Problem Books in Mathematics

Books in this series are devoted exclusively to problems - challenging, difficult, but accessible problems. They are intended to help at all levels - in college, in graduate school, and in the profession. Arthur Engels "Problem-Solving Strategies" is good for elementary students and Richard Guys "Unsolved Problems in Number Theory" is the classical advanced prototype. The series also features a number of successful titles that prepare students for problem-solving competitions.

Volodymyr Brayman • Andrii Chaikovskyi •
Oleksii Konstantinov • Alexander Kukush •
Yuliya Mishura • Oleksii Nesterenko

Functional Analysis and Operator Theory

Volodymyr Brayman
Mathematical Analysis
Taras Shevchenko National University of Kyiv
Kyiv, Ukraine

Andrii Chaikovskyi
Mathematical Analysis
Taras Shevchenko National University of Kyiv
Kyiv, Ukraine

Oleksii Konstantinov
Mathematical Analysis
Taras Shevchenko National University of Kyiv
Kyiv, Ukraine

Alexander Kukush
Mathematical Analysis
Taras Shevchenko National University of Kyiv
Kyiv, Ukraine

Yuliya Mishura
Probability, Stat. and Actuarial Maths
Taras Shevchenko National University of Kyiv
Kyiv, Ukraine

Oleksii Nesterenko
Mathematical Analysis
Taras Shevchenko National University of Kyiv
Kyiv, Ukraine

ISSN 0941-3502 ISSN 2197-8506 (electronic)
Problem Books in Mathematics
ISBN 978-3-031-56426-0 ISBN 978-3-031-56427-7 (eBook)
https://doi.org/10.1007/978-3-031-56427-7

Mathematics Subject Classification: 00A07, 46-01, 47-01, 45-01

Translation from the Ukrainian language edition: "ЗБІРНИК ЗАДАЧ З ФУНКщОНАЛЬНОГО АНАЛІЗУ" by Volodymyr Brayman et al., © Taras Shevchenko National University of Kyiv 2023. Published by Taras Shevchenko National University of Kyiv. All Rights Reserved.

This Springer imprint is published by the registered company Springer Nature Switzerland AG
The registered company address is: Gewerbestrasse 11, 6330 Cham, Switzerland

Preface

Functional analysis is one of the most abstract and difficult disciplines in the curricula of the specialties "Mathematics," "Applied Mathematics," "Statistics," "Engineering Mathematics," and many others related to the teaching of mathematical methods. At the same time, the elements of functional analysis are used in all modern courses of mathematical physics, quantum theory, integral equations, calculus of variations, approximation theory, harmonic and wavelet analysis, stochastic analysis, theory of stochastic processes, stochastic differential equations, financial mathematics, etc. A deep acquaintance with this discipline is absolutely necessary for every researcher who works in the mentioned areas of mathematics and its applications. A particular reason is that even the problem statements and obtained results are often formulated in the language of functional analysis. Therefore, every qualified mathematician should master the basic concepts and methods of functional analysis. Undoubtedly, the studying of this deep and extensive course by students and postgraduates should be accompanied by a large number of problems and exercises.

The proposed collection of problems summarizes more than 30 years of experience in teaching functional analysis, in particular, the experience of conducting practical classes in the courses "Functional Analysis" and "Operator Theory and Integral Equations" at the Faculty of Mechanics and Mathematics of Taras Shevchenko National University of Kyiv. First of all, this book is intended for undergraduate students who are starting to study the course of functional analysis. However, the authors hope that it will also be useful for graduate and postgraduate students and researchers who wish to refresh their knowledge and deepen their understanding of the subject, as well as for teachers of functional analysis and related disciplines. It can be used for the independent study as well. It is assumed that the reader has mastered standard courses of calculus and measure theory and has basic knowledge of linear algebra, analytic geometry, and differential equations.

The main goal of the authors was to create a collection of problems that could help students at different levels of training and various areas of specialization to learn how to solve problems in functional analysis. In order to realize this ambitious goal, the authors were guided by the following principles in their preparation of the problem book. Firstly, the book should contain examples of typical problems with detailed solutions; this is relevant not only for those students who have significant difficulties in studying this subject but also for other students who, due to various circumstances, can be deprived of communication with a teacher. Secondly, there should be problems for independent solving, and the corresponding selection of problems should reflect all main plot lines that relate to a given topic. Thirdly, the number of problems should be sufficient both for a teacher to conduct practical classes, to set homework, and to prepare tasks for various forms of control and for those students who seek to study the discipline more deeply. Fourthly, problems of computational nature should be provided with answers, while "theoretical" problems, the solutions of which require nontrivial ideas or new techniques, should be provided with detailed hints or solutions to introduce a reader to the corresponding ideas or techniques. Fifthly, the book should contain theoretical preliminaries to ensure that a reader understands the statements of the problems and is able to solve them successfully, and amount of these preliminaries should be minimal.

In order to implement these principles, each of the first 14 chapters of the book has the same structure and consists of three sections: *Theoretical Background*, *Examples of Problems with Solutions*, and *Problems to Solve*. In the first of these sections, the theoretical material is presented, which contains all necessary definitions, as well as standard theorems needed to solve the proposed problems. Some more specific concepts are introduced later in the statements of the problems. The majority of this material is presented in Chaps. 6–11 of the textbook [8], but the reader can use any theoretical course which covers the corresponding theory, for example, [10, 11, 21, 25, 26]. The second section contains several problems with complete solutions. These problems provide a reader with standard tools and techniques. Some of them are typical algorithmic problems which every student should know how to solve in order to pass the exam. Other problems in this section contain statements and proofs of useful lemmas which are not always included as separate results in theoretical courses (e.g., problems 1.2, 2.2, 2.4, 3.2, 5.1, etc.). The third section contains statements of a rather large spectrum of problems of varying difficulty for students to work on independently. It contains a significant number of typical algorithmic problems similar to those solved in the second section. However, the main attention is paid to simple meaningful problems that explain the most important concepts and facts of the course and teach students to operate with them fluently. Some of independent study problems that make up a necessary minimum are marked with small circles (°). "Strong" students will see some of those problems as trivial exercises. However, these students will find more challenging problems for themselves as well, the most difficult of which are marked with asterisks (*). It is not necessary but highly preferable to read and solve the problems in the order in which they are presented in the book: the solutions of some rather difficult problems become much easier in the light of the statements of the previous problems or the ideas from their solutions. At the end of the book there is a chapter which contains answers and detailed hints or solutions to almost all the problems, allowing students to check their work. It goes without saying that students are strongly advised to work on the problems by themselves prior to looking at the solutions. On the other hand, it might be useful to compare your own solutions with the hints and solutions provided in the book which in many cases contain typical ideas and methods of reasoning. In total, the collection contains more than 800 problems, and more than 1500 problems taking into account that many problems contain several items. More than 60 problems with complete and detailed solutions, which contain more than 110 individual items, are presented in the "Examples of problems with solutions" sections.

Chapter 1 (*Banach Spaces*) deals with the problems on normed linear and Banach spaces. Special attention is paid to classical spaces $C([a, b])$, l_p, $L_p(T, \mu)$. Chapter 2 (*Hilbert Spaces*) studies the geometry of Hilbert space. In particular, problems on orthogonality, projection theorems, orthonormal systems and bases are considered. Chapters 3–5 (*Continuous Linear Functionals, Hahn–Banach Theorem, Weak and Weak* Convergence*) deal with dual spaces and the properties of continuous linear functionals on normed spaces. Chapters 6–8 (*Bounded Linear Operators, Uniform, Strong and Weak Operator Convergences, Inverse Operators*) are devoted to the study of general properties of bounded linear operators in Banach spaces. Chapter 9 (*Classes of Linear Operators in Hilbert Space*) contains problems on the properties of operators in Hilbert spaces. In particular the properties of adjoint operators and the classes of self-adjoint, normal, isometric, unitary operators, and orthoprojectors are studied. Chapter 10 (*Compact Sets and Operators*) is devoted to the study of compact operators in Banach spaces. In Chap. 11 (*Spectrum of Linear Operators*) the spectrum and spectral properties of bounded linear operators in Banach and Hilbert spaces are studied. Chapter 12 (*Spectral Theory of Compact Operators*) deals with the spectral properties of compact operators. The spectral theorem for self-adjoint compact operators, singular numbers, minimax principle, ideals of Hilbert–Schmidt operators, and trace-class operators are discussed as well. Chapter 13 (*Integral Equations*) deals with solving of Fredholm integral equation of the second kind with a degenerate kernel, constructing the resolvent of the Volterra or Fredholm integral equation, as well as problems on the application of the Fredholm alternative. Chapter 14 (*Generalized Functions*) is devoted to the spaces of generalized functions $\mathcal{D}'$ and $\mathcal{S}'$. In particular, we study convergence, derivatives, support, and Fourier transform of distributions.

Despite the fact that a very large number of both classic and modern textbooks on functional analysis and its applications have been written and published, among which we shall mention [2, 4, 5, 6, 7, 8, 9, 10, 11, 12, 16, 17, 18, 19, 21, 23, 26, 27, 28, 29, 30, 33, 35, 36, 37], this list is by no means comprehensive; the number of problem books on functional analysis and its applications is relatively small. We shall mention, without pretending to make a complete list again, problem books [1, 3, 13, 14, 15, 20, 22, 24, 31, 32, 34]. Many problems or problem ideas in our book are taken from well-known educational and monographic literature. In particular, a significant number of problems are taken from books [8, 20, 26, 33, 34]. Origins of some other problems can be traced to journal papers. The source base of the problems is very wide, and some of the sources were not translated into English, or became a bibliographical rarity. At the same time, a significant part of our collection contains original material. We gave a lot of thought to presenting the problems in each chapter in right order to ensure a smooth transition from easy problems to hard ones. Many solutions were significantly simplified, in particular, due to application of statements and ideas presented in previous problems.

The theorems and problems in the book are numbered chapter-wise. For example, problem 2 of Chap. 1 is referred to as problem 1.2 everywhere.

Concluding the preface, the authors consider it their pleasant duty to mention with words of gratitude the bright memory of Academician of the National Academy of Sciences of Ukraine, Yu. M. Berezanskyi; Corresponding Member of the National Academy of Sciences of Ukraine, M. L. Horbachuk; former Head of the Department of Mathematical Analysis, Professor A. Ya. Dorogovtsev; and professors S. D. Ivasyshen and Yu.G. Kondratiev. With their unsurpassed lectures on various sections of analysis, which most of the authors of this problem book were lucky to listen to, they established strong traditions of teaching the modern course of functional analysis at the Faculty of Mechanics and Mathematics of Taras Shevchenko National University of Kyiv. The development of these traditions at the University became possible thanks to the activities of the Head of the Department of Mathematical Analysis, Corresponding Member of the National Academy of Sciences of Ukraine I. O. Shevchuk, professors M. F. Horodnii and V. L. Ostrovskyi, and Associate Professor G. F. Us who taught mandatory and special courses on functional analysis. Discussions of scientific and pedagogical ideas and other forms of cooperation with these outstanding mathematicians and wonderful people had a positive impact both on the professional activity of the authors in general and on the content and structure of this book in particular, for which the authors express their sincere gratitude.

Kyiv, Ukraine
January 2024

Volodymyr Brayman
Andrii Chaikovskyi
Oleksii Konstantinov
Alexander Kukush
Yuliya Mishura
Oleksii Nesterenko

Contents

List of Notations

$\mathbb{N}$	Set of positive integers
$\mathbb{Z}$	Set of integers
$\mathbb{Q}$	Set of rational numbers
$\mathbb{R}$	Set of real numbers
$\mathbb{C}$	Set of complex numbers
$\mathbb{K}$	A field of numbers, $\mathbb{K} = \mathbb{R}$ or $\mathbb{K} = \mathbb{C}$
$C(M)$	The set of $\mathbb{K}$-valued continuous functions on the set M
$C^n(M)$	The set of $\mathbb{K}$-valued functions that have n continuous derivatives on the set M
$C^\infty(M)$	The set of $\mathbb{K}$-valued functions that are infinitely differentiable on the set M
$R([a, b])$	Class of Riemann integrable functions on the interval $[a, b]$
$BV([a, b])$	Class of functions of bounded variation on the interval $[a, b]$
$BV_0([a, b])$	Class of functions $f \in BV([a, b])$ such that f is right continuous on (a, b) and $f(a) = 0$
$V(f, [a, b])$	Variation of function f on the interval $[a, b]$
$\chi_A(x)$	Characteristic function of the set A, which takes the value 1, if $x \in A$, and the value 0, if $x \notin A$
m	Lebesgue measure in the space $\mathbb{R}^k$, $k \geq 1$
$f_n \to f \pmod{\mu}$	The sequence of functions $\{f_n : n \geq 1\}$ converges to function f almost everywhere with respect to measure μ as $n \to \infty$
$f_n \overset{\mu}{\to} f$	The sequence of functions $\{f_n : n \geq 1\}$ converges to function f in measure μ
$f_n \rightrightarrows f$	The sequence of functions $\{f_n : n \geq 1\}$ converges uniformly to the function f on a given set
$\int\limits_A f(x) d\mu(x)$	The Lebesgue integral of the function f over the set A with respect to the measure μ
$\int\limits_A f(x) dx$	The Lebesgue integral of the function f over the set A with respect to the Lebesgue measure m

$\int_a^b f(x)dx$	The Lebesgue integral of the function f over the interval $[a, b]$ with respect to the Lebesgue measure m; if f is Riemann integrable, then this integral coincides with the Riemann integral
$L(A, \mu)$	The set of measurable functions that are Lebesgue integrable over the set A with respect to the measure μ
$\\|\cdot\\|$	(Chapter 1) The norm in a normed linear space (NLS)
$B(x_0, r)$	(Chapter 1) Open ball in an NLS with center at point x_0 and radius $r > 0$
$\overline{B}(x_0, r)$	(Chapter 1) Closed ball in an NLS with center at point x_0 and radius $r > 0$
$S(x_0, r)$	(Chapter 1) Sphere in an NLS with center at point x_0 and radius $r > 0$
$\rho(x, A)$	(Chapter 1) Distance from point x to set A
$\mathrm{span}(M)$	(Chapter 1) Linear span of set M
$\overline{\mathrm{span}(M)}$	(Chapter 1) Closed linear span of set M
$\dim X$	(Chapter 1) Dimension of space X
$A + B$	(Chapter 1) Sum of sets A and B in an NLS
$L_p(T, \mu)$	(Chapter 1) Space of measurable functions, the absolute value of which is integrable in p-th power on T with respect to measure μ, $1 \le p < +\infty$
$L_p(T)$	(Chapter 1) Space of Lebesgue measurable functions, the absolute value of which is integrable in the p-th power on the Borel set $T \subset \mathbb{R}^n$ with respect to the Lebesgue measure m, $1 \le p < +\infty$
$L_\infty(T, \mu)$	(Chapter 1) Space of measurable essentially bounded functions with respect to the measure μ on T
$L_\infty(T)$	(Chapter 1) Space of Lebesgue measurable essentially bounded functions with respect to the Lebesgue measure m on the Borel set $T \subset \mathbb{R}^n$
l_p	(Chapter 1) Space of sequences with a finite p-norm $\\|\cdot\\|_p$
$\mathbb{R}^m_p$	(Chapter 1) Space $\mathbb{R}^m$ with the norm $\\|\cdot\\|_p$
$\mathbb{C}^m_p$	(Chapter 1) Space $\mathbb{C}^m$ with the norm $\\|\cdot\\|_p$
$\\|\cdot\\|_p$	(Chapter 1) A norm on $L_p(T), l_p, \mathbb{R}^m$ or $\mathbb{C}^m$, $1 \le p \le +\infty$
c_0	(Chapter 1) Space of sequences that converge to zero
c	(Chapter 1) Space of sequences that converge
$(\cdot, \cdot)$	(Chapter 2) Scalar product
$\overline{z}$	Conjugate number to the complex number z
$x \perp y$	(Chapter 2) Vector x is orthogonal to vector y
$M^\perp$	(Chapter 2) Orthogonal complement of a set M in a Hilbert space
$\mathrm{pr}_M x$	(Chapter 2) Orthogonal projection of a vector x onto a subspace M
$M \oplus N$	(Chapter 2) Orthogonal sum of sets M and N

X^*	(Chapter 3) Dual space of an NLS X
$\operatorname{Ker} f$	(Chapter 3) Kernel of a functional f
$x_n \overset{w}{\to} x$	(Chapter 5) The sequence $\{x_n : n \geq 1\}$ of elements of an NLS converges weakly to the element x
$f_n \overset{w-*}{\longrightarrow} f$	(Chapter 5) The sequence $\{f_n : n \geq 1\}$ of continuous linear functionals converges weakly-$*$ to the linear functional f
$\mathcal{L}(X_1, X_2)$	(Chapter 6) Space of continuous linear operators acting from an NLS X_1 to an NLS X_2
$\mathcal{L}(X)$	(Chapter 6) Space of continuous linear operators acting in an NLS X
$\operatorname{Ker} A$	(Chapter 6) The kernel of a linear operator A
$R(A)$	(Chapter 6) The range of a linear operator A
$A_n \rightrightarrows A$	(Chapter 7) The sequence of continuous linear operators $\{A_n : n \geq 1\}$ converges uniformly to the linear operator A
$A_n \overset{s}{\to} A$	(Chapter 7) The sequence of continuous linear operators $\{A_n : n \geq 1\}$ converges strongly to the linear operator A
$A_n \overset{w}{\to} A$	(Chapter 7) The sequence of continuous linear operators $\{A_n : n \geq 1\}$ converges weakly to the linear operator A
A^{-1}	(Chapter 8) Inverse operator of the operator A
A^*	(Chapter 9) Adjoint operator to the operator A, acting in a Hilbert space
A'	(Chapter 9) Adjoint operator to the operator A, acting in an NLS
$A \geq 0$	(Chapter 9) Operator A, acting in the Hilbert space, is nonnegative
$A \geq B$	(Chapter 9) Operators A, B, acting in the Hilbert space, are self-adjoint and $A - B \geq 0$
P_G	(Chapter 9) Orthogonal projector in the Hilbert space onto a subspace G
$\sqrt{A} = A^{\frac{1}{2}}$	(problem 9.52*) Square root of a nonnegative operator A
$S_\infty(X_1, X_2)$	(Chapter 10) Space of compact operators, acting from an NLS X_1 to an NLS X_2
$S_\infty(X)$	(Chapter 10) Space of compact operators, acting in an NLS X
$S_0(X_1, X_2)$	(Chapter 10) Space of finite-dimensional linear operators acting from an NLS X_1 to an NLS X_2
$S_0(X)$	(Chapter 10) Space of finite-dimensional linear operators acting in an NLS X
$\rho(A)$	(Chapter 11) Resolvent set of an operator A
$\sigma(A)$	(Chapter 11) Spectrum of an operator A
$\sigma_p(A)$	(Chapter 11) Point spectrum of an operator A
$\sigma_c(A)$	(Chapter 11) Continuous spectrum of an operator A
$\sigma_r(A)$	(Chapter 11) Residual spectrum of an operator A
$r(A)$	(Chapter 11) Spectral radius of the operator A
$R_\lambda(A)$	(Chapter 11) Resolvent of the operator A

$\operatorname{supp}\varphi$ ($\operatorname{supp} f$)	(Chapter 14) Support of the function φ (the generalized function f)
$\mathcal{D}(\mathbb{R}^m)$	(Chapter 14) Space of all infinitely differentiable functions with bounded support from $\mathbb{R}^m$ to $\mathbb{C}$ (space of test functions)
$\mathcal{D}'(\mathbb{R}^m)$	(Chapter 14) Space of generalized functions (distributions)
$L_1^{loc}(\mathbb{R})$	(Chapter 14) Space of locally integrable functions on $\mathbb{R}$ (i.e., functions integrable on every segment in $\mathbb{R}$)
δ, δ_0	(Chapter 14) Dirac delta function
δ_a	(Chapter 14) Delta function shifted by a, $a \in \mathbb{R}$
$f * g$	(Chapter 14) Convolution of functions f and g
$\theta = \chi_{[0,+\infty)}$	(Chapter 14) Heaviside function
$\mathcal{S}(\mathbb{R}^m)$	(Chapter 14) Space of rapidly decreasing functions
$\mathcal{S}'(\mathbb{R}^m)$	(Chapter 14) Space of tempered distributions
$F[\varphi]$	(Chapter 14) The Fourier transform of the function φ

Chapter 1
Banach Spaces

Theoretical Background

Let X be a linear space over a field $\mathbb{K}$ ($\mathbb{K} = \mathbb{R}$ or $\mathbb{K} = \mathbb{C}$), and 0 be the zero element of X.

A function $\|\cdot\| : X \to \mathbb{R}$ is called a *norm* on X, if the following conditions (*axioms of norm*) hold:

(a) $\forall x \in X : \|x\| \geq 0$, moreover $\|x\| = 0 \Leftrightarrow x = 0$ (nonnegativity);
(b) $\forall \alpha \in \mathbb{K}\, \forall x \in X : \|\alpha x\| = |\alpha| \cdot \|x\|$ (homogeneity);
(c) $\forall x, y \in X : \|x + y\| \leq \|x\| + \|y\|$ (the triangle inequality).

A linear space equipped with a norm is called a *normed linear space (NLS)*. If $\mathbb{K} = \mathbb{R}$ then the space is called *real* NLS, and if $\mathbb{K} = \mathbb{C}$ then it is called *complex* NLS.

If $\|\cdot\|$ is a norm on X then

(1) $\forall x, y \in X : |\, \|x\| - \|y\|\, | \leq \|x - y\|$ (the second triangle inequality);
(2) the function $\rho(x, y) = \|x - y\|,\ x, y \in X$, is a metric in X. It is said that ρ is the metric generated by the norm on X.

A sequence $\{x_n : n \geq 1\} \subset X$ is said to be *convergent* to an element $x \in X$, if $\|x_n - x\| \to 0, n \to \infty$.

A sequence $\{x_n : n \geq 1\} \subset X$ is called a *Cauchy sequence*, if
$\forall \varepsilon > 0\, \exists N \in \mathbb{N}\, \forall n \geq N\, \forall m \geq N : \|x_n - x_m\| < \varepsilon$.

A normed linear space X is called a *Banach space*, if every Cauchy sequence $\{x_n : n \geq 1\} \subset X$ converges to some element $x \in X$, i.e. if the space X is complete in the metric generated by the norm.

An open ball of radius $r > 0$ with the center at a point $x_0 \in X$ is the set $B(x_0, r) = \{x \in X \mid \|x - x_0\| < r\}$.

A closed ball of radius $r > 0$ with the center at a point $x_0 \in X$ is the set $\overline{B}(x_0, r) = \{x \in X \mid \|x - x_0\| \leq r\}$.

V. Brayman et al., *Functional Analysis and Operator Theory*, Problem Books in Mathematics, https://doi.org/10.1007/978-3-031-56427-7_1

A sphere of radius $r > 0$ with the center at a point $x_0 \in X$ is the set $S(x_0, r) = \{x \in X \mid \|x - x_0\| = r\}$.

A set $A \subset X$ is said to be *bounded* if it is contained in some ball.

A distance from a point $x \in X$ to a set $A \subset X$, $A \neq \varnothing$, is the number $\rho(x, A) = \inf\limits_{y\in A} \|x - y\|$. If there exists $y^* \in A$ such that $\rho(x, A) = \|x - y^*\|$, then the element y^* is said to be *the element of the best approximation* for the element $x \in X$ in the set A.

A point $x_0 \in X$ is called an *interior point* of a set $A \subset X$ if
$\exists r > 0 \ : \ B(x_0, r) \subset A$.

The set of all interior points of a set A is denoted by A°.

A set $A \subset X$ is called *an open set* if every point in it is an interior point.

A point $x_0 \in X$ is called *a limit point* of a set $A \subset X$ if
$\forall r > 0 \, \exists x \in B(x_0, r) \backslash \{x_0\} \ : \ x \in A$.

The set of all limit points of a set A is denoted by A'.

A set $A \subset X$ is called a *closed set* if it contains all its limit points.

The set $\overline{A} = A \cup A'$ is called the *closure* of a set A.

A set $L \subset X$ is called a *linear set* if

$$\forall x, y \in L \, \forall \alpha, \beta \in \mathbb{K} \ : \ \alpha x + \beta y \in L.$$

A closed linear subset of a space X is called a *subspace*.

A linear span of a set $M \subset X$ is the smallest linear set which contains M:

$$\operatorname{span}(M) = \left\{ \sum_{k=1}^{n} \alpha_k x_k \;\middle|\; n \in \mathbb{N}, \ \alpha_k \in \mathbb{K}, \ x_k \in M, \ 1 \leq k \leq n \right\}.$$

The closure of the set $\operatorname{span}(M)$ is called a *closed linear span* of a set M. The set $\overline{\operatorname{span}(M)}$ is the smallest subspace containing M. A set is called a *total set* in X if

$$\overline{\operatorname{span}(M)} = X.$$

Vectors $x_1, \ldots, x_n \in X, \ n \in \mathbb{N}$, are said to be *linearly independent* if $\sum\limits_{k=1}^{n} \alpha_k x_k = 0$ implies that $\alpha_k = 0, \ 1 \leq k \leq n$. In the opposite case these vectors are said to be *linearly dependent*.

A space is called an *n-dimensional space* if there are n linearly independent vectors in it and every $(n + 1)$ vectors are linearly dependent. An infinite set of vectors is *linearly independent* if every nonempty finite subset of this set is linearly independent.

A space X is called *infinite-dimensional* if there exists an infinite linearly independent subset. *The dimension of a space* is the maximum number of linearly independent vectors in this space. The dimension of a space X is denoted by $\dim X$.

The set $\{tx + (1 - t)y \mid t \in [0, 1]\}$ is called the *segment* connecting points $x, y \in X$.

A set $A \subset X$ is called a *convex set* if, for any two points in A, the segment connecting them belongs to A.

The sum of sets $A \subset X$ and $B \subset X$ is the set $A + B = \{x + y \mid x \in A,\ y \in B\}$.

A set $M \subset X$ is said to be *dense* (or *everywhere dense*) in X if
$\forall x \in X\ \forall \varepsilon > 0\ \exists y \in M\ :\ \|x - y\| < \varepsilon$, i.e. $\overline{M} = X$.

A normed linear space is said to be *separable* if it contains a countable dense set.

Two norms $\|\cdot\|_1$ and $\|\cdot\|_2$ on a linear space X are said to be *equivalent* if $\exists C_1 > 0\ \exists C_2 > 0\ \forall x \in X\ :\ C_1\|x\|_1 \leq \|x\|_2 \leq C_2\|x\|_1$.

Linear spaces X_1 and X_2 are said to be *algebraically isomorphic* if there exists a linear bijection $U : X_1 \to X_2$; U is called an *algebraic isomorphism*.

Normed linear spaces X_1 and X_2 are said to be *isomorphic* if they are algebraically isomorphic and this isomorphism U is a homeomorphism, i.e. both mappings $U : X_1 \to X_2$ and $U^{-1} : X_2 \to X_1$ are continuous. An isomorphism U is called an *isometric isomorphism* if $\|x\|_{X_1} = \|Ux\|_{X_2}$ for all $x \in X_1$.

Theorem 1.1 *Finite-dimensional normed linear spaces over the same field are isomorphic if and only if they have the same dimension.*

For every NLS X there exists a Banach space $\widetilde{X}$ (completion of X) and an isometric isomorphism $U : X \to U(X) \subset \widetilde{X}$ such that $U(X)$ is dense in $\widetilde{X}$ (see problem 1.88).

Basic Examples of Banach Spaces

1. $\mathbb{R}^m = \{x = (x_1, \ldots, x_m) \mid x_k \in \mathbb{R},\ 1 \leq k \leq m\}$ is a real m-dimensional separable Banach space with each of the norms

$$\|x\|_p := \left(\sum_{k=1}^{m} |x_k|^p\right)^{\frac{1}{p}},\ 1 \leq p < +\infty,\ \|x\|_\infty := \max_{1 \leq k \leq m} |x_k|.$$

Convergence in $(\mathbb{R}^m, \|\cdot\|_p)$ is equivalent to coordinate-wise convergence. If $p = 2$, the corresponding norm is called the Euclidean norm.

2. $\mathbb{C}^m = \{x = (x_1, \ldots, x_m) \mid x_k \in \mathbb{C},\ 1 \leq k \leq m\}$ is a complex m-dimensional separable Banach space with each of the norms

$$\|x\|_p := \left(\sum_{k=1}^{m} |x_k|^p\right)^{\frac{1}{p}},\ 1 \leq p < +\infty,\ \|x\|_\infty := \max_{1 \leq k \leq m} |x_k|.$$

Convergence in $(\mathbb{C}^m, \|\cdot\|_p)$ is equivalent to coordinate-wise convergence. If $p = 2$, the corresponding norm is called the Euclidean norm.

In what follows, $\mathbb{K} = \mathbb{R}$ or $\mathbb{K} = \mathbb{C}$.

3. Let Q be a compact metric space (for example, $Q = [a, b]$ with the usual distance), and $C(Q)$ be the set of all continuous functions $x : Q \to \mathbb{K}$. $C(Q)$ will become a linear space if we put

$$(\alpha x + \beta y)(t) = \alpha x(t) + \beta y(t),\ t \in Q,\ x, y \in C(Q),\ \alpha, \beta \in \mathbb{K}.$$

The space $C(Q)$ is a Banach space with the norm $\|x\| = \sup\limits_{t \in Q} |x(t)|,\ x \in C(Q)$, moreover it is a separable space (see problem 1.58*). If Q is an infinite set, then $C(Q)$ is an infinite-dimensional space (see problem 1.58*). Convergence in $C(Q)$ is equivalent to uniform convergence in Q.
4. Let $n \in \mathbb{N}$. The space
$C^n([a, b]) := \{x : [a, b] \to \mathbb{K} \mid \forall t \in [a, b]\ \exists x^{(n)}(t),\ x^{(n)} \in C([a, b])\}$ is a Banach space with the norm $\|x\| := \sum\limits_{k=0}^{n} \sup\limits_{t \in [a,b]} |x^{(k)}(t)|,\ x \in C^n([a, b])$.
5. Let $1 \le p < +\infty$, $(T, \mathcal{F}, \mu)$ be a measure space,
$\mathcal{L}_p(T, \mu) := \{x : T \to \mathbb{K} \mid x$ is $\mathcal{F}$-measurable function and $\int\limits_T |x(t)|^p d\mu(t) < +\infty\}$,
$\mathcal{L} := \{x : T \to \mathbb{K} \mid x$ is $\mathcal{F}$-measurable function and $x(t) = 0 \pmod{\mu}\}$. Then the factor space $L_p(T, \mu) = \mathcal{L}_p(T, \mu)/\mathcal{L}$ is a Banach space with the norm $\|x\|_p = \left(\int\limits_T |x(t)|^p d\mu(t)\right)^{\frac{1}{p}},\ x \in L_p(T, \mu)$.
If $T = \mathbb{N},\ \mathcal{F} = 2^{\mathbb{N}},\ \mu(\{n\}) = 1,\ n \in \mathbb{N}$, then $L_p(T, \mu) = l_p :=$

$$= \left\{x = (x_1, \ldots, x_n, \ldots) \mid x_n \in \mathbb{K}, n \in \mathbb{N}, \sum_{n=1}^{\infty} |x_n|^p < +\infty\right\}$$

is a separable Banach space with the norm

$$\|x\|_p := \left(\sum_{n=1}^{\infty} |x_n|^p\right)^{\frac{1}{p}},\ x = (x_1, \ldots, x_n, \ldots) \in l_p.$$

If T is a Borel subset of $\mathbb{R}^n$ and $\mu = m$ is the Lebesgue measure on T, then $L_p(T)$ is usually written instead of $L_p(T, m)$.
6. Let $(T, \mathcal{F}, \mu)$ be a measure space,
$\mathcal{L}_\infty(T, \mu) := \{x : T \to \mathbb{K} \mid x$ is $\mathcal{F}$-measurable function and $\operatorname{ess\,sup}\limits_{t \in T} |x(t)| < +\infty\}$, $\mathcal{L} := \{x : T \to \mathbb{K} \mid x$ is $\mathcal{F}$-measurable function and $x(t) = 0 \pmod{\mu}\}$. Then the factor space $L_\infty(T, \mu) = \mathcal{L}_\infty(T, \mu)/\mathcal{L}$ is a Banach space with the norm

$$\|x\|_\infty := \operatorname*{ess\,sup}_{t \in T} |x(t)| := \inf\left\{C \ge 0 \,\middle|\, |x(t)| \le C \pmod{\mu}\right\}.$$

It can be proved that $\operatorname*{ess\,sup}_{t\in T} |x(t)| = \min\left\{C \geq 0 \,\middle|\, |x(t)| \leq C \,(\text{mod}\,\mu)\right\}$ (see problem 1.20).

If $T = \mathbb{N}$, $\mathcal{F} = 2^{\mathbb{N}}$, $\mu(\{n\}) = 1$, $n \in \mathbb{N}$, then $L_\infty(T, \mu) = l_\infty :=$

$$= \{x = (x_1, \ldots, x_n, \ldots) \mid x_n \in \mathbb{K}, n \in \mathbb{N}, \sup_{n\in\mathbb{N}} |x_n| < +\infty\}$$

is a nonseparable Banach space with the norm

$$\|x\|_\infty := \sup_{n\in\mathbb{N}} |x_n|, \; x = (x_1, \ldots, x_n, \ldots) \in l_\infty.$$

7. $c := \{x = (x_1, \ldots, x_n, \ldots) \mid x_n \in \mathbb{K}, n \in \mathbb{N}, \exists \lim_{n\to\infty} x_n \in \mathbb{K}\}$ and

$$c_0 := \{x = (x_1, \ldots, x_n, \ldots) \mid x_n \in \mathbb{K}, \; n \in \mathbb{N}, \; \lim_{n\to\infty} x_n = 0\}$$

are Banach spaces with the norm $\|x\|_\infty = \sup_{n\in\mathbb{N}} |x_n|$.

In the spaces l_p, $1 \leq p \leq +\infty$, c and c_0 the notation

$$e_n = (0, \ldots, 0, \underbrace{1}_{n}, 0, 0, \ldots), \; n \geq 1,$$

is used.

We also present two important inequalities.

(1) Let $1 < p < +\infty$, $\frac{1}{p} + \frac{1}{q} = 1$, $f \in L_p(T, \mu)$, $g \in L_q(T, \mu)$. Then $fg \in L_1(T, \mu)$ and *Hölder's inequality* holds:

$$\int_T |fg| d\mu \leq \left(\int_T |f|^p d\mu\right)^{\frac{1}{p}} \left(\int_T |g|^q d\mu\right)^{\frac{1}{q}},$$

i.e. $\|fg\|_1 \leq \|f\|_p \cdot \|g\|_q$.

(2) Let $1 \leq p < +\infty$, $f, g \in L_p(T, \mu)$. Then $f + g \in L_p(T, \mu)$ and the *Minkowski inequality* holds:

$$\left(\int_T |f+g|^p d\mu\right)^{\frac{1}{p}} \leq \left(\int_T |f|^p d\mu\right)^{\frac{1}{p}} + \left(\int_T |g|^p d\mu\right)^{\frac{1}{p}},$$

i.e. $\|f + g\|_p \leq \|f\|_p + \|g\|_p$.

Examples of Problems with Solutions

1.1 Check whether the given functions are norms on the corresponding spaces:

(1) $C([a,b]) \ni x \mapsto \varphi(x) = \max\limits_{\frac{1}{2}(a+b) \le t \le b} |x(t)|$;

(2) $C^1([a,b]) \ni x \mapsto \varphi(x) = \int\limits_a^b |x'(t)|dt$;

(3) $C^1([a,b]) \ni x \mapsto \varphi(x) = |x(a)| + \int\limits_a^b |x'(t)|dt$.

Solution

(1) No, since $\varphi(x) = 0$ does not imply that $x = 0$. For example, if

$$x_0(t) = \begin{cases} t - \frac{1}{2}(a+b), & t \in [a, \frac{1}{2}(a+b)], \\ 0, & t \in [\frac{1}{2}(a+b), b], \end{cases}$$

then $x_0 \in C([a,b]),\ x_0 \neq 0$, but $\varphi(x_0) = \max\limits_{\frac{1}{2}(a+b) \le t \le b} |x_0(t)| = 0$.

(2) No, since $\varphi(x) = 0$ does not imply that $x = 0$. Indeed, if $x_0(t) = 1,\ t \in [a,b]$, then $x_0 \in C^1([a,b]),\ x_0 \neq 0$, but $\varphi(x_0) = 0$.

(3) Yes. The function φ is correctly defined, because the integrand function is continuous and therefore it is Riemann integrable. Let us check the axioms of norm.

(a) $\varphi(x) \ge 0,\ x \in C^1([a,b])$, because both terms are nonnegative (since the absolute value of a number is nonnegative and the integral of a nonnegative function is nonnegative). If $x = 0$, then $\varphi(x) = 0$. On the contrary, if $\varphi(x) = 0$, then $|x(a)| = 0$ and $\int\limits_a^b |x'(t)|dt = 0$. Since the integrand function is continuous and nonnegative, the last equality implies that $|x'(t)| = 0,\ t \in [a,b]$, and $x(t) = c,\ t \in [a,b]$, where c is some constant. But $x(a) = 0$, therefore $c = 0$, i.e. $x(t) = 0,\ t \in [a,b]$. The first axiom holds.

(b) For all $\alpha \in \mathbb{C}$ and $x \in C^1([a,b])$ we have

$$\varphi(\alpha x) = |\alpha x(a)| + \int\limits_a^b |(\alpha x)'(t)|dt = |\alpha| \cdot |x(a)| + |\alpha| \int\limits_a^b |x'(t)|dt = |\alpha| \varphi(x).$$

The second axiom holds.

(c) For all $x, y \in C^1([a,b])$ we have

$$\varphi(x+y) = |x(a)+y(a)| + \int_a^b |x'(t)+y'(t)|dt \leq$$

$$\leq |x(a)| + |y(a)| + \int_a^b (|x'(t)| + |y'(t)|)dt =$$

$$= |x(a)| + \int_a^b |x'(t)|dt + |y(a)| + \int_a^b |y'(t)|dt = \varphi(x) + \varphi(y).$$

The third axiom holds.

1.2

(1) Let a sequence $\{x_n : n \geq 1\} \subset C([a,b])$ be such that $x_n \to x,\ n \to \infty$, in $C([a,b])$. Prove that $x_n(t) \to x(t), n \to \infty$, for every $t \in [a,b]$ (i.e. pointwise convergence takes place).
(2) Let a sequence $\{x^{(n)} = (x_1^{(n)}, \ldots, x_k^{(n)}, \ldots) : n \geq 1\} \subset l_p,\ 1 \leq p \leq +\infty$, be such that $x^{(n)} \to x, n \to \infty$, in l_p, where $x = (x_1, \ldots, x_k, \ldots) \in l_p$. Prove that $x_k^{(n)} \to x_k, n \to \infty$, for every $k \in \mathbb{N}$ (i.e. coordinate-wise convergence takes place).
(3) Let a sequence $\{x_n : n \geq 1\} \subset L_p(T,\mu),\ 1 \leq p < +\infty$, be such that $x_n \to x,\ n \to \infty$ in $L_p(T,\mu)$, and $x_n \to y,\ n \to \infty$ μ-almost everywhere on T. Prove that $x = y$ μ-almost everywhere on T.

Solution

(1) For each fixed $t_0 \in [a,b]$ we have

$$|x_n(t_0) - x(t_0)| \leq \max_{t \in [a,b]} |x_n(t) - x(t)|,$$

whence $x_n(t_0) \to x(t_0),\ n \to \infty$.
(2) For each fixed $k_0 \in \mathbb{N}$, the inequality $|x_{k_0}^{(n)} - x_{k_0}| \leq \|x^{(n)} - x\|_p$ holds (indeed, for $1 \leq p < +\infty$ we have

$$|x_{k_0}^{(n)} - x_{k_0}| = (|x_{k_0}^{(n)} - x_{k_0}|^p)^{\frac{1}{p}} \leq \left(\sum_{k=1}^{\infty} |x_k^{(n)} - x_k|^p\right)^{\frac{1}{p}} = \|x^{(n)} - x\|_p,$$

and for $p = +\infty$ we have $|x^{(n)}_{k_0} - x_{k_0}| \le \sup_{k\ge1} |x^{(n)}_k - x_k| = \|x^{(n)} - x\|_\infty$). Since by the condition of the problem we have $x^{(n)} \to x,\ n \to \infty$, it follows from the last inequality that $x^{(n)}_{k_0} \to x_{k_0},\ n \to \infty$.

(3) Let us show that convergence in $L_p(T, \mu)$ implies convergence in measure μ. Using Chebyshev's inequality, for any $\varepsilon > 0$ we get

$$\mu(\{t \in T \mid |x_n(t) - x(t)| \ge \varepsilon\}) \le \frac{1}{\varepsilon^p} \int_T |x_n(t) - x(t)|^p d\mu(t) \to 0,$$

$n \to \infty$, i.e. $x_n \xrightarrow[n\to\infty]{\mu} x,\ n \to \infty$. Now the Riesz theorem implies the existence of a subsequence $\{x_{n_k} : k \ge 1\}$ such that $x_{n_k} \to x,\ k \to \infty$, μ-almost everywhere on T. By the condition of the problem we have $x_{n_k} \to y,\ k \to \infty$, μ-almost everywhere on T. Thus $x = y$ μ-almost everywhere on T.

1.3 Examine the convergence of the following sequences in a normed linear space X. Find limits for convergent sequences.

(1) $X = C([0, 1]),\ x_n(t) = \sin^n \pi t,\ t \in [0, 1]$;
(2) $X = C([0, 1]),\ x_n(t) = \frac{nt}{1+n^\alpha t^2}$, where $\alpha \ge 1$ is fixed;
(3) $X = l_p,\ 1 \le p \le +\infty,\ x^{(n)} = (\underbrace{0, \dots, 0}_{n-1}, 1, 0, \dots)$;
(4) $X = l_p,\ 1 \le p \le +\infty,\ x^{(n)} = (1, 0, \frac{1}{3}, 0, \frac{1}{5}, \dots, \frac{1}{2n-1}, 0, 0, 0, \dots)$;
(5) $X = L_p([0, 1]),\ 1 \le p < +\infty,\ x_n(t) = \sin^n \pi t,\ t \in [0, 1]$;
(6) $X = L_p(\mathbb{R}),\ 1 \le p < +\infty,\ x_n(t) = \frac{1}{\sqrt{n}} \chi_{[n,2n]}(t),\ t \in \mathbb{R}$.

Solution

(1) According to problem 1.2, item (1), the only "candidate" for the limit of the sequence in $C([0, 1])$ is its pointwise limit, so let us find this limit first. For each $t \in [0, 1]$ we have

$$x_n(t) = \sin^n \pi t \to x_0(t) = \begin{cases} 1, & t = \frac{1}{2}, \\ 0, & t \in [0, 1]\setminus\{\frac{1}{2}\}, \end{cases}$$

(because $|\sin \pi t| < 1$ for $t \in [0, 1]\setminus\{\frac{1}{2}\}$). However $x_0 \notin C([0, 1])$, hence $x_n \not\to x_0$ in $C([0, 1])$, i.e. the sequence $\{x_n : n \ge 1\}$ diverges in $C([0, 1])$.

(2) Let us find the pointwise limit: if $\alpha > 1$, then for all $t \in [0, 1]$ we have $x_n(t) = \frac{nt}{1+n^\alpha t^2} \to x_0(t) = 0$, and if $\alpha = 1$, then

$$x_n(t) = \frac{nt}{1 + nt^2} \to x_0(t) = \begin{cases} 0, & t = 0, \\ \frac{1}{t}, & t \in (0, 1]. \end{cases}$$

Since for $\alpha = 1$ the pointwise limit $x_0 \notin C([0,1])$, then by problem 1.2, item (1) the sequence $\{x_n : n \geq 1\}$ diverges. Let now $\alpha > 1$. It is necessary to check whether $d_n = \max\limits_{t\in[0,1]} |x_n(t) - x(t)| \to 0,\ n \to \infty$. Let us find $d_n = \max\limits_{t\in[0,1]} \frac{nt}{1+n^\alpha t^2}$. We have $x_n'(t) = \frac{n-n^{1+\alpha}t^2}{(1+n^\alpha t^2)^2} = 0,\ n - n^{1+\alpha}t^2 = 0$, whence $t = n^{-\frac{\alpha}{2}}$. Since $x_n(0) = 0,\ x_n\big(n^{-\frac{\alpha}{2}}\big) = \frac{1}{2}n^{1-\frac{\alpha}{2}},\ x_n(1) = \frac{n}{1+n^\alpha} \leq \frac{n}{2n^{\frac{\alpha}{2}}},\ n \geq 1$, then $d_n = x_n\big(n^{-\frac{\alpha}{2}}\big) = \frac{1}{2}n^{1-\frac{\alpha}{2}}$. If $1 - \frac{\alpha}{2} \geq 0$, i.e. $\alpha \leq 2$, then $d_n = \frac{1}{2}n^{1-\frac{\alpha}{2}} \nrightarrow 0,\ n \to \infty$. If $1 - \frac{\alpha}{2} < 0$, i.e. $\alpha > 2$, then $d_n = \frac{1}{2}n^{1-\frac{\alpha}{2}} \to 0,\ n \to \infty$. So, according to problem 1.2, item (1), for $1 \leq \alpha \leq 2$ the sequence $\{x_n : n \geq 1\}$ diverges, and for $\alpha > 2$ the sequence converges to the element $x_0(t) = 0,\ t \in [0,1]$.

Note that another way of finding d_n without using the derivative is to notice that by the AM-GM inequality

$$\frac{nt}{1+n^\alpha t^2} = \frac{1}{\frac{1}{nt} + n^{\alpha-1}t} \leq \frac{1}{2\sqrt{\frac{1}{nt}n^{\alpha-1}t}} = \frac{1}{2}n^{1-\frac{\alpha}{2}},\ t \in (0,1].$$

Equality is achieved when $\frac{1}{nt} = n^{\alpha-1}t$, i.e. for $t = n^{-\frac{\alpha}{2}}$. Hence $d_n = \frac{1}{2}n^{1-\frac{\alpha}{2}}$.

(3) According to problem 1.2, item (2), the only "candidate" for the limit of the sequence in l_p is its coordinate-wise limit. Let us find this limit. For each fixed $k \in \mathbb{N}$ we have $x_k^{(n)} = 0$ for all n, starting from some number, therefore $x_k^{(n)} \to 0,\ n \to \infty$. Put $\tilde{x} = (0,\ldots,0,\ldots)$, i.e. $\tilde{x} = 0$. Let us check whether $x^{(n)} \to \tilde{x},\ n \to \infty$ in l_p. We have $\|x^{(n)} - \tilde{x}\|_p = 1 \nrightarrow 0,\ n \to \infty$, therefore the sequence $\{x^{(n)} : n \geq 1\}$ diverges in l_p.

(4) Let us find the coordinate-wise limit. Fix an arbitrary $k \in \mathbb{N}$. If k is an even number, then $x_k^{(n)} = 0$ for all $n \in \mathbb{N}$, therefore $x_k^{(n)} \to 0 = x_k,\ n \to \infty$. If k is an odd number, then $x_k^{(n)} = \frac{1}{k}$ for all $n \in \mathbb{N}$, starting from some number, therefore $x_k^{(n)} \to \frac{1}{k} = x_k,\ n \to \infty$. Put $\tilde{x} = (x_1,\ldots,x_k,\ldots)$, i.e. $\tilde{x} = (1, 0, \frac{1}{3}, 0, \frac{1}{5}, 0, \ldots)$. Since the series $\sum\limits_{k=1}^{\infty} \frac{1}{2k-1}$ diverges, $\tilde{x} \notin l_1$ and it follows from problem 1.2, item (2) that the sequence $\{x^{(n)} : n \geq 1\}$ diverges in l_1. For $1 < p < +\infty$ the series $\sum\limits_{k=1}^{\infty} \frac{1}{(2k-1)^p}$ converges, moreover $\sup\limits_{k\geq 1} |\frac{1}{2k-1}| = 1 < +\infty$, therefore $\tilde{x} \in l_p$ for $1 < p \leq +\infty$.

Let us check whether $x^{(n)} \to \tilde{x}$ in l_p. For $1 < p < +\infty$ we have

$$\|x^{(n)} - \tilde{x}\|_p^p = \sum_{k=n+1}^{+\infty} \frac{1}{(2k-1)^p} \to 0,\ n \to \infty,$$

as a remainder of a convergent series, therefore $x^{(n)} \to \tilde{x},\ n \to \infty$, in l_p. For $p = +\infty$ we have

$$\|x^{(n)} - \tilde{x}\|_\infty = \sup_{k \geq n+1} \frac{1}{2k-1} = \frac{1}{2n+1} \to 0,\ n \to \infty,$$

therefore $x^{(n)} \to \tilde{x},\ n \to \infty$, in l_∞.

(5) According to problem 1.2, item (3), if the sequence $\{x_n : n \geq 1\} \subset L_p([0,1]),\ 1 \leq p < +\infty$, converges almost everywhere with respect to the Lebesgue measure m to the function x_0, then this sequence can converge in $L_p([0,1])$ only to x_0. We have $x_n(t) = \sin^n \pi t \to 0 = x_0(t)$ m-almost everywhere on $[0,1]$. Let us check whether $x_n \to x_0,\ n \to \infty$, in $L_p([0,1])$. We have $\|x_n - x_0\|_p^p = \int\limits_0^1 \sin^{np} \pi t dt \to 0,\ n \to \infty$, according to Lebesgue's dominated convergence theorem (indeed, $y_n(t) = \sin^{np} \pi t \to 0 \,(\mathrm{mod}\, m)$, $n \to \infty$, on $[0,1]$, as well as $|y_n(t)| \leq 1 = g(t),\ t \in [0,1],\ n \geq 1,\ g \in L_1([0,1], m)$). Thus $x_n \to x_0,\ n \to \infty$, in $L_p([0,1])$.

(6) Let us find an almost everywhere limit. For each $t \leq 0 : x_n(t) = 0,\ n \geq 1$, and for $t \geq 0 : x_n(t) = 0$ for all $n \geq n_0 = [t] + 1$, hence,

$$\forall t \in \mathbb{R} : x_n(t) \to 0 = x_0(t),\ n \to \infty.$$

Let us check whether $x_n \to x_0,\ n \to \infty$, in $L_p(\mathbb{R})$. We have

$$\|x_n - x_0\|_p^p = \int\limits_{\mathbb{R}} |x_n(t) - x_0(t)|^p dt = \int\limits_n^{2n} \frac{1}{n^{\frac{p}{2}}} dt = n^{1-\frac{p}{2}} \to 0,\ n \to \infty,$$

if and only if $1 - \frac{p}{2} < 0$, i.e. for $p > 2$. Thus $\{x_n : n \geq 1\}$ converges in $L_p(\mathbb{R})$ only for $p > 2$.

1.4 Let X be a normed linear space over the field $\mathbb{K}$, G be a subspace in X, $y \notin G$. Prove that the set $G_1 = \{x + \lambda y \mid x \in G,\ \lambda \in \mathbb{K}\}$ is a subspace in X.

Solution Let us check the linearity of G_1. Let $z_1, z_2 \in G_1,\ \alpha_1, \alpha_2 \in \mathbb{K}$. Then $z_1 = x_1 + \lambda_1 y,\ z_2 = x_2 + \lambda_2 y$, where $x_1, x_2 \in G,\ \lambda_1, \lambda_2 \in \mathbb{K}$. Therefore

$$\alpha_1 z_1 + \alpha_2 z_2 = (\alpha_1 x_1 + \alpha_2 x_2) + (\alpha_1 \lambda_1 + \alpha_2 \lambda_2) y = x + \lambda y,$$

where $x = \alpha_1 x_1 + \alpha_2 x_2 \in G,\ \lambda = \alpha_1 \lambda_1 + \alpha_2 \lambda_2 \in \mathbb{K}$, hence $\alpha_1 z_1 + \alpha_2 z_2 \in G_1$.

Now we shall verify whether G_1 is closed. Let $\{z_n : n \geq 1\} \subset G_1$, and $z_n \to z,\ n \to \infty$. By definition of the set G_1 we have $z_n = x_n + \lambda_n y$, where $x_n \in G,\ \lambda_n \in \mathbb{K},\ n \geq 1$. Let us show that the sequence $\{\lambda_n : n \geq 1\}$ is bounded. Indeed, if it is not true, then there exists a subsequence $\{\lambda_{n_k} : k \geq 1\}$ such that $|\lambda_{n_k}| \to +\infty,\ k \to$

$+\infty$. Since $x_n + \lambda_n y - z \to 0,\ n \to \infty$, then $\frac{x_{n_k}}{\lambda_{n_k}} + y - \frac{z}{\lambda_{n_k}} \to 0,\ k \to \infty$. Taking into account $\frac{z}{\lambda_{n_k}} \to 0,\ k \to \infty$, we get that $\frac{x_{n_k}}{\lambda_{n_k}} \to -y,\ k \to \infty$. Since $\frac{x_{n_k}}{\lambda_{n_k}} \in G,\ k \geq 1$, and G is a closed set, then $-y \in G$, whence $y \in G$, because G is a linear set. We have got a contradiction with the condition of the problem, so the sequence $\{\lambda_n : n \geq 1\}$ is bounded. Therefore it contains a convergent subsequence $\{\lambda_{n_k} : k \geq 1\}$. Let $\lambda_{n_k} \to \lambda \in \mathbb{K},\ k \to \infty$. Since $x_n + \lambda_n y - z \to 0,\ n \to \infty$, we get $x_{n_k} \to z - \lambda y,\ k \to \infty$. Hence $z - \lambda y \in G$, because G is a closed set. Thus $z = x + \lambda y$, where $x = z - \lambda y \in G$. Consequently, $z \in G_1$, so G_1 is a closed set.

Problems to Solve

1.5° Find the norm of an element in a given space:

(1) $x(t) = t^n$ in $C([0, 1]),\ n \geq 1$;
(2) $x(t) = e^{-t}$ in $C([0, 1])$;
(3) $x = (1, \frac{1}{2}, \frac{1}{3}, \ldots)$ in l_2; in l_∞;
(4) $x(t) = t$ in $L_4([0, 1])$;
(5) $x(t) = t^2$ in $C^1([0, 1])$;
(6) $x(t) = \chi_{\mathbb{Q}}(t)$ in $L_1([0, 1])$;
(7) $x(t) = \chi_{\mathbb{R}\setminus\mathbb{Q}}(t)$ in $L_2([0, 1])$;
(8) $x(t) = 1 - e^t$ in $L_3([0, 1])$;
(9) $x(t) = \sin t + \cos t$ in $L_2([0, 1])$;
(10) $x(t) = t \ln t$ in $L_1([1, e])$.

1.6° Are the following functions norms on $C([a, b])$?

(1) $\|x\| = \max\limits_{a \leq t \leq \frac{1}{2}(a+b)} |x(t)| + |x(b)|$;

(2) $\|x\| = \left(\int\limits_a^b \alpha(t)|x(t)|^2 dt \right)^{\frac{1}{2}}$, where $\alpha \in C([a, b])$ is such that $\alpha(t) > 0,\ t \in [a, b]$;

(3) $\|x\| = \max\limits_{a \leq t \leq \frac{1}{2}(a+b)} |x(t)| + \int\limits_{\frac{1}{2}(a+b)}^{b} |x(t)| dt$.

1.7° Are the following functions norms on $C^1([a, b])$?

(1) $\|x\| = \max\limits_{a \leq t \leq b} |x(t)|$;
(2) $\|x\| = \max\limits_{a \leq t \leq b} |x'(t)|$;
(3) $\|x\| = \max\limits_{a \leq t \leq \frac{1}{2}(a+b)} |x(t)| + \max\limits_{\frac{1}{2}(a+b) \leq t \leq b} |x'(t)|$;
(4) $\|x\| = |x(a)| + \max\limits_{a \leq t \leq b} |x'(t)|$;
(5) $\|x\| = |x(b) - x(a)| + \max\limits_{a \leq t \leq b} |x'(t)|$;
(6) $\|x\| = \int\limits_a^b |x(t)| dt + \max\limits_{a \leq t \leq b} |x'(t)|$.

1.8° Are the following functions norms on $C^2([a,b])$?

(1) $\|x\| = |x(a)| + |x'(a)| + \max\limits_{a\le t\le b} |x''(t)|$;

(2) $\|x\| = |x(a)| + |x(b)| + \max\limits_{a\le t\le b} |x''(t)|$;

(3) $\|x\| = \int\limits_a^b |x(t)|dt + |x'(a)| + \max\limits_{a\le t\le b} |x''(t)|$.

1.9° Is the sequence of elements convergent in the given space? If so, find its limit.

I. In the space $C([0,1])$:

(1) $x_n(t) = t^n$;
(2) $x_n(t) = t^n - t^{n+1}$;
(3) $x_n(t) = t^n - t^{2n}$;
(4) $x_n(t) = e^{-\frac{t}{n}}$;
(5) $x_n(t) = \sin t - \sin \frac{t}{n}$;
(6) $x_n(t) = \frac{nt}{\sqrt{n^2+1}}$;
(7) $x_n(t) = nte^{-nt}$;
(8) $x_n(t) = n \sin \frac{t}{n}$;
(9) $x_n(t) = n \ln(1 + \frac{t}{n})$;
(10) $x_n(t) = t^n - t^{3n}$;
(11) $x_n(t) = \varphi(t + \frac{1}{n})$, where $\varphi \in C(\mathbb{R})$ is a fixed function.

II. In the space $C^1([0,1])$:

(1) $x_n(t) = \frac{t^{n+1}}{n+1} - \frac{t^{n+2}}{n+2}$;
(2) $x_n(t) = \int\limits_0^t e^{-\frac{u^2}{2n}} du$.

III. In the space $l_p,\ 1 \le p \le +\infty$:

(1) $x^{(n)} = (\underbrace{0,\dots,0,\tfrac{1}{n}}_{n},0,\dots)$;
(2) $x^{(n)} = (1,\underbrace{0,\dots,0,\tfrac{1}{n}}_{n},0,\dots)$;
(3) $x^{(n)} = (\tfrac{1}{n},\underbrace{0,\dots,0,\tfrac{1}{n}}_{n},0,\dots)$;
(4) $x^{(n)} = (\underbrace{0,\dots,0}_{n-1},1,\tfrac{1}{2},\tfrac{1}{3},\dots)$;
(5) $x^{(n)} = (\underbrace{0,\dots,0}_{n-1},\tfrac{1}{n},\tfrac{1}{n+1},\dots)$;
(6) $x^{(n)} = (1,\tfrac{1}{2},\dots,\tfrac{1}{n},0,\dots)$;
(7) $x^{(n)} = (1,\tfrac{1}{\sqrt{2}},\dots,\tfrac{1}{\sqrt{n}},0,\dots)$;
(8) $x^{(n)} = (\underbrace{\tfrac{1}{n},\dots,\tfrac{1}{n}}_{n},0,\dots)$;
(9) $x^{(n)} = (\underbrace{\tfrac{1}{\ln n},\dots,\tfrac{1}{\ln n}}_{n},0,\dots)$;
(10) $x^{(n)} = (\underbrace{1,\dots,1}_{n},0,\dots)$;
(11) $x^{(n)} = (\tfrac{1}{n},\tfrac{1}{n-1},\dots,\tfrac{1}{2},1,0,\dots)$;
(12) $x^{(n)} = (\sqrt[n]{1},\sqrt[n]{2},\dots,\sqrt[n]{n},0,\dots)$;
(13) $x^{(n)} = (\underbrace{\left(\tfrac{1}{2}\right)^n,\left(\tfrac{2}{3}\right)^n,\dots,\left(\tfrac{n}{n+1}\right)^n}_{n},0,0,\dots)$.

IV. In the space $L_p([0,1]),\ 1 \le p < +\infty$:

(1) $x_n(t) = t^n$;
(2) $x_n(t) = t^n - t^{n+1}$;
(3) $x_n(t) = n \cdot \chi_{[0,\frac{1}{n^2}]}(t)$;
(4) $x_n(t) = ne^{-nt},\ p = 1$;
(5) $x_n(t) = 1 - \frac{t}{n},\ p = 2$;
(6) $x_n(t) = (\sqrt{n} - n\sqrt{n}\,t)\chi_{[0,\frac{1}{n}]}(t)$, $p = 2$.

1.10° Prove directly that the following spaces are complete:

(1) $\mathbb{C}^m$ with the norm $\|x\|_\infty = \max\limits_{1\le k\le m} |x_k|$;

(2) l_p, $1 \le p < +\infty$, with the norm $\|x\|_p = \left(\sum\limits_{k=1}^{\infty} |x_k|^p\right)^{\frac{1}{p}}$;

(3) l_∞ with the norm $\|x\|_\infty = \sup\limits_{k\ge 1} |x_k|$.

1.11 Let X be a Banach space. Prove that the space of continuous bounded functions $C_b(\mathbb{R}, X)$ with uniform norm $\|x\| = \sup\limits_{t\in\mathbb{R}} \|x(t)\|_X$, $x \in C_b(\mathbb{R}, X)$, is a Banach space.

1.12°

(1) Let $X = \mathbb{R}^2$ with the norm $\|\cdot\|_p$, where $p = 1, 2, +\infty$. Draw unit balls centered at the origin.
(2) Let $X = \mathbb{R}^3$ with the norm $\|\cdot\|_p$, where $p = 1, 2, +\infty$. Draw unit balls centered at the origin.

1.13° Prove that $C([a, b]) \subset L_p([a, b])$ for all $1 \le p \le +\infty$ and $[a, b] \subset \mathbb{R}$; moreover, $\|x\|_{L_p([a,b])} \le (b-a)^{\frac{1}{p}} \|x\|_{C([a,b])}$ for $1 \le p < +\infty$ and $\|x\|_{L_\infty([a,b])} = \|x\|_{C([a,b])}$, $x \in C([a, b])$.

1.14

(1) Show that every norm on the space $\mathbb{R}$ has form $\|x\| = \alpha|x|$, $x \in \mathbb{R}$, where $\alpha > 0$ is a fixed constant.
(2)* Let $f : \mathbb{R} \to (0, +\infty)$ be a π-periodic function, for which the set $A_f = \{(\rho\cos\varphi, \rho\sin\varphi) \mid \varphi \in \mathbb{R}, 0 \le \rho \le f(\varphi)\} \subset \mathbb{R}^2$ is convex. Put $\|x\| = \frac{\rho}{f(\varphi)}$, where (ρ, φ) are the polar coordinates of a point $x \in \mathbb{R}^2$. Prove that $\|\cdot\|$ is a norm on $\mathbb{R}^2$.
(3) Prove that each norm on $\mathbb{R}^2$ can be obtained in the way described in (2)*. Find the functions f that correspond to norms on the spaces $\mathbb{R}^2_p$, $1 \le p \le +\infty$.

1.15 Prove that for every $1 \le p_1 < p_2 \le +\infty$ the inclusion $l_{p_1} \subset l_{p_2}$ holds, and $\|x\|_\infty = \lim\limits_{p\to\infty} \|x\|_p$ for all $x \in l_1$.

1.16 Let $(T, \mathcal{F}, \mu)$ be a space with a finite measure. Prove that for every $1 \le p_1 < p_2 \le +\infty$ the inclusion $L_{p_2}(T, \mu) \subset L_{p_1}(T, \mu)$ holds, and $\|x\|_\infty = \lim\limits_{p\to\infty} \|x\|_p$ for all $x \in L_\infty(T, \mu)$.

1.17° Prove that the spaces $L_{p_1}(\mathbb{R})$ and $L_{p_2}(\mathbb{R})$ do not contain each other for any $1 \le p_1 < p_2 \le +\infty$.

1.18* Let $(T, \mathcal{F}, \mu)$ be a space with a measure, $1 \le p_1 < p_2 < +\infty$. Prove that

(1) $L_{p_2}(T, \mu) \subset L_{p_1}(T, \mu)$ if and only if there exists $C > 0$ such that $A \in \mathcal{F}$, $\mu(A) < \infty \Rightarrow \mu(A) \le C$.

(2) $L_{p_1}(T,\mu) \subset L_{p_2}(T,\mu)$ if and only if there exists $\delta > 0$ such that $A \in \mathcal{F},\ \mu(A) > 0 \Rightarrow \mu(A) \geq \delta$.

1.19 Let $1 \leq p < r < s$. Prove that $\big(L_p(\mathbb{R}) \cap L_s(\mathbb{R})\big) \subset L_r(\mathbb{R})$.

1.20 Let $x \in L_\infty(T,\mu)$. Prove that

$$\operatorname*{ess\,sup}_{t\in T} |x(t)| = \min\{C \geq 0 \mid |x(t)| \leq C \ (\mathrm{mod}\,\mu)\} =$$

$$= \inf\left\{ \sup_{t\in T\setminus A} |x(t)| \,\Big|\, A \in \mathcal{F},\ \mu(A) = 0 \right\}.$$

1.21°

(1) Is the function $x \mapsto \|x\| = \left(\sum\limits_{i=1}^{\infty} |x_i|^p\right)^{\frac{1}{p}}$ a norm on l_p for $0 < p < 1$, where l_p for $0 < p < 1$ is formally defined in the same way as for $1 \leq p < +\infty$?

(2) Is the function $x \mapsto \|x\| = \left(\int\limits_a^b |x(t)|^p dt\right)^{\frac{1}{p}}$ a norm on $C([a,b])$ for $0 < p < 1$?

1.22 Under what condition on a sequence $\{\alpha_n \ : \ n \geq 1\} \subset [0,+\infty)$ the function $l_2 \ni x \mapsto \left(\sum\limits_{k=1}^{\infty} \alpha_k |x_k|^2\right)^{1/2}$ is a norm on l_2?

1.23 When is equality achieved (1) in Hölder's inequality; (2) in the Minkowski inequality?

1.24 Let $f \in L_2(\mathbb{R}),\ tf(t) \in L_2(\mathbb{R})$. Prove that $f \in L_1(\mathbb{R})$.

1.25 Prove that for $1 \leq p \leq 2$ and for each function $f \in L_p(\mathbb{R})$ there exist $f_1 \in L_1(\mathbb{R})$ and $f_2 \in L_2(\mathbb{R})$ such that $f = f_1 + f_2$.

1.26 Let $f \in L_1(\mathbb{R})$ be such that for some $\sigma > 0$

$$\exists A > 0 \ : \ \int\limits_{\mathbb{R}} |x^k f(x)| dx \leq A\sigma^k, \ k \geq 1.$$

Prove that $f = 0 \,(\mathrm{mod}\, m)$ on $(-\infty, -\sigma) \cup (\sigma, +\infty)$.

1.27° Let $0 < \alpha < 1$. For which $1 \leq p < +\infty$ the function

$$x(t) = \begin{cases} n^\alpha, & \frac{1}{n+1} \leq t < \frac{1}{n},\ n \geq 1, \\ 0, & t = 0,\ t = 1, \end{cases}$$

belongs to the space $L_p([0,1])$?

1.28 Give examples of infinite-dimensional subspaces L in $L_1(\mathbb{R})$ such that

(1) L consists of continuous functions;
(2) L does not contain any nonzero continuous function.

1.29 Let

$$X = \left\{ x = (x_1, x_2, \ldots) \mid \sup_{n\geq 1} \left| \sum_{k=1}^{n} x_k \right| < +\infty,\ x_k \in \mathbb{K}, k \geq 1 \right\}.$$

Define a norm on X as $\|x\| = \sup_{n\geq 1} \left| \sum_{k=1}^{n} x_k \right|$. Prove that $(X, \|\cdot\|)$ is a Banach space.

1.30° Prove that the functions $1,\ \cos t,\ \cos^2 t$ are linearly independent in $C([0, \pi])$, and the functions $1,\ \cos^2 t,\ \cos 2t$ are linearly dependent in $C([0, \pi])$.

1.31° Define the function $\|x\| = \max_{t\in[a,b]} |x'(t) - x(t)|,\ x \in C^1([a, b])$. Is it a norm (1) on the set P_n of polynomials with powers at most n, where $n \in \mathbb{N}$ is fixed? (2) on $C^1([a, b])$?

1.32 Let $\|\cdot\|_1, \|\cdot\|_2$ be norms on a linear space X. Prove that the following conditions are equivalent:

(1) the norms $\|\cdot\|_1$ and $\|\cdot\|_2$ are equivalent;
(2) $x_n \to x,\ n \to \infty$, in $(X, \|\cdot\|_1)$ if and only if $x_n \to x,\ n \to \infty$, in $(X, \|\cdot\|_2)$;
(3) the topologies (i.e. classes of open sets) in the spaces $(X, \|\cdot\|_1)$ and $(X, \|\cdot\|_2)$ coincide.

1.33° Let $\|\cdot\|_1$ and $\|\cdot\|_2$ be equivalent norms on a linear space X. Prove that

(1) if $(X, \|\cdot\|_1)$ is a Banach space, then $(X, \|\cdot\|_2)$ is also a Banach space;
(2) if $(X, \|\cdot\|_1)$ is a separable space, then the space $(X, \|\cdot\|_2)$ is also separable.

1.34° Let the spaces $(X_1, \|\cdot\|_1)$ and $(X_2, \|\cdot\|_2)$ be isomorphic, and $(X_1, \|\cdot\|_1)$ be a Banach space. Prove that $(X_2, \|\cdot\|_2)$ is also a Banach space. As a corollary, show that every finite-dimensional NLS is a Banach space.

1.35° Prove that the norms $\|x\| = \max_{a\leq t\leq b} |x(t)|$ and $\|x\| = \left(\int_a^b |x(t)|^p dt \right)^{\frac{1}{p}}$, $1 \leq p < +\infty$, in $C([a, b])$ are not equivalent.

1.36° Is it true that the norms $\|x\| = |x(a)| + \max_{a\leq t\leq b} |x'(t)|$ and $\|x\| = \max_{a\leq t\leq b} |x(t)| + \max_{a\leq t\leq b} |x'(t)|$ in $C^1([a, b])$ are equivalent?

1.37 Let $\{r_n : n \geq 1\} = \mathbb{Q} \cap [0, 1]$. Prove that $\|x\| = \sum_{n=1}^{\infty} \frac{|x(r_n)|}{n^2}$, $x \in C([0, 1])$, is a norm on $C([0, 1])$. Is it equivalent to the uniform norm?

1.38 Let $\|\cdot\|_1$, $\|\cdot\|_2$ be norms on the linear space X, and $\overline{B}_i(0,1)$ be the unit closed ball in $(X, \|\cdot\|_i)$, $i = 1, 2$. Prove that if $\overline{B}_1(0,1) = \overline{B}_2(0,1)$, then $\|x\|_1 = \|x\|_2$, $x \in X$.

1.39 Let X be an NLS, $\{x_n : n \geq 1\} \subset X$. Which of the following conditions are equivalent:

(1) $\{x_n : n \geq 1\}$ is a Cauchy sequence;
(2) $\|x_n - x_m\| \to 0$, $m, n \to \infty$;
(3) $\lim\limits_{m\to\infty} \lim\limits_{n\to\infty} \|x_m - x_n\| = 0$?

1.40 Prove that an NLS X is a Banach space if and only if any series $\sum\limits_{k=1}^{\infty} x_k$, for which $\sum\limits_{k=1}^{\infty} \|x_k\| < +\infty$, converges in X.

1.41 Let $n \geq 1$. Prove that $C^n([a,b])$ is a Banach space with the norm

$$\|x\|_n = \sum_{k=0}^{n} \max_{a\leq t\leq b} |x^{(k)}(t)|,$$

but it is not complete with the norm $\|x\|_{n-j} = \sum\limits_{k=0}^{n-j} \max\limits_{a\leq t\leq b} |x^{(k)}(t)|$ for every j, $1 \leq j \leq n$.

1.42 Let X be an NLS, $x \in X$, and $M \subset X$ be some nonempty set.

(1) Prove that the function $\varphi(x) = \rho(x, M)$ is continuous on X.
(2) Prove that $\rho(x, M) = \rho(x, \overline{M})$, where $\overline{M}$ is the closure of the set M.
(3) Prove statements (1) and (2) for the case when (X, ρ) is a metric space and $\rho(x, M) := \inf\limits_{y\in M} \rho(x, y)$.

1.43 Let Q be a metric space, and the Banach space $C_b(Q)$ of bounded continuous functions on Q with norm $\|x\| = \sup\limits_{t\in Q} |x(t)|$ be separable. Prove that Q is a compact set.

1.44 Let T be some set, and $B(T)$ be the space of bounded functions on T with the norm $\|x\| = \sup\limits_{t\in T} |x(t)|$.

(1) Prove that $B(T)$ is a Banach space.
(2) Under what condition on T is the space $B(T)$ separable?

1.45 Prove that the space $BV_0([a,b])$ of functions of bounded variation on $[a,b]$ such that $x(a) = 0$, with the norm $\|x\| = V(x, [a,b])$ is a Banach space. Is it separable?

1.46 Let $H^{\lambda}([a,b])$ denote the set of all functions that satisfy the Hölder condition on $[a,b]$ with the exponent $\lambda \in (0,1]$:

$$\varphi_{\lambda}(x) = \sup_{a \le s < t \le b} \frac{|x(t) - x(s)|}{(t-s)^{\lambda}} < +\infty.$$

Prove that $H^{\lambda}([a,b])$ is a Banach space with respect to the norm

$$\|x\| = \max_{a \le t \le b} |x(t)| + \varphi_{\lambda}(x), \ x \in H^{\lambda}([a,b]).$$

Is it separable?

1.47 Let X be a linear metric space with a metric ρ having the following properties:

(i) $\forall x, y, z \in X : \rho(x+z, y+z) = \rho(x,y)$;
(ii) $\forall x \in X \ \forall \lambda \in \mathbb{C} : \rho(0, \lambda x) = |\lambda| \rho(0,x)$.
(1) Show that X is an NLS with the norm $\|x\| = \rho(0,x)$.
(2) Give an example of a metric in $\mathbb{R}$ that does not have any of the properties (i), (ii).

1.48 Let X be an NLS, $x, y \in X$ be such that $\|x+y\| = \|x\| + \|y\|$.

(1) Let $X = \mathbb{R}^m$ with Euclidean norm. Prove that x and y are linearly dependent. Is this correct in other normed spaces?
(2) Prove that $\|\alpha x + \beta y\| = \alpha\|x\| + \beta\|y\|$ for all $\alpha, \beta \ge 0$.

1.49 Let X be an NLS. Prove that

(1) $\forall x, y \in X : \|x\| \le \max\{\|x+y\|, \|x-y\|\}$;
(2) $\forall x, y \in X : \|x\| + \|y\| \le \|x+y\| + \|x-y\|$.

1.50 Let X be the linear space of all complex numerical sequences. Prove that

(1) there is no norm on X such that convergence in the space $(X, \|\cdot\|)$ is equivalent to coordinate-wise convergence;
(2)* there exists a metric on X such that convergence in the space (X, ρ) is equivalent to coordinate-wise convergence.

1.51 Let $x \in L_1([a,b])$. Prove that $\lim\limits_{h \to 0+} \int\limits_a^{b-h} |x(t+h) - x(t)| dt = 0$.

1.52° Prove that

(1) the system of functions $\{1, t, t^2, \ldots\}$ is total in $C([a,b])$ and in $L_p([a,b])$, $1 \le p < +\infty$;
(2) the system of functions $\{t^{2k} : k \ge 0\}$ is total in $C([a,b])$ and in $L_p([a,b])$, $1 \le p < +\infty$, for $a \ge 0$;

(3) the system of functions $\{1, \cos nt, \sin nt : n \geq 1\}$ is total in $L_p([0, 2\pi]),\ 1 \leq p < +\infty$;
(4) the system of functions $\{e^{int} : n \in \mathbb{Z}\}$ is total in $L_p([0, 2\pi]),\ 1 \leq p < +\infty$;
(5) the system of functions $\{e^{int} : n \in \mathbb{Z}\}$ is total in $C([a, b])$ if and only if $|b-a| < 2\pi$;
(6) the system of functions $\{1, \sin nt : n \geq 1\}$ is total in $C([0, b])$ if and only if $0 < b < \pi$;
(7) the system of functions $\{1, \cos nt : n \geq 1\}$ is total in $C([0, b])$ if and only if $0 < b \leq \pi$;
(8) the system of elements $\{e_n = (0, \ldots, 0, \underbrace{1}_{n}, 0, \ldots) : n \geq 1\}$ is total in $l_p,\ 1 \leq p < +\infty$.

1.53° Prove that the following sets are dense in $L_p([a, b]),\ 1 \leq p < +\infty$:

(1) simple functions;
(2) step functions;
(3) continuous functions;
(4) polynomials;
(5) polynomials with rational coefficients;
(6) polynomials with zero sum of coefficients;
(7) even polynomials if $a \geq 0$;
(8) continuous functions x such that $x(a) = 0$;
(9) continuous functions x such that $x(a) = x(b) = 0$;
(10) polynomials of e^t.

1.54°

(1) Prove that the set $C_0(\mathbb{R})$ of continuous functions on $\mathbb{R}$ which vanish outside some finite interval is dense in $L_p(\mathbb{R}),\ 1 \leq p < +\infty$.
(2) Prove that the space $L_p(\mathbb{R}),\ 1 \leq p < +\infty$ is separable.

1.55° Prove that the set $C_b(\mathbb{R})$ of bounded continuous functions on $\mathbb{R}$ is not dense in $L_\infty(\mathbb{R})$.

1.56 Prove the nonseparability of the space $L_\infty([a, b])$.

1.57 Let K be a compact set in the metric space $(\mathbb{R}^m, \rho)$. Prove that the space $C(K)$ with uniform norm is separable.

1.58* Let (K, ρ) be a compact metric space. Prove that

(1) if K is an infinite set, then $C(K)$ is an infinite-dimensional space;
(2) if $\{U_i : 1 \leq i \leq n\}$ is a finite open cover of K, then there exists a system of real continuous functions $\{\varphi_i : 1 \leq i \leq n\}$ on K such that
(i) $\varphi_i(t) \geq 0,\ 1 \leq i \leq n,\ t \in K$,
(ii) $\varphi_i(t) = 0,\ t \notin U_i,\ i = 1, \ldots, n$,

(iii) $\sum\limits_{i=1}^{n} \varphi_i(t) = 1,\ t \in K$
(such a system is called *a partition of unity*);

(3) the Banach space $C(K)$ is separable.

1.59 Let X be an NLS. Prove that X is separable if and only if there exists $r > 0$ such that the sphere $S(0, r)$ is a separable metric space.

1.60 Let $p \in [1, \infty)$. The space of *locally p-integrable* functions on $\mathbb{R}$ is defined as

$$L_p^{loc}(\mathbb{R}) := \{f : \mathbb{R} \to \mathbb{C} \,|\, \forall a, b \in \mathbb{R},\ a < b\ :\ f \cdot \chi_{[a,b]} \in L_p(\mathbb{R})\}.$$

Under what condition on a function $g \in L_p^{loc}(\mathbb{R})$ the set

$$\{g \cdot \chi_{(\alpha,\beta]} \,|\, -\infty < \alpha < \beta < +\infty\}$$

is total in $L_p(\mathbb{R})$?

1.61° Are the following subsets subspaces in $C([-1, 1])$:

(1) monotonic functions;
(2) nondecreasing functions;
(3) even functions;
(4) odd functions;
(5) polynomials;
(6) polynomials of degree at most m, where $m \in \mathbb{N}$ is fixed;
(7) continuous piecewise smooth functions;
(8) continuously differentiable functions;
(9) continuous functions of bounded variation;
(10) functions x for which $x(0) = 0$;
(11) functions x for which $\int\limits_{-1}^{1} x(t)dt = 0$;
(12) functions satisfying the Lipschitz condition.

Remark Piecewise smooth functions are the functions that have a continuous derivative everywhere except for a finite number of points.

1.62° Is the set

$$M = \left\{ x \in C^1([-1, 1]) \,\middle|\, \int_{-1}^{1} x(t)dt = 0 \right\}$$

a subspace (1) in the space $C^1([-1, 1])$ with the norm

$$\|x\| = \max_{-1\le t\le 1} |x(t)| + \max_{-1\le t\le 1} |x'(t)|?$$

(2) in the space $C([-1, 1])$ with the norm $\|x\| = \max\limits_{-1\le t\le 1} |x(t)|$?

1.63 Consider l_1 as a subset of l_∞. Find its closure in l_∞.

1.64 Let $M = \left\{x = (x_1, x_2, \ldots) \in X \;\middle|\; \sum\limits_{k=1}^{\infty} x_k = 0\right\}$. Is M a subspace of X if

(1) $X = l_1$; (2) $X = l_p,\ 1 < p < +\infty$; (3) $X = l_\infty$?

1.65 Is the space l_1 complete with the norm of the space l_2?

1.66 Let us define in l_∞ the set $c = \left\{x \in l_\infty \;\middle|\; \exists \lim\limits_{i\to\infty} x_i \in \mathbb{K}\right\}$. Prove that

(1) c is a subspace in l_∞;
(2) c with the norm $\|x\| = \|x\|_\infty = \sup\limits_{i\ge 1} |x_i|$ is a separable Banach space.

1.67 Let us define in l_∞ the set $c_0 = \left\{x \in l_\infty \;\middle|\; \lim\limits_{k\to\infty} x_k = 0\right\}$. Prove that

(1) c_0 is a subspace in l_∞;
(2) c_0 with the norm $\|x\| = \|x\|_\infty = \sup\limits_{i\ge 1} |x_i|$ is a separable Banach space.

1.68° Let A be a set in NLS. Is it always true that $\operatorname{span}(\overline{A}) = \overline{\operatorname{span}(A)}$?

1.69° Prove that a subspace in NLS is a convex set.

1.70 Prove that the intersection of any family of convex sets in a linear space is a convex set.

1.71 Let a function φ on a linear space X satisfy the first two axioms of the norm. Prove that φ is a norm on X if and only if the set $B = \{x \in X \mid \varphi(x) \le 1\}$ is convex.

1.72 Prove that in any finite-dimensional NLS all norms are equivalent.

1.73° Prove that in any finite-dimensional NLS every bounded closed set is compact.

1.74

(1) Prove that in a finite-dimensional NLS every bounded sequence has at least one partial limit (i.e. the limit of some subsequence).
(2) Give an example of a bounded sequence in l_2 which does not have any partial limits.

1.75° Prove that in every NLS X the distance from any fixed point x to any finite-dimensional subspace M is attained.

1.76° Let X be an NLS, $x_1, \ldots, x_n \in X$. Prove that

$$\operatorname{span}(\{x_1, \ldots, x_n\}) = \overline{\operatorname{span}(\{x_1, \ldots, x_n\})}.$$

1.77° Let X be an NLS, $x_1, \ldots, x_n \in X$ be linearly independent elements. Prove that $\rho\big(x_n, \operatorname{span}(\{x_1, \ldots, x_{n-1}\})\big) > 0$.

1.78 Let L_1, L_2 be subspaces in NLS X, and at least one of these subspaces is finite-dimensional. Prove that $L_1 + L_2$ is a subspace in X.

1.79° Let $(X_1, \|\cdot\|_{X_1})$ and $(X_2, \|\cdot\|_{X_2})$ be NLS over a field $\mathbb{K}$. The Cartesian product $X := X_1 \times X_2$, in which linear operations are defined as follows:

$$(x_1, x_2) + (y_1, y_2) = (x_1 + y_1, x_2 + y_2), \quad \alpha\,(x_1, x_2) = (\alpha x_1, \alpha x_2),$$

$(x_1, x_2) \in X$, $(y_1, y_2) \in X$, $\alpha \in \mathbb{K}$, is called the direct sum of X_1 and X_2. It is denoted by $X_1 \bigoplus X_2 = X$.

(1) Prove that X is a linear space.
(2) Prove that the functions

$$\|(x_1, x_2)\|_p = \left(\|x_1\|_{X_1}^p + \|x_2\|_{X_2}^p\right)^{\frac{1}{p}}, \quad 1 \le p < +\infty,$$

$$\|(x_1, x_2)\|_\infty = \max\{\|x_1\|_{X_1}, \|x_2\|_{X_2}\}$$

are norms in X.

1.80 Let X be an NLS over a field $\mathbb{K}$. Prove that a set $M \subset X$ is bounded if and only if for each sequence $\{x_n : n \ge 1\} \subset M$ and each sequence $\{\alpha_n : n \ge 1\} \subset \mathbb{K}$ such that $\alpha_n \to 0,\ n \to \infty$, the convergence $\alpha_n x_n \to 0,\ n \to \infty$, in X holds.

1.81 Let X be an NLS, and $M \subset X$ be a linear set, $M \ne X$. Prove that M has no interior points.

1.82 A linearly independent system $\{x_\alpha : \alpha \in A\}$ of elements of a linear space X is called a *Hamel basis* if $\operatorname{span}(\{x_\alpha : \alpha \in A\}) = X$.

(1) Prove that in every linear space there exists a Hamel basis.
(2) Prove that each element of the space X is uniquely represented as a linear combination of a finite number of elements from the Hamel basis of the space X.
(3)* Based on the Baire category theorem (see, for example, [26], Theorem III.8 or [21], Ch. 2, Sec. 7, Theorem 2), prove that in an infinite-dimensional Banach space, there is no countable Hamel basis.
(4) Show that in item (3)* completeness of the space is essential.

1.83 Prove that an NLS is nonseparable if and only if it contains an uncountable number of balls of some fixed radius $r > 0$ that do not intersect pairwise.

1.84 Let X be a Banach space, $\left\{\overline{B}(x_n, r_n) : n \geq 1\right\}$ be a sequence of nested closed balls, and $\lim\limits_{n\to\infty} r_n = 0$. Prove that

$$\exists!\, x \in \bigcap_{n=1}^{\infty} \overline{B}(x_n, r_n).$$

1.85 Let X be an NLS in which any sequence $\left\{\overline{B}(x_n, r_n) : n \geq 1\right\}$ of nested closed balls, for which $\lim\limits_{n\to\infty} r_n = 0$, has a nonempty intersection. Prove that X is a Banach space.

1.86

(1) Give an example of a Banach space X and a sequence of nested nonempty closed sets that has an empty intersection.
(2) Prove that every sequence of nested closed balls in a Banach space has a nonempty intersection.

1.87 Let X be a Banach space, $M \subset X$ be a linear set. Prove that the completion of M in the norm of the space X coincides with the closure of M.

1.88 Let X be an NLS. Consider the set $\overline{X}$, which consists of all Cauchy sequences of the space X, i.e. $\overline{x} \in \overline{X}$, if $\overline{x} = (x_1, x_2, \ldots)$, $\{x_n : n \geq 1\}$ is a Cauchy sequence in X.

(1) Two sequences $\overline{x} = \{x_n : n \geq 1\}$ and $\overline{y} = \{y_n : n \geq 1\}$ are said to be equivalent ($\overline{x} \sim \overline{y}$), if $\|x_n - y_n\| \to 0,\ n \to \infty$. Prove that the relation $\overline{x} \sim \overline{y}$ is an equivalence relation, i.e. it is reflexive, symmetric, and transitive.
(2) Thus the set $\overline{X}$ is split into classes of equivalent sequences. Let us denote the set of these classes by $\widetilde{X}$. Prove that it is possible to introduce the structure of a linear space in $\widetilde{X}$.
(3) For every $\tilde{x} \in \widetilde{X}$, we set $\|\tilde{x}\|_{\widetilde{X}} = \lim\limits_{n\to\infty} \|x_n\|_X$. Prove that this formula determines a norm on $\widetilde{X}$.
(4) Prove that $(\widetilde{X}, \|\tilde{x}\|_{\widetilde{X}})$ is a complete space.
(5) Prove that X is dense in $\widetilde{X}$ if an element $x \in X$ is identified with the equivalence class $\tilde{x} \in \widetilde{X}$ that contains the sequence $(x, x, \ldots)$.

1.89 Let us call the sequence $\{e_1, \ldots, e_n, \ldots\} \subset X$ a *Schauder basis* in the NLS X if for each $x \in X$ there exists a unique representation $x = \sum\limits_{n=1}^{\infty} \alpha_n e_n,\ \alpha_n \in \mathbb{K},\ n \geq 1$, where the series converges in norm.

(1) Prove that any space with a Schauder basis is separable.
(2) Prove that there exists a Schauder basis in the spaces $l_p,\ 1 \leq p < +\infty,\ c_0,\ c$.

1.90* Let $\{e_n : n \geq 1\} \subset X$ be the sequence of nonzero elements in a Banach space X with the following properties:

(a) $\overline{\text{span}(\{e_n : n \geq 1\})} = X$;

(b) $\exists C > 0 \, \forall m > k \, \forall \lambda_1, \dots, \lambda_m \in \mathbb{K} : \left\| \sum\limits_{i=1}^{k} \lambda_i e_i \right\| \leq C \left\| \sum\limits_{i=1}^{m} \lambda_i e_i \right\|$.

Prove that $\{e_n : n \geq 1\}$ is a Schauder basis in the space X.

Remark The sequence $\{e_n : n \geq 1\} \subset X \setminus \{0\}$ is a Schauder basis in the Banach space X if and only if it satisfies conditions (a) and (b) (see problem 8.30*).

1.91* Consider the sequence of functions on $[0, 1]$ defined as follows:

$$h_{2^k+l}(t) = \begin{cases} 1, & \frac{2l-2}{2^{k+1}} \leq t < \frac{2l-1}{2^{k+1}}, \\ -1, & \frac{2l-1}{2^{k+1}} \leq t < \frac{2l}{2^{k+1}}, \\ 0, & \text{otherwise}, \end{cases}$$

where $k = 0, 1, 2, \dots$ and $1 \leq l \leq 2^k$, $h_1(t) = 1$, $t \in [0, 1]$. The sequence $\{h_n : n \geq 1\}$ is called the *Haar system*. Prove that

(1) $\overline{\text{span}(\{h_n : n \geq 1\})} = L_p([0, 1])$ in the norm of $L_p([0, 1])$, $1 \leq p < +\infty$;
(2) the Haar system is a Schauder basis in $L_p([0, 1])$ for every $1 \leq p < +\infty$;

(3)* $\forall n \geq 1 \, \forall \lambda_1, \lambda_2, \dots, \lambda_{n+1} \in \mathbb{K} : \left\| \sum\limits_{i=1}^{n} \lambda_i h_i \right\|_p \leq \left\| \sum\limits_{i=1}^{n+1} \lambda_i h_i \right\|_p$.

A basis satisfying this condition is said to be *monotone*.

1.92* Let $\{h_n : n \geq 1\}$ be the Haar system from the previous problem. Consider the sequence of functions $\{s_n : n \geq 0\}$ in $C([0, 1])$ defined as follows: $s_0(t) = 1$, $t \in [0, 1]$, and $s_n(t) = \int\limits_0^t h_n(\tau)d\tau$, $t \in [0, 1]$, $n \geq 1$. The sequence $\{s_n : n \geq 0\}$ is called the *Schauder system*. Prove that the Schauder system is a monotone Schauder basis in $C([0, 1])$.

1.93 Let A be a set in an NLS X, and let A contains a countable subset which is dense in A. Prove that $\overline{\text{span}(A)}$ is a separable subspace in X.

1.94 An NLS X is said to be *strictly normed* if for any $x_1, x_2 \in X$ equalities $\|x_1\| = \|x_2\| = \|\frac{1}{2}(x_1 + x_2)\| = 1$ imply that $x_1 = x_2$. The norm on such a space is said to be *strictly convex*. Prove that

(1) X is strictly normed if and only if for all $x_1, x_2 \in X$ and $\alpha \in [0, 1]$ equalities $\|x_1\| = \|x_2\| = \|\alpha x_1 + (1 - \alpha)x_2)\| = 1$ imply that $x_1 = x_2$;
(2) X is strictly normed if and only if the unit ball $\overline{B}(0, 1)$ is a strictly convex set (for an open set A, the set $\overline{A}$ is said to be *strictly convex* if $\alpha x_1 + (1 - \alpha)x_2 \in A$ for all $x_1, x_2 \in \overline{A}$ and $\alpha \in (0, 1)$).

1.95 For which $1 \le p \le +\infty$ the space $\mathbb{R}^m$ with the norm $\|\cdot\|_p$ is strictly normed?

1.96 Prove that

(1) the spaces $L_p(T,\mu)$, $1 < p < +\infty$, are strictly normed;
(2) the spaces $C([a,b])$, $L_1([a,b])$, $L_\infty([a,b])$ are not strictly normed.

1.97* Prove that a real NLS X is strictly normed if and only if for an arbitrary convex set $L \subset X$ and for any $x \in X$ there exists at most one element of the best approximation (i.e. if the element of the best approximation exists, then it is unique).

In $(\mathbb{R}^2, \|\cdot\|_1)$, give examples of the subspace $L \subset \mathbb{R}^2$ and the element $x \in \mathbb{R}^2$, for which

(1) there exists a unique element of the best approximation in L;
(2) there exist several elements of the best approximation in L.

1.98 A normed linear space X is said to be *uniformly convex* if

$$\forall \varepsilon > 0\, \exists \delta > 0\, \forall x_1, x_2 \in X,\ \|x_1\| = \|x_2\| = 1,$$
$$\|\tfrac{1}{2}(x_1 + x_2)\| > 1 - \delta\ :\ \|x_1 - x_2\| < \varepsilon.$$

Prove that (1) a uniformly convex space is strictly normed; (2) a finite-dimensional strictly normed space is uniformly convex.

1.99*

(1) Let $1 < p < \infty$. Prove that for any $0 < c < 1$ there exists $a_c \ge 1$ such that

$$\frac{|z_1|^p + |z_2|^p}{2} \le a_c \left(\frac{|z_1|^p + |z_2|^p}{2} - \left|\frac{z_1 + z_2}{2}\right|^p \right)$$

for all $z_1, z_2 \in \mathbb{K}$, for which $\left|\frac{z_1 - z_2}{2}\right|^p \ge c\left(\frac{|z_1|^p + |z_2|^p}{2}\right)$.
(2) Prove that the space $L_p(T,\mu)$, $1 < p < +\infty$, is uniformly convex.

1.100* Let X be a uniformly convex Banach space. Prove that for every element $x \in X$ in an arbitrary convex closed set $L \subset X$ there exists a unique element of the best approximation.

1.101 Let X be a uniformly convex Banach space, $K \subset X$ be a nonempty closed convex set. Prove that the function $f(x) = \|x\|$, $x \in X$, attains its minimum on K exactly once.

1.102* Let $(T, \mathcal{F}, \mu)$ be a measurable space with a measure μ, $\mu(T) < +\infty$. Let Φ denote the set of all $n \in \mathbb{N}$ for which there exists a partition of the space $T = \bigcup_{k=1}^{n} A_k$, where $A_i \in \mathcal{F}$, $\mu(A_i) > 0$, $1 \le i \le n$, $A_i \cap A_j = \varnothing$, $i \ne j$. Prove that for every $1 \le p < +\infty$ the dimension of the space $L_p(T, \mu)$ is given by $\dim L_p(T, \mu) = \sup \Phi$. At the same time,

$$\dim L_p(T, \mu) = n < +\infty \Leftrightarrow \text{there exists a partition of } T \text{ into } n \text{ atoms.}$$

(A set $A \in \mathcal{F}$ is called an atom if $\mu(A) > 0$ and for every subset $B \subset A$, $B \in \mathcal{F}$, either $\mu(B) = 0$, or $\mu(A \backslash B) = 0$.)

Chapter 2
Hilbert Spaces

Theoretical Background

Let H be a complex linear space. A function $(\cdot, \cdot) : H \times H \to \mathbb{C}$ is called a *scalar product* (or an *inner product*), if

(a) $\forall x \in H \ : \ (x, x) \geq 0$, moreover, $(x, x) = 0 \ \Leftrightarrow \ x = 0$;
(b) $\forall x, y, z \in H \ \forall \alpha, \beta \in \mathbb{C} \ : \ (\alpha x + \beta y, z) = \alpha(x, z) + \beta(y, z)$ (linearity in the first argument);
(c) $\forall x, y \in H \ : \ (y, x) = \overline{(x, y)}$ (conjugate symmetry).

In a real linear space H, a scalar product is a function $(\cdot, \cdot) : H \times H \to \mathbb{R}$, which satisfies conditions (a), (b) for $\alpha, \beta \in \mathbb{R}$ and the condition

(c)' $\forall x, y \in H \ : \ (y, x) = (x, y)$ (symmetry).

A linear space with a scalar product is called a *pre-Hilbert* space (or *inner product space*).

If $(\cdot, \cdot)$ is a scalar product in a complex linear space H, then:

(1) $\forall \ x, y, z \in H \ \forall \ \alpha, \beta \in \mathbb{C} \ : \ (x, \alpha y + \beta z) = \overline{\alpha}(x, y) + \overline{\beta}(x, z)$ (anti-linearity in the second argument).
(2) $\forall \ x, y \in H \ : \ |(x, y)|^2 \leq (x, x) \cdot (y, y)$ (the Cauchy–Schwarz inequality).

In a pre-Hilbert space H, the function $x \mapsto \|x\| = \sqrt{(x, x)}$ is a norm on H. If the space H is complete with respect to convergence in this norm, then H is called *Hilbert space*.

Theorem 2.1 (Completion of a Pre-Hilbert Space) *For every pre-Hilbert space H there exists a Hilbert space $\widetilde{H}$ (completion of H) such that $H \subset \widetilde{H}$, H is dense in $\widetilde{H}$ and $(x, y)_{\widetilde{H}} = (x, y)_H$, $x, y \in H$.*

The elements $x, y \in H$ are said to be *orthogonal*, if $(x, y) = 0$ (the orthogonality is denoted by $x \perp y$). An element $x \in H$ is said to be orthogonal to a set $M \subset H$,

V. Brayman et al., *Functional Analysis and Operator Theory*, Problem Books in Mathematics, https://doi.org/10.1007/978-3-031-56427-7_2

if $(x, y) = 0$ for all $y \in M$ (denoted as $x \perp M$). The sets $M \subset H$ and $N \subset H$ are said to be *orthogonal*, if $(x, y) = 0$ for all $x \in M$ and all $y \in N$ (denoted as $M \perp N$).

The orthogonal complement $M^{\perp}$ of a set $M \subset H$ is the set of vectors from H which are orthogonal to M, i.e.

$$M^{\perp} = \{x \in H \mid \forall\, y \in M \;:\; (x, y) = 0\}.$$

Note that for every set $M \subset H$ the orthogonal complement $M^{\perp}$ is a subspace (see problem 2.22°).

A vector $y \in M$ such that $(x - y) \perp M$ is called the *projection* of a vector $x \in H$ onto the subspace M. It is denoted as $y = \mathrm{pr}_M x$.

Theorem 2.2 (Projection onto a Subspace) *The projection* $\mathrm{pr}_M x$ *of a vector* x *onto a subspace* M *exists and is unique, moreover*

$$\|x - \mathrm{pr}_M x\| = \inf_{y \in M} \|x - y\| = \rho(x, M).$$

Theorem 2.3 (Decomposition of a Hilbert Space) *Let* M *be a subspace of a Hilbert space* H. *Then*

$$\forall x \in H\; \exists!\, x' \in M\; \exists!\, x'' \in M^{\perp} \;:\; x = x' + x'',$$

where $x' = \mathrm{pr}_M x$, *and* $x'' = \mathrm{pr}_{M^{\perp}} x$.

If $M \subset H$ and $N \subset H$ are orthogonal sets, then their *orthogonal sum* is the set $M \bigoplus N = \{x + y \mid x \in M,\ y \in N\}$. If M and N are orthogonal subspaces of a Hilbert space then $M \bigoplus N$ is a subspace. If M is a subspace of a Hilbert space H then Theorem 2.3 implies that $H = M \bigoplus M^{\perp}$.

A system $\{x_\alpha : \alpha \in A\} \subset H$ is said to be *orthogonal*, if $(x_\alpha, x_\beta) = 0,\ \alpha, \beta \in A,\ \alpha \neq \beta$. An orthogonal system $\{x_\alpha : \alpha \in A\} \subset H$ is said to be *orthonormal*, if $(x_\alpha, x_\alpha) = 1,\ \alpha \in A$. An orthonormal system in a pre-Hilbert space H is called an orthonormal basis in H if its closed linear span coincides with H.

Theorem 2.4 *In an arbitrary Hilbert space there exists an orthonormal basis, and in a separable Hilbert space such a basis is countable.*

Theorem 2.5 *Let* $\{e_n : n \geq 1\}$ *be a countable orthonormal system in a Hilbert space* H. *Then:*

(1) the series $\sum\limits_{n=1}^{\infty} c_n e_n$ $(c_n \in \mathbb{K})$ *converges in* H *if and only if the series* $\sum\limits_{n=1}^{\infty} |c_n|^2$ *converges;*

(2) for every $x \in H$ *the series* $\sum\limits_{n=1}^{\infty} (x, e_n) e_n$ *converges in* H, *moreover* $\sum\limits_{n=1}^{\infty} |(x, e_n)|^2 \leq \|x\|^2$ *(Bessel inequality);*

(3) if $L_n = \text{span}(\{e_1, \ldots, e_n\}),\ n \geq 1$, *then for every* $x \in H$

$$\text{pr}_{L_n} x = \sum_{k=1}^{n} (x, e_k) e_k,$$

$$\rho^2(x, L_n) = \|x - \text{pr}_{L_n} x\|^2 = \|x\|^2 - \sum_{k=1}^{n} |(x, e_k)|^2.$$

If $\{e_n : n \geq 1\}$ is a countable orthonormal system in H, then the numbers $(x, e_n),\ n \geq 1$, are called *Fourier coefficients* of the element $x \in H$ with respect to the system $\{e_n : n \geq 1\}$, and the series $\sum_{n=1}^{\infty} (x, e_n) e_n$ is called the Fourier series of the element $x \in H$ with respect to the system $\{e_n : n \geq 1\}$.

A system $\{e_n : n \geq 1\}$ is said to be *complete*, if there exist no $x \in H$ such that $x \neq 0$ and $(x, e_n) = 0$ for all $n \geq 1$.

Theorem 2.6 *Let* $\{e_n : n \geq 1\}$ *be an orthonormal sequence in a Hilbert space* H. *Then the following conditions are equivalent:*

(1) $\{e_n : n \geq 1\}$ *is an orthonormal basis;*
(2) $\forall x \in H : x = \sum_{n=1}^{\infty} (x, e_n) e_n$;
(3) $\forall x \in H : \|x\|^2 = \sum_{n=1}^{\infty} |(x, e_n)|^2$ *(Parseval's identity);*
(4) $\{e_n : n \geq 1\}$ *is a complete system.*

Examples of Hilbert Spaces

1. l_2 with the scalar product $(x, y) = \sum_{n=1}^{\infty} x_n \overline{y_n}$. This space has a basis $\{e_n = (0, \ldots, 0, \underbrace{1}_{n}, 0, 0, \ldots) : n \geq 1\}$.
2. $L_2(T, \mu)$ with the scalar product $(x, y) = \int_T x(t)\overline{y(t)} d\mu(t)$. The space $L_2([-\pi, \pi])$ has an orthonormal basis $\left\{\frac{1}{\sqrt{2\pi}} e^{int},\ t \in [-\pi, \pi] : n \in \mathbb{Z}\right\}$.

Remark For $1 \leq p \leq +\infty,\ p \neq 2$, the space l_p is not a Hilbert space, and $L_p(T, \mu)$ is a Hilbert space only if it is one-dimensional.

Examples of Problems with Solutions

2.1 Prove that a scalar product in a Hilbert space is a continuous function in each of the arguments.

Solution Let us fix $x \in H$ and consider the function $f_x(y) = (x, y),\ y \in H$. According to the Cauchy–Schwarz inequality for all $y, z \in H$ we have $|f_x(y) - f_x(z)| = |(x, y - z)| \le \|x\| \cdot \|y - z\|$, so the function f_x is uniformly continuous on H. Continuity in the first argument can be proved similarly.

2.2 Let L and M be linear sets in a Hilbert space H, and $L \perp M$.

(1) Prove that for all $x \in L \bigoplus M$ the representation $x = x' + x'',\ x' \in L,\ x'' \in M$, is unique.
(2) Let $H = L \bigoplus M$. Prove that $M^{\perp} = L$.

Solution

(1) Suppose that $x = x' + x'' = y' + y''$, where $x', y' \in L,\ x'', y'' \in M$. Then $x'' - y'' = y' - x' \in L \cap M$. Hence $x' = y', x'' = y''$ because $\|x' - y'\|^2 = (x' - y', y'' - x'') = 0$.
(2) Since $L \perp M$, then $L \subset M^{\perp}$. Let us prove that $M^{\perp} \subset L$. Consider any $x \in M^{\perp}$. Since $H = L \bigoplus M$, we have $x = x' + x''$, where $x' \in L,\ x'' \in M$. It follows that $0 = (x, x'') = (x', x'') + (x'', x'') = \|x''\|^2$, i.e. $x'' = 0$ and $x = x' \in L$. Thus $M^{\perp} \subset L$.

Remark The orthogonal complement of an arbitrary subset of H is a subspace (see problem 2.22°), therefore the sets $L,\ M$ in item (2) are subspaces.

2.3 Find the orthogonal complement in $L_2([-1, 1])$ of the set of all even functions.

Remark A function $x \in L_2([-1, 1])$ is said to be even if $x(t) = x(-t)$ for almost all $t \in [-1, 1]$. The notion of an odd function is introduced in a similar way.

Solution Let M be a set of even functions in $L_2([-1, 1])$, and L be a set of odd functions. It is clear that M and L are linear sets and $M \perp L$ (the product of an even and an odd function is an odd function). Note that for each $x \in H$ there exist $x' \in M,\ x'' \in L$ such that $x = x' + x''$. Indeed, we can put $x'(t) = \frac{1}{2}(x(t) + x(-t)),\ x''(t) = \frac{1}{2}(x(t) - x(-t))$. Therefore according to problem 2.2, item (2) $M^{\perp} = L$, i.e. the complement is the set of all odd functions.

2.4 Let L be a total set in a Hilbert space H. Prove that $L^{\perp} = \{0\}$.

Solution Let $x \perp L$. Let us check that $x \perp \operatorname{span}(L)$. Indeed, if $y = \sum_{k=1}^{n} \alpha_k x_k \in \operatorname{span}(L)$, where $\alpha_k \in \mathbb{K},\ x_k \in L,\ 1 \le k \le n$, then $(y, x) = \sum_{k=1}^{n} \alpha_k (x_k, x) =$

$\sum_{k=1}^{n} 0 = 0$. Since L is a total set, there exists a sequence $\{x_n : n \geq 1\} \subset \text{span}(L)$ such that $x_n \to x$ in H. According to problem 2.1, $0 = (x_n, x) \to (x, x),\ n \to \infty$. Thus $(x, x) = \|x\|^2 = 0$, i.e. $x = 0$.

2.5 Find the orthogonal complement of the set $M = \{e^{kt} \mid k \geq 10\}$ in $L_2([0, 1])$.

Solution Let $x \perp M$, i.e. $\int_0^1 x(t)e^{kt}\,dt = 0$ for all $k \geq 10$. It follows that the function $y(t) = x(t)e^{10t}$ is orthogonal to the set $M_0 = \{e^{kt} \mid k \geq 0\}$. By the Stone-Weierstrass theorem $\text{span}(M_0)$ is dense in $C([0, 1])$ with the uniform norm, and therefore also in $C([0, 1])$ with the norm of $L_2([0, 1])$. By problem 1.53°, item (3) the set $C([0, 1])$ is dense in $L_2([0, 1])$, therefore $\text{span}(M_0)$ is dense in $L_2([0, 1])$. According to problem 2.4, $M_0^{\perp} = \{0\}$. It follows that $x = 0$ almost everywhere with respect to the Lebesgue measure. Therefore $M^{\perp} = \{0\}$.

Problems to Solve

2.6° Let $\alpha = (\alpha_1, \alpha_2, \ldots)$, where $\alpha_k > 0,\ k \geq 1$. Consider the set $l_{2,\alpha}$ of all numerical sequences $x = (x_1, x_2, \ldots)$ for which $\sum_{k=1}^{\infty} \alpha_k |x_k|^2 < \infty$. Verify that $l_{2,\alpha}$ is a Hilbert space with the scalar product $(x, y) = \sum_{k=1}^{\infty} \alpha_k x_k \overline{y_k},\ x, y \in l_{2,\alpha}$.

Construct an orthonormal basis in the space $l_{2,\alpha}$ in the following cases: (1) $\alpha_k = k,\ k \geq 1$; (2) $\alpha_k = k^2,\ k \geq 1$; (3) $\alpha_k = e^{-k},\ k \geq 1$.

2.7 Determine for which $\alpha = (\alpha_1, \alpha_2, \ldots)$, where $\alpha_k > 0,\ k \geq 1$, the inclusion $l_{2,\alpha} \subset l_2$ holds.

2.8°

(1) Let H be a Hilbert space, $\|\cdot\|$ be a norm in H, generated by a scalar product. Check that

$$\forall x, y \in H : \|x + y\|^2 + \|x - y\|^2 = 2(\|x\|^2 + \|y\|^2)$$

(parallelogram law).

(2) Prove that in the Banach space $C([0, 1])$ the norm is not generated by a scalar product.

2.9° Prove that in the Banach spaces $c_0,\ l_p,\ L_p([a, b]),\ p \neq 2$, the norm is not generated by a scalar product.

2.10 Prove that for arbitrary x, y, z in a Hilbert space Apollonius identity holds:

$$\|x-z\|^2+\|y-z\|^2=\frac{1}{2}\|x-y\|^2+2\left\|z-\frac{x+y}{2}\right\|^2.$$

2.11 Prove that for arbitrary x, y, z, u in a Hilbert space Ptolemy's inequality holds:

$$\|x-z\|\cdot\|y-u\|\le\|x-y\|\cdot\|z-u\|+\|y-z\|\cdot\|x-u\|.$$

2.12° Prove that in a Hilbert space over a field $\mathbb{K}$ the elements x and y are orthogonal if and only if

(1) $\|x+y\|^2=\|x\|^2+\|y\|^2$ in the case of $\mathbb{K}=\mathbb{R}$;
(2) $\|\lambda x+\mu y\|^2=\|\lambda x\|^2+\|\mu y\|^2$ for any $\lambda, \mu\in\mathbb{C}$ in the case of $\mathbb{K}=\mathbb{C}$.

2.13° Prove that the *polarization identity* holds in a complex Hilbert space:

$$(x,y)=\frac{1}{4}(\|x+y\|^2-\|x-y\|^2+i\|x+iy\|^2-i\|x-iy\|^2).$$

What does it look like in a real Hilbert space?

2.14 Let X be a complex NLS, and let the function $S: X\times X\to\mathbb{C}$ has the following properties:

(1) $\forall x\in X: S(x,x)\ge 0$, and $S(x,x)=0 \Leftrightarrow x=0$;
(2) $\forall x,y\in X: S(y,x)=\overline{S(x,y)}$;
(3) $\forall x_1,x_2,y\in X: S(x_1+x_2,y)=S(x_1,y)+S(x_2,y)$;
(4) $\forall x,y\in X: S(ix,y)=iS(x,y)$;
(5) $\forall\{x_n: n\ge 1\}\subset X,\ y\in X,\ x_n\to x,\ n\to\infty: S(x_n,y)\to S(x,y),\ n\to\infty$.

Prove that $S(\cdot,\cdot)$ is a scalar product in the space X. Is X necessarily a Hilbert space with this scalar product if X is a Banach space?

2.15* Let X be a complex NLS in which the norm satisfies the parallelogram law $\|x+y\|^2+\|x-y\|^2=2(\|x\|^2+\|y\|^2)$ for all $x,y\in X$. Prove that the formula

$$S(x,y)=\frac{1}{4}\left(\|x+y\|^2-\|x-y\|^2+i\|x+iy\|^2-i\|x-iy\|^2\right),$$

$x,y\in X$, defines a scalar product in the space X, and $\|\cdot\|$ is the norm generated by this scalar product.

2.16 Let p be a Lebesgue-measurable function on (a,b) such that $p(t)>0$ for almost all $t\in(a,b)$, and $L_{2,p}([a,b])=\left\{x \mid p^{\frac{1}{2}}x\in L_2([a,b])\right\}$. Prove that

$L_{2,p}([a,b])$ is a Hilbert space with scalar product $(x, y) = \int\limits_a^b x(t)\overline{y(t)}p(t)dt$. For which p the inclusion $L_{2,p}([a,b]) \subset L_2([a,b])$ holds?

2.17 Let $\{x_k : k \geq 1\}$ be an orthogonal system in a Hilbert space H. Prove that the series $\sum\limits_{k=1}^{\infty} x_k$ converges in H if and only if $\sum\limits_{k=1}^{\infty} \|x_k\|^2 < \infty$.

2.18 Prove that a finite system $\{x_1, \ldots, x_n\}$ of elements of a Hilbert space is linearly independent if and only if its Gram determinant

$$\begin{vmatrix} (x_1, x_1) & (x_1, x_2) & \ldots & (x_1, x_n) \\ (x_2, x_1) & (x_2, x_2) & \ldots & (x_2, x_n) \\ \ldots & \ldots & \ldots & \ldots \\ (x_n, x_1) & (x_n, x_2) & \cdots & (x_n, x_n) \end{vmatrix}$$

is nonzero.

2.19° Let H be a Hilbert space, $x, y \in H$, $\|x\| = \|y\| = 1$, $a, b \in \mathbb{R}$. Prove that $\|ax + by\| = \|bx + ay\|$.

2.20° Let H be a Hilbert space, $\{x_n\} \subset H$, $\{y_n\} \subset H$, and $\|x_n\| = \|y_n\| = 1$, $n \geq 1$. Prove that

(1) $(x_n, y_n) \to 1,\ n \to \infty \Rightarrow \|x_n - y_n\| \to 0,\ n \to \infty$;
(2) $\|x_n + y_n\| \to 2,\ n \to \infty \Rightarrow \|x_n - y_n\| \to 0,\ n \to \infty$.

2.21 Prove that an element x of a Hilbert space H is orthogonal to a subspace L if and only if $\|x\| \leq \|x - y\|$ for all $y \in L$.

2.22° Prove that for an arbitrary set $M \subset H$ the orthogonal complement $M^{\perp}$ is a subspace.

2.23° Let $M \subset N \subset H$. Prove that $M^{\perp} \supset N^{\perp}$.

2.24 Prove that (1) $M \subset (M^{\perp})^{\perp}$ for any $M \subset H$; (2) the equality $M = (M^{\perp})^{\perp}$ holds if and only if M is a subspace.

2.25 Let H be a Hilbert space. Prove that

(1) $M^{\perp} = \big(\mathrm{span}(M)\big)^{\perp} = \big(\overline{\mathrm{span}(M)}\big)^{\perp}$ for all subsets $M \subset H$;

(2) $\left(\bigcup\limits_{\alpha \in A} M_{\alpha}\right)^{\perp} = \bigcap\limits_{\alpha \in A} M_{\alpha}^{\perp}$, where $\{M_{\alpha} \mid \alpha \in A\}$ are subsets of H;

(3) $\left(\bigcap\limits_{\alpha \in A} L_{\alpha}\right)^{\perp} = \overline{\mathrm{span}\left(\bigcup\limits_{\alpha \in A} L_{\alpha}^{\perp}\right)}$, where $\{L_{\alpha} \mid \alpha \in A\}$ are subspaces of H.

2.26° Let $L = \{x \in L_2(\mathbb{R}) \mid x(t) = 0,\ t \geq 0\ (\mathrm{mod}\, m)\}$. Prove that L is a subspace of $L_2(\mathbb{R})$ and find the orthogonal complement of L.

2.27° Let $L \subset H$. Prove that $\overline{\mathrm{span}(L)} = H$ if and only if $L^{\perp} = \{0\}$.

2.28° Find in $L_2([0, 1])$ the orthogonal complement of the set

(1) $C([0, 1])$;
(2) all polynomials;
(3) all polynomials of t^2;
(4) all polynomials $x(t)$, for which $x(0) = 0$;
(5) all polynomials of e^t;
(6) $\{t^k,\ t \in [0, 1] \mid k \geq 10\}$;
(7) $\{t^{3k},\ t \in [0, 1] \mid k \geq 1\}$.

2.29° Find in $L_2([-1, 1])$ the orthogonal complement of the set

(1) $\{x(t) = t^k,\ t \in [-1, 1] \mid k \geq 13\}$;
(2) $\{t^{2k},\ t \in [-1, 1] \mid k \geq 0\}$;
(3) $\{t^{2k+1},\ t \in [-1, 1] \mid k \geq 0\}$;
(4) $\{t^{2k},\ t \in [-1, 1] \mid k \geq 7\}$.

2.30° Find in $L_2([-\pi, \pi])$ the orthogonal complement of the set

(1) $\{\sin kt \mid k \geq 1\}$;
(2) $\{\cos kt \mid k \geq 1\}$;
(3) $\{e^{ikt} \mid k \geq 5\}$;
(4) $\{e^{-ikt} \mid k \geq 3\}$.

2.31° Find in l_2 the orthogonal complement of the set

(1) $\{(1, 1, 0, \ldots)\}$; (2) $\{e_k, \mid 1 \leq k \leq n\},\ n \in \mathbb{N}$; (3) $\{e_{2k} \mid k \in \mathbb{N}\}$.

2.32° Prove that the function

$$(x, y) = \int_a^b x(t)\overline{y(t)}dt + \int_a^b x'(t)\overline{y'(t)}dt,\ x, y \in C^1([a, b]),$$

defines a scalar product in the linear space $C^1([a, b])$. The completion of the space $C^1([a, b])$ in the norm generated by this scalar product is a Hilbert space called Sobolev space $W_2^1([a, b])$.

2.33* Find the orthogonal complement of the set $\{e^{ikt} \mid k \in \mathbb{Z}\}$ in the space $W_2^1([-\pi, \pi])$.

2.34 Consider the space H that consists of the functions $x : \mathbb{R} \to \mathbb{R}$, for which $\{x \in \mathbb{R} \mid x(t) \neq 0\}$ is a countable set and $\sum\limits_{t \in \mathbb{R}} |x(t)|^2 < +\infty$. Prove that H is a nonseparable Hilbert space with the scalar product $(x, y) = \sum\limits_{t \in \mathbb{R}} x(t)y(t),\ x, y \in H$.

2.35 Prove that the system of functions $\{e^{-int} \mid n \in \mathbb{Z}\}$ is complete in $L_2([a, b])$ if and only if $b - a \leq 2\pi$.

2.36 Let L be a linear span of the set $\{e^{i\lambda t} \mid \lambda \in \mathbb{R}\}$ equipped with the scalar product $(x, y) = \lim\limits_{T\to\infty} \frac{1}{2T} \int\limits_{-T}^{T} x(t)\overline{y(t)}dt,\ x, y \in L$. Prove that

(1) $(e^{i\lambda t}, e^{i\mu t}) = \delta_{\lambda\mu} = \begin{cases} 1, & \lambda = \mu, \\ 0, & \lambda \neq \mu, \end{cases} \quad \lambda, \mu \in \mathbb{R};$

(2) the completion of the set L is nonseparable.

2.37 Prove that the system of Haar functions

$$\{x_{kn} \mid 1 \le k \le 2^n,\ n \ge 0\} \cup \{x_0\},$$

where $x_{kn}(t) = \begin{cases} 2^{\frac{n}{2}}, & t \in \left[\frac{k-1}{2^n}, \frac{k-1/2}{2^n}\right), \\ -2^{\frac{n}{2}}, & t \in \left[\frac{k-1/2}{2^n}, \frac{k}{2^n}\right), \\ 0, & t \in [0, 1] \setminus [\frac{k-1}{2^n}, \frac{k}{2^n}], \end{cases} \quad x_0(t) = 1,\ t \in [0, 1],$

is an orthonormal basis in the space $L_2([0, 1])$.

2.38 Verify that the following systems are orthogonal in H:

(1) Rademacher's system of functions

$$x_n(t) = \begin{cases} (-1)^m, & t \in [\frac{m}{2^n}, \frac{m+1}{2^n}),\ m = 0, 1, \dots, 2^n - 1, \\ -1, & t = 1, \end{cases}$$

$n \ge 1,\ H = L_2([0, 1])$;

(2) Legendre polynomials $P_n(t) = \frac{1}{2^n n!} \frac{d^n}{dt^n}(t^2 - 1)^n,\ n \ge 0,\ H = L_2([-1, 1])$;

(3) Hermite functions $H_n(t) = e^{t^2/2} \frac{d^n}{dt^n} e^{-t^2},\ n \ge 0,\ H = L_2(\mathbb{R})$;

(4) $x_n(t) = \frac{d^n}{dt^n}[(t-a)(t-b)]^n,\ n \ge 0, H = L_2([a, b])$;

(5) Laguerre functions $x_n(t) = e^{t/2} \frac{d^n}{dt^n}(t^n e^{-t}),\ n \ge 0,\ H = L_2([0, +\infty))$.

2.39 Prove that in problem 2.38

(1) Rademacher's system of functions $\{x_n(t) : n \ge 1\}$ is not an orthonormal basis in $L_2([0, 1])$;

(2) the functions $\left\{\sqrt{\frac{2n+1}{2}} P_n(t) : n \ge 0\right\}$ form an orthonormal basis in $L_2([-1, 1])$.

2.40 In the space $L_2(\mathbb{R})$, consider the set

$$M = \left\{ x \in C_0(\mathbb{R}) \,\middle|\, \int\limits_{-\infty}^{\infty} x(t)dt = 0 \right\},$$

where $C_0(\mathbb{R})$ is the set of continuous functions on $\mathbb{R}$ which vanish outside some finite interval, see problem 1.54°. Prove that M is dense in $L_2(\mathbb{R})$.

2.41 In the space l_2, find the orthogonal complement of the set

(1) $\{x = (1, \alpha, \alpha^2, \dots, \alpha^k, \dots) \mid \alpha \in (0, 1)\}$;
(2) $\{x = (1, \frac{1}{n}, \frac{1}{n^2}, \dots, \frac{1}{n^k}, \dots) \mid n \geq 2\}$.

2.42° Let $M \subset H$, where H is a Hilbert space. Is the formula $H = M \bigoplus M^\perp$ always true?

2.43 Let $\{e_n : n \geq 1\}$ be a standard basis in l_2, $x_1 = \sum\limits_{k=1}^{\infty} \frac{e_k}{k}$ and $H = \text{span}(\{x_1, e_k : k \geq 2\}) \subset l_2$. Prove that in the pre-Hilbert space H the set $M = \{e_n : n \geq 2\}$ is a complete orthonormal system, but $\overline{\text{span}(M)} \neq H$.

2.44* Prove that a separable space equipped with a scalar product is complete if and only if every complete orthonormal system in this space is a basis.

2.45 Prove that in a Hilbert space there exists a dense linear subset, which does not coincide with the entire space, if and only if the space is infinite-dimensional.

2.46 Let $\{e_n : n \geq 1\}$ be linearly independent system of elements of a Hilbert space H such that $\sum\limits_{n=1}^{\infty} c_n e_n = 0$ for some sequence $\{c_n : n \geq 1\} \subset \mathbb{K}$. Is it true that $c_n = 0, \ n \geq 1$?

2.47 Let H be a Hilbert space over a field $\mathbb{K}$, $\{e_n : n \geq 1\} \subset H$ be an orthonormal system and $x = \sum\limits_{n=1}^{\infty} x_n e_n$, where $x_n \in \mathbb{K}, \ n \geq 1$. Prove that $x_n = (x, e_n), \ n \geq 1$.

2.48 Prove that the system $\{t^n, \ t \in [0, 1] : n \geq 0\}$ is linearly independent, and its closed linear span is equal to $L_2([0, 1])$, but not for all $x \in L_2([0, 1])$ there exist coefficients $\{c_n : n \geq 0\}$ such that $x(t) = \sum\limits_{n=0}^{\infty} c_n t^n$ (the convergence here is in the sense of convergence in $L_2([0, 1])$.) Does this contradict to the statement of Theorem 2.6? Why?

2.49 Let $D = \overline{B}(0, 1) \subset \mathbb{C}$ be the closed unit circle in the complex plane, $A_2(D)$ be the set of functions $x \in L_2(D)$, for which $x(z) = \sum\limits_{n=0}^{\infty} a_n z^n$ for $|z| < 1$. Prove that $A_2(D)$ is the completion of the set $P(D)$ of all polynomials defined on D, in the norm of $L_2(D)$.

2.50 Apply the orthogonalization process to the sequence $\{z^n : n \geq 0\}$, $z = t + is$, with respect to the following scalar product on the set of all polynomials $P(\mathbb{C})$:

$$(x, y) = \int\limits_{\mathbb{C}} x(z)\overline{y(z)} e^{-|z|^2} dt\, ds, \ x, y \in P(\mathbb{C}).$$

Describe the completion of the set $P(\mathbb{C})$.

2.51 Let $\{\varphi_n : n \geq 1\}$ and $\{\psi_n : n \geq 1\}$ be complete orthonormal systems in $L_2(X_1, \mathcal{F}_1, \mu_1)$ and $L_2(X_2, \mathcal{F}_2, \mu_2)$, respectively, and μ_i be σ-finite measures on $\mathcal{F}_i$, $i = 1, 2$. Prove that $e_{n,m}(x, y) = \varphi_n(x)\psi_m(y)$, $(x, y) \in X = X_1 \times X_2$, $n, m \geq 1$, is a complete orthonormal system in $L_2(X, \mathcal{F}, \mu)$, where $\mu = \mu_1 \times \mu_2$, $\mathcal{F} = \sigma(\mathcal{F}_1 \times \mathcal{F}_2)$.

2.52 Prove that a Hilbert space is uniformly convex (see problem 1.98).

2.53 Let M be a nonempty closed convex set in a Hilbert space H. Prove that in M there exists a unique element with the smallest norm. Show that the convexity condition is essential.

2.54 Let H be a Hilbert space, $x_0 \in H$, $r > 0$. For $x \in H \setminus S(x_0, r)$ find $y \in S(x_0, r)$ such that $\rho(x, S(x_0, r)) = \|x - y\|$.

2.55* Let $x \in H$ and $M \subset H$ be a closed convex set in a real Hilbert space H. Prove that $y \in M$ satisfies the condition $\rho(x, M) = \|x - y\|$ if and only if $(x - y, y - z) \geq 0$ for all $z \in M$.

2.56 Let L be a one-dimensional subspace of a Hilbert space H, $a \in L$, $a \neq 0$. Prove that $\rho(x, L^\perp) = \frac{|(x,a)|}{\|a\|}$ for all $x \in H$.

2.57 For the set $M_n = \left\{ x \in l_2 \,\middle|\, \sum\limits_{k=1}^{n} x_k = 0 \right\}$, $n \in \mathbb{N}$, find $M_n^\perp$, $\rho(x_0, M_n)$ and $\lim\limits_{n\to\infty} \rho(x_0, M_n)$, where $x_0 = (1, 0, \ldots)$.

2.58 Let M, N be subspaces of a Hilbert space H and for each $x \in H$ there exist unique $x_1 \in M$ and $x_2 \in N$ such that $x = x_1 + x_2$. Is it true that $N = M^\perp$?

2.59 Let

$$M = \big\{(x_1, 0, x_3, 0, \ldots) \mid x = (x_1, x_2, x_3, \ldots) \in l_2\big\},$$

$$N = \big\{(x_1, x_1, x_3, \frac{x_3}{3}, x_5, \frac{x_5}{5}, \ldots) \mid x = (x_1, x_2, x_3, \ldots) \in l_2\big\}.$$

Prove that M and N are subspaces of l_2 and the set $M + N$ is dense in l_2, but $M + N \neq l_2$, i.e. $M + N$ is not a subspace of l_2.

2.60 Let $\{e_n : n \geq 1\}$ be an orthonormal basis in H, $u_n = \cos\frac{1}{n} \cdot e_{2n} + \sin\frac{1}{n} \cdot e_{2n+1}$. Prove that $\{u_n : n \geq 1\}$ is an orthonormal system. Put $M = \overline{\text{span}(\{e_{2n} : n \geq 1\})}$ and $N = \overline{\text{span}(\{u_n : n \geq 1\})}$. Prove that the linear set $M + N$ is not closed.

2.61 Let M and N be subspaces of a Hilbert space H. Prove that $M + N$ is a subspace of H, if the following condition holds:

$$\exists \varepsilon \in (0, 1] \,:\, \sup\{|(x, y)| \mid \|x\| = \|y\| = 1,\ x \in M,\ y \in N\} \leq 1 - \varepsilon,$$

in particular if $M \perp N$.

2.62 Let $\{x_n : n \geq 1\}$, $\{y_n : n \geq 1\}$ be orthonormal systems in $L_2([a, b])$ and $L_2([b, c])$ respectively, $\lambda > 0$, $\mu > 0$, $\lambda^2 + \mu^2 = 1$. Put

$$z_n(t) = \begin{cases} \lambda x_n(t), & a \leq t \leq b, \\ \mu y_n(t), & b \leq t \leq c. \end{cases}$$

Prove that $\{z_n : n \geq 1\}$ is an orthonormal system in $L_2([a, c])$. Is $\{z_n : n \geq 1\}$ necessarily an orthonormal basis in $L_2([a, c])$, if $\{x_n : n \geq 1\}$ and $\{y_n : n \geq 1\}$ are orthonormal bases in $L_2([a, b])$ and $L_2([b, c])$, respectively?

2.63 Let $a < b < c$ and $\{x_n : n \geq 1\}$ be an arbitrary countable system of functions from $L_2([a, b])$. Prove that the functions x_n can be extended to the segment $[b, c]$ to obtain an orthogonal system in $L_2([a, c])$.

2.64 Show that the functions $x_{n,k}(t) = \sqrt{2}\chi_{[k-1,k]}(t) \sin \pi n t$ for $n \in \mathbb{N}$, $k \in \mathbb{N}$ form an orthonormal basis in $L_2([0, +\infty))$, and for $n \in \mathbb{N}$, $k \in \mathbb{Z}$ they form an orthonormal basis in $L_2(\mathbb{R})$.

Chapter 3
Continuous Linear Functionals

Theoretical Background

Let X be an NLS over a field $\mathbb{K}$ ($\mathbb{K} = \mathbb{R}$ or $\mathbb{K} = \mathbb{C}$). Mapping $f : X \to \mathbb{K}$ is called a functional. A functional f is called *continuous*, if the mapping $f : X \to \mathbb{K}$ is continuous on X, i.e.

$\forall\, x_0 \in X\ \forall\, \{x_n : n \geq 1\} \subset X, x_n \to x_0 \ :\ f(x_n) \to f(x_0),\ n \to \infty.$

A functional f is called *linear* if

$\forall\, x, y \in X\ \forall\, \alpha, \beta \in \mathbb{K} \ :\ f(\alpha x + \beta y) = \alpha f(x) + \beta f(y).$

A linear functional f is called *bounded* if

$\exists\, C \geq 0\ \forall\, x \in X \ :\ |f(x)| \leq C\|x\|.$

Note that for every linear functional f the equality $f(0) = 0$ holds.

Theorem 3.1 *Let f be a linear functional on X. Then the following statements are equivalent:*

(1) f is continuous on X;
(2) f is continuous at one point;
(3) f is bounded.

The set of all continuous linear functionals on an NLS X is denoted by X^*. The set X^* is endowed naturally with the structure of a linear space:

$$\forall\, f, g \in X^*\ \forall\, \alpha, \beta \in \mathbb{K} \ :\ (\alpha f + \beta g)(x) := \alpha f(x) + \beta g(x),\ x \in X;$$

$\alpha f + \beta g$ is also a continuous linear functional (see problem 3.6°).

The *norm of a functional* $f \in X^*$ is defined as the number $\|f\| = \sup\limits_{x \neq 0} \frac{|f(x)|}{\|x\|}$.

From the definition of the norm of a functional, it is easy to derive the following result.

V. Brayman et al., *Functional Analysis and Operator Theory*, Problem Books in Mathematics, https://doi.org/10.1007/978-3-031-56427-7_3

Lemma 3.1 *Let $f \in X^*$. Then*

$$\|f\| = \sup_{\|x\|=1} |f(x)| = \sup_{\|x\|\le 1} |f(x)| = \min\{C \ge 0 \mid \forall x \in X \,:\, |f(x)| \le C\|x\|\}.$$

In particular, if $|f(x)| \le C\|x\|$ for every $x \in X$ then $\|f\| \le C$.

Theorem 3.2 *(1) The norm of a functional satisfies axioms of norm on X^*, i.e. X^* is an NLS; (2) X^* is a Banach space.*

The Banach space X^* is called the *dual space* of the NLS X.

The set $\text{Ker}\, f = \{x \in X \mid f(x) = 0\}$ is called the *kernel* of a functional f.

Let us describe the general form of continuous linear functionals in some classical normed linear spaces and provide the description of dual spaces.

The number $q \in (1, +\infty)$ such that $\frac{1}{p} + \frac{1}{q} = 1$ is called the conjugate index to the number $p \in (1, +\infty)$. For $p = 1$ set $q = +\infty$, and for $p = +\infty$ set $q = 1$. In what follows, q is the conjugate index to p.

Theorem 3.3 *Let $(T, \mathcal{F}, \mu)$ be a space with a σ-finite measure and $1 \le p < +\infty$. For every functional $f \in (L_p(T, \mu))^*$ there exists a unique element $h \in L_q(T, \mu)$ such that $f(x) = \int\limits_T h(t)x(t)d\mu(t)$, $x \in L_p(T, \mu)$. Conversely, for every element $h \in L_q(T, \mu)$ the last formula defines a functional $f \in (L_p(T, \mu))^*$. Moreover, $\|f\| = \|h\|_q$.*

The function $\varphi : (L_p(T, \mu))^* \to L_q(T, \mu)$, which maps each functional to the element $h \in L_q(T, \mu)$ from Theorem 3.3, is an isometric isomorphism, i.e. a linear bijection that preserves the norm. The notation of the isometric isomorphism between $(L_p(T, \mu))^*$ and $L_q(T, \mu)$ is written simply as the equality $(L_p(T, \mu))^* = L_q(T, \mu)$, $1 \le p < +\infty$.

For $T = \mathbb{N}$ (or $T = \{1, \ldots, m\}$, $m \in \mathbb{N}$), $\mathcal{F} = 2^T$, $\mu(\{n\}) = 1$, $n \in T$, we get the following corollaries.

Corollary 3.1 *Let $1 \le p < +\infty$. Then $l_p^* = l_q$, and the isometric isomorphism $\varphi : l_p^* \to l_q$ is defined by the formula $f(x) = \sum\limits_{n=1}^{\infty} a_n x_n$, $x \in l_p$, $a = \varphi(f) \in l_q$. Here $\|f\| = \|a\|_q$.*

Denote by $\mathbb{R}_p^m$ ($\mathbb{C}_p^m$) the space $\mathbb{R}^m$($\mathbb{C}^m$) with the norm $\|\cdot\|_p$.

Corollary 3.2 *Let $1 \le p < +\infty$. Then $(\mathbb{R}_p^m)^* = \mathbb{R}_q^m$, and the isometric isomorphism $\varphi : (\mathbb{R}_p^m)^* \to \mathbb{R}_q^m$ is defined by the formula $f(x) = \sum\limits_{n=1}^{m} a_n x_n$, $x \in \mathbb{R}_p^m$, $a = \varphi(f) \in \mathbb{R}_q^m$. Here $\|f\| = \|a\|_q$.*

Similarly, $(\mathbb{C}_p^m)^* = \mathbb{C}_q^m$.

Note that the Corollary 3.2 remains valid for $p = +\infty$. However, in general case $(L_\infty(T, \mu))^* \ne L_1(T, \mu)$. More precisely, if the measure μ is not concentrated in a finite number of atoms (see the definition of atom in problem 1.102*), then there is a strict inclusion $L_1(T, \mu) \subset (L_\infty(T, \mu))^*$, in particular, $l_1 \subset l_\infty^*$.

Let us consider the general form of a continuous linear functional on the space $C([a, b])$ over the field $\mathbb{K}$. Denote by $BV_0([a, b])$ the Banach space of functions of bounded variation $g : [a, b] \to \mathbb{K}$, which are right-continuous on (a, b) and for which $g(a) = 0$. This space is equipped with the norm $\|g\|_V = V(g, [a, b])$, where $V(g, [a, b])$ is the variation of a function g on the interval $[a, b]$.

Theorem 3.4 (F. Riesz) *For every functional $f \in (C([a, b]))^*$ there exists a unique function $g \in BV_0([a, b])$ such that $f(x) = \int_a^b x(t)dg(t)$, for all $x \in C([a, b])$. Conversely, for any function $g \in BV_0([a, b])$ the last formula defines a functional $f \in (C([a, b]))^*$. Moreover, $\|f\| = \|g\|_V$. Thus $(C([a, b]))^* = BV_0([a, b])$.*

The integral in Theorem 3.4 is the Riemann-Stieltjes integral. Note that for an arbitrary function $g : [a, b] \to \mathbb{K}$, which has bounded variation on $[a, b]$, the formula in Theorem 3.4 defines a continuous linear functional on $C([a, b])$, but the equality $\|f\| = V(g, [a, b])$ may fail (e.g., if g is discontinuous at some point $t_0 \in (a, b)$ and the value $g(t_0)$ does not belong to the interval with endpoints $g(t_0-)$ and $g(t_0+)$). Also the correspondence between functions of bounded variation on $[a, b]$ and continuous linear functionals on $C([a, b])$ is not bijective (e.g., if we replace g with $g +$ const or change the values of the function g on a finite subset of (a, b), we will get the same functional defined in Theorem 3.4).

Let us reformulate Theorem 3.4 in terms of the Lebesgue integral with respect to a finite complex measure. Recall the necessary definitions. Let $(T, \mathcal{F})$ be a measurable space; a σ-additive function $\nu : \mathcal{F} \to \mathbb{K}$ is called a *finite signed measure* if $\mathbb{K} = \mathbb{R}$, and a *finite complex measure* if $\mathbb{K} = \mathbb{C}$. A finite signed measure is clearly a complex measure. The *variation of the complex measure ν on a set $A \in \mathcal{F}$* is defined as the number $|\nu|(A) = \sup \sum_{k=1}^{\infty} |\nu(A_k)|$, where the supremum is taken over all countable partitions of A into pairwise disjoint subsets $A_k \in \mathcal{F}$; $|\nu|$ is a finite measure on $\mathcal{F}$. If ν is a finite signed measure, then $|\nu|(A) = \nu_+(A) + \nu_-(A)$, $A \in \mathcal{F}$, where ν_+, ν_- are measures from Jordan decomposition $\nu = \nu_+ - \nu_-$ of the signed measure ν, and the formula $\int_A x d\nu = \int_A x d\nu_+ - \int_A x d\nu_-$ defines *the Lebesgue integral of a function $x \in L(A, |\nu|)$ with respect to the signed measure ν over the set A*. If ν is a finite complex measure, then $\nu_R(A) = \operatorname{Re} \nu(A)$ and $\nu_I(A) = \operatorname{Im} \nu(A)$, $A \in \mathcal{F}$, are finite signed measures, $\nu(A) = \nu_R(A) + i\nu_I(A)$, $A \in \mathcal{F}$, and the formula $\int_A x d\nu = \int_A x d\nu_R + i \int_A x d\nu_I$ defines *the Lebesgue integral of a function $x \in L(A, |\nu|)$ with respect to the complex measure ν over the set A*.

Theorem 3.5 (F. Riesz) *In case of a complex space $C([a, b])$, for every functional $f \in (C([a, b]))^*$ there exists a unique finite complex measure ν on the σ-algebra $\mathcal{B}$ of Borel subsets of $[a, b]$ such that $f(x) = \int_{[a,b]} x d\nu$, $x \in C([a, b])$. Conversely,* *for any finite complex measure* ν *on* $\mathcal{B}$ *the last formula defines a functional* $f \in (C([a, b]))^*$. *Moreover,* $\|f\| = |\nu|([a, b])$.

In case of a real space $C([a, b])$, Theorem 3.5 remains true if complex measures are replaced with signed measures.

Theorem 3.6 (F. Riesz) *Let* H *be a Hilbert space. Then for every functional* $f \in H^*$ *there exists a unique element* $a \in H$ *such that* $f(x) = (x, a)$, $x \in H$. *Conversely, for every element* $a \in H$ *this formula defines a functional* $f \in H^*$. *Moreover,* $\|f\| = \|a\|$. *Thus* $H^* = H$.

Note that in complex Hilbert space the function $\varphi : H^* \to H$, which maps each functional $f \in H^*$ to the respective element $a \in H$, is antilinear, i.e.

$$\forall f, g \in H^* \; \forall \alpha, \beta \in \mathbb{C} \; : \; \varphi(\alpha f + \beta g) = \overline{\alpha}\varphi(f) + \overline{\beta}\varphi(g).$$

Examples of Problems with Solutions

3.1 Determine whether the given functionals are linear and continuous. For continuous linear functionals, find their norms using the definition.

(1) $X = C([a, b]),\; f(x) = x(\frac{a+b}{2})$;
(2) $X = C([a, b]),\; f(x) = x(a) - x(b)$;
(3) $X = C([-1, 1]),\; f(x) = \int\limits_{-1}^{1} x(t)\, \mathrm{sign}\, t dt$;
(4) $X = C([0, 1]),\; f(x) = \int\limits_{0}^{1} t|x(t)|dt$;
(5) $X = C([0, 1])$ with the norm $\|x\| = \int\limits_{0}^{1} |x(t)|dt,\; f(x) = x(1)$;
(6) $X = L_p([-1, 1]),\; 1 \le p \le +\infty,\; f(x) = \int\limits_{-1}^{1} t^2 x(t)\, \mathrm{sign}\, t dt$;
(7) $X = l_2,\; f(x) = \sum\limits_{n=1}^{\infty} \frac{1}{2^n}(x_n + x_{n+1})$;
(8) $X = \left\{ x \in l_2 \;\middle|\; \sum\limits_{n=1}^{\infty} x_n \cos n \text{ is a convergent series} \right\},\; \|\cdot\|_X = \|\cdot\|_{l_2}$,

$$f(x) = \sum_{n=1}^{\infty} x_n \cos n,\; x \in X.$$

Solution

(1) Since for all $x, y \in C([a, b])$ and $\alpha, \beta \in \mathbb{K}$ we have

$$f(\alpha x + \beta y) = (\alpha x + \beta y)(\tfrac{a+b}{2}) = \alpha x(\tfrac{a+b}{2}) + \beta y(\tfrac{a+b}{2}) = \alpha f(x) + \beta f(y),$$

the functional f is linear. For a linear functional, continuity is equivalent to boundedness. The functional f is bounded, because

$$\forall x \in C([a, b]) \; : \; |f(x)| = |x(\tfrac{a+b}{2})| \leq \|x\|.$$

Therefore f is a continuous linear functional, and by Lemma 3.1 we have $\|f\| \leq 1$. To show that $\|f\| = 1$, set $x_0(t) = 1,\ t \in [a, b]$. Then $\|x_0\| = 1$, $f(x_0) = x_0(\frac{a+b}{2}) = 1$, hence $\|f\| \geq \frac{|f(x_0)|}{\|x_0\|} = 1$. Therefore $\|f\| = 1$.

(2) The functional f is linear, because for all $x, y \in C([a, b])$ and $\alpha, \beta \in \mathbb{K}$ we have

$$f(\alpha x + \beta y) = (\alpha x + \beta y)(a) - (\alpha x + \beta y)(b) =$$
$$= \alpha(x(a) - x(b)) + \beta(y(a) - y(b)) = \alpha f(x) + \beta f(y).$$

For a linear functional, continuity is equivalent to boundedness. The functional f is bounded, because

$$\forall x \in C([a, b]) \; : \; |f(x)| = |x(a) - x(b)| \leq |x(a)| + |x(b)| \leq 2\|x\|.$$

Therefore f is a continuous linear functional and $\|f\| \leq 2$. To show that $\|f\| = 2$, set $x_0(t) = t - \frac{a+b}{2},\ t \in [a, b]$. Since $\|x_0\| = \frac{b-a}{2}$ and $f(x_0) = x_0(a) - x_0(b) = a - b$, then $\|f\| \geq \frac{|f(x_0)|}{\|x_0\|} = 2$. Therefore $\|f\| = 2$.

(3) The functional f is defined correctly, because the integrand is bounded and has at most one point of discontinuity, so it is Riemann integrable. The linearity of the functional f follows from the linearity of the Riemann integral. The functional is bounded because

$$\forall x \in C([-1, 1]) \; : \; |f(x)| \leq \int_{-1}^{1} |x(t) \operatorname{sign} t| dt =$$
$$= \int_{-1}^{1} |x(t)| dt \leq \int_{-1}^{1} \max_{s \in [-1,1]} |x(s)| dt = \|x\| \int_{-1}^{1} dt = 2\|x\|,$$

hence, taking into account the linearity of f, we get that f is a continuous linear functional and $\|f\| \leq 2$. The equality $|f(x)| = 2\|x\|$ is achieved, for example, on the function $x_0(t) = \operatorname{sign} t,\ t \in [-1, 1]$, but $x_0 \notin C([-1, 1])$. Consider a sequence of continuous functions which converge to x_0 almost everywhere with respect to the Lebesgue measure on $[-1, 1]$. For $n \geq 1$ set

$$x_n(t) = \begin{cases} 1, & t \in [\frac{1}{n}, 1], \\ -1, & t \in [-1, -\frac{1}{n}], \end{cases}$$

and let the function x_n be linear on the interval $[-\frac{1}{n}, \frac{1}{n}]$ (draw the graph of the function x_n). Since $x_n(t) \to \operatorname{sign} t,\ n \to \infty$, for all $t \in [-1, 1]$, then $y_n(t) = x_n(t) \operatorname{sign} t \to 1 \,(\operatorname{mod} m),\ n \to \infty$, on $[-1, 1]$. Since $|y_n(t)| \leq y_0(t) = 1$, $t \in [-1, 1]$, and the function y_0 is Lebesgue integrable on $[-1, 1]$, by the Lebesgue's dominated convergence theorem

$$f(x_n) = \int_{-1}^{1} y_n(t)dt \to \int_{-1}^{1} dt = 2,\ n \to \infty.$$

Consequently, given that $\|x_n\| = 1$, we have $\|f\| \geq |f(x_n)|,\ n \geq 1$. Passing to the limit as $n \to \infty$ we obtain that $\|f\| \geq 2$. Hence $\|f\| = 2$.

(4) The functional f is not linear. Indeed, if $x(t) = 1,\ t \in [0, 1]$, then $f(x) = f(-x) = \frac{1}{2}$ and $f(-x) \neq -f(x)$. The functional f is continuous. Indeed, if $\{x,\ x_n,\ n \geq 1\} \subset C([0, 1])$ and $\|x_n - x\| \to 0,\ n \to \infty$, then

$$|f(x_n) - f(x)| = \left| \int_0^1 t|x_n(t)|dt - \int_0^1 t|x(t)|dt \right| \leq \int_0^1 t\big||x_n(t)| - |x(t)|\big|dt \leq$$

$$\leq \int_0^1 t\|x_n - x\|dt = \frac{1}{2}\|x_n - x\| \to 0,\ n \to \infty.$$

(5) Since for all $x, y \in C([0, 1])$ and $\alpha, \beta \in \mathbb{K}$

$$f(\alpha x + \beta y) = (\alpha x + \beta y)(1) = \alpha x(1) + \beta y(1) = \alpha f(x) + \beta f(y),$$

f is a linear functional. The functional f is not continuous. Indeed, if $x(t) = 0$ and $x_n(t) = t^n,\ t \in [0, 1],\ n \geq 1$, then

$$\|x_n - x\| = \int_0^1 |t^n - 0|dt = \frac{1}{n+1} \to 0,\ n \to \infty,$$

but $f(x_n) = x_n(1) = 1 \not\to 0 = x(1) = f(x), n \to \infty$.

(6) The functional f is defined correctly, this follows from the Hölder inequality. The linearity of f follows from the linearity of the Lebesgue integral. Let us verify that f is bounded and calculate $\|f\|$.

Let $p = 1$. Then

$$|f(x)| = \left| \int_{-1}^{1} t^2 x(t) \operatorname{sign} t dt \right| \le \int_{-1}^{1} t^2 |x(t)| dt \le \int_{-1}^{1} |x(t)| dt = \|x\|_1.$$

Hence f is bounded and $\|f\| \le 1$. If

$$x_n(t) = \begin{cases} 0, & t \in [-1, 1 - \frac{1}{n}], \\ n, & t \in (1 - \frac{1}{n}, 1], \end{cases}$$

then $\|x_n\| = n \int_{1-\frac{1}{n}}^{1} dt = 1$ and $f(x_n) = n \int_{1-\frac{1}{n}}^{1} t^2 dt \ge \left(1 - \frac{1}{n}\right)^2 \to 1,\ n \to \infty$. Since $\|f\| \ge |f(x_n)|,\ n \ge 1$, we have $\|f\| \ge 1$ and finally $\|f\| = 1$.

Let $1 < p < +\infty$. By the Hölder inequality,

$$|f(x)| \le \int_{-1}^{1} t^2 |x(t)| dt \le \left(\int_{-1}^{1} t^{2q} dt \right)^{\frac{1}{q}} \cdot \left(\int_{-1}^{1} |x(t)|^p dt \right)^{\frac{1}{p}} = \left(\frac{2}{2q+1} \right)^{\frac{1}{q}} \cdot \|x\|_p.$$

On the other hand, if $x_0(t) = (t^2)^{q-1} \operatorname{sign} t,\ t \in [-1, 1]$, then

$$\|x_0\|_p = \left(\int_{-1}^{1} t^{2(q-1)p} dt \right)^{\frac{1}{p}} = \left(\int_{-1}^{1} t^{2q} dt \right)^{\frac{1}{p}},$$

$$f(x_0) = \int_{-1}^{1} t^2 \cdot t^{2(q-1)} dt = \int_{-1}^{1} t^{2q} dt,$$

hence

$$\|f\| \ge \frac{|f(x_0)|}{\|x_0\|_p} = \left(\int_{-1}^{1} t^{2q} dt \right)^{1-\frac{1}{p}} = \left(\frac{2}{2q+1} \right)^{\frac{1}{q}}.$$

Therefore $\|f\| = \left(\frac{2}{2q+1} \right)^{\frac{1}{q}}$.

Let $p = +\infty$. Then

$$|f(x)| \le \int_{-1}^{1} t^2 |x(t)| dt \le \|x\|_\infty \int_{-1}^{1} t^2 dt = \frac{2}{3} \|x\|_\infty.$$

If $x_0(t) = \operatorname{sign} t,\ t \in [-1, 1]$, then $x_0 \in L_\infty([-1, 1]),\ \|x_0\|_\infty = 1,\ f(x_0) = \int_{-1}^{1} t^2 dt = \frac{2}{3}$, Hence $\|f\| = \frac{2}{3}$.

(7) We present f in the form

$$f(x) = \sum_{n=1}^{\infty} \frac{x_n}{2^n} + \sum_{n=1}^{\infty} \frac{x_{n+1}}{2^n} = \sum_{n=1}^{\infty} \frac{x_n}{2^n} + \sum_{n=2}^{\infty} \frac{x_n}{2^{n-1}} =$$

$$= \frac{x_1}{2} + \sum_{n=2}^{\infty} x_n \left(\frac{1}{2^n} + \frac{1}{2^{n-1}} \right) = \frac{x_1}{2} + \sum_{n=2}^{\infty} \frac{3x_n}{2^n}.$$

The linearity of f follows from the linearity of the sum of the series. Let us prove that f is bounded. Indeed, for every $x \in l_2$ by the Cauchy–Schwarz inequality

$$|f(x)| \le \frac{|x_1|}{2} + \sum_{k=2}^{\infty} \frac{3|x_n|}{2^n} \le \left(\frac{1}{4} + \sum_{n=2}^{\infty} \frac{9}{4^n} \right)^{1/2} \cdot \left(\sum_{n=1}^{\infty} |x_n|^2 \right)^{1/2} = \|x\|_2.$$

It follows that f is bounded, and therefore continuous, and $\|f\| \le 1$. For the element $\tilde{x} = (\frac{1}{2}, \frac{3}{2^2}, \ldots, \frac{3}{2^n}, \ldots) \in l_2$ equality is achieved in the Cauchy–Schwarz inequality, therefore $\|f\| \ge \frac{|f(\tilde{x})|}{\|\tilde{x}\|} = 1$. Hence $\|f\| = 1$.

(8) The linearity of f follows from the linearity of the sum of the series. Let us prove that the functional f is not bounded. Set $x^{(n)} = (\cos 1, \ldots, \cos n, 0, \ldots)$, $n \ge 1$. Then $\|x^{(n)}\|^2 = f(x^{(n)}) = \sum_{k=1}^{n} \cos^2 k,\ n \ge 1$, hence

$$\frac{f(x^{(n)})}{\|x^{(n)}\|} = \left(\sum_{k=1}^{n} \cos^2 k \right)^{1/2} \to \left(\sum_{k=1}^{\infty} \cos^2 k \right)^{1/2} = +\infty,\ n \to \infty,$$

because $\cos k \nrightarrow 0,\ k \to \infty$. Since f is a linear unbounded functional, it is not continuous.

3.2 Let X be a Banach space, f be a linear functional on X. Prove that

(1) $\text{Ker } f$ is a linear set, moreover if $f \in X^*$ then $\text{Ker } f$ is a subspace of X;
(2) if $z \in X \setminus \text{Ker } f$ then

$$\forall x \in X \; \exists!\, y \in \text{Ker } f \; \exists!\, \lambda \in \mathbb{K} \; : \; x = y + \lambda z.$$

Solution

(1) For all $x, y \in \text{Ker } f$ and $\alpha, \beta \in \mathbb{K}$ it follows from the linearity of the functional f that $f(\alpha x + \beta y) = \alpha f(x) + \beta f(y) = 0$, hence $\alpha x + \beta y \in \text{Ker } f$. Therefore $\text{Ker } f$ is a linear set. If $f \in X^*$, $\{x_n : n \geq 1\} \subset \text{Ker } f$ and $x_n \to x, \; n \to \infty$, then $0 = f(x_n) \to f(x), \; n \to \infty$. Hence $f(x) = 0$, so $x \in \text{Ker } f$. Therefore $\text{Ker } f$ is a closed linear set under the condition that $f \in X^*$.
(2) Let $x \in X$ be a fixed element and $y = x - \lambda z$, where $\lambda \in \mathbb{K}$. Then $y \in \text{Ker } f$, i.e. $f(y) = f(x) - \lambda f(z) = 0$, if and only if $\lambda = \frac{f(x)}{f(z)}$ (here $f(z) \neq 0$, because $z \notin \text{Ker } f$).

3.3 Using the description of the respective dual spaces, prove the linearity and continuity of the given functionals. Find the norms of continuous linear functionals.

(1) $X = C([-1, 1]), \; f(x) = \int\limits_{-1}^{1} t x(t) dt - x(0)$;
(2) $X = C([a, b]), \; f(x) = x(\frac{a+b}{2})$;
(3) $X = C([0, 3]), \; f(x) = \int\limits_{0}^{1} x(t) dt + \int\limits_{2}^{3} t x(t) dt$;
(4) $X = l_2, \; f(x) = \sum\limits_{k=1}^{\infty} \frac{1}{2^k}(x_k + x_{k+1})$;
(5) $X = l_p, \; 1 \leq p \leq +\infty, \; f(x) = x_j + x_{j+1}, \; j$ is fixed;
(6) $X = l_p, \; 1 \leq p \leq +\infty, \; f(x) = \sum\limits_{k=1}^{\infty} \frac{x_k}{3^k}$;
(7) $X = L_p([0, 1]), \; f(x) = \int\limits_{0}^{\frac{1}{2}} x(t) dt - \int\limits_{\frac{2}{3}}^{1} x(t) dt$.

Solution Recall a formula for calculation of the Riemann–Stieltjes integral. Let $g \, : \, [a, b] \to \mathbb{K}$ be a function of bounded variation, which is continuous on $[a, b] \setminus \{t_1, \ldots, t_n\}$, there exists a derivative g' on $[a, b]$ except for a finite number of points, and $g' \in R([a, b])$. Then for every function $x \in C([a, b])$ we have $\int\limits_{a}^{b} x(t) dg(t) = \int\limits_{a}^{b} x(t) g'(t) dt + \sum\limits_{k=1}^{n} x(t_k)(g(t_k+) - g(t_k-))$, where $g(b+) := g(b), \; g(a-) := g(a)$. If $g(t_k)$ belongs to the interval with endpoints $g(t_k-)$ and $g(t_k+), \; 1 \leq k \leq n$, then the variation of function g on $[a, b]$ equals $V(g, [a, b]) = \int\limits_{a}^{b} |g'(t)| dt + \sum\limits_{k=1}^{n} |g(t_k+) - g(t_k-)|$.

(1) Construct a function $g \in BV_0([-1,1])$ such that $f(x) = \int_{-1}^{1} x(t)dg(t),\ x \in C([-1,1])$. From the formula that defines the functional, it is clear that $g'(t) = t,\ t \neq 0$, and the function g has a jump of size -1 at the point 0. Therefore

$$g(t) = \begin{cases} \frac{t^2}{2} + C, & t \in [-1, 0), \\ \frac{t^2}{2} + C - 1, & t \in [0, 1]. \end{cases}$$

In particular, $g \in BV_0([-1,1])$, if $g(-1) = 0$, i.e. if $C = -\frac{1}{2}$. For any choice of C we have $f(x) = \int_{-1}^{1} x(t)dg(t),\ x \in C([-1,1])$, therefore $f \in (C([-1,1]))^*$ and $\|f\| = \|g\|_V = V(g, [-1,1]) = \int_{-1}^{1} |t|dt + 1 = 2$.

(2) We have $g'(t) = 0,\ t \neq \frac{a+b}{2}$, because there is no integral in the formula which defines the functional. The function g has a jump of size 1 at the point $t = \frac{a+b}{2}$. Therefore

$$g(t) = \begin{cases} 0, & t \in [a, \frac{a+b}{2}), \\ 1, & t \in [\frac{a+b}{2}, b]. \end{cases}$$

Hence $g \in BV_0([a,b])$ and $f(x) = \int_{a}^{b} x(t)dg(t),\ x \in C([a,b])$. Consequently $f \in (C([a,b]))^*$ and $\|f\| = V(g, [a,b]) = 1$.

(3) From the formula that defines the functional, it is clear that the function g is continuous and

$$g'(t) = \begin{cases} 1, & t \in [0, 1), \\ 0, & t \in (1, 2), \\ t, & t \in (2, 3], \end{cases} \text{ whence } g(t) = \begin{cases} t, & t \in [0, 1), \\ 1, & t \in [1, 2), \\ \frac{t^2}{2} - 1, & t \in [2, 3]. \end{cases}$$

Therefore $g \in BV_0([0,3])$ and $f(x) = \int_{0}^{3} x(t)dg(t),\ x \in C([0,3])$. Thus $f \in (C([0,3]))^*$ and $\|f\| = V(g, [0,3]) = g(3) - g(0) = \frac{7}{2}$ (since g is nondecreasing).

(4) Taking into account Corollary 3.1, we need to choose $y \in l_2$ such that $f(x) = (x, y),\ x \in l_2$. Rewrite the functional f as

$$f(x) = \sum_{n=1}^{\infty} \frac{1}{2^n}(x_n + x_{n+1}) = \sum_{n=1}^{\infty} \frac{1}{2^n} x_n + \sum_{n=2}^{\infty} \frac{1}{2^{n-1}} x_n = \frac{1}{2} x_1 + \sum_{n=2}^{\infty} \frac{3}{2^n} x_n.$$

Hence for $y = (\frac{1}{2}, \frac{3}{4}, \frac{3}{8}, \ldots) \in l_2$ we have $f(x) = (x, y),\ x \in l_2$. Therefore $f \in l_2^*$ and $\|f\| = \|y\| = \sqrt{\frac{1}{4} + \sum\limits_{n=2}^{\infty} \frac{9}{4^n}} = 1$.

(5) By Corollary 3.1 and problem 3.13°, we have to find $a \in l_q$ such that $f(x) = \sum\limits_{n=1}^{\infty} x_n a_n,\ x \in l_p$. Put $a = e_j + e_{j+1}$. Since $a \in l_q$ for every $1 \leq q \leq +\infty$ then $f \in l_p^*$. If $p = 1$, then $\|f\| = \|a\|_\infty = 1$; if $1 < p < +\infty$, then $\|f\| = \|a\|_q = 2^{\frac{1}{q}}$; if $p = +\infty$, then $\|f\| = \|a\|_1 = 2$.

(6) Put $a = (\frac{1}{3}, \frac{1}{3^2}, \ldots)$. Then $f(x) = \sum\limits_{n=1}^{\infty} x_n a_n,\ x \in l_p$. Since $a \in l_q,\ 1 \leq q \leq +\infty$, then $f \in l_p^*$ for every $1 \leq p \leq +\infty$. If $p = 1$, then

$$\|f\| = \|a\|_\infty = \sup_{n \geq 1} \left|\frac{1}{3^n}\right| = \frac{1}{3};$$

if $1 < p < +\infty$, then

$$\|f\| = \|a\|_q = \left(\sum_{n=1}^{\infty} \frac{1}{3^{qn}}\right)^{\frac{1}{q}} = \left(\frac{1}{3^q - 1}\right)^{\frac{1}{q}};$$

finally, if $p = +\infty$, then $\|f\| = \|a\|_1 = \sum\limits_{n=1}^{\infty} \frac{1}{3^n} = \frac{1}{2}$ (we used Corollary 3.1 and problem 3.13° here).

(7) Construct a function h such that $f(x) = \int\limits_0^1 x(t)h(t)dt,\ x \in L_p([0, 1])$. Obviously,

$$h(t) = \begin{cases} 1, & t \in [0, \frac{1}{2}], \\ 0, & t \in (\frac{1}{2}, \frac{2}{3}), \\ -1, & t \in [\frac{2}{3}, 1]. \end{cases}$$

Since $h \in L_q([0, 1])$ for all $1 \leq q \leq +\infty$, then by Theorem 3.3 and problem 3.14° we have $f \in (L_p([0, 1]))^*$ and $\|f\| = \|h\|_q$. If $p = 1$, then

$$\|f\| = \|h\|_\infty = \operatorname*{ess\,sup}_{t \in [0,1]} |h(t)| = 1;$$

if $1 < p < +\infty$, then

$$\|f\| = \|h\|_q = \left(\int_0^1 |h(t)|^q dt\right)^{\frac{1}{q}} = \left(\frac{5}{6}\right)^{\frac{1}{q}};$$

if $p = +\infty$, then $\|f\| = \|h\|_1 = \int\limits_0^1 |h(t)|dt = \frac{5}{6}$.

3.4 Let $1 \le p < +\infty$ be fixed. For which $\alpha \in \mathbb{R}$ do the functionals belong to the respective dual spaces? Calculate the norms of continuous linear functionals.

(1) $X = L_p([1, +\infty)),\ f(x) = \int\limits_1^{+\infty} \frac{x(t)}{t^\alpha} dt;$

(2) $X = l_p,\ f(x) = \sum\limits_{n=1}^{\infty} \alpha^n x_n.$

Solution

(1) Put $h(t) = \frac{1}{t^\alpha},\ t \ge 1$. Then

$$f(x) = \int_1^{+\infty} x(t)h(t)dt,\ x \in L_p([1, +\infty)).$$

By Theorem 3.3 we have $f \in (L_p([1, +\infty)))^* \Leftrightarrow h \in L_q([1, +\infty])$. If $1 < p < +\infty$, then $1 < q < +\infty$ and $h \in L_q([1, +\infty))$ under the condition that $\int\limits_1^{+\infty} \frac{dt}{t^{\alpha q}} < +\infty$. This is true when $\alpha q > 1$, i.e. for $\alpha > \frac{1}{q}$. If $p = 1$, then $q = +\infty$ and $h \in L_\infty([1, +\infty))$ under the condition that this function is bounded, i.e. for $\alpha \ge 0$.

(2) Put $a = (\alpha, \alpha^2, \ldots, \alpha^k, \ldots)$. Then $f(x) = \sum\limits_{n=1}^{\infty} x_n a_n,\ x \in l_p$. By Corollary 3.1 we have $f \in l_p^* \Leftrightarrow a \in l_q$. If $1 < p < +\infty$, then $1 < q < +\infty$ and $a \in l_q$ under the condition that $\sum\limits_{n=1}^{\infty} |\alpha|^{nq} < +\infty$. This is a geometric progression, which converges if and only if $|\alpha| < 1$. If $p = 1$, then $q = +\infty$ and $a \in l_\infty$ under the condition that the sequence $a = (\alpha, \alpha^2, \ldots)$ is bounded, i.e. for $|\alpha| \le 1$.

Remark One can deduce the answer to this problem in the case of $p = +\infty$ from problems 3.13°, 3.14° and 5.31: (1) $\alpha > 1$; (2) $|\alpha| < 1$.

Problems to Solve

3.5° Let X be an NLS over the field $\mathbb{K}$. Prove that the set of values of a nonzero linear functional on X coincides with $\mathbb{K}$.

3.6° Let X be an NLS, $f, g \in X^*$. Prove that for any $\alpha, \beta \in \mathbb{K}$ the functional $\alpha f + \beta g$, where $(\alpha f + \beta g)(x) := \alpha f(x) + \beta g(x),\ x \in X$, also belongs to X^*.

3.7° Let X be an NLS, $f \in X^*$. Prove that

(1) $\|f\| = \sup\limits_{\|x\| \le 1} |f(x)|$;
(2) $\|f\| = \sup\limits_{\|x\| = 1} |f(x)|$;
(3) $\|f\| = \min\{C \ge 0 \mid |f(x)| \le C\|x\|,\ x \in X\}$;
(4) $\|f\| = \sup\limits_{\|x\| \le 1} f(x)$, if the space X is real;
(5) $\|f\| = \sup\limits_{x \in M, x \neq 0} \frac{|f(x)|}{\|x\|}$, where M is a dense subset of X.

3.8° Let X be an NLS, $f_n \in X^*$ and $\|f_n\| \le 1,\ n \ge 1$. Set

$$f(x) = \sum_{n=1}^{\infty} \frac{1}{n^2} f_n(x),\ x \in X.$$

Prove that $f \in X^*$.

3.9° Determine whether the given functionals on $C([0, 1])$ are linear and continuous. For continuous linear functionals, calculate their norms using the definition:

(1) $f(x) = x(0)$;
(2) $f(x) = \int\limits_0^1 x(t)dt$;
(3) $f(x) = \int\limits_0^1 |x(t)|dt$;
(4) $f(x) = \|x\|$;
(5) $f(x) = \sum\limits_{n=1}^{\infty} \frac{1}{2^n} x(\frac{1}{2^n})$;
(6) $f(x) = \int\limits_0^1 tx(t)dt$;
(7) $f(x) = \int\limits_0^1 x(t)\,\mathrm{sign}(t - \frac{1}{2})dt$;
(8) $f(x) = \int\limits_0^1 \mathrm{sign}\, x(t)dt$;
(9) $f(x) = \int\limits_0^1 tx^2(t)dt$;
(10) $f(x) = \int\limits_0^1 (1 - 2t)x(t)dt$;
(11) $f(x) = \int\limits_0^1 x(t)\cos \pi t dt$;
(12) $f(x) = \int\limits_0^1 x(t)\sin \pi t dt$;
(13) $f(x) = \int\limits_0^1 |x(t)|\cos \pi t dt$;
(14) $f(x) = \int\limits_0^1 x(t)dt - x(0)$;
(15) $f(x) = \int\limits_0^{\frac{1}{3}} x(t)dt - \int\limits_{\frac{1}{3}}^1 x(t)dt$;
(16) $f(x) = \sum\limits_{k=1}^{n} \lambda_k x(t_k)$, where $t_1, \ldots, t_n$ is a set of different points from $[0, 1]$, $\lambda_1, \ldots, \lambda_n$ is a set of real numbers.

3.10° Determine whether the given functionals are linear and continuous. For continuous linear functionals, calculate their norms using the definition:

(1) $X = l_1,\ f(x) = \sum\limits_{n=1}^{\infty} x_n;$

(2) $X = l_1,\ f(x) = \sum\limits_{n=1}^{\infty} (1 - \frac{1}{n})x_n;$

(3) $X = l_1,\ f(x) = \sum\limits_{n=1}^{\infty} x_n y_n$, where $y \in l_\infty$ is fixed;

(4) $X = l_2,\ f(x) = \sum\limits_{n=1}^{\infty} |x_n|^2;$

(5) $X = l_2,\ f(x) = \sum\limits_{n=1}^{\infty} \frac{x_n}{\sqrt{n(n+1)}};$

(6) $X = l_2,\ f(x) = \sum\limits_{n=1}^{\infty} (-1)^n \frac{x_n}{n};$

(7) $X = l_{\frac{4}{3}},\ f(x) = \sum\limits_{n=1}^{\infty} \frac{x_{2n}}{3^{n/4}};$

(8) $X = l_1,\ f(x) = \sum\limits_{n=1}^{\infty} (1 - (-1)^n)\frac{n-1}{n} x_n;$

(9) $X = l_p,\ 1 \le p < +\infty,\ f(x) = \sum\limits_{n=1}^{\infty} \frac{x_n}{n};$

(10) $X = l_p,\ 1 \le p \le +\infty,\ f(x) = \sum\limits_{n=1}^{\infty} \frac{x_n}{2^n};$

(11) $X = l_p,\ 1 \le p \le +\infty,\ f(x) = \operatorname{sign}(x_1);$

(12) $X = l_p,\ 1 \le p < +\infty,\ f(x) = \sum\limits_{n:\, x_n \neq 0} e^{-\frac{1}{|x_n|}};$

(13) $X = l_p,\ 1 \le p \le +\infty,\ f(x) = x_j,\ j \in \mathbb{N}$ is fixed;

(14) $X = l_p,\ 1 \le p \le +\infty,\ f(x) = x_j + x_{j+1},\ j \in \mathbb{N}$ is fixed;

(15) $X = l_p,\ 1 \le p \le +\infty,\ f(x) = x_1 + 2x_2.$

3.11° Determine whether the given functionals are linear and continuous. For continuous linear functionals, calculate their norms using the definition:

(1) $X = L_2([0, 1]),\ f(x) = \int\limits_0^1 t x(t) dt;$

(2) $X = L_2([0, 1]),\ f(x) = \int\limits_0^1 x(t) \operatorname{sign}(t - \frac{1}{2}) dt;$

(3) $X = L_2([0, 1]),\ f(x) = \int\limits_0^1 \frac{x(t)}{\sqrt[4]{t}} dt;$

(4) $X = L_2([0, 1]),\ f(x) = \int\limits_0^1 |x(t)|^{\frac{3}{4}} dt;$

(5) $X = L_p([0, 1]),\ 1 \le p \le +\infty,\ f(x) = \int\limits_0^1 t x(t) dt;$

(6) $X = L_p([0, \frac{\pi}{2}])$, $1 \le p \le +\infty$, $f(x) = \int\limits_0^{\frac{\pi}{2}} x(t)\sin t dt$;

(7) $X = L_p([0, 1])$, $1 \le p \le +\infty$, $f(x) = \int\limits_0^{\frac{1}{2}} t x(t) dt$;

(8) $X = L_p([0, 2])$, $1 \le p \le +\infty$, $f(x) = \int\limits_0^1 x(t)dt - 3\int\limits_1^2 x(t)dt$;

(9) $X = L_1(\mathbb{R})$, $f(x) = \int\limits_{\mathbb{R}} x(t)\mathrm{arctg}\, t dt$.

3.12° Determine whether the given functionals are linear and continuous. For continuous linear functionals, calculate their norms using the definition:

(1) $X = C^1([0, 1])$ with the norm $\|x\| = \max\limits_{t\in[0,1]} |x(t)| + \max\limits_{t\in[0,1]} |x'(t)|$, $f(x) = x'(0)$;

(2) $X = C^1([0, 1])$ with the uniform norm, $f(x) = x'(0)$;

(3) $X = C^1([0, 1])$ with the uniform norm, $f(x) = \int\limits_0^1 x'(t)\sin \pi t dt$;

(4) $X = C^1([0, 1])$ with norm $\|x\| = \max\limits_{t\in[0,1]} |x(t)| + \max\limits_{t\in[0,1]} |x'(t)|$, $f(x) = x'(0) + x(1)$;

(5) $X = C([0, 1])$, $f(x) = \int\limits_0^1 x(\sqrt{t})dt$;

(6) $X = C([0, 1])$, $f(x) = \int\limits_0^1 x(t^2)dt$;

(7) $X = \left\{ x \in L_2([0, 1]) \;\middle|\; \int\limits_0^1 |x(t^\alpha)|dt < +\infty \right\}$, $\alpha > 0$, $\|\cdot\|_X = \|\cdot\|_{L_2([0,1])}$,
$f(x) = \int\limits_0^1 x(t^\alpha)dt$;

(8) $X = C([0, 1])$, $f(x) = \lim\limits_{n\to\infty} \int\limits_0^1 x(t^n)dt$;

(9) $X = c$, $f(x) = \lim\limits_{n\to\infty} x_n$;

(10) $X = \left\{ x \in l_2 \;\middle|\; \sum\limits_{n=1}^{\infty} x_n \sin n \text{ converges} \right\}$, $\|\cdot\|_X = \|\cdot\|_{l_2}$, $f(x) = \sum\limits_{n=1}^{\infty} x_n \sin n$;

(11) $X = \left\{ x \in l_2 \;\middle|\; \sum\limits_{n=1}^{\infty} x_{n^2} \text{ converges} \right\}$, $\|\cdot\|_X = \|\cdot\|_{l_2}$, $f(x) = \sum\limits_{n=1}^{\infty} x_{n^2}$;

(12) $X = \left\{ x \in l_\infty \;\middle|\; \sum\limits_{n=1}^{\infty} x_n y_n \text{ converges} \right\}$, where $\{y_n : n \ge 1\}$ is a fixed sequence such that the series $\sum\limits_{n=1}^{\infty} y_n$ converges conditionally, $\|\cdot\|_X = \|\cdot\|_{l_\infty}$, $f(x) = \sum\limits_{n=1}^{\infty} x_n y_n$.

3.13° Prove that $l_1 \subset l_\infty^*$, i.e. for every element $a = (a_1, \dots, a_n, \dots) \in l_1$ the functional

$$f_a(x) = \sum_{n=1}^{\infty} a_n x_n, \ x = (x_1, \dots, x_n, \dots) \in l_\infty,$$

is linear and continuous on l_∞, and $\|f_a\| = \|a\|_1$.

3.14° Prove that $L_1(T, \mu) \subset (L_\infty(T, \mu))^*$, i.e. for every element $h \in L_1(T, \mu)$ the functional $f(x) = \int_T h(t)x(t)d\mu(t)$, $x \in L_\infty(T, \mu)$, is linear and continuous on $L_\infty(T, \mu)$, and $\|f\| = \|h\|_1$.

3.15° Using the description of the dual space $(C([a, b]))^*$, prove the linearity and continuity of the given functionals on $C([a, b])$ and calculate their norms:

(1) $X = C([a, b]),\ f(x) = \frac{1}{2}(x(a) + x(b))$;
(2) $X = C([0, 1]),\ f(x) = x(0)$;
(3) $X = C([-1, 1]),\ f(x) = x(0) - 2ix(-1) + (4 + 3i)x(1)$;
(4) $X = C([0, 1]),\ f(x) = x(0) + \int_0^1 t^2 x(t)dt$;
(5) $X = C([-1, 1]),\ f(x) = \int_{-1}^0 x(t)dt - \int_0^1 x(t)dt$;
(6) $X = C([0, 2]),\ f(x) = \int_0^1 tx(t)dt$;
(7) $X = C([0, 2]),\ f(x) = \int_0^1 (1 - 2t)x(t)dt - ix(0) + 2x(2)$;
(8) $X = C([0, \pi]),\ f(x) = \int_0^\pi x(t)\cos t dt$;
(9) $X = C([0, \pi]),\ f(x) = \int_0^\pi x(t)|\cos t| dt$;
(10) $X = C([0, \pi]),\ f(x) = \int_0^\pi x(t)\cos t dt + i\int_0^\pi x(t)\sin t dt$;
(11) $X = C([0, 1]),\ f(x) = \int_0^{\ln 2} e^t x(t)dt + 2ix(1)$;
(12) $X = C([-1, 2]),\ f(x) = \int_{-1}^1 |t|x(t)dt + 3\int_1^2 x(t)dt - 2x(0)$;
(13) $X = C([0, 1]),\ f(x) = \sum_{n=1}^{\infty} \frac{(-1)^n}{n^2} x(\frac{1}{n})$;
(14) $X = C([0, 1]),\ f(x) = \sum_{n=1}^{\infty} \frac{1}{(4n+1)(4n+5)} x(\frac{1}{n^2})$;

(15) $X = C([0,1]),\ f(x) = \sum\limits_{n=1}^{\infty} \frac{(-1)^n}{2^n} x(1 - \frac{1}{n})$;

(16) $X = C([a,b]),\ f(x) = \sum\limits_{k=1}^{n} \lambda_k x(t_k)$, where $t_1, \dots, t_n$ is a set of different points from $[a,b]$, $\lambda_1, \dots, \lambda_n$ is a set of complex numbers;

(17) $X = C([0,1]),\ f(x) = \int\limits_0^1 x(t^2)dt$.

3.16° Let $1 \le p \le +\infty$ be fixed. Using the descriptions of the respective dual spaces, prove the linearity and continuity of the given functionals and calculate their norms:

(1) $X = l_1,\ f(x) = \sum\limits_{n=1}^{\infty} \frac{x_n}{n}$;

(2) $X = l_1,\ f(x) = \sum\limits_{n=1}^{\infty} x_n \operatorname{arctg} n$;

(3) $X = l_2,\ f(x) = \sum\limits_{n=1}^{\infty} \frac{x_n}{\sqrt{n(n+1)}}$;

(4) $X = l_\infty,\ f(x) = \sum\limits_{n=2}^{\infty} \frac{x_n}{n^2-1}$;

(5) $X = l_{\frac{4}{3}},\ f(x) = x_1 + \sqrt[4]{3} i x_3$;

(6) $X = l_{\frac{7}{5}},\ f(x) = \sum\limits_{n=1}^{\infty} \frac{x_{3n}}{\sqrt[7]{n^4}}$;

(7) $X = l_p,\ f(x) = x_j$, where $j \in \mathbb{N}$ is fixed;

(8) $X = l_p,\ f(x) = 2ix_1 + 4x_2$;

(9) $X = l_p,\ f(x) = \sum\limits_{n=1}^{\infty} \frac{x_n}{2^n}$;

(10) $X = l_2,\ f(x) = \sum\limits_{n=1}^{\infty} \frac{\sqrt{n}}{2^{n/2}} x_n$;

(11) $X = L_1(\mathbb{R}),\ f(x) = \int\limits_{\mathbb{R}} x(t) \operatorname{arcctg} t\,dt$;

(12) $X = L_2([0, 2\pi]),\ f(x) = \int\limits_0^{\pi} x(t) e^{it} \sin t\,dt$;

(13) $X = L_p([1,4]),\ f(x) = \int\limits_1^2 t x(t)dt - 2\int\limits_3^4 x(t)dt$;

(14) $X = L_p(\mathbb{R}),\ f(x) = \int\limits_{\mathbb{R}} e^{-t^2} x(t)dt$;

(15) $X = L_\infty([0, +\infty)),\ f(x) = \int\limits_0^{+\infty} \frac{\sin^2 t}{t^2} x(t)dt$.

3.17° Let $1 \leq p < +\infty$ be fixed. For which $\alpha \in \mathbb{R}$ the following functionals belong to the respective dual spaces? Calculate the norms of these functionals.

(1) $X = l_p,\ f(x) = \sum\limits_{n=1}^{\infty} \frac{x_n}{n^\alpha};$

(2) $X = l_p,\ f(x) = \sum\limits_{n=1}^{\infty} x_n((n+1)^\alpha - n^\alpha);$

(3) $X = L_p([0, 1]),\ f(x) = \int\limits_0^1 \frac{x(t)}{t^\alpha} dt;$

(4) $X = L_p([0, +\infty)),\ f(x) = \int\limits_0^{+\infty} t^\alpha e^{-t} x(t) dt;$

(5) $X = L_p([0, \frac{\pi}{2}]),\ f(x) = \int\limits_0^{\frac{\pi}{2}} x(t) \sin^\alpha t dt.$

3.18° (1) Establish the general form of a linear functional on a finite-dimensional NLS and prove that every linear functional on this space is continuous.
(2) Describe all subsets $M \subset \mathbb{R}^2$, for which there exists a nonzero linear functional f on $\mathbb{R}^2$ such that $\operatorname{Ker} f = M$. Also solve this problem for subsets of $\mathbb{R}^3$.

3.19 Prove that $c_0^* = l_1$, and each continuous linear functional on c_0 is given by the formula $f(x) = \sum\limits_{n=1}^{\infty} x_n a_n,\ x \in c_0$, where $a = (a_1, \ldots, a_n, \ldots) \in l_1$ is a fixed element, $\|f\| = \|a\|_1$.

3.20 Prove that $c^* = l_1$, and each continuous linear functional on c is given by the formula $f(x) = a_0 \lim\limits_{n\to\infty} x_n + \sum\limits_{n=1}^{\infty} a_n x_n,\ x \in c$, where $a = (a_0, a_1, \ldots, a_n, \ldots) \in l_1$ is a fixed element, $\|f\| = \|a\|_1$.

3.21 Describe all continuous linear functionals on c_0, which achieve their norm on $\overline{B}(0, 1)$.

3.22*

(1) Find the general form of a continuous linear functional in a real Banach space $C^1([a, b])$.
(2) Find the general form of the continuous linear functional in a real Banach space $C^n([a, b])$, where $n \in \mathbb{N}$ is fixed.

3.23 Prove that $f(x) = \int\limits_a^b p(t)x(t)dt,\ x \in C([a, b])$ is a continuous linear functional, and calculate its norm, if

(1) $p \in C([a, b])$ is a fixed function;
(2)* $p \in L_1([a, b])$ is a fixed function.

3.24 Give an example of a functional $f \in C([0,1])^*$, for which

(1) there is no element $x_0 \in C([0,1])$ such that $\|x_0\| = 1$ and $f(x_0) = \|f\|$;
(2) there are infinitely many such elements;
(3) there is exactly one such element.

3.25 In the spaces l_1 and c give examples of continuous linear functionals, which achieve and which do not achieve their norm on the closed unit ball.

3.26 Let X be an NLS.

(1) Can there exist a functional $f \in X^*$, for which

(a) $\exists!\, x_0 \in \overline{B}(0,1) \ : \ |f(x_0)| = \|f\|$?
(b) $\exists!\, x_0 \in \overline{B}(0,1) \ : \ f(x_0) = \|f\|$?

(2) Let $f \in X^* \setminus \{0\}$. Is it correct that

(a) if $x \in X,\ |f(x)| = \|f\|$, then $\|x\| = 1$?
(b) if $x \in \overline{B}(0,1),\ |f(x)| = \|f\|$, then $\|x\| = 1$?

(3) Let $f \in X^*$. Consider the following statements:

(a) $\exists\, x \in X,\ x \neq 0 \ : \ |f(x)| = \|f\| \cdot \|x\|$;
(b) $\exists\, x \in \overline{B}(0,1) \ : \ |f(x)| = \|f\|$;
(c) $\exists\, x \in \overline{B}(0,1) \ : \ f(x) = \|f\|$.

Which of these statements are equivalent?

3.27° Let X_1, X_2 be two normed linear spaces such that $X_1 \subset X_2$ and convergence $x_n \to x$ in X_1 implies convergence $x_n \to x$ in X_2. For a given $f \in X_2^*$ set $g(x) = f(x),\ x \in X_1$. Prove that $g \in X_1^*$.

3.28 Let X be an NLS, $f \in X^* \backslash \{0\}$, $L := \{x \in X \mid f(x) = 1\}$. Prove that $\frac{1}{\|f\|} = \inf\limits_{x \in L} \|x\|$.

3.29 Let X be an NLS over the field $\mathbb{K}$, f be a linear functional on X. Prove that each of the following conditions is equivalent to the continuity of f:

(1) there exists $R > 0$ such that the set $f\big(\overline{B}(0,R)\big)$ is bounded;
(2) for any sequence $\{x_n : n \geq 1\} \subset X$ which converges to zero, the set $\{f(x_n) : n \geq 1\}$ is bounded;
(3) for any Cauchy sequence $\{x_n : n \geq 1\} \subset X$, the sequence of numbers $\{f(x_n) : n \geq 1\}$ converges in $\mathbb{K}$.
(4) for any real $c \in \mathbb{R}$, the set $\{x \mid f(x) < c\}$ is an open set (in the case $\mathbb{K} = \mathbb{R}$);
(5) there exists $C > 0$ such that the set $\{x \mid |f(x)| \leq C\}$ has interior points.

3.30 Let X be an NLS over the field $\mathbb{K}$, and f be a linear unbounded functional on X. Prove that f attains all values from the field $\mathbb{K}$ in any neighborhood of zero. Is it true that f attains all values from the field $\mathbb{K}$ in any neighborhood of any point?

3.31 Let X be an NLS over the field $\mathbb{K}$, and $f, g \in X^*$ be such that $\operatorname{Ker} f \subset \operatorname{Ker} g$. Prove that there exists $\alpha \in \mathbb{K}$ such that $g = \alpha f$.

3.32 Let X be an NLS, $f, f_1, \ldots, f_n$ be linear functionals on X. Prove that f is a linear combination of functionals $f_1, \ldots, f_n$ if and only if

$$\bigcap_{k=1}^{n} \operatorname{Ker} f_k \subset \operatorname{Ker} f.$$

3.33° In which normed linear spaces X does there exist a functional $f \in X^*$ (1) such that $|f(x)| = \|f\| \cdot \|x\|$, $x \in X$? (2) such that $\operatorname{Ker} f = \{0\}$?

3.34 Let there be functionals $f, g \in X^*$ on an NLS X such that $\operatorname{Ker} f \cap \operatorname{Ker} g = \{0\}$. Prove that $\dim X \leq 2$.

3.35 Let X be an NLS over the field $\mathbb{K}$, f be a nonzero linear functional on X. Prove that f is an open mapping, i.e. the image of any open set in X under the mapping f is an open set in $\mathbb{K}$.

3.36* Let X be an NLS, f be a linear functional on X. Prove that $f \in X^*$ if and only if $\operatorname{Ker} f$ is a closed set.

3.37 Let X be an NLS over the field $\mathbb{K}$, f be a linear functional on X. Prove that

(1) $\operatorname{Ker} f$ is closed in X or is a dense set in X;
(2) if the set $\{x \in X \mid f(x) = c_0\}$ is closed for some $c_0 \in \mathbb{K}$, then the set $\{x \in X \mid f(x) = c\}$ is closed for every $c \in \mathbb{K}$.

3.38* Let X be an NLS, $f \in X^*$. Prove that $|f(x)| = \|f\| \cdot \rho(x, \operatorname{Ker} f)$ for all $x \in X$, where $\rho(x, \operatorname{Ker} f) = \inf\limits_{y \in \operatorname{Ker} f} \|x - y\|$.

3.39 Let X be an NLS over the field $\mathbb{K} = \mathbb{R}$, $f \in X^*$, and elements $x_0, y_0 \in X$ be such that $f(x_0) \cdot f(y_0) < 0$. Prove that

$$\|x_0 - y_0\| \geq \rho(x_0, \operatorname{Ker} f) + \rho(y_0, \operatorname{Ker} f).$$

3.40* Let X be an NLS, $f \in X^*$, $f \neq 0$.

(1) Prove that the following statements are equivalent:
   (a) there exists an element $z \in \overline{B}(0, 1)$ such that $|f(z)| = \|f\|$;
   (b) for every $x \in X$ there exists $y \in \operatorname{Ker} f$ such that $\rho(x, \operatorname{Ker} f) = \|x - y\|$ (i.e. y is the element of the best approximation for x in $\operatorname{Ker} f$);
   (c) for some $x \in X \setminus \operatorname{Ker} f$ there exists $y \in \operatorname{Ker} f$ such that $\rho(x, \operatorname{Ker} f) = \|x - y\|$.
(2) Under what conditions on the functional f for every element $x \in X$ there exists no more than one element of the best approximation in $\operatorname{Ker} f$?

3.41 Give examples of subspaces L in $X = C([0, 1])$ and in $X = c_0$ such that

(1) for any $x_0 \notin L$ there is no element of the best approximation in L;
(2) for any $x_0 \notin L$ there exist infinitely many elements of the best approximation in L.

3.42 A linear functional f on the space $C([a, b])$ is called nonnegative, if for every $x \in C([a, b])$ such that $x(t) \geq 0,\ t \in [a, b]$, the inequality $f(x) \geq 0$ holds. Prove the following statements.

(1) A nonnegative functional f on a real space $C([a, b])$ is continuous, and $\|f\| = f(e_0)$, where $e_0(t) = 1,\ t \in [a, b]$.
(2) The statement of item (1) is valid in a complex space $C([a, b])$ as well.
(3) If a functional $f \in (C([a, b]))^*$ is such that $\|f\| = f(e_0)$, then f is nonnegative.

3.43 Establish the necessary and sufficient condition on the function $g \in BV_0([a, b])$, under which the functional $f(x) = \int\limits_a^b x(t)dg(t)$ is nonnegative.

3.44 Prove that for every functional $f \in (C([a, b]))^*$, where $C([a, b])$ is a real space, there exist nonnegative functionals f_1 and f_2 such that $f = f_1 - f_2$.

3.45 A linear functional f on a real space $C([a, b])$ is called multiplicative if $f(xy) = f(x)f(y),\ x, y \in C([a, b])$. Prove that

(1) for a fixed $\tau \in [a, b]$ the functional $f(x) = x(\tau),\ x \in C([a, b])$, is multiplicative;
(2) every nonzero multiplicative functional is continuous and has a norm equal to one;
(3)* if f is a nonzero multiplicative functional, then there exists $\tau \in [a, b]$ such that $f(x) = x(\tau),\ x \in C([a, b])$.

3.46 Let X_1, X_2 be two normed linear spaces, $X := X_1 \bigoplus X_2$ be their direct sum with the norm

$$\|(x_1, x_2)\|_p = \left(\|x_1\|_{X_1}^p + \|x_2\|_{X_2}^p\right)^{\frac{1}{p}},\ 1 \leq p < +\infty.$$

Prove that every functional $h \in X^*$ is uniquely represented in the form

$$h((x, y)) = f_1(x) + f_2(y),\ (x, y) \in X,$$

where $f_i \in X_i^*,\ i = 1, 2$.

3.47 Prove that in an infinite-dimensional NLS there exist linear discontinuous functionals.

3.48 Let X be an NLS, $M \subset X$ be a convex dense subset in X and $f \in X^*$. Prove that the set $M \cap \operatorname{Ker} f$ is dense in $\operatorname{Ker} f$.

3.49* **(Yamabe)** Let X be an NLS, $M \subset X$ be a convex dense subset in X, $x \in X$, $\varepsilon > 0$, $f_1, \dots, f_n \in X^*$. Prove that
$\exists\, y \in M,\ \|x - y\| < \varepsilon\ :\ f_i(y) = f_i(x),\ 1 \le i \le n.$

3.50 Prove that for any function $x \in C([a, b])$, any points $t_1, \dots, t_n \in [a, b]$ and any $\varepsilon > 0$ there exists a polynomial p such that $\max\limits_{t \in [a,b]} |x(t) - p(t)| < \varepsilon$, and $p(t_i) = x(t_i)$, $1 \le i \le n$.

Chapter 4
Hahn–Banach Theorem

Theoretical Background

Let X be a normed linear space over the field $\mathbb{K}$ ($\mathbb{K} = \mathbb{R}$ or $\mathbb{K} = \mathbb{C}$), G be a linear subset in X, $f : G \to \mathbb{K}$ be a linear functional. A linear functional $F : X \to \mathbb{K}$ is called an *extension* of the functional f to X, if $F(x) = f(x)$, $x \in G$. In this case, the functional f is called a *restriction* of the functional F to G (notation: $F \upharpoonright G = f$). Further, G is considered as an NLS with the norm induced by the norm in X, i.e., $\|x\|_G = \|x\|_X$, $x \in G$.

If X is an NLS, $f \in G^*$, $F \in X^*$ and $F \upharpoonright G = f$, then $\|F\| \geq \|f\|$.

Theorem 4.1 (Extension by Continuity) *Let X be an NLS, $G \subset X$ be a linear set which is dense in X. Then for every functional $f \in G^*$ there exists a unique functional $F \in X^*$ such that $F \upharpoonright G = f$. Moreover, $\|F\| = \|f\|$.*

Theorem 4.2 (Hahn–Banach) *Let X be an NLS, $G \subset X$ be a linear set. Then for every functional $f \in G^*$ there exists a functional $F \in X^*$ such that $F \upharpoonright G = f$ and $\|F\| = \|f\|$.*

Corollary 4.1 *Let X be an NLS, $G \subset X$ be a subspace. Then for every element $y \notin G$ there exists a functional $f \in X^*$ such that $\|f\| = 1$, $f \upharpoonright G = 0$ and $f(y) = \rho(y, G)$.*

Corollary 4.2 *Let X be an NLS, $y \in X$, $y \neq 0$. Then there exists a functional $f \in X^*$ such that $\|f\| = 1$, $f(y) = \|y\|$.*

Corollary 4.3 *Let X be an NLS, $x, y \in X$, $x \neq y$. Then there exists a functional $f \in X^*$ such that $f(x) \neq f(y)$.*

Corollary 4.4 *A subset $M \subset X$ is total in X if and only if there exists no nonzero $f \in X^*$ such that $f \upharpoonright M = 0$.*

V. Brayman et al., *Functional Analysis and Operator Theory*, Problem Books in Mathematics, https://doi.org/10.1007/978-3-031-56427-7_4

Let X be an NLS, X^* be its dual space, $X^{**} = (X^*)^*$ be the dual space to X^*, which is called the *second dual space* to X.

Note that for every $x \in X$ the mapping $F_x : X^* \to \mathbb{K}$ defined as $F_x(f) = f(x),\ f \in X^*$, is a continuous linear functional on X^*, i.e. $F_x \in X^{**}$.

Theorem 4.3 *Let X be an NLS space. The mapping $\varphi : X \to X^{**}$, where $\varphi(x) = F_x,\ x \in X$, is linear, injective and preserves the norm, i.e., $\|F_x\|_{X^{**}} = \|x\|_X$.*

It follows from Theorem 4.3 that X is isometrically isomorphic to a subspace of X^{**}, or to the whole X^{**}, i.e., $\varphi(X) \subset X^{**}$, which means $X \subset X^{**}$. The mapping φ is called the natural (or canonical) embedding of X into X^{**}. If $\varphi(X) = X^{**}$, then the space X is called *reflexive*.

Any finite-dimensional space is reflexive. Spaces $L_p(T, \mu)$ and $l_p,\ 1 < p < +\infty$, are also reflexive. Spaces $l_1,\ l_\infty,\ c_0,\ c,\ C([a, b])$, are nonreflexive. Spaces $L_1(T, \mu)$ and $L_\infty(T, \mu)$ are nonreflexive if the measure μ is not concentrated in a finite number of atoms (see the definition of atom in problem 1.102*).

A subspace $G \subset X$ is called a *hypersubspace*, if $G \neq X$ and there exists an element $x_0 \in X$ such that $X = \operatorname{span}(G \cup \{x_0\})$. A *hyperplane* in X is a set of the form $x_0 + G = \{x_0 + y \mid y \in G\}$, where $x_0 \in X$ and $G \subset X$ are some element and hypersubspace, respectively.

The Hahn–Banach Theorem remains valid in a more general setting. Let X be a linear space over the field $\mathbb{K}$. A functional $p : X \to \mathbb{R}$ is said to be

semi-additive, if $p(x + y) \leq p(x) + p(y)$ for all $x, y \in X$;
positively homogeneous, if $p(\lambda x) = \lambda p(x)$ for all $x \in X$ and $\lambda \geq 0$;
homogeneous, if $p(\lambda x) = |\lambda| p(x)$ for all $x \in X$ and $\lambda \in \mathbb{K}$.

A semi-additive homogeneous functional $p : X \to \mathbb{R}$ is called a *seminorm*. If the functional $p : X \to \mathbb{R}$ is a seminorm, then $p(x) \geq 0,\ x \in X$ (see item (1) of problem 4.52).

Theorem 4.2 admits the following generalizations.

Theorem 4.4 (Hahn–Banach Theorem for Real Space) *Let X be a real linear space, $p : X \to \mathbb{R}$ be a semi-additive positively homogeneous functional, $G \subset X$ be a linear set. Then for every linear functional $f : G \to \mathbb{R}$ such that $f(x) \leq p(x),\ x \in G$, there exists a linear functional $F : X \to \mathbb{R}$ for which $F \upharpoonright G = f$ and $F(x) \leq p(x),\ x \in X$.*

Theorem 4.5 (Hahn–Banach Theorem for Complex Space) *Let X be a complex linear space, $p : X \to \mathbb{R}$ be a seminorm, $G \subset X$ be a linear set. Then for every linear functional $f : G \to \mathbb{C}$ such that $|f(x)| \leq p(x),\ x \in G$, there exists a linear functional $F : X \to \mathbb{C}$, for which $F \upharpoonright G = f$ and $|F(x)| \leq p(x),\ x \in X$.*

Examples of Problems with Solutions

4.1 A functional f on the subspace $G = \{(x_1, 0) \mid x_1 \in \mathbb{R}\}$ of the space $\mathbb{R}^2$ is defined by the formula $f(x) = \alpha x_1,\ x = (x_1, 0) \in G$, where $\alpha \in \mathbb{R}$ is fixed.

Describe all linear extensions F of the functional f to $\mathbb{R}^2$. For which of these extensions $\|F\| = \|f\|$, if $\mathbb{R}^2$ is endowed with the norm $\|\cdot\|_p$, $1 \le p \le +\infty$? In which cases the extension that preserves the norm is unique?

Solution Every linear functional on $\mathbb{R}^2$ has the form

$$F(x) = a_1 x_1 + a_2 x_2, \; x = (x_1, x_2) \in \mathbb{R}^2,$$

where $a_1, a_2 \in \mathbb{R}$ are fixed numbers (which are different for different functionals F). If F is an extension of f, then $F((x_1, 0)) = f((x_1, 0))$, $(x_1, 0) \in G$, i.e. $a_1 x_1 = \alpha x_1$, $x_1 \in \mathbb{R}$. Hence any extension F to $\mathbb{R}^2$ has the form $F(x) = \alpha x_1 + a_2 x_2$, $x = (x_1, x_2) \in \mathbb{R}^2$, where $a_2 \in \mathbb{R}$ is an arbitrary number. It follows from Corollary 3.2 that $\|F\| = \|a\|_q$, where $a = (\alpha, a_2) \in \mathbb{R}^2$, q is the conjugate index to p.

Let us find $\|f\|$. We have

$$\|f\| = \sup_{(x_1,0)\in G\setminus\{0\}} \frac{|f((x_1, 0))|}{\|(x_1, 0)\|_p} = \sup_{x_1 \ne 0} \frac{|\alpha x_1|}{|x_1|} = |\alpha|.$$

Now for every $1 \le p \le +\infty$ we will determine when $\|F\| = \|f\|$ and when such an extension is unique.

If $p = 1$, then $\|F\| = \|f\|$ if and only if $\max\{|\alpha|, |a_2|\} = |\alpha|$. Therefore the extension F of the functional f that preserves the norm has the form

$$F(x) = \alpha x_1 + a_2 x_2, \; x = (x_1, x_2) \in \mathbb{R}^2,$$

where $a_2 \in [-|\alpha|, |\alpha|]$ is an arbitrary number. Such an extension is unique if and only if $\alpha = 0$.

If $1 < p \le +\infty$, then $1 \le q < +\infty$ and $\|F\| = \|f\|$ if and only if $|\alpha|^q + |a_2|^q = |\alpha|^q$, hence $a_2 = 0$. Therefore the extension of the functional f that preserves the norm is unique and has the form $F(x) = \alpha x_1$, $x = (x_1, x_2) \in \mathbb{R}^2$.

4.2 Let $G = \left\{ x \in C([0, 1]) \;\middle|\; \int_0^1 x(t)dt = 0, \; x(1) = 0 \right\}$. Construct a functional $f \in (C([0, 1]))^*$ such that $f(x) = 0$, $x \in G$, and $f(x_1) = f(x_2) = 1$, where $x_1(t) = 1$, $x_2(t) = 1 - t$, $t \in [0, 1]$.

Solution We will try to find the functional f of the form

$$f(x) = \alpha \int_0^1 x(t)dt + \beta x(1), \; x \in C([0, 1]),$$

where $\alpha, \beta \in \mathbb{R}$ are some coefficients to be determined. Obviously, if $x \in G$, then $f(x) = 0$. Since $f(x_1) = \alpha + \beta = 1$, $f(x_2) = \frac{\alpha}{2} = 1$, then $\alpha = 2$, $\beta = -1$, so $f(x) = 2\int\limits_0^1 x(t)dt - x(1)$, $x \in C([0, 1])$.

Remark Functional with the prescribed properties is unique. Indeed, for every $y \in C([0, 1])$ it is easy to check that $y - y(1)x_1 - 2\left(\int\limits_0^1 y(t)dt - y(1)\right)x_2 \in G$, hence

$$f(y) = y(1)f(x_1) + 2\left(\int\limits_0^1 y(t)dt - y(1)\right)f(x_2) = 2\int\limits_0^1 y(t)dt - y(1).$$

4.3 Prove that there exists a nonzero functional $F \in (L_\infty([a, b]))^*$ such that

(1) $\forall x \in C([a, b])\ :\ F(x) = x(\frac{a+b}{2})$;
(2) $\forall x \in C([a, b])\ :\ F(x) = 0$.

Solution

(1) Set $f(x) = x(\frac{a+b}{2})$, $x \in C([a, b])$. One can verify that $f \in (C([a, b]))^*$ either directly or by Riesz's theorem (Theorem 3.4), considering the representation $f(x) = \int\limits_a^b x(t)dg(t)$, $x \in C([a, b])$, where $g(t) = \chi_{[\frac{a+b}{2}, b]}(t)$, $t \in [a, b]$, $g \in BV_0([a, b])$. Since $C([a, b])$ is a linear subset in $L_\infty([a, b])$, by Hahn–Banach Theorem there exists a functional $F \in (L_\infty([a, b]))^*$ such that $F(x) = f(x)$ for all $x \in C([a, b])$. The functional F is nonzero, because if $x_0(t) = 1$, $t \in [a, b]$, then $F(x_0) = f(x_0) = x_0(\frac{a+b}{2}) = 1 \neq 0$.
(2) Note that $C([a, b])$ is a subspace in $L_\infty([a, b])$. Indeed, it is obvious that the set $C([a, b])$ is linear. Also the set $C([a, b])$ is closed in $L_\infty([a, b])$, it follows from the completeness of $C([a, b])$ with the uniform norm and the equality

$$\operatorname*{ess\,sup}_{t\in[a,b]} |x(t)| = \max_{t\in[a,b]} |x(t)|,\ x \in C([a, b]).$$

Consider $y(t) = \chi_{[\frac{a+b}{2}, b]}(t)$, $t \in [a, b]$. Then $y \in L_\infty([a, b])$, but y does not belong to the subspace $C([a, b])$ in $L_\infty([a, b])$, since there is no function $x \in C([a, b])$ such that $y(t) = x(t)\,(\text{mod}\, m)$ on $[a, b]$. By Corollary 4.1 of Hahn–Banach Theorem there exists a functional $F \in (L_\infty([a, b]))^*$ such that $F(x) = 0$ for every $x \in C([a, b])$ and $F(y) \neq 0$.

Problems to Solve

4.4° A functional f is defined on the subspace $G = \{(x_1, x_2, 0) \mid x_1, x_2 \in \mathbb{R}\}$ of the space $\mathbb{R}^3$ by the formula $f(x) = \alpha x_1 + \beta x_2$, $x = (x_1, x_2, 0) \in G$, where

$\alpha, \beta \in \mathbb{R}$ are some fixed numbers. Describe all linear extensions F of the functional f to $\mathbb{R}^3$. For which of these extensions $\|F\| = \|f\|$, if $\mathbb{R}^3$ is endowed with the norm $\|\cdot\|_p$, $1 \le p \le +\infty$? In which cases the extension that preserves the norm is unique?

4.5° Let $\alpha > 0$ be given, $G = \{(x_1, x_1) \mid x_1 \in \mathbb{R}\} \subset \mathbb{R}^2$, $f(x) = \alpha x_1$, $x \in G$. Describe all linear extensions F of the functional f to $\mathbb{R}^2$. For which of these extensions $\|F\| = \|f\|$, if $\mathbb{R}^2$ is endowed with the norm $\|\cdot\|_p$, $1 \le p \le +\infty$? In which cases the extension that preserves the norm is unique?

4.6° Let $f(x) = x_1$, $x \in L = \{x = (x_1, x_2) \in \mathbb{R}^2 \mid 2x_1 - x_2 = 0\}$. Prove that there exists a unique extension F of the functional f to the Euclidean space $\mathbb{R}^2$, which preserves the norm. Find this extension.

4.7°

(1) Let G be a subspace of a Hilbert space H and $f \in G^*$. Describe all continuous linear extensions F of the functional f to H. Which of them preserve the norm?
(2) Solve problem (1) for $G = \{x \in H \mid (x, h) = 0\}$ and $f(x) = (x, a)$, $x \in G$, where $a, h \in H$ are fixed elements.
(3) Solve problem (1) for $G = \{x \in H \mid (x, h_i) = 0,\ 1 \le i \le n\}$ and $f(x) = (x, a)$, $x \in G$, where $a \in H$ is a fixed element, and $\{h_i : 1 \le i \le n\} \subset H$ is a fixed orthonormal system.

4.8° Let $(T, \mathcal{F}, \mu)$ be a space with a σ-finite measure, $1 \le p < +\infty$, $A \in \mathcal{F}$, $G = \{x \in L_p(T, \mu) \mid x(t) = 0$ for almost all $t \in T \setminus A\}$, $a \in L_q(A, \mu)$, q is the conjugate index to p. Describe all continuous linear extensions F to $L_p(T, \mu)$ of the functional $f(x) = \int\limits_A a(t)x(t)d\mu(t)$, $x \in G$. For which of these extensions $\|F\| = \|f\|$?

4.9 Consider c_0 as a subspace of the space c. Describe all continuous linear extensions of the functional $f \in c_0^*$ to c. Which of them preserve the norm?

4.10° Let $G = \{x \in C([0, 1]) \mid x(0) = 0\}$. Construct a continuous linear functional on $C([0, 1])$, which equals zero on G and takes the value 2 on the function $x_0(t) = t + 1$, $t \in [0, 1]$.

4.11° Let $G = \{x \in C([0, 1]) \mid x(0) = x(1) = 0\}$, $x_1(t) = 1$, $x_2(t) = t$, $t \in [0, 1]$. Construct a continuous linear functional on $C([0, 1])$, which equals zero on G and such that (1) $f(x_1) = 2$, $f(x_2) = 3$; (2) $f(x_1) = 2$, $\|f\| = 2$, $f(x_2) \le 0$.

4.12° Let $A \subset [0, 1]$ be a Lebesgue measurable set of positive Lebesgue measure and $G = \{x \in L_2([0, 1]) \mid x = 0$ almost everywhere on $A\}$. Construct a functional $f \in (L_2([0, 1]))^*$, which equals zero on G and such that $f(x_0) = 1$, where $x_0(t) = t$, $t \in [0, 1]$. Is such functional unique?

4.13 Let X be an NLS and G be a linear subset in X, which we will consider as an NLS with the norm induced by the norm of X.

(1) Let G be dense in X. Prove that $G^* = X^*$.
(2) Is the converse statement correct?

4.14 Let $\Phi = \{x \in l_2 \mid \exists n \in \mathbb{N} : x_k = 0,\ k \geq n\}$ (the set of elements of l_2 which have finitely many nonzero coordinates) with the norm $\|x\| = \left(\sum\limits_{n=1}^{\infty} |x_n|^2\right)^{1/2}$. Describe the dual space Φ^*.

4.15° Let $P = \left\{x(t) = \sum\limits_{k=0}^{n} a_k t^k \mid n \in \mathbb{N} \cup \{0\},\ a_0, \ldots, a_n \in \mathbb{C}\right\}$ be the set of polynomials. Determine which of the given linear functionals on P admit a continuous linear extension to $C([0, 1])$ and, if such extensions exist, find them.

(1) $f(x) = a_0$;
(2) $f(x) = \sum\limits_{k=0}^{n} a_k$;
(3) $f(x) = \sum\limits_{k=0}^{n} (-1)^k a_k$;
(4) $f(x) = \sum\limits_{k=0}^{N} c_k a_k$, where $N \in \mathbb{N}$ and $c_0, \ldots, c_N \in \mathbb{R}$ are fixed numbers.

4.16° Prove that there exist nonzero functionals $F \in (L_\infty([-1, 1]))^*$ such that

(1) $F(x) = x(0)$ for every $x \in C([-1, 1])$;
(2) $F(x) = x(0)$ for every $x \in L_\infty([-1, 1])$ such that x is continuous at point 0 (i.e. there exists a continuous at point 0 function $\tilde{x}$ such that $x(t) = \tilde{x}(t)$ almost everywhere on $[-1, 1]$);
(3) $F(x) = 0$ for every $x \in L_\infty([-1, 1])$ such that x is continuous at point 0;
(4) $F(x) = 0$ for every $x \in L_\infty([-1, 1])$ such that $x = 0$ almost everywhere on $[-1, 0]$ or the function x is continuous at point 0.

4.17°

(1) Prove that there exists a functional $F \in l_\infty^*$ such that $F(x) = \lim\limits_{n\to\infty} x_n$ for every $x \in c$;
(2) Prove that the inclusion $l_1 \subset l_\infty^*$ is strict.

4.18 Prove that there exists a functional $F \in l_\infty^*$ such that $F(x) = \lim\limits_{n\to\infty} x_n$ for every $x \in c$ and $F(x) = 0$ for every $x \in l_\infty$ of the form $x = (0, x_2, 0, x_4, 0, \ldots)$.

4.19 Prove that the inclusion $L_1([a, b]) \subset (L_\infty([a, b]))^*$ is strict.

4.20

(1) Let l_∞ be a real Banach space. Prove that there exists a functional $f \in l_\infty^*$ such that

(a) $\|f\| = 1$;
(b) $f(e) = 1$ for $e = (1, 1, 1, \ldots)$;
(c) $f(Tx) = f(x)$ for all $x \in l_\infty$, where $T(x_1, x_2, \ldots) = (x_2, x_3, \ldots)$. Any functional $f \in l_\infty^*$ with properties (a)–(c) is called a *Banach limit* and is denoted by $\operatorname*{Lim}_{n\to\infty} x_n = f(x),\ x \in l_\infty$.

(2) Prove that $\operatorname*{Lim}_{n\to\infty} x_n$ has the following properties:

(d) if $x_n \geq 0,\ n \geq 1$, then $\operatorname*{Lim}_{n\to\infty} x_n \geq 0$;
(e) $\liminf_{n\to\infty} x_n \leq \operatorname*{Lim}_{n\to\infty} x_n \leq \limsup_{n\to\infty} x_n$;
(f) if $x_n \to a,\ n \to \infty$, then $\operatorname*{Lim}_{n\to\infty} x_n = a$.

4.21 Prove that for any Banach limit (see problem 4.20):

(1) There exists $x \in l_\infty$ such that $x_n = 1$ for an infinite number of indices n, but $\operatorname*{Lim}_{n\to\infty} x_n = 0$.
(2) There exist $x, y \in l_\infty$ such that $\operatorname*{Lim}_{n\to\infty} x_n y_n \neq \operatorname*{Lim}_{n\to\infty} x_n \cdot \operatorname*{Lim}_{n\to\infty} y_n$.

4.22

(1) Let $B([0, +\infty))$ be a linear space of bounded real functions on $[0, +\infty)$ with the norm $\|x\| = \sup_{t\geq 0} |x(t)|,\quad x \in B([0, +\infty))$. Prove that there exists a functional $f \in (B([0, +\infty)))^*$ such that

(a) $\|f\| = 1$;
(b) $f(e) = 1$ for $e(t) = 1,\ t \geq 0$;
(c) $f(T_s x) = f(x)$ for all $x \in B([0, +\infty))$ and $s > 0$, where $T_s x(t) = x(t + s),\ t \geq 0$.

(2) For any functional $f \in (B([0, +\infty)))^*$ with properties (a)–(c) denote $\operatorname*{Lim}_{t\to+\infty} x(t) = f(x),\ x \in B([0, +\infty))$. Prove that $\operatorname*{Lim}_{t\to+\infty} x(t)$ has the following properties:

(d) if $x(t) \geq 0,\ t \geq 0$, then $\operatorname*{Lim}_{t\to+\infty} x(t) \geq 0$;
(e) $\liminf_{t\to+\infty} x(t) \leq \operatorname*{Lim}_{t\to+\infty} x(t) \leq \limsup_{t\to+\infty} x(t)$;
(f) if $x(t) \to a,\ t \to +\infty$, then $\operatorname*{Lim}_{t\to+\infty} x(t) = a$.

4.23 Let X be an NLS, $G \subset X$ be a subspace, X^* be a strictly normed space (see problem 1.94). Prove that for every functional from G^* there exists a unique extension to X^* that preserves the norm.

4.24* Let f be a continuous linear functional on the subspace $c_0 \subset l_\infty$. Prove that there exists a unique continuous linear extension of f to l_∞ that preserves the norm of the functional.

4.25 Let X be a Banach space, A be a set of indices. The systems

$$\{x_\alpha : \alpha \in A\} \subset X, \ \{f_\alpha : \alpha \in A\} \subset X^*$$

are called *biorthogonal*, if $f_\alpha(x_\beta) = \delta_{\alpha\beta}, \ \alpha, \beta \in A$, where $\delta_{\alpha\beta}$ is the Kronecker symbol ($\delta_{\alpha\beta} = 0, \ \alpha \neq \beta, \ \delta_{\alpha\alpha} = 1$).

(1) Let $\{x_1, \ldots, x_n\} \subset X$ be a linearly independent system. Prove that there exists a system biorthogonal to it.
(2) Is the statement of item (1) correct for an infinite system of linearly independent elements of space X?
(3) Let $\{x_\alpha : \alpha \in A\} \subset X, \ \{f_\alpha : \alpha \in A\} \subset X^*$ be biorthogonal systems. Is it correct that the system $\{x_\alpha : \alpha \in A\} \subset X$ is linearly independent? Is it correct that the system $\{f_\alpha : \alpha \in A\} \subset X^*$ is linearly independent?
(4)* Under what conditions on the linearly independent system $\{x_\alpha : \alpha \in A\} \subset X$ does a biorthogonal system exist?
(5)* Let $\{f_1, \ldots, f_n\} \subset X^*$ be a linearly independent system. Prove that there exists a system in X that is biorthogonal to it.

4.26° Let X be an NLS, $\{x_1, \ldots, x_n\} \subset X$ be a linearly independent system and $c_1, \ldots, c_n \in \mathbb{K}$. Prove that there exists a functional $f \in X^*$ such that $f(x_k) = c_k, \ k = 1, \ldots, n$.

4.27 Let X be an NLS, $\{x_n : n \geq 1\} \subset X, \ \{c_n : n \geq 1\} \subset \mathbb{K}, \ M > 0$. Show that a functional $f \in X^*$ such that $f(x_n) = c_n, \ n \geq 1$, and $\|f\| \leq M$, exists if and only if for any finite system $\{\lambda_1, \ldots, \lambda_m\} \subset \mathbb{K}$ the inequality

$$\left|\sum_{k=1}^{m} \lambda_k c_k\right| \leq M \left\|\sum_{k=1}^{m} \lambda_k x_k\right\|$$

holds.

4.28° Is it true that in an NLS X the elements x and y are equal, if the equality $f(x) = f(y)$ holds (1) for all $f \in X^*$? (2) for all f from a total subset in X^*?

4.29

(1) Let X be an infinite-dimensional NLS. Show that X^* is an infinite-dimensional space.
(2) Is the converse statement to statement (1) true?
(3) How are the dimensions of spaces X and X^* related, if both these spaces are finite-dimensional?

4.30° Let $\{x_n : n \geq 1\}$ be a fixed sequence of elements in an NLS X, and L be its linear span. Prove that $x \in \overline{L}$ if and only if for any functional $f \in X^*$ the equalities $f(x_n) = 0,\ n \geq 1$, imply that $f(x) = 0$.

4.31 Prove that the system of functions $\{t^{100k} : k \geq k_0\},\ k_0 \in \mathbb{N}$, is total in $L_p([a, b]),\ 1 \leq p < +\infty$ for $a \geq 0$.

4.32° Let X be an NLS, $x \in X$. Prove that

$$\|x\| = \max\big\{|f(x)| \mid f \in X^*,\ \|f\| = 1\big\}.$$

4.33 Let X be a reflexive Banach space. Prove that each functional $f \in X^*$ attains maximum of its absolute value on $\overline{B}(0, 1)$.

4.34 Let X be an NLS, $G \subset X$ be its subspace, $x \in X$. Prove that the element $y_0 \in G$ satisfies condition $\rho(x, G) = \|x - y_0\|$ if and only if there exists a functional $f \in X^*$ such that $\|f\| = 1,\ f(x) = \|x - y_0\|$ and $f(y) = 0,\ y \in G$.

4.35 Prove that a finite-dimensional NLS is reflexive.

4.36 Find the image $\varphi(y) \in l_\infty$ of the element $y \in c$ for the canonical embedding from Theorem 4.3.

4.37 Prove that the spaces c and c_0 are nonreflexive.

4.38* Prove that the subspace of a reflexive space is reflexive.

4.39* Prove that a Banach space X is reflexive if and only if its dual space X^* is reflexive.

4.40* Let X be a Banach space and the space X^* is separable. Prove that the space X is separable. Is the converse statement true?

4.41 Is it true that

(1) Every reflexive NLS is a Banach space?
(2) If for a Banach space X the set $\varphi(X)$ is dense in X^{**}, where $\varphi : X \to X^{**}$ is the canonical embedding from Theorem 4.3, then X is reflexive?

4.42 Prove that the functional $F(g) = \int\limits_0^{1/2} dg(t),\ g \in BV_0([0, 1])$, is linear and continuous on $BV_0([0, 1])$. Deduce that the space $C([0, 1])$ is nonreflexive.

4.43 Prove the nonreflexivity of the spaces $C([0, 1])$ and l_1 (1) with the help of the statement of problem 4.33; (2) with the help of the statement of problem 4.40*.

4.44° Explain the geometric content (1) of Corollary 4.1 of the Hahn–Banach Theorem in $\mathbb{R}^3$, if G is a line that passes through the origin; (2) of Corollary 4.2 of the Hahn–Banach Theorem in $\mathbb{R}^2$ and $\mathbb{R}^3$.

4.45 Let X be an NLS and $M \subset X$ be a closed set. Put

$$M^{\perp} = \{f \in X^* \mid f(x) = 0 \text{ for all } x \in M\}.$$

(1) Prove that $M^{\perp}$ is a subspace in X.
(2) What is $M^{\perp}$, if X is a Hilbert space?
(3) Prove that $\{x \in X \mid f(x) = 0 \text{ for all } f \in M^{\perp}\} = \overline{\text{span}(M)}$.
(4) Prove that if X is a reflexive space and $M \subset X$ is a subspace, then $M^{\perp\perp} = M$ (more precisely, $M^{\perp\perp} = \varphi(M)$, where $\varphi : X \to X^{**}$ is the canonical embedding from Theorem 4.3).
(5) Is the statement of item (4) valid for a nonreflexive space X?
(6) Prove that $M^{\perp\perp} \cap X = \overline{\text{span}(M)}$ (more precisely, $M^{\perp\perp} \cap \varphi(X) = \varphi(\overline{\text{span}(M)})$).

4.46 Let X be an NLS. Is it true that for an arbitrary subspace $L \subset X^*$ there exists a subspace $M \subset X$ such that $L = M^{\perp}$?

4.47 Let X be a Banach space, $M \subset X^*$. Consider the following statements: (a) M is a total set in X^*; (b) if $x \in X$ is such that $f(x) = 0$ for all $f \in M$, then $x = 0$.

(1) Prove that (a) implies (b).
(2) Prove that if X is a reflexive Banach space then (b) implies (a).
(3) Let X be a dual space of some Banach space and (b) implies (a). Prove that the space X is reflexive.

4.48 Let $M = \{\delta_t \mid t \in [a, b]\} \subset (C([a, b]))^*$, where $\delta_t(x) = x(t)$, $x \in C([a, b])$.

(1) Prove that if $x \in C([a, b])$ and $f(x) = 0$ for all $f \in M$, then $x = 0$, but the set M is not total in $(C([a, b]))^*$.
(2) Find $\overline{\text{span}(M)}$ in $(C([a, b]))^*$.

4.49 Prove that in a real NLS X every closed ball with center at point 0 is the intersection of some family of sets of the form $\{x \in X \mid f(x) \le a\}$, where $f \in X^*$, $a \in \mathbb{R}$.

4.50

(1) Prove that in a separable NLS X every closed ball is the intersection of a countable family of sets of the form $\{x \in X \mid\ |f(x) - b| \le a\}$, where $f \in X^*$, $a \in \mathbb{R}$, $b \in \mathbb{K}$.
(2) Prove that in a real separable NLS X every closed ball is the intersection of a countable family of sets of the form $\{x \in X \mid f(x) \le a\}$, where $f \in X^*$, $a \in \mathbb{R}$.

4.51 Let X be an NLS. Prove that

(1) A subset $G \subset X$ is a hypersubspace if and only if there exists a functional $f \in X^* \setminus \{0\}$ such that $G = \text{Ker}\, f$.
(2) A subset $M \subset X$ is a hyperplane if and only if there exists a functional $f \in X^* \setminus \{0\}$ and $\lambda \in \mathbb{K}$ such that $M = \{x \in X \mid f(x) = \lambda\}$.

4.52 Let X be a linear space. Prove that

(1) If a functional $p : X \to \mathbb{R}$ is a seminorm, then $p(x) \geq 0,\ x \in X$;
(2) If p is a seminorm, then $|p(x) - p(y)| \leq p(x - y),\ x, y \in X$;
(3) If f is a linear functional on X, then the functional $p(x) = |f(x)|,\ x \in X$, is a seminorm;
(4) If p is a seminorm and f is a linear functional on X such that $f(x) \leq p(x),\ x \in X$, then $|f(x)| \leq p(x),\ x \in X$.

Under what conditions a seminorm is a norm?

4.53 Let p be a semi-additive positively homogeneous functional on a real linear space X. Prove that (1) p is not necessarily a seminorm; (2) the statement of problem 4.52, item (4) may be false for p.

4.54 For a complex linear space X the same space endowed with operations of addition and multiplication by real numbers becomes a real linear space. It is denoted by $X_{\mathbb{R}}$ and called *real space associated with* X. Let p be a seminorm on X. Prove that

(1) If f is a linear functional on X for which $|f(x)| \leq p(x),\ x \in X$, and $g(x) = \mathrm{Re} f(x),\ h(x) = \mathrm{Im} f(x),\ x \in X$, then g, h are linear functionals on $X_{\mathbb{R}}$ such that $|g(x)| \leq p(x),\ |h(x)| \leq p(x),\ x \in X$, and also $h(x) = -g(ix),\ x \in X$.
(2) If g is a linear functional on $X_{\mathbb{R}}$, for which $|g(x)| \leq p(x),\ x \in X$, and $f(x) = g(x) - ig(ix),\ x \in X$, then f is a linear functional on X such that $|f(x)| \leq p(x),\ x \in X$.

4.55 Let X be a complex NLS, $X_{\mathbb{R}}$ be a real space associated with X (see problem 4.54). The hyperplane M_0 in $X_{\mathbb{R}}$ is called a real hyperplane in X. Prove that

(1) If M_0 is a real hyperplane, then iM_0 is also a real hyperplane in X.
(2) The set $M \subset X$ is a hypersubspace in X if and only if there exists a real hypersubspace M_0 in X such that $M = M_0 \cap (iM_0)$.
(3) If the set $M \subset X$ is a hyperplane in X, then there exist real hyperplanes M_1, M_2 in X such that $M = M_1 \cap M_2$.

4.56* Prove that an NLS X is strictly normed (see problem 1.94) if and only if for every functional $f \in X^* \setminus \{0\}$ there exists no more than one element $x \in \overline{B}(0, 1)$ for which $f(x) = \|f\|$.

4.57 Let X be a real linear space, $p : X \to \mathbb{R}$ be a semi-additive positively homogeneous functional. Prove that there exists a linear functional $F : X \to \mathbb{R}$ such that $F(x) \leq p(x),\ x \in X$.

4.58 Let $M(\mathbb{R})$ be a linear space of bounded real functions on $\mathbb{R}$ with period 1. For $x \in M(\mathbb{R})$ define

$$\pi(x; \alpha_1, \dots, \alpha_k) = \sup_{t \in \mathbb{R}} \frac{1}{k} \sum_{j=1}^{k} x(t + \alpha_j), \;\; p(x) = \inf \pi(x; \alpha_1, \dots, \alpha_k),$$

where the infimum is taken over all finite sets of real numbers $\alpha_1, \dots, \alpha_k$. Prove that p is a semi-additive positively homogeneous functional on $M(\mathbb{R})$.

4.59*

(1) Let $M(\mathbb{R})$ be a linear space of bounded real functions on $\mathbb{R}$ with period 1, p be a semi-additive positively homogeneous functional on $M(\mathbb{R})$, defined in problem 4.58, and $F : M(\mathbb{R}) \to \mathbb{R}$ be a linear functional such that $F(x) \le p(x), \; x \in M(\mathbb{R})$ (such a functional exists according to problem 4.57). Prove that

(a) if $x(t) \ge 0, \; t \in \mathbb{R}$, then $F(x) \ge 0$;
(b) $F(T_s x) = F(x)$ for all $x \in M(\mathbb{R})$ and $s \in \mathbb{R}$, where

$$T_s x(t) = x(t + s), \; t \in \mathbb{R}.$$

(c) $F(e) = 1$ for $e(t) = 1, \; t \in \mathbb{R}$;
(d) if $x(t) = \chi_{[a,b]}(t)$ for $t \in [0, 1], \; 0 \le a < b \le 1$, then $F(x) = b - a$.

(2) Prove that every linear functional F on $M(\mathbb{R})$, which satisfies conditions (a)–(c), coincides with the Riemann integral $\int_0^1 x(t)dt$ for Riemann integrable on [0, 1] functions x.

(3) Prove that the functional F which satisfies conditions (a)–(c) can be chosen in such a way that it coincides with the Lebesgue integral $\int_0^1 x(t)dt$ for Lebesgue integrable on [0, 1] functions x.

4.60

(1) Prove that it is possible to assign to every set $A \subset [0, 1]$ a number $\mu(A)$, the "measure" of the set A, in such a way that for any sets $A \subset [0, 1], \; B \subset [0, 1]$ the following conditions hold true:

(a) if $A \cap B = \varnothing$, then $\mu(A \cup B) = \mu(A) + \mu(B)$;
(b) $\mu(A) \ge 0$;
(c) if a set A can be transformed to a set B by parallel translation, then $\mu(A) = \mu(B)$;
(d) $\mu([0, 1]) = 1$.

(2) Let the set $A \subset [0, 1]$ be Jordan measurable. Prove that $\mu(A)$ equals to the Jordan measure of the set A.
(3) Prove that the functional $\mu : 2^{[0,1]} \to \mathbb{R}$, which satisfies conditions (a)–(d), can be chosen in such a way that it coincides with the Lebesgue measure for the Lebesgue measurable sets.

Chapter 5
Weak and Weak-∗ Convergence

Theoretical Background

Theorem 5.1 (Banach–Steinhaus Theorem; Uniform Boundedness Principle) *Let X be a Banach space, $\{f_\alpha : \alpha \in A\}$ be a family of functionals from X^* which is bounded at each point, i.e.* $\forall\, x \in X\ \exists\, C_x > 0\ \forall\, \alpha \in A\ :\ |f_\alpha(x)| \le C_x$.

Then the family of norms of these functionals is bounded, i.e.
$\exists\, C > 0\ \forall\, \alpha \in A\ :\ \|f_\alpha\| \le C$.

Let X be an NLS. A sequence of functionals $\{f_n : n \ge 1\} \subset X^*$ *converges weakly-∗* to a functional $f \in X^*$, if $f_n(x) \to f(x)$, $n \to \infty$, for each $x \in X$. This is denoted as $f_n \stackrel{w-*}{\longrightarrow} f$.

A sequence of functionals $\{f_n : n \ge 1\} \subset X^*$ *converges strongly* to a functional $f \in X^*$, if $f_n \to f$, $n \to \infty$, in norm of the Banach space X^*, i.e. $\|f_n - f\| \to 0$, $n \to \infty$. Strong convergence of functionals implies the weak-∗ convergence: if $f_n \to f$ in X^*, then $f_n \stackrel{w-*}{\longrightarrow} f$, $n \to \infty$. The converse statement, generally speaking, is incorrect.

Theorem 5.2 (Criterion of Weak-∗ Convergence of Functionals) *Let X be a Banach space, $\{f_n : n \ge 1\} \subset X^*$, $f \in X^*$. Then $f_n \stackrel{w-*}{\longrightarrow} f$, $n \to \infty$, if and only if the following conditions are fulfilled:*

(a) $\exists\, C > 0\ \forall\, n \ge 1\ :\ \|f_n\| \le C$;
(b) there exists a total set M in X such that $f_n(x) \to f(x)$, $n \to \infty$, for all $x \in M$.

Theorem 5.3 (Weak-∗ Completeness of a Dual Space) *Let X be a Banach space, and let the sequence of functionals $\{f_n : n \ge 1\} \subset X^*$ be weakly-∗ Cauchy sequence, i.e., for every $x \in X$ the sequence $\{f_n(x) : n \ge 1\}$ is a Cauchy sequence. Then there exists a functional $f \in X^*$ such that $f_n \stackrel{w-*}{\longrightarrow} f$, $n \to \infty$.*

V. Brayman et al., *Functional Analysis and Operator Theory*, Problem Books in Mathematics, https://doi.org/10.1007/978-3-031-56427-7_5

Theorem 5.4 (Weak-∗ Compactness of the Ball in a Dual Space) *Let X be a separable NLS, $R > 0$. Then any sequence of functionals from the ball $\overline{B}(0, R) = \{f \in X^* \mid \|f\| \leq R\}$ in X^* has a subsequence that converges weakly-∗ to some functional from X^*.*

A sequence $\{x_n : n \geq 1\}$ of elements of an NLS X *converges weakly* to an element $x \in X$, if $f(x_n) \to f(x)$, $n \to \infty$, for every functional $f \in X^*$. This is denoted as $x_n \overset{w}{\to} x$.

A sequence $\{x_n : n \geq 1\} \subset X$ *converges strongly* to $x \in X$, if $x_n \to x$, $n \to \infty$, in norm of X, i.e. $\|x_n - x\| \to 0$, $n \to \infty$.

Strong convergence of a sequence of elements of an NLS implies weak convergence: if $x_n \to x$ in X, then $x_n \overset{w}{\to} x$, $n \to \infty$. The converse statement, generally speaking, is incorrect (see problem 5.2). The sequence $\{x_n : n \geq 1\} \subset X$ is called *weakly bounded*, if for every bounded functional $f \in X^*$ the sequence $\{f(x_n) : n \geq 1\}$ is bounded.

The sequence $\{x_n : n \geq 1\} \subset X$ is called *a weakly Cauchy sequence*, if for every bounded functional $f \in X^*$ the sequence $\{f(x_n) : n \geq 1\}$ is a Cauchy sequence.

Theorem 5.5 (Criterion of Weak Convergence of Elements) *Let X be an NLS, $\{x_n : n \geq 1\} \subset X$, $x \in X$. Then $x_n \overset{w}{\to} x$ if and only if the following conditions are fulfilled:*

(a) $\exists C > 0 \, \forall n \geq 1 : \|x_n\| \leq C$;
(b) there exists a total set M in X^* such that $f(x_n) \to f(x)$, $n \to \infty$, for all $f \in M$.

Examples of Problems with Solutions

5.1 Prove the following statements:

(1) (**Criterion of weak convergence of elements in $C([a, b])$.**) The sequence $\{x_n : n \geq 1\} \subset C([a, b])$ converges weakly to the element $x \in C([a, b])$ if and only if the following conditions are fulfilled:

  (a) $\exists C > 0 \, \forall n \geq 1 : \|x_n\| \leq C$;
  (b) $\forall t \in [a, b] : x_n(t) \to x(t), \ n \to \infty$.

(2) (**Criterion of weak convergence of elements in l_p, $1 < p < +\infty$.**) The sequence $\{x^{(n)} = (x_1^{(n)}, x_2^{(n)}, \ldots) : n \geq 1\} \subset l_p$, $1 < p < +\infty$, converges weakly to the element $x = (x_1, x_2, \ldots) \in l_p$ if and only if the following conditions are fulfilled:

  (a) $\exists C > 0 \, \forall n \geq 1 : \|x^{(n)}\| \leq C$;
  (b) $\forall k \geq 1 : x_k^{(n)} \to x_k, \ n \to \infty$.

(3) **(Criterion of weak convergence of elements in $L_p([a,b])$, $1 < p < +\infty$.)** The sequence $\{x_n : n \geq 1\} \subset L_p([a,b])$, $1 < p < +\infty$, converges weakly to the element $x \in L_p([a,b])$ if and only if the following conditions are fulfilled:

(a) $\exists\, C > 0\ \forall\, n \geq 1 \ :\ \|x_n\| \leq C$;

(b) $\forall\, \tau \in (a,b] \ :\ \int_a^\tau x_n(t)dt \to \int_a^\tau x(t)dt,\ n \to \infty.$

Solution

(1) *Necessity.* Condition (a) is fulfilled according to the criterion of weak convergence (Theorem 5.5). For any fixed $t \in [a,b]$ consider the functional $f(x) = x(t)$, $f \in (C([a,b]))^*$. Since

$$f(x_n) = x_n(t) \to f(x) = x(t),\ n \to \infty,$$

condition (b) is also fulfilled. *Sufficiency.* By Theorem 3.4 any functional $f \in (C([a,b]))^*$ has the form $f(x) = \int_a^b x(t)dg(t)$, where $g \in BV_0([a,b])$. Let λ_v be the Lebesgue–Stieltjes measure associated with the function $v(t) = V(g,[a,t])$, $t \in [a,b]$. Then

$$|f(x_n) - f(x)| = \left| \int_a^b (x_n(t) - x(t))dg(t) \right| \leq$$

$$\leq \int_a^b |x_n(t) - x(t)|dv(t) = \int_{[a,b]} |x_n(t) - x(t)|d\lambda_v(t).$$

It follows from conditions (a), (b) and the Lebesgue's dominated convergence theorem that $f(x_n) \to f(x)$, $n \to \infty$.

(2) Apply the criterion of weak convergence (Theorem 5.5). Consider the functionals $f_k(x) = x_k$, $x = (x_1, x_2, \ldots) \in l_p$, $k \geq 1$. The condition (b) of item (2) implies that $f_k(x^{(n)}) \to f_k(x)$, $n \to \infty$. Therefore it is enough to check that the set $M = \{f_k : k \geq 1\}$ is total in l_p^*. This is true because the spaces l_p^* and l_q, where $\frac{1}{p} + \frac{1}{q} = 1$, are isometrically isomorphic (see Corollary 3.1 of Theorem 3.3), and the image of the set M under this isomorphism is the set $\{e_k : k \geq 1\}$ which is total in l_q.

(3) Apply the criterion of weak convergence (Theorem 5.5). Consider the functionals $f_\tau(x) = \int_a^\tau x(t)dt$, $x \in L_p([a,b])$, $\tau \in (a,b]$. The condition (b) of item (3) implies that $f_\tau(x_n) \to f_\tau(x)$, $n \to \infty$. Therefore it is enough to check that the set $M = \{f_\tau \mid \tau \in (a,b]\}$ is total in $(L_p([a,b]))^*$. This is true because the spaces $(L_p([a,b]))^*$ and $L_q([a,b])$, where $\frac{1}{p} + \frac{1}{q} = 1$, are

isometrically isomorphic (see Theorem 3.3), and the image of the set M under this isomorphism is the set $\{\chi_{[a,\tau]} \mid \tau \in (a, b]\}$ which is total in $L_q([a, b])$.

Remark The criterion of weak convergence in $L_p(\mathbb{R})$, $1 < p < +\infty$, is similar (see problem 5.7).

5.2 Prove that every strongly convergent in NLS sequence converges weakly. Is the converse statement true?

Solution Let X be an NLS, the sequence $\{x_n : n \geq 1\}$ converges in norm to $x \in X$. Then for any $f \in X^*$ we have

$$|f(x_n) - f(x)| = |f(x_n - x)| \leq \|f\| \cdot \|x_n - x\| \to 0, \ n \to \infty,$$

so $f(x_n) \to f(x)$, $n \to \infty$. Hence $x_n \xrightarrow{w} x$, $n \to \infty$.

The converse statement is false, as shown in the following example. Let $X = l_2$, $x^{(n)} = e_n$, $n \geq 1$. Since $\|e_n - e_m\| = \sqrt{2}$ for all $n \neq m$, the sequence $\{x^{(n)} : n \geq 1\}$ is not a Cauchy sequence, hence it does not converge in l_2. Now let us establish that $x^{(n)} \xrightarrow{w} 0$, $n \to \infty$.

Method I Use the criterion of weak convergence in l_2 (item (2) of problem 5.1). (a) We have $\|x^{(n)}\| = 1$, $n \geq 1$. (b) For all $k \geq 1$ we have $x_k^{(n)} = 0$, $n \geq k + 1$, hence $x_k^{(n)} \to 0$, $n \to \infty$. Since conditions a) and b) are fulfilled, it follows that $x^{(n)} \xrightarrow{w} 0$, $n \to \infty$.

Method II Use the definition of weak convergence. If $f \in l_2^*$ then there exists $a \in l_2$ such that $f(x) = \sum\limits_{n=1}^{\infty} x_n \overline{a_n}$, $x \in l_2$. Hence

$$f\big(x^{(n)}\big) = \overline{a_n} \to 0 = f(0), \ n \to \infty,$$

because the series $\sum\limits_{n=1}^{\infty} |a_n|^2$ converges. Therefore $x^{(n)} \xrightarrow{w} 0$, $n \to \infty$.

5.3 Examine the sequences for strong and weak convergence in an NLS X. For convergent sequences find their limits.

(1) $X = l_p$, $1 \leq p < +\infty$, $x^{(n)} = (\underbrace{0, \ldots, 0, \tfrac{n}{n+1}}_{n}, 0, \ldots)$;

(2) $X = l_p$, $1 \leq p \leq +\infty$, $x^{(n)} = (1, \frac{1}{2}, \ldots, \frac{1}{n}, 0, \ldots)$;

(3) $X = L_p([0, 1])$, $1 \leq p \leq +\infty$, $x_n(t) = \begin{cases} n, & t \in [0, \frac{1}{n^2}], \\ 0, & t \in (\frac{1}{n^2}, 1]; \end{cases}$

(4) $X = L_p([0, 1])$, $1 < p < +\infty$, $x_n(t) = \sin \pi n t$, $t \in [0, 1]$;

(5) $X = C([0, 1])$, $x_n(t) = \frac{nt^2}{1+n^2t^4}$, $t \in [0, 1]$.

Solution

(1) If the sequence converges weakly (in particular if it converges strongly) in l_p, $1 \le p < +\infty$, then it converges coordinate-wise. For $1 < p < +\infty$ it follows from problem 5.1, item (2), and for $p = 1$ one can directly consider the functionals $f_k(x) = x_k$, $x \in l_1$, $k \ge 1$. Therefore the coordinate-wise limit of the sequence is the unique "candidate" to be the weak or strong limit. For each $k \ge 1$ we have $x_k^{(n)} = 0$, $n \ge k + 1$, hence $x_k^{(n)} \to 0$, $n \to \infty$. Now let us examine the sequence for strong and weak convergence to $x = 0$.

For all $1 \le p < +\infty$ we have

$$\|x^{(n)} - 0\|_p^p = \sum_{k=1}^{\infty}(x_k^{(n)} - 0)^p = \left(\frac{n}{n+1}\right)^p \to 1 \ne 0, \ n \to \infty,$$

hence there is no strong convergence.

For $1 < p < +\infty$ we apply the criterion of weak convergence in l_p (item (2) of problem 5.1). (a) We have $\|x^{(n)}\| = \frac{n}{n+1} \le 1$, $n \ge 1$. (b) Coordinate-wise convergence to zero has already been established. Therefore $x^{(n)} \xrightarrow{w} 0$, $n \to \infty$.

It remains to deal with $p = 1$. Consider $f(x) = \sum\limits_{n=1}^{\infty} x_n$, $x \in l_1$. Then $f \in l_1^*$, because $(1, 1, \ldots) \in l_\infty$, and $f(x^{(n)}) = \frac{n}{n+1} \to 1 \ne 0 = f(0)$, $n \to \infty$. Hence $x^{(n)} \overset{w}{\not\to} 0$, $n \to \infty$, in l_1.

Remark For $p = +\infty$ the sequence does not converge strongly. Since both the sequence members and the "candidate" to be a limit belong to the subspace c_0 of the space l_∞, the sequence converges weakly in l_∞ if and only if it converges weakly in c_0. One can apply the criterion of weak convergence in c_0 (see item (1) of problem 5.8) to verify that this sequence converges weakly.

(2) Let us find the coordinate-wise limit. For each $k \ge 1$ we have $x_k^{(n)} = \frac{1}{k}$, $n \ge k + 1$, thus $x_k^{(n)} \to \frac{1}{k}$, $n \to \infty$. Put $x = (1, \frac{1}{2}, \frac{1}{3}, \ldots)$ and verify whether $x^{(n)} \to x$, $n \to \infty$, in l_p.

For $p = 1$ we have $x \notin l_1$, because the series $\sum\limits_{n=1}^{\infty} \frac{1}{n}$ diverges. Therefore for $p = 1$ the sequence does not converge either weakly or strongly.

Let $1 < p < +\infty$. Then $x \in l_p$, because the series $\sum\limits_{n=1}^{\infty} \frac{1}{n^p}$ converges. We have $\|x^{(n)} - x\|_p^p = \sum\limits_{k=n+1}^{\infty} \frac{1}{k^p} \to 0$, $n \to \infty$, as a remainder of a convergent series. Thus $x^{(n)} \to x$, $n \to \infty$, in l_p, consequently $x^{(n)} \xrightarrow{w} x$, $n \to \infty$.

For $p = +\infty$ we have $x \in l_\infty$, because the sequence $\{\frac{1}{k} : k \geq 1\}$ is bounded. Since $\|x^{(n)} - x\|_\infty = \sup\limits_{k \geq n+1} \frac{1}{k} = \frac{1}{n+1} \to 0,\ n \to \infty$, then $x^{(n)} \to x,\ n \to \infty$, in l_∞, and therefore $x^{(n)} \xrightarrow{w} x,\ n \to \infty$, in l_∞.

(3) Neither strong nor weak convergence in $L_p([0, 1])$, $1 \leq p \leq +\infty$, implies convergence almost everywhere. However, if for a given sequence there exists an almost everywhere limit, it is natural to consider it as a "candidate" for weak and strong limits.

For each $t \in (0, 1]$ there exists $N \in \mathbb{N}$ such that $\frac{1}{n^2} < t,\ n \geq N$, therefore $x_n(t) = 0,\ n \geq N$. Thus $x_n(t) \to 0 = x_0(t),\ n \to \infty$. Hence $x_n(t) \to x_0(t)\ (\mathrm{mod}\, m)$ on $[0, 1]$.

Let $1 \leq p < +\infty$. Then

$$\|x_n - x_0\|_p^p = \int\limits_0^1 |x_n(t) - x_0(t)|^p dt = \int\limits_0^{1/n^2} n^p dt = \frac{1}{n^{2-p}},\quad n \geq 1.$$

Since $\frac{1}{n^{2-p}} \to 0,\ n \to \infty$, if and only if $p < 2$, then $x_n \to x_0$ in $L_p([0, 1])$ for $1 \leq p < 2$ and $x_n \not\to x_0$ in $L_p([0, 1])$ for $2 \leq p < +\infty$. Moreover, for $2 < p < +\infty$ we have $\frac{1}{n^{2-p}} \to +\infty$, hence the sequence $\{x_n\}$ is not bounded in $L_p([0, 1])$. The criterion of weak convergence (Theorem 5.5) implies that for $2 < p < +\infty$ the sequence does not converge weakly, and therefore does not converge strongly. In case $p = +\infty$ we have

$$\|x_n - x_0\|_\infty = \operatorname*{ess\,sup}_{t \in [0,1]} |x_n(t) - x_0(t)| = \sup_{t \in [0,1]} |x_n(t)| = n,\quad n \geq 1.$$

Thus the sequence $\{x_n : n \geq 1\}$ is not bounded in $L_\infty([0, 1])$, and therefore does not converge in $L_\infty([0, 1])$ either strongly or weakly.

It remains to deal with the case $p = 2$. We have already established that $x_n \not\to x_0,\ n \to \infty$, in $L_2([0, 1])$. Now we will show that $x_n \xrightarrow{w} x_0,\ n \to \infty$, in $L_2([0, 1])$. From here we can deduce that the sequence $\{x_n : n \geq 1\}$ cannot have another strong limit, and therefore does not converge strongly in $L_2([0, 1])$.

Method I Apply the criterion of weak convergence in $L_2([0, 1])$ (item (3) of problem 5.1). (a) We have $\|x_n\|_2^2 = \int\limits_0^{1/n^2} n^2 dt = n^2 \cdot \frac{1}{n^2} = 1,\ n \geq 1$. (b) For each $\tau \in (0, 1]$ there exists $N \in \mathbb{N}$ such that $\frac{1}{n^2} < \tau,\ n \geq N$. Therefore for $n \geq N$ we have $\int\limits_0^\tau x_n(t)dt = \int\limits_0^{1/n^2} n dt = \frac{1}{n} \to 0 = \int\limits_0^\tau x_0(t)dt,\ n \to \infty$.

Since conditions (a) and (b) are fulfilled, it follows that $x_n \xrightarrow{w} x_0,\ n \to \infty$, in $L_2([0, 1])$.

Method II Use the definition of weak convergence. For each function $g \in L_2([0, 1])$ by the Cauchy–Schwarz inequality we have

$$\left|\int_0^1 x_n(t)g(t)dt\right| = \left|\int_0^{1/n^2} ng(t)dt\right| \leq \left(\int_0^{1/n^2} n^2 dt \cdot \int_0^{1/n^2} |g(t)|^2 dt\right)^{1/2} =$$

$$= \left(\int_0^{1/n^2} |g(t)|^2 dt\right)^{1/2} \to 0,\ n \to \infty,$$

therefore $\int_0^1 x_n(t)g(t)dt \to 0 = \int_0^1 x_0(t)g(t)dt,\ n \to \infty$. Hence $x_n \overset{w}{\to} x_0$, $n \to \infty$, in $L_2([0, 1])$.

Thus if $1 \leq p < 2$, then $x_n \to x_0$, $n \to \infty$, in $L_p([0, 1])$, if $p = 2$, then $x_n \overset{w}{\to} x_0$, $n \to \infty$, but there is no strong convergence, and if $2 < p \leq +\infty$, then the sequence $\{x_n : n \geq 1\}$ does not converge either weakly or strongly.

(4) Set $x_0(t) = 0$, $t \in [0, 1]$. We will apply the criterion of weak convergence in $L_p([0, 1])$, $1 < p < +\infty$ (item (3) of problem 5.1) to show that $x_n \overset{w}{\to} x_0$, $n \to \infty$.

(a) $\|x_n\|_p^p = \int_0^1 |\sin \pi nt|^p dt \leq \int_0^1 dt = 1$, $n \geq 1$. (b) For each $\tau \in (0, 1]$ we have

$$\int_0^\tau x_n(t)dt = \int_0^\tau \sin \pi nt dt = \frac{1}{\pi n}(-\cos \pi nt)\Big|_0^\tau$$

$$= \frac{1 - \cos \pi n\tau}{\pi n} \to 0 = \int_0^\tau x_0(t)dt,\ n \to \infty.$$

It follows from conditions (a) and (b) that $x_n \overset{w}{\to} x_0,\ n \to \infty$.

Let us prove that $x_n \not\to x_0$, $n \to \infty$, in $L_p([0, 1])$. Indeed,

$$\|x_n\|_p^p = \int_0^1 |\sin \pi nt|^p dt = |s = \pi nt| = \frac{1}{\pi n}\int_0^{\pi n} |\sin s|^p ds =$$

$$= \frac{1}{\pi}\int_0^{\pi} |\sin s|^p ds \not\to 0,\ n \to \infty.$$

(5) Apply the criterion of weak convergence in $C([0, 1])$ (item (1) of problem 5.1). If a sequence converges weakly (in particular if it converges strongly) in $C([0, 1])$, then it converges pointwise. Therefore the pointwise limit $x_0(t) = \lim\limits_{n\to\infty} \frac{nt^2}{1+n^2t^4} = 0$, $t \in [0, 1]$, is the only "candidate" to be the weak or strong limit Let us show that $x_n \not\to x_0$, $n \to \infty$, in $C([0, 1])$. One can use the derivative to find that

$$\|x_n - x_0\| = \max_{t\in[0,1]} |x_n(t) - 0| = x_n\Big(\frac{1}{\sqrt{n}}\Big) = \frac{1}{2}, \quad n \geq 1,$$

hence, $x_n \not\to x_0$, $n \to \infty$.

Since $\|x_n\| = \frac{1}{2}$, $n \geq 1$, and also $x_n(t) \to x_0(t)$, $n \to \infty$, for all $t \in [0, 1]$, then $x_n \xrightarrow{w} x_0$, $n \to \infty$, by the criterion of weak convergence in $C([0, 1])$.

5.4 Prove that the sequence of functionals $\{f_n : n \geq 1\} \subset X^*$ is weakly-∗ convergent, if

(1) $X = L_1([a, b])$, $f_n(x) = \int\limits_a^b x(t)e^{int}dt$;

(2) $X = L_p([0, 1])$, $2 \leq p < +\infty$, $f_n(x) = \int\limits_0^{1/n^2} nx(t)dt$;

(3) $X = C([0, 1])$, $f_n(x) = \sqrt{n} \int\limits_{1-\frac{1}{\sqrt{n}}}^{1} x(t)dt$.

Solution

(1) Apply the criterion of weak-∗ convergence of functionals (Theorem 5.2). Let us check its conditions. (a) Let $h_n(t) = e^{int}$, $t \in [a, b]$. Since $h_n \in L_\infty([a, b])$, then by Theorem 3.3 we have $f_n \in X^*$ and $\|f_n\| = \|h_n\|_\infty = 1$. (b) The set $M = \{\chi_{[a,s]} \mid s \in [a, b]\}$ is total in $L_1([a, b])$. For arbitrary $x = \chi_{[a,s]} \in M$ we have

$$|f_n(x)| = \left|\int_a^s e^{int}dt\right| = \left|\frac{e^{ins} - e^{ina}}{in}\right| \leq \frac{2}{n} \to 0, \; n \to \infty.$$

Hence $f_n \xrightarrow{w-*} 0$, $n \to \infty$, in $(L_1([a, b]))^*$.

Remark The convergence means that $\int\limits_a^b x(t)e^{int}dt \to 0$, $n \to \infty$, for all $x \in L_1([a, b])$ (Riemann–Lebesgue lemma).

(2) If $\{f_n : n \ge 0\} \subset (L_p([a,b]))^*$, then $f_n(x) = \int\limits_a^b h_n(t)x(t)dt$, $n \ge 0$, where $h_n \in L_q([a,b])$ and $\frac{1}{p} + \frac{1}{q} = 1$. By definition $f_n \overset{w-*}{\longrightarrow} f_0$, $n \to \infty$, if

$$\forall x \in L_p([a,b]) \ : \ \int\limits_a^b h_n(t)x(t)dt \to \int\limits_a^b h_0(t)x(t)dt, \ n \to \infty.$$

This is equivalent to weak convergence $h_n \overset{w}{\to} h_0$, $n \to \infty$, in $L_q([a,b])$. In our case $[a,b] = [0,1]$ and

$$h_n(t) = \begin{cases} n, & t \in [0, \frac{1}{n^2}], \\ 0, & t \in (\frac{1}{n^2}, 1], \end{cases} \quad n \ge 1.$$

According to problem 5.3, item (3) we have $h_n \overset{w}{\to} 0 = h_0$, $n \to \infty$, in $L_q([0, 2\pi])$ for $1 < q \le 2$. Therefore $f_n \overset{w-*}{\longrightarrow} 0$, $n \to \infty$, in $(L_p([0, 2\pi]))^*$ for $2 \le p < +\infty$.

(3) Set $f(x) = x(1)$, $x \in C([0,1])$. Let us show that $f_n \overset{w-*}{\longrightarrow} f$, $n \to \infty$. Consider the space $C([0,1])$ of real-valued functions first. By the mean value theorem for the Riemann integral, for each function $x \in C([0,1])$ and for each $n \ge 1$ there exists $\theta_n \in \left[1 - \frac{1}{\sqrt{n}}, 1\right]$ such that $f_n(x) = x(\theta_n)$, $n \ge 1$. Since $\theta_n \to 1$, $n \to \infty$, and $x \in C([0,1])$, then $f_n(x) = x(\theta_n) \to x(1) = f(x)$, $n \to \infty$. In a similar way, in the case of complex space $C([0,1])$ for every function $x \in C([0,1])$ there exist $\theta_n', \theta_n'' \in \left[1 - \frac{1}{\sqrt{n}}, 1\right]$ such that

$$f_n(x) = \operatorname{Re} x(\theta_n') + i \operatorname{Im} x(\theta_n'') \to x(1) = f(x), \ n \to \infty.$$

By definition, we have $f_n \overset{w-*}{\longrightarrow} f$, $n \to \infty$, in $(C([0,1]))^*$.

Problems to Solve

5.5° Prove the uniqueness of the limit

(1) for a sequence of continuous linear functionals that converges weakly-$*$;
(2) for a sequence of elements of an NLS that converges weakly.

5.6° Prove that in a finite-dimensional space strong convergence is equivalent to weak convergence.

5.7 Prove that the sequence $\{x_n : n \geq 1\} \subset L_p(\mathbb{R})$, $1 < p < +\infty$, converges weakly to an element $x \in L_p([a, b])$ if and only if the following conditions are fulfilled:

(a) $\exists C > 0 \, \forall n \geq 1 : \|x_n\| \leq C$;

(b) $\forall s, t \in \mathbb{R}, s < t : \int\limits_s^t x_n(u)du \to \int\limits_s^t x(u)du, \; n \to \infty$.

5.8 (1) Prove that the sequence $\left\{x^{(n)} = (x_1^{(n)}, x_2^{(n)}, \ldots) : n \geq 1\right\} \subset c_0$ converges weakly to the element $x = (x_1, x_2, \ldots) \in c_0$ if and only if the following conditions are fulfilled:

(a) $\exists C > 0 \, \forall n \geq 1 : \|x^{(n)}\| \leq C$;

(b) $\forall k \geq 1 : x_k^{(n)} \to x_k, \; n \to \infty$.

(2) Formulate and prove the criterion of weak convergence in the space c.

5.9° Examine the sequences for strong and weak convergence in an NLS X. For convergent sequences find their limits.

(1) $X = l_p$, $1 < p < +\infty$, $x^{(n)} = (\underbrace{0, \ldots, 0}_{n}, 1, \frac{1}{2}, \frac{1}{3}, \ldots)$;

(2) $X = l_p$, $1 < p < +\infty$, $x^{(n)} = (\underbrace{0, \ldots, 0}_{n}, \frac{1}{n+1}, \frac{1}{n+2}, \ldots)$;

(3) $X = l_p$, $1 < p < +\infty$, $x^{(n)} = (\underbrace{1, \ldots, 1}_{n-1}, \frac{1}{n}, \frac{1}{n+1}, \ldots)$;

(4) $X = l_p$, $1 < p < +\infty$, $x^{(n)} = (\frac{1}{n}, \frac{1}{n+1}, \ldots)$;

(5) $X = l_p$, $1 < p < +\infty$, $x^{(n)} = (\frac{1+n}{1+n}, \frac{1+n}{1+2n}, \ldots, \frac{1+n}{1+kn}, \ldots)$;

(6) $X = l_p$, $1 \leq p < +\infty$, $x^{(n)} = (\frac{1}{2}, \frac{2}{2^2}, \ldots, \frac{n}{2^n}, \frac{1}{(n+1)^2}, \frac{1}{(n+2)^2}, \ldots)$;

(7) $X = l_p$, $1 \leq p < +\infty$, $x^{(n)} = (1, 0, \frac{1}{3^2}, 0, \frac{1}{5^2}, \ldots, 0, \frac{1}{(2n-1)^2}, \frac{1}{(2n)^2}, \frac{1}{(2n+1)^2}, \ldots)$;

(8) $X = l_\infty$, $x^{(n)} = (\underbrace{0, \ldots, 0}_{n-1}, \ln n, 0, 0, \ldots)$;

(9) $X = L_p([0, 1])$, $1 \leq p < +\infty$, $x_n(t) = t^n$;

(10) $X = L_p([0, 1])$, $1 \leq p \leq +\infty$, $x_n(t) = \begin{cases} \sqrt{n}, & t \in [0, \frac{1}{n}), \\ 0, & t \in [\frac{1}{n}, 1]; \end{cases}$

(11) $X = L_p([0, 1])$, $1 < p < +\infty$, $x_n(t) = e^{int}$;

(12) $X = L_p([0, 1])$, $1 < p < +\infty$, $x_n(t) = \sin(t \cdot \ln n)$;

(13) $X = L_p([0, 1])$, $1 < p < +\infty$, $x_n(t) = \cos(n^2 t)$;

(14) $X = L_p([0, 1])$, $1 < p < +\infty$, $x_n(t) = \sin(2^n t)$;

(15) $X = L_p([0, 1])$, $1 < p < +\infty$, $x_n(t) = \begin{cases} \sqrt{n} - n\sqrt{nt}, & t \in [0, \frac{1}{n}), \\ 0, & t \in [\frac{1}{n}, 1]; \end{cases}$

(16) $X = L_p([0, 1]),\ 1 < p \le +\infty,\ x_n(t) = \begin{cases} 2n(1 - nt), & t \in [0, \frac{1}{n}), \\ 0, & t \in [\frac{1}{n}, 1]; \end{cases}$
(17) $X = L_p([0, 1]),\ 1 \le p < +\infty,\ x_n(t) = \sin(t^n)$;
(18) $X = L_p(\mathbb{R}),\ 1 \le p \le +\infty,\ x_n(t) = \sqrt[4]{n}\chi_{[n,n+\frac{1}{n}]}(t)$;
(19) $X = L_p(\mathbb{R}),\ 1 < p < +\infty,\ x_n(t) = \chi_{[n,n+1]}(t)$;
(20) $X = L_p(\mathbb{R}),\ 1 \le p \le +\infty,\ x_n(t) = \frac{1}{\sqrt{n}}\chi_{[n,2n]}(t)$.

5.10° Examine the given sequences in $C([0, 1])$ for weak and strong convergence.

(1) $x_n(t) = t^n - t^{3n}$;
(2) $x_n(t) = e^{-\frac{t}{n}}$;
(3) $x_n(t) = e^{-nt}$;
(4) $x_n(t) = te^{-nt}$;
(5) $x_n(t) = nte^{-nt}$;
(6) $x_n(t) = \frac{n^2 t}{1+n^4t^2}$;
(7) $x_n(t) = \frac{n^3t^2}{1+4n^4t^4}$.

5.11° Examine the given sequences $\{x_n : n \ge 1\}$ for weak and strong convergence in an NLS X.

(1) $X = L_2([0, 1]),\ x_n(t) = \sin(nt^2)$;
(2) $X = L_2([0, 1]),\ x_n(t) = \sin(ne^t)$;
(3) $X = L_p([0, 1]),\ 1 < p < +\infty,\ x_n(t) = x_0(2^n t),\ t \in [0, 1]$, where $x_0(t) = (-1)^k$ for $t \in [k, k + 1),\ k \in \mathbb{N} \cup \{0\}$;
(4) $X = L_1(\mathbb{R}),\ x_n(t) = \chi_{[n,n+1]}(t)$;
(5) $X = l_1,\ x_n = e_n$.

5.12° Prove that the sequence of functionals $\{f_n : n \ge 1\} \subset X^*$ converges weakly-$*$, and determine whether it converges strongly.

(1) $X = L_2([0, 1]),\ f_n(x) = \int\limits_0^1 \cos 2\pi ntx(t)dt$;
(2) $X = L_2([0, 1]),\ f_n(x) = \int\limits_0^1 e^{2\pi int}x(t)dt$;
(3) $X = C([0, 1]),\ f_n(x) = n\int\limits_0^{\frac{1}{n}} (1 - 2nt)x(t)dt$;
(4) $X = C([0, 1]),\ f_n(x) = n\int\limits_0^1 x(t)t^n dt$;
(5) $X = C^1([-1, 1]),\ f_n(x) = \frac{n}{2}(x(\frac{1}{n}) - x(-\frac{1}{n}))$.

5.13 Let X be the space $C^1([-1, 1])$ with the norm $\|x\| = \max\limits_{t\in[0,1]} |x(t)|$. Consider

$$f_n(x) = \frac{n}{2}\left(x\left(\frac{1}{n}\right) - x\left(-\frac{1}{n}\right)\right),\quad x \in X.$$

(1) Prove that for each $x \in C^1([-1, 1])$ the sequence $\{f_n(x) : n \geq 1\}$ is convergent, and find its limit.
(2) Prove that the sequence $\{f_n : n \geq 1\}$ does not converge weakly-$*$.
(3) Prove that the sequence $\{\|f_n\| : n \geq 1\}$ is not bounded.

Do these statements contradict the Uniform boundedness principle?

5.14 Construct an example of an NLS X and of a weakly-$*$ convergent sequence $\{f_n : n \geq 1\} \subset X^*$, for which the sequence $\{\|f_n\| : n \geq 1\}$ is unbounded.

5.15° Let X be a Banach space, $\{f_n : n \geq 1\} \subset X^*$, $f, g \in X^*$, $f_n \xrightarrow{w} f$, $f_n \xrightarrow{w-*} g$, $n \to \infty$. Prove that $f = g$.

5.16

(1) Let the functionals $\{f_n : n \geq 1\} \subset c_0^*$ be defined as follows: $f_n(x) = x_n$, $x \in c_0$, $n \in \mathbb{N}$. Prove that $f_n \xrightarrow{w-*} 0$, but $f_n \overset{w}{\not\to} 0$, $n \to \infty$.
(2) Let the functionals $\{f_n : n \geq 1\} \subset l_1^*$ be defined as follows: $f_n(x) = \sum\limits_{k=n+1}^{\infty} x_k$, $x \in l_1$, $n \in \mathbb{N}$. Prove that $f_n \xrightarrow{w-*} 0$, but $f_n \overset{w}{\not\to} 0$, $n \to \infty$.
(3) Let the functionals $\{f_n : n \geq 1\} \subset (C([0, 1]))^*$, $f_0 \in (C([0, 1]))^*$ be defined as follows: $f_n(x) = x(\frac{1}{n})$, $n \in \mathbb{N}$, $f_0(x) = x(0)$, $x \in C([0, 1])$. Prove that $f_n \xrightarrow{w-*} f_0$, but $f_n \overset{w}{\not\to} f_0$, $n \to \infty$.

5.17° Let H be a Hilbert space, $x, x_n \in H$, $n \geq 1$. Prove that

(1) $x_n \xrightarrow{w} x$ if and only if $(x_n, a) \to (x, a)$, $n \to \infty$, for all $a \in H$;
(2) if $x_n \xrightarrow{w} x$ and $y_n \to y$ in norm, then $(x_n, y_n) \to (x, y)$, $n \to \infty$;
(3) every orthonormal sequence converges weakly to zero;
(4) if $x_n \xrightarrow{w} x$ and $\|x_n\| \to \|x\|$ (or $\limsup\limits_{n\to\infty} \|x_n\| \leq \|x\|$), then $x_n \to x$ in norm.

5.18° Let H be an infinite-dimensional Hilbert space. Construct examples of sequences $\{x_n : n \geq 1\} \subset H$, $\{y_n : n \geq 1\} \subset H$ such that

(1) $x_n \xrightarrow{w} x$, $y_n \xrightarrow{w} y$, where $x, y \in H$, but $(x_n, y_n) \not\to (x, y)$;
(2) $x_n \xrightarrow{w} x$, $y_n \xrightarrow{w} y$, where $x, y \in H$, $x_n \not\to x$, $y_n \not\to y$ in norm, but $(x_n, y_n) \to (x, y)$.

5.19° Let $\{x_n : n \geq 1\}$ be an orthogonal system in a Hilbert space H. Prove the equivalence of the following statements:

(1) the series $\sum\limits_{n=1}^{\infty} x_n$ converges strongly;
(2) the series $\sum\limits_{n=1}^{\infty} x_n$ converges weakly;
(3) the series $\sum\limits_{n=1}^{\infty} \|x_n\|^2$ converges.

5.20 (1) Let X be a Banach space, and the sequence $\{f_n : n \geq 1\} \subset X^*$ is such that for all $x \in X$ there exists $\lim_{n\to\infty} f_n(x) = f(x)$. Prove that $f \in X^*$ and $\|f\| \leq \liminf_{n\to\infty} \|f_n\|$.

(2) Let X be an NLS, and the sequence $\{x_n : n \geq 1\} \subset X$ converges weakly to $x \in X$. Prove that $\|x\| \leq \liminf_{n\to\infty} \|x_n\|$.

5.21° Let X be a Banach space. Prove that a sequence $\{f_n : n \geq 1\} \subset X^*$ is weakly-$*$ convergent if and only if the following conditions are fulfilled:

(a) $\exists C > 0 \, \forall n \geq 1 : \|f_n\| \leq C$;
(b) there exists a total set M in X such that for each $x \in M$ the sequence $\{f_n(x) : n \geq 1\}$ is a Cauchy sequence.

5.22° Let X be a reflexive Banach space, $\{x_n : n \geq 1\} \subset X$. Prove that if for each $f \in X^*$ the sequence $\{f(x_n) : n \geq 1\}$ is a Cauchy sequence, then there exists $x \in X$ such that $x_n \overset{w}{\to} x$.

5.23 Let X be a reflexive Banach space. Prove that if the sequence $\{x_n : n \geq 1\} \subset X$ is bounded in norm, then there exist a subsequence $\{x_{n_k} : k \geq 1\}$ and an element $x \in X$ such that $x_{n_k} \overset{w}{\to} x$.

5.24 Prove that the following Banach spaces are not complete with respect to the weak convergence: (1) $C([0, 1])$; (2) c_0; (3) c.

5.25° Let X be a Banach space, $x, x_n \in X$, $f, f_n \in X^*$, $n \geq 1$. Prove that $f_n(x_n) \to f(x)$, if one of the following conditions is satisfied:

(1) $x_n \to x$, $f_n \to f$ in norm;
(2) $x_n \overset{w}{\to} x$, $f_n \to f$ in norm;
(3) $x_n \to x$ in norm, $f_n \overset{w-*}{\longrightarrow} f$.

5.26 Let X be a separable NLS, the sequence $\{x_n : n \geq 1\} \subset X$ converges weakly to $x \in X$. Also, let $f_n(x_n) \to f(x)$, $n \to \infty$, for each functional $f \in X^*$ and each sequence $\{f_n : n \geq 1\} \subset X^*$, which converges weakly-$*$ to f. Prove that $x_n \to x$ in norm.

5.27 Let X be an NLS. A set $M \subset X$ is called *weakly closed* if for any sequence $\{x_n : n \geq 1\} \subset M$ such that $x_n \overset{w}{\to} x \in X$, it follows that $x \in M$. Prove that

(1) every weakly closed set in X is strongly closed;
(2) the converse statement to statement (1) is false;
(3) every subspace in X is a weakly closed set;
(4) every closed ball in X is a weakly closed set.

5.28 Let X be a uniformly convex Banach space (see problem 1.98), a sequence $\{x_n : n \geq 1\} \subset X$ converges weakly to $x \in X$, and $\|x_n\| \to \|x\|$, $n \to \infty$. Prove that $x_n \to x$ in norm.

5.29* (Schur) Prove that in the space l_1 weak convergence is equivalent to convergence in norm.

5.30 (Singularity Fixation Principle) Let X be a Banach space, $\{f_n : n \geq 1\} \subset X^*$, $\sup\limits_{n\geq 1} \|f_n\| = +\infty$. Prove that there exists $x \in X$, for which $\sup\limits_{n\geq 1} |f_n(x)| = +\infty$.

5.31 Let $1 \leq p \leq +\infty$, q be the conjugate index to p.

(1) Assume that a sequence $a = \{a_n : n \geq 1\}$ is such that for every $x \in l_p$ the series $\sum\limits_{n=1}^{\infty} a_n x_n$ converges. Prove that $a \in l_q$.
(2) Assume that a measurable function h is such that for every $x \in L_p([a, b])$ the Lebesgue integral $\int\limits_a^b h(t)x(t)dt$ is finite. Prove that $h \in L_q([a, b])$.

5.32 Formulate and prove statements similar to problem 5.31 for the spaces c and c_0.

5.33 Let X be an NLS. Prove that the sequence $\{x_n : n \geq 1\} \subset X$ is weakly bounded if and only if it is bounded in norm in X.

5.34 Let $f_n(x) = \frac{1}{n} \sum\limits_{k=0}^{n-1} x(\frac{k}{n})$, $x \in C([0, 1])$, $n \geq 1$. Prove that the sequence of functionals $\{f_n : n \geq 1\} \subset (C([0, 1]))^*$ converges weakly-$*$, and find its limit. Does this sequence converge in norm?

5.35 Let $f_n(x) = \sum\limits_{k=1}^{n} A_{nk} x(t_{nk})$, $x \in C([a, b])$, where

$$a \leq t_{n1} \leq \ldots \leq t_{nn} \leq b, \ n \geq 1, \ \{A_{nk} \,|\, 1 \leq k \leq n, \ n \geq 1\} \subset \mathbb{K}.$$

Prove that $f_n(x) \to \int\limits_a^b x(t)dt$, $n \to \infty$, for all $x \in C([a, b])$ if and only if the following conditions are fulfilled:

(a) $\sup\limits_{n\geq 1} \sum\limits_{k=1}^{n} |A_{nk}| < +\infty$;
(b) $f_n(p) \to \int\limits_a^b p(t)dt$, $n \to \infty$, for every polynomial p.

Verify that (b) implies (a), if $A_{nk} \geq 0$ for all $n \geq 1$ and $1 \leq k \leq n$.

5.36 Let $\mathbb{K} = \mathbb{R}$, $f_n(x) = \int\limits_a^b x(t)dF_n(t)$, $x \in C([a, b])$, where $F_n \in BV_0([a, b])$, $n \geq 0$. Prove that $f_n \xrightarrow{w-*} f_0$ if and only if the following conditions are fulfilled: $\sup\limits_{n\geq 1} V(F_n, [a, b]) < +\infty$, $F_n(t) \to F_0(t)$, $n \to \infty$, at points of continuity of F_0 and $F_n(b) \to F_0(b)$, $n \to \infty$.

5.37 Let $\mathbb{K} = \mathbb{R}$, $\{p_n : n \geq 1\} \subset L_1([0,1])$, $f_n(x) = \int\limits_0^1 p_n(t)x(t)dt$, $x \in C([0,1])$, $n \geq 1$.

(1) Let $p_n(t) \leq p_{n+1}(t)$ almost everywhere on $[0,1]$ with respect to the Lebesgue measure for all $n \in \mathbb{N}$, and also for each $x \in C([0,1])$ the sequence $\{f_n(x) : n \geq 1\}$ is bounded. Show that the sequence $\{f_n : n \geq 1\}$ converges weakly-$*$.
(2) Let $f_n \overset{w-*}{\longrightarrow} 0$. Is it necessarily true that there exists a subsequence of the sequence $\{p_n : n \geq 1\}$, which converges almost everywhere on $[0,1]$?
(3) The same question as in item (2), if $f_n \to 0$ in norm.

5.38 Let X be an NLS, $f \in X^*$. Under what condition on the sequence $\{\alpha_n : n \geq 1\} \subset \mathbb{R}$ does the sequence of functionals $f_n(x) = f(\alpha_n x)$, $x \in X$, $n \geq 1$, converge (1) weakly-$*$? (2) in norm?

5.39 Let $\{f_n : n \geq 1\}$ be a sequence of linear nonnegative functionals on a real space $C([a,b])$ (see problem 3.42), and $f(x) = x(c)$, $x \in C([a,b])$, where $c \in [a,b]$ is fixed. Denote $e_k(t) = t^k$, $t \in [a,b]$, $k = 0, 1, 2$, $y(t) = (t-c)^2$, $t \in [a,b]$. Prove that

(1) for each $x \in C([a,b])$ and each $\varepsilon > 0$ the following inequality holds:

$$-\frac{2\|x\|}{\delta^2}y(t) - \varepsilon \leq x(t) - x(c) \leq \frac{2\|x\|}{\delta^2}y(t) + \varepsilon, \ t \in [a,b],$$

where the number $\delta = \delta(x,\varepsilon) > 0$ is such that $|x(t) - x(c)| < \varepsilon$ for all $t \in [a,b]$, for which $|t-c| < \delta$.

(2) **(Korovkin)** $f_n \overset{w-*}{\longrightarrow} f$, if one of the following conditions is fulfilled:

(a) $f_n(e_0) \to f(e_0)$ and $f_n(y) \to f(y)$, $n \to \infty$;
(b) $f_n(e_k) \to f(e_k)$, $n \to \infty$, for $k = 0, 1, 2$.

5.40 Let functionals $f, f_n \in (C([a,b]))^*$ be such that $f_n(e_0) \to f(e_0)$, $n \to \infty$, where $e_0(t) = 1$, $t \in [a,b]$, and for each $n \in \mathbb{N}$ the functional $f_n - f$ is nonnegative. Prove that $f_n \to f$ in norm, as $n \to \infty$.

5.41 (Toeplitz) An infinite set of numbers $\{a_{nk} : k \geq 1, \ n \geq 1\} \subset \mathbb{K}$ is given. Prove that for any convergent sequence $\{x_n : n \geq 1\} \subset \mathbb{K}$ the sequence $\{\sum\limits_{k=1}^{\infty} a_{nk}x_k, n \geq 1\}$ converges to the number $\lim\limits_{n\to\infty} x_n$ if and only if the following conditions are fulfilled:

(1) $\forall k \geq 1 : \lim\limits_{n\to\infty} a_{nk} = 0$;

(2) $\lim\limits_{n\to\infty} \sum\limits_{k=1}^{\infty} a_{nk} = 1$;

(3) $\exists C > 0 \ \forall n \geq 1 : \sum\limits_{k=1}^{\infty} |a_{nk}| \leq C$.

5.42* **(Mertens–Schur)** Let $\{a_n : n \geq 0\} \subset \mathbb{C}$, $\{x_n : n \geq 0\} \subset \mathbb{C}$, $y_n = \sum\limits_{k=0}^{n} a_{n-k}x_k$, $n \geq 0$. Prove that the series $\sum\limits_{n=0}^{\infty} y_n$ converges for any convergent series $\sum\limits_{n=0}^{\infty} x_n$ if and only if the series $\sum\limits_{n=0}^{\infty} |a_n|$ converges.

5.43 Prove that

(1) The mapping $\sum\limits_{n=1}^{\infty} a_n \mapsto \sum\limits_{n=1}^{\infty} a_n b_n$, defined by the sequence $\{b_n : n \geq 1\} \subset \mathbb{C}$, transforms convergent series into convergent ones if and only if $\sum\limits_{n=1}^{\infty} |b_{n+1} - b_n| < +\infty$;

(2) The series $\sum\limits_{n=1}^{\infty} a_n b_n$ is convergent for all sequences $\{a_n : n \geq 1\} \subset \mathbb{C}$, for which the partial sums $\sum\limits_{k=1}^{n} a_k$, $n \geq 1$, are uniformly bounded, if and only if $b_n \to 0$, $n \to \infty$, and $\sum\limits_{n=1}^{\infty} |b_{n+1} - b_n| < +\infty$.

Chapter 6
Bounded Linear Operators

Theoretical Background

Let X_1, X_2 be normed linear spaces with norms $\|\cdot\|_1$ and $\|\cdot\|_2$, respectively. An *operator* is an arbitrary mapping $A : X_1 \to X_2$. The image of an element $x \in X_1$ under an operator A is denoted by Ax.

An operator A is called *continuous* if the mapping $A : X_1 \to X_2$ is continuous on X_1, i.e.

$\forall x_0 \in X_1 \ \forall \{x_n : n \geq 1\} \subset X_1, x_n \to x_0 \ : \ Ax_n \to Ax_0, \ n \to \infty.$

An operator $A : X_1 \to X_2$ is called *linear* if

$\forall x, y \in X_1 \ \forall \alpha, \beta \in \mathbb{K} \ : \ A(\alpha x + \beta y) = \alpha Ax + \beta Ay.$

The *kernel* of a linear operator $A : X_1 \to X_2$ is the set $\operatorname{Ker} A = \{x \in X_1 \mid Ax = 0\}$.

The *range* of the operator $A : X_1 \to X_2$ is the set $R(A) = \{Ax \mid x \in X_1\}$.

A linear operator $A : X_1 \to X_2$ is said to be *bounded* if

$\exists C \geq 0 \ \forall x \in X_1 \ : \ \|Ax\|_2 \leq C\|x\|_1.$

A linear operator is bounded if and only if it is continuous.

Denote by $\mathcal{L}(X_1, X_2)$ the class of all continuous linear operators from X_1 to X_2. In case $X_2 = X_1$ the notation $\mathcal{L}(X_1)$ is often used instead of $\mathcal{L}(X_1, X_1)$.

For $A \in \mathcal{L}(X_1, X_2)$ the number $\|A\| = \sup\limits_{x \in X_1, \ x \neq 0} \frac{\|Ax\|_2}{\|x\|_1}$ is called *the norm of the operator* A. The space $\mathcal{L}(X_1, X_2)$ endowed with standard linear operations and this norm is an NLS.

If X_2 is a complete NLS (i.e., Banach space), then $\mathcal{L}(X_1, X_2)$ is also a Banach space.

Let X, Y, Z be normed linear spaces. The *product* of the operator $A \in \mathcal{L}(Y, Z)$ and the operator $B \in \mathcal{L}(X, Y)$ is the operator $AB \in \mathcal{L}(X, Z)$ for which $(AB)x = A(Bx)$. The space $\mathcal{L}(X)$ with the introduced multiplication is an algebra with unity (unity is the identity operator I).

Operators $A, B \in \mathcal{L}(X)$ *commute* if $AB = BA$.

V. Brayman et al., *Functional Analysis and Operator Theory*, Problem Books in Mathematics, https://doi.org/10.1007/978-3-031-56427-7_6

The *graph* of a linear operator A is the set

$$\Gamma(A) = \{(x, Ax) \mid x \in X_1\} \subset X_1 \times X_2.$$

For every $A \in \mathcal{L}(X_1, X_2)$ the set $\Gamma(A)$ is a subspace in the space $X_1 \times X_2$, which is endowed with the norm $\|(x_1, x_2)\| = \|x_1\|_{X_1} + \|x_2\|_{X_2}$. In the case of Banach spaces X_1, X_2 the converse result holds.

Theorem 6.1 (Closed Graph Theorem) *Let X_1, X_2 be Banach spaces, A is a linear operator from X_1 to X_2. The graph of the operator $\Gamma(A)$ is a subspace in $X_1 \times X_2$ if and only if $A \in \mathcal{L}(X_1, X_2)$.*

Examples of Problems with Solutions

6.1 Are the given operators $A : l_2 \to l_2$ linear? Continuous? For continuous linear operators, find their norms.

(1) $Ax = (0, 2x_3, 0, 2x_4, 0, 2x_5, 0, \ldots),\ x \in l_2$;
(2) $Ax = (x_1^2, x_2^2, x_3^2, \ldots),\ x \in l_2$.

Solution

(1) For all $x, y \in l_2$ and $\alpha, \beta \in \mathbb{K}$ we have

$$A(\alpha x + \beta y) = A(\alpha x_1 + \beta y_1, \alpha x_2 + \beta y_2, \ldots) =$$
$$= (0, 2\alpha x_3 + 2\beta y_3, 0, 2\alpha x_4 + 2\beta y_4, 0, \ldots) =$$
$$= \alpha(0, 2x_3, 0, 2x_4, 0, \ldots) + \beta(0, 2y_3, 0, 2y_4, 0, \ldots) = \alpha Ax + \beta Ay.$$

Hence A is a linear operator. Therefore instead of continuity we can verify whether A is bounded. For all $x \in l_2$ we have

$$\|Ax\|^2 = \sum_{n=3}^{\infty} 4|x_n|^2 \leq \sum_{n=1}^{\infty} 4|x_n|^2 = 4\|x\|^2,$$

so $\|Ax\| \leq 2\|x\|$. Hence linear operator A is bounded which implies A is continuous, moreover $\|A\| \leq 2$. To show that $\|A\| = 2$, it is enough to find a vector $y \in l_2$ such that $y \neq 0$ and $\|Ay\| = 2\|y\|$. Take $y = e_3$. Then $Ay = 2e_2$, $\|y\| = 1$ and $\|Ay\| = 2$. Hence $\|A\| = 2$.

(2) Note that there exist vectors for which linearity fails. For example, for $x = y = e_1$ we have $A(x+y) = 4e_1 \neq Ax+Ay = 2e_1$. Thus the operator A is nonlinear. Let us verify its continuity. Let $x^{(n)} \to x$, $n \to \infty$, in l_2. Then

$$\|Ax^{(n)} - Ax\|_2^2 = \sum_{k=1}^{\infty} \left|(x_k^{(n)})^2 - x_k^2\right|^2 = \sum_{k=1}^{\infty} \left|x_k^{(n)} + x_k\right|^2 \left|x_k^{(n)} - x_k\right|^2 \leq$$

$$\leq \|x^{(n)} + x\|_\infty^2 \cdot \|x^{(n)} - x\|_2^2 \leq (\|x^{(n)}\|_2 + \|x\|_2)^2 \cdot \|x^{(n)} - x\|_2^2 \to 0,$$

$n \to \infty$, because the sequence is bounded by the norm. Therefore A is a continuous operator.

6.2 Let $X = L_2(T, \mu)$, $\alpha \in L_\infty(T, \mu)$, μ is σ-finite. The operator of multiplication by α is defined as $(Ax)(t) = \alpha(t)x(t)$. Prove that A belongs to $\mathcal{L}(X)$ and find its norm.

Solution It is obvious that the operator A is linear. Moreover,

$$\|Ax\|_2 = \left(\int_T |\alpha(t)x(t)|^2 d\mu(t)\right)^{\frac{1}{2}} \leq \left(\int_T \|\alpha\|_\infty^2 |x(t)|^2 d\mu(t)\right)^{\frac{1}{2}} = \|\alpha\|_\infty \|x\|_2.$$

(Here $\|\alpha\|_\infty = \operatorname{ess\,sup}_{t\in T} |\alpha(t)|$ is the essential supremum of the function $|\alpha|$.) For arbitrary $\varepsilon > 0$ consider the set $B_\varepsilon = \left\{t \in T \mid |\alpha(t)| > \|\alpha\|_\infty - \varepsilon\right\}$. According to the definition of $\|\alpha\|_\infty$ we have $\mu(B_\varepsilon) > 0$. Consider a measurable subset $B'_\varepsilon \subset B_\varepsilon$ such that $0 < \mu(B'_\varepsilon) < +\infty$ (such a set exists due to the σ-finiteness of the measure μ). Put $x_\varepsilon = \frac{1}{\sqrt{\mu(B'_\varepsilon)}} \chi_{B'_\varepsilon}$. It is easy to verify that $\|x_\varepsilon\| = 1$ and $\|Ax_\varepsilon\| \geq \|\alpha\|_\infty - \varepsilon$. Hence $\|A\| = \|\alpha\|_\infty$.

6.3 Let X, Y be normed linear spaces, $f \in X^*$, $y \in Y$ and $Ax = f(x)y$, $x \in X$. Prove that $A \in \mathcal{L}(X, Y)$ and find $\|A\|$.

Solution Due to the linearity of f, for all $x_1, x_2 \in l_2$ and $\alpha_1, \alpha_2 \in \mathbb{K}$ we have

$$A(\alpha_1 x_1 + \alpha_2 x_2) = f(\alpha_1 x_1 + \alpha_2 x_2)y = \alpha_1 f(x_1)y + \alpha_2 f(x_2)y = \alpha_1 Ax_1 + \alpha_2 Ax_2.$$

Since

$$\sup_{x\in X,\, x\neq 0} \frac{\|Ax\|_Y}{\|x\|_X} = \sup_{x\in X,\, x\neq 0} \frac{|f(x)| \cdot \|y\|_Y}{\|x\|_X} = \|f\| \cdot \|y\|_Y < +\infty,$$

then $A \in \mathcal{L}(X, Y)$ and $\|A\| = \|f\| \cdot \|y\|_Y$.

6.4 Find the norm of the operator $(Ax)(t) = \int_0^{2\pi} \sin(t-s)x(s)ds$ in $L_2([0, 2\pi])$.

Solution Set $e_1(t) = \frac{1}{\sqrt{\pi}} \cos t$, $e_2(t) = \frac{1}{\sqrt{\pi}} \sin t$. It is clear that $\{e_1, e_2\}$ is an orthonormal system in $L_2([0, 2\pi])$. We have

$$Ax = \pi\big((x, e_1)e_2 - (x, e_2)e_1\big).$$

It follows from the Pythagorean theorem and Bessel's inequality that

$$\|Ax\|^2 = \pi^2\big(|(x, e_1)|^2 + |(x, e_2)|^2\big) \le \pi^2\|x\|^2,$$

hence $\|A\| \le \pi$. On the other hand, $Ae_1 = \pi e_2$ and $\|Ae_1\| = \pi$. Therefore $\|A\| = \pi$.

Problems to Solve

6.5° Let X_1, X_2 be normed linear spaces, $A \in \mathcal{L}(X_1, X_2)$. Prove the equalities

$$\|A\| = \sup_{\|x\|_1=1} \|Ax\|_2 = \sup_{\|x\|_1\le 1} \|Ax\|_2 =$$
$$= \min\{C \ge 0 \,|\, \forall x \in X_1 \;:\; \|Ax\|_2 \le C\|x\|_1.\}$$

6.6° Prove that a continuous linear operator $A : X \to Y$ remains continuous if the norms in X and Y are replaced with equivalent ones.

6.7° Let $\|\cdot\|_1$ and $\|\cdot\|_2$ be two equivalent norms on a linear space X. Denote by X_i the linear space X endowed with the norm $\|\cdot\|_i$, $i = 1, 2$. Prove that the norms in $\mathcal{L}(X_1)$ and $\mathcal{L}(X_2)$ are equivalent.

6.8° Let X_1 be a finite-dimensional NLS, X_2 be an arbitrary NLS. Prove that every linear operator $A : X_1 \to X_2$ is continuous. In particular, any linear operator in a finite-dimensional NLS is continuous.

6.9° Find the general form of a linear operator $A : X \to Y$ (which is continuous according to problem 6.8°) and calculate its norm in the following cases:

(1) $X = \mathbb{R}_1^m,\ Y = \mathbb{R}_1^n$;
(2) $X = \mathbb{R}_\infty^m,\ Y = \mathbb{R}_\infty^n$;
(3) $X = \mathbb{R}_1^m,\ Y = \mathbb{R}_\infty^n$;
(4) $X = \mathbb{R}_\infty^m,\ Y = \mathbb{R}_1^n$;
(5) $X = \mathbb{R}_2^m,\ Y = \mathbb{R}_\infty^n$.

Here $\mathbb{R}_p^m$ is the space $\mathbb{R}^m$ endowed with the norm $\|x\|_p = \left(\sum_{k=1}^m |x_k|^p\right)^{1/p}$ for $1 \le p < +\infty$, and $\mathbb{R}_\infty^m$ is the space $\mathbb{R}^m$ endowed with the norm $\|x\|_\infty = \max_{1\le k\le m} |x_k|$.

6.10° Let $\{\alpha_n : n \geq 1\}$ be a bounded sequence. Find the norm of the diagonal operator $Ax = (\alpha_1 x_1, \alpha_2 x_2, \ldots)$ in l_p, $1 \leq p \leq \infty$.

6.11° Let $\alpha = \alpha(t)$ be a fixed function from $C([a, b])$, A is the operator of multiplication by the function α: $(Ax)(t) = \alpha(t)x(t)$.

(1) Prove that A is a continuous linear operator in $C([a, b])$, and find the norm of A.
(2) Prove that A is a continuous linear operator in $L_p([a, b])$, $1 \leq p \leq +\infty$, and find the norm of A.

6.12 Let A be the operator of multiplication by a Lebesgue measurable function α in $L_p([a, b])$, $1 \leq p \leq +\infty$: $(Ax)(t) = \alpha(t)x(t)$. Prove that A is bounded if and only if $\alpha \in L_\infty([a, b])$. Find the norm of A.

6.13° Find the norm of the operator $A : \mathbb{R}^m \to l_2$, if

$$Ax = (x_1, \ldots, x_m, \frac{x_1}{2}, \ldots, \frac{x_m}{2}, \frac{x_1}{3}, \ldots, \frac{x_m}{3}, \ldots),$$

$x = (x_1, \ldots x_m) \in \mathbb{R}^m$, if $\mathbb{R}^m$ is endowed with the Euclidean norm.

6.14° Let $(\alpha_{jk})_{j,k=1}^{\infty}$ be an infinite matrix with real or complex entries such that $\sum\limits_{j,k=1}^{\infty} |\alpha_{jk}|^2 < \infty$. Prove that the operator of multiplication by this matrix, defined by the equalities $(Ax)_j = \sum\limits_{k=1}^{\infty} \alpha_{jk} x_k$, $j \geq 1$, is linear and continuous in the space l_2. Estimate the norm of A.

6.15 The integral operator $A : C([a, b]) \to C([a, b])$ with a continuous kernel $K \in C([a, b]^2)$ is defined by the formula $(Ax)(t) = \int\limits_a^b K(t, s)x(s)ds$, $t \in [a, b]$, $x \in C([a, b])$. Prove that A is linear and continuous. Also prove that $\|A\| = \max\limits_{t \in [a,b]} \int\limits_a^b |K(t, s)|ds$.

6.16 The integral operator $A : L_1([a, b]) \to L_1([a, b])$ with a continuous kernel $K \in C([a, b]^2)$ is defined by the formula $(Ax)(t) = \int\limits_a^b K(t, s)x(s)ds$, $t \in [a, b]$, $x \in L_1([a, b])$. Prove that A is linear and continuous. Also prove that $\|A\| = \max\limits_{s \in [a,b]} \int\limits_a^b |K(t, s)|dt$.

6.17° The integral operator $A : L_2([a,b]) \to L_2([a,b])$ with kernel $K \in L_2([a,b]^2)$ is defined by the formula $(Ax)(t) = \int_{[a,b]} K(t,s)x(s)ds,\ t \in [a,b]$. Prove that A is linear and continuous. Also prove the estimate

$$\|A\| \le \|K\|_{L_2([a,b]^2)} = \left(\int_{[a,b]^2} |K(t,s)|^2 dt ds \right)^{\frac{1}{2}}.$$

6.18 The Volterra integral operator $A : C([a,b]) \to C([a,b])$ with continuous kernel K is defined by the formula $(Ax)(t) = \int_a^t K(t,s)x(s)ds,\ t \in [a,b]$, where $K \in C(\{(t,s) \in \mathbb{R}^2 \mid a \le s \le t \le b\})$. Prove that A is linear and continuous. Find $\|A\|$.

6.19 Let u be a continuous 2π-periodic function on $\mathbb{R}$. The integral operator $A : L_1([-\pi,\pi]) \to L_1([-\pi,\pi])$ is defined by the formula $(Ax)(t) = \int_{-\pi}^{\pi} u(t-s)x(s)ds,\ t \in [-\pi,\pi]$. Prove that A is linear and continuous. Find $\|A\|$.

6.20

(1) Prove that the operator $A : C^1([a,b]) \to C([a,b])$, which is defined by the formula $(Ax)(t) = \frac{dx(t)}{dt},\ t \in [a,b]$, is linear and continuous. Find its norm.

(2) Let X, Y be linear normed spaces. An operator $A : D(A) \subset X \to Y$ is said to be *densely defined* if $\overline{D(A)} = X$, and *closed*, if for any sequence $\{x_n : n \ge 1\} \subset X$ such that $x_n \to x$ in X and $Ax_n \to y$ in Y it follows that $x \in D(A)$ and $y = Ax$.

Prove that the operator $A : C^1([a,b]) \subset C([a,b]) \to C([a,b])$, where $(Ax)(t) = \frac{dx(t)}{dt},\ t \in [a,b],\ x \in C^1([a,b])$, is densely defined and closed, but not continuous.

6.21° Let $u, v \in L_2([a,b])$. Prove that the integral operator $A : L_2([a,b]) \to L_2([a,b])$, which is defined by the formula

$$(Ax)(t) = \int_{[a,b]} u(t)v(s)x(s)ds,\ t \in [a,b],$$

is linear and continuous. Also prove that $\|A\| = \|u\| \cdot \|v\|$.

6.22 Find the norm of the operator defined as $Ax = x$, which acts:

(1) from $C^1([a,b])$ to $C([a,b])$;

(2) from $L_p([a,b])$ to $L_r([a,b])$, $1 \le r \le p \le \infty$.

6.23 Let $K \in C([a,b]^2)$, $\alpha < 1$. Prove that the integral operator $A : C([a,b]) \to C([a,b])$, where $(Ax)(t) = \int\limits_a^b \frac{K(t,s)}{|t-s|^\alpha} x(s)ds$, $t \in [a,b]$, is linear and continuous.

6.24 (1) Find the nth power of the Fredholm integral operator $A : L_2([a,b]) \to L_2([a,b])$, which is defined by the formula

$$(Ax)(t) = \int\limits_a^b K(t,s)x(s)ds, \; t \in [a,b], \; K \in L_2([a,b]^2).$$

(2) Find the nth power of the Volterra integral operator $A : L_2([a,b]) \to L_2([a,b])$, which is defined by the formula

$$(Ax)(t) = \int\limits_a^t K(t,s)x(s)ds, \; t \in [a,b], \; K \in L_2(\{(t,s) \in [a,b]^2 \mid s \le t\}).$$

6.25 Let the numbers α, β, γ be such that $0 \le \gamma \le \alpha < \beta$. Denote by C_α the Banach space of functions $x \in C([0,+\infty))$ for which

$$\sup_{0 \le t < \infty} e^{\alpha t}|x(t)| < \infty,$$

with norm $\|x\|_\alpha = \sup\limits_{0 \le t < \infty} e^{\alpha t}|x(t)|$. Show that the operator $A : C_\alpha \to C_\gamma$, defined by the formula $(Ax)(t) = \int\limits_0^t e^{-\beta(t-s)}x(s)ds$, $t \ge 0$, is linear and continuous. Find $\|A\|$.

6.26° Let H be a Hilbert space, $y, z \in H$, $Ax = (x,y)z$, $x \in H$. Show that $A \in \mathcal{L}(H)$, and find $\|A\|$.

6.27° Let $u, v \in C([a,b])$ and $(Ax)(t) = \int\limits_a^b u(t)v(s)x(s)ds$. Show that $A \in \mathcal{L}(X)$, and find the norm $\|A\|$. Consider the cases (1) $X = L_p([a,b])$; (2) $X = C([a,b])$.

6.28° An operator $A \in \mathcal{L}(X,Y)$ is called *finite-dimensional* if $\dim R(A) < \infty$. Show that A is finite-dimensional if and only if there exist $f_1, \dots, f_n \in X^*$ and $y_1, \dots, y_n \in Y$ such that

$$Ax = \sum_{k=1}^n f_k(x)y_k, \; x \in X.$$

6.29° Establish the general form of a finite-dimensional operator (1) in a Hilbert space; (2) in $L_p([a,b])$, $1 \le p < \infty$; (3) in $C([a,b])$.

6.30 Let $\{y_k \mid 1 \le k \le n\}$ and $\{z_k \mid 1 \le k \le n\}$ be two orthonormal systems in a Hilbert space H. Find the norm of the operator A, where $Ax = \sum_{k=1}^{n} c_k(x, y_k)z_k$, $c_k \in \mathbb{C}$.

Remark It follows from problem 12.32 that every finite-dimensional operator in a Hilbert space H has form described in problem 6.30.

6.31° Prove that the following integral operators are linear and continuous. Find their norms.

(1) $X = C([0, 1])$, $(Ax)(t) = \int_0^1 t^\alpha s^\beta x(s)ds$, $\alpha \ge 0, \beta > -1$;

(2) $X = C([0, 1])$, $(Ax)(t) = \int_0^1 e^{3t-2s} x(s)ds$;

(3) $X = C([0, 1])$, $(Ax)(t) = \int_0^1 \sin \pi(t - s)x(s)ds$;

(4) $X = C([0, 1])$, $(Ax)(t) = \int_0^\pi \cos t \sin sx(s)ds$;

(5) $X = L_2([0, 2\pi])$, $(Ax)(t) = \int_0^{2\pi} \cos(t - s)x(s)ds$;

(6) $X = L_2([0, 1])$, $(Ax)(t) = \int_0^1 t^\alpha s^\beta x(s)ds$, $\alpha > -\frac{1}{2}$, $\beta > -\frac{1}{2}$;

(7) $X = L_2([0, 2\pi])$, $(Ax)(t) = \int_0^{2\pi} \sin(t + s)x(s)ds$;

(8) $X = L_2([0, 2\pi])$, $(Ax)(t) = \int_0^{2\pi} \cos(2t + 2s)x(s)ds$;

(9) $X = C([0, 1])$, $(Ax)(t) = \int_0^t x(s)ds$.

6.32° Prove that the following operators $A : l_2 \to l_2$ are linear and continuous. Find their norms.

(1) $Ax = (\underbrace{0, 0, \dots, 0}_{k}, x_1, x_2, \dots)$;

(2) $Ax = (\underbrace{0, 0, \dots, 0}_{k}, x_{k+1}, x_{k+2}, \dots)$;

(3) $Ax = (x_{k+1}, x_{k+2}, \dots)$;

(4) $Ax = (x_1, 0, x_2, 0, x_3, 0, \dots)$;

(5) $Ax = (x_1, 0, x_3, 0, x_5, 0, \dots)$;

(6) $Ax = (0, x_2, 0, x_4, 0, x_6, \dots)$.

6.33° Prove that the following operators are continuous and not linear:

(1) $X = C([0, 1])$, $(Ax)(t) = x^2(t)$;
(2) $X = C([0, 1])$, $(Ax)(t) = \sin x^2(t)$;
(3) $X = C([0, 1])$, $(Ax)(t) = \int_0^1 x^2(t)dt$.

6.34°

(1) Let X, Y be normed linear spaces. Prove that the kernel $\operatorname{Ker} A$ of every continuous linear operator $A : X \to Y$ is a subspace in X.
(2) Let $A : X \to Y$ be a linear operator, for which $\operatorname{Ker} A$ is a subspace in X. Does it follow that A is a bounded operator?
(3) Provide an example of a continuous linear operator A the range of which (a) is a closed set; (b) is not a closed set.

6.35° Let X, Y be Banach spaces, $A \in \mathcal{L}(X, Y)$ be an operator, for which there exists $c > 0$ such that $\|Ax\| \geq c\|x\|$, $x \in X$. Prove that

(1) $R(A)$ is a subspace in Y;
(2) $\operatorname{Ker} A = \{0\}$.

6.36* Let X, Y be normed linear spaces, and $A : X \to Y$ be a linear operator, for which $\dim R(A) < +\infty$, and $\operatorname{Ker} A$ is a closed set in X. Prove that $A \in \mathcal{L}(X, Y)$.

6.37° Let H be a Hilbert space, $G \subset H$ be a subspace, $Px = \operatorname{pr}_G x$. Prove that the projection operator P is linear and continuous. Find the norm of P.

6.38 Let A be a linear operator in an NLS X. Prove that A is continuous if and only if the set $\{x \in X \mid \|Ax\| < 1\}$ has interior points.

6.39 Let A be a linear operator in an NLS X, which satisfies one of the following conditions:

(1) A maps every bounded sequence into a bounded sequence;
(2) A maps a closed unit ball into a bounded set;
(3) A maps every strongly convergent sequence into a weakly convergent sequence.

Prove that A is a bounded operator.

6.40 Let X be an NLS, $A : X \to X$. Prove that $A \in \mathcal{L}(X)$ if and only if for every $f \in X^*$ the functional $f_0(x) = f(Ax)$, $x \in X$, is linear and continuous.

6.41° Let X_1, X_2 be normed linear spaces, $\{x, x_n : n \geq 1\} \subset X_1$ and $A \in \mathcal{L}(X_1, X_2)$. Prove that if $x_n \xrightarrow{w} x$, $n \to \infty$, then $Ax_n \xrightarrow{w} Ax$, $n \to \infty$.

6.42 (1) For which $\alpha \geq 0$ is the operator $(Ax)(t) = x(t^\alpha)$ linear and continuous in $C([0, 1])$? Find its norm.
(2) For which $\alpha > 0$ is the operator $(Ax)(t) = x(t^\alpha)$ linear and continuous in $L_2([0, 1])$? Find its norm.
(3) For which $\alpha > 0$ and $\beta \in \mathbb{R}$ is the operator $(Ax)(t) = t^\beta x(t^\alpha)$ linear and continuous in $L_2([0, 1])$? Find its norm.

6.43°

(1) Prove that the operators $Ax = (x_2, 0)$, $Bx = (x_2, x_1)$, $x \in \mathbb{R}^2$, are linear, continuous and do not commute;
(2) Prove that the operators $Ax = (x_2, 0)$, $Bx = (0, x_1)$, $x \in \mathbb{R}^2$, are linear, continuous and do not commute.
(3) Prove that the operators $(Ax)(t) = tx(t)$, $(Bx)(t) = \int_0^t x(s)ds$, $t \in [0, 1]$, are linear, continuous in $L_2([0, 1])$ and do not commute.

6.44

(1) Let $X = l_p$, $1 \le p \le +\infty$. Prove that the space $\mathcal{L}(X)$ is nonseparable.
(2) Let $X = L_p([a, b])$, $1 \le p \le +\infty$. Prove that the space $\mathcal{L}(X)$ is nonseparable.

6.45 Let X be a nonseparable NLS. Prove that the space $\mathcal{L}(X)$ is nonseparable.

6.46° Let X be a Banach space, $A \in \mathcal{L}(X)$. Are the following functions norms on X: (1) $\|x\|_1 = \|Ax\|$? (2) $\|x\|_2 = \|x\| + \|Ax\|$? Is it true that X endowed with the norm $\|\cdot\|_2$ is a Banach space?

6.47 Let X be an NLS, Y be a Banach space, the linear set $M \subset X$ is dense in X, and $A \in \mathcal{L}(M, Y)$. Prove that the operator A admits a unique extension $\overline{A} \in \mathcal{L}(X, Y)$, and $\|\overline{A}\| = \|A\|$.

6.48° Let H be a Hilbert space, $A : H \to H$ be a linear operator. Prove that $A \in \mathcal{L}(H)$ if and only if
$\exists C > 0\, \forall x, y \in H : |(Ax, y)| \le C\|x\| \cdot \|y\|$.
Moreover, $\|A\| = \sup\limits_{\|x\|=1, \|y\|=1} |(Ax, y)|$.

6.49°

(1) Let the operator $A : L_2(T, \mu) \to L_2(T, \mu)$ be an integral operator with kernel K, $(Ax)(t) = \int_T K(t, s)x(s)d\mu(s)$, $t \in T$, where $c_1 = \operatorname{ess\,sup}_{t \in T} \int_T |K(t, s)|d\mu(s) < \infty$, $c_2 = \operatorname{ess\,sup}_{s \in T} \int_T |K(t, s)|d\mu(t) < \infty$. Prove that $A \in \mathcal{L}(L_2(T, \mu))$ and $\|A\| \le (c_1c_2)^{\frac{1}{2}}$.
(2) Let $K \in L_\infty([a, b]^2)$, $\alpha < 1$. Prove that the operator $A : L_2([a, b]) \to L_2([a, b])$, where $(Ax)(t) = \int_a^b \frac{K(t,s)}{|t-s|^\alpha} x(s)ds$, $t \in [a, b]$, is linear and continuous.
(3) Let $u \in L_1(\mathbb{R})$. Prove that the integral operator $A : L_2(\mathbb{R}) \to L_2(\mathbb{R})$, defined by the formula

$$(Ax)(t) = \int_{\mathbb{R}} u(t - s)x(s)ds, \; t \in \mathbb{R},$$

(the operator of convolution with function u) is linear and continuous.

6.50* A system of sets $\mathcal{A} = \{A_\alpha : \alpha \in T\}$ is said to be *linked*, if every pair of sets from $\mathcal{A}$ has a nonempty intersection. A Banach space X is a space of *type* $\mathcal{M}$, if every linked system of closed balls in X has a nonempty intersection.

Prove that

(1) The space $\mathbb{C}$ is not a space of type $\mathcal{M}$, while the real spaces $\mathbb{R}$, $L_\infty(T, \mu)$ are spaces of type $\mathcal{M}$.
(2) Every space of type $\mathcal{M}$ is a Banach space.
(3) Let X be a real NLS, $G \subset X$ be a subspace in X, Y be a real space of type $\mathcal{M}$, and $A : G \to Y$ be a continuous linear operator. Then there exists a continuous linear extension $\overline{A}$ of operator A on X that preserves the norm, i.e., an extension for which $\|\overline{A}\| = \|A\|$.

6.51° Let A, B be linear operators in the Hilbert space H such that $(Ax, y) = (x, By)$ for all $x, y \in H$. Prove that $A, B \in \mathcal{L}(H)$.

Chapter 7
Uniform, Strong and Weak Operator Convergence

Theoretical Background

Let X_1, X_2 be normed linear spaces.

A sequence $\{A_n : n \geq 1\} \subset \mathcal{L}(X_1, X_2)$ is called *uniformly convergent* (or convergent in norm) to an operator $A \in \mathcal{L}(X_1, X_2)$, if $\|A_n - A\| \to 0$, $n \to \infty$. Uniform convergence is denoted as $A_n \rightrightarrows A$, $n \to \infty$.

A sequence $\{A_n : n \geq 1\} \subset \mathcal{L}(X_1, X_2)$ is called *strongly convergent* (or pointwise convergent) to an operator $A \in \mathcal{L}(X_1, X_2)$, if

$\forall x \in X_1 \ : \ A_n x \to Ax, \ n \to \infty$, in X_2.

Strong convergence is denoted as $A_n \overset{s}{\to} A, \ n \to \infty$, or $A = s\text{-}\lim\limits_{n\to\infty} A_n$.

A sequence $\{A_n : n \geq 1\} \subset \mathcal{L}(X_1, X_2)$ is called *weakly convergent* to an operator $A \in \mathcal{L}(X_1, X_2)$, if

$\forall x \in X_1 \ : \ A_n x \overset{w}{\to} Ax, \ n \to \infty$, in X_2.

Weak convergence is denoted as $A_n \overset{w}{\to} A, \ n \to \infty$, or $A = w\text{-}\lim\limits_{n\to\infty} A_n$.

Uniform convergence of the sequence of operators implies strong convergence, and strong convergence implies weak convergence. The converse statements, generally speaking, are incorrect.

Theorem 7.1 (Banach–Steinhaus theorem; Uniform Boundedness Principle) *Let X_1 be a Banach space, X_2 be a normed linear space, and the set of operators $\{A_\alpha : \alpha \in T\} \subset \mathcal{L}(X_1, X_2)$ be such that*

$\forall x \in X_1 \ \exists C_x > 0 \ \forall \alpha \in T \ : \ \|A_\alpha x\| \leq C_x.$

Then

$\exists C > 0 \ \forall \alpha \in T \ : \ \|A_\alpha\| \leq C.$

The criterion for strong convergence of operators is given in problem 7.6°.

V. Brayman et al., *Functional Analysis and Operator Theory*, Problem Books in Mathematics, https://doi.org/10.1007/978-3-031-56427-7_7

Examples of Problems with Solutions

7.1 Examine the sequence of operators $A_n : X \to X$, $n \geq 1$ for uniform, strong and weak convergence, if

(1) $X = l_2$, $A_n x = (\underbrace{0, 0, \ldots, 0}_{n-1}, \frac{x_n}{n}, \frac{x_{n+1}}{n+1}, \ldots, \frac{x_{2n}}{2n}, 0, 0, \ldots)$;
(2) $X = l_2$, $A_n x = (x_1, \ldots, x_n, 0, 0, \ldots)$;
(3) $X = l_2$, $A_n x = (\underbrace{0, 0, \ldots, 0}_{n-1}, x_1, 0, 0, \ldots)$;
(4) $X = C([0, 1])$, $(A_n x)(t) = \int\limits_0^1 \sqrt{(t-s)^2 + \frac{1}{n}}\, x(s)ds$;
(5) $X = C([0, 1])$, $(A_n x)(t) = \int\limits_0^1 t^n x(s)ds$;
(6) $X = L_p(\mathbb{R})$, $1 \leq p < +\infty$, $(A_n x)(t) = e^{-(t-n)^2} x(t)$;
(7) $X = L_p(\mathbb{R})$, $1 \leq p < +\infty$, $(A_n x)(t) = \begin{cases} x(t-n), & t \geq n, \\ 0, & t < n. \end{cases}$

Solution

(1) To begin with, we will determine an operator to which the sequence of operators can converge. If the sequence $\{A_n : n \geq 1\}$ converges weakly (in particular, if it converges strongly or uniformly), then for every $x \in l_2$ the sequence of elements $\{A_n x : n \geq 1\}$ converges weakly in l_2, and therefore converges coordinate-wise in l_2. Let us fix any $x \in l_2$ and find the coordinate-wise limit of the sequence $\{A_n x : n \geq 1\}$. Since for every $k \in \mathbb{N}$ we have $(A_n x)_k = 0$, $n > k$, then $(A_n x)_k \to 0$, $n \to \infty$. Thus if the sequence $\{A_n : n \geq 1\}$ converges uniformly, strongly or weakly, then the limit is the zero operator $A = 0$. Now we will verify whether the sequence $\{A_n x : n \geq 1\}$ converges to the operator A uniformly, strongly, weakly, or it does not converge in any of the above senses.

Since for all $x \in l_2$ we have

$$\|A_n x - Ax\|^2 = \sum_{k=n}^{2n} \frac{|x_k|^2}{k^2} \leq \frac{1}{n^2}\|x\|^2,$$

then $\|A_n - A\| \leq \frac{1}{n}$, $n \geq 1$. Therefore $A_n \rightrightarrows A$, so $A_n \xrightarrow{s} A$ and $A_n \xrightarrow{w} A$.

(2) Let us determine an operator A to which the sequence can converge. For each $x \in l_2$ we find the coordinate-wise limit of the sequence $\{A_n x : n \geq 1\}$. Since for each $k \in \mathbb{N}$ we have $(A_n x)_k = x_k$, $n \geq k$, then $(A_n x)_k \to x_k$, $n \to \infty$. Therefore if the sequence converges in some sense, then the limit is the identity operator $A = I$. Now let us study in which sense (if any) the sequence $\{A_n : n \geq 1\}$ converges to the operator A.

Firstly, we will verify whether the sequence converges strongly. If it happens, then the sequence also converges weakly, and it remains to verify whether the sequence converges uniformly. Otherwise the sequence does not converge uniformly, and it remains to verify whether the sequence converges weakly.

For every $x \in l_2$ we have $A_n x - Ax = (0, \ldots, 0, -x_{n+1}, -x_{n+2}, \ldots)$, so

$$\|A_n x - Ax\|^2 = \sum_{k=n+1}^{\infty} |x_k|^2 \to 0, \ n \to \infty,$$

as a remainder of the convergent series $\sum_{n=1}^{\infty} |x_n|^2$. Therefore $A_n \xrightarrow{s} A$, hence $A_n \xrightarrow{w} A$.

Now we will show that $A_n \not\rightrightarrows A$. Put $x^{(n)} = e_{n+1}$, $n \ge 1$. Then $\|x^{(n)}\| = 1$, $\|A_n - A\| = \sup_{\|x\| \le 1} \|A_n x - Ax\| \ge \|A_n x^{(n)} - Ax^{(n)}\| = 1$. Hence $\|A_n - A\| \not\to 0$, $n \to \infty$.

(3) Determine an operator A to which the sequence can converge. For each $x \in l_2$ we will find the coordinate-wise limit of the sequence $\{A_n x : n \ge 1\}$. Since for each $k \in \mathbb{N}$ we have $(A_n x)_k = 0$, $n > k$, then $(A_n x)_k \to 0$, $n \to \infty$. Therefore $A = 0$. Now let us study in which sense (if any) the sequence $\{A_n : n \ge 1\}$ converges to the operator A.

For every $x \in l_2$ we have $\|A_n x - Ax\|^2 = |x_1|^2 \not\to 0$, $n \to \infty$, if $x_1 \ne 0$. Therefore $A_n \overset{s}{\not\to} A$, hence $A_n \not\rightrightarrows A$.

Let us prove that $A_n \xrightarrow{w} A$, i.e. $A_n x \xrightarrow{w} Ax = 0$ for each $x \in l_2$. We will check the conditions of the weak convergence criterion in the space l_2 (item (2) of problem 5.1): (a) $\|A_n x\| = |x_1|$, $n \ge 1$, hence the sequence $\{\|A_n x\| \mid n \ge 1\}$ is bounded; (b) We already checked that $(A_n x)_k \to 0$, $n \to \infty$, for each $k \in \mathbb{N}$. By the criterion of weak convergence in l_2 we have $A_n x \xrightarrow{w} 0$, $n \to \infty$, hence $A_n \xrightarrow{w} A$.

(4) Determine an operator A to which the sequence can converge. If the sequence $\{A_n : n \ge 1\}$ converges weakly (in particular, if it converges strongly or uniformly), then for every $x \in C([0, 1])$ the sequence of elements $\{A_n x : n \ge 1\}$ converges weakly in $C([0, 1])$, and therefore converges pointwise. Let us fix any $x \in C([0, 1])$ and $t \in [0, 1]$ and find the limit $\lim_{n\to\infty} \int_0^1 \sqrt{(t-s)^2 + \frac{1}{n}}\, x(s) ds$. It is natural to try pass to the limit under the integral. Since $\sqrt{(t-s)^2 + \frac{1}{n}} \to |t - s|$, $n \to \infty$, $t, s \in [0, 1]$, let us consider $(Ax)(t) = \int_0^1 |t - s| x(s) ds$, $t \in [0, 1]$, $x \in C([0, 1])$.

Let us verify that $A_n \rightrightarrows A$. For all $t, s \in [0, 1]$ we have

$$\left|\sqrt{(t-s)^2+\frac{1}{n}}-|t-s|\right| = \frac{(t-s)^2+\frac{1}{n}-|t-s|^2}{\sqrt{(t-s)^2+\frac{1}{n}}+|t-s|} \leq \frac{1}{\sqrt{n}}.$$

Therefore $\|A_n x - Ax\| \leq \frac{1}{\sqrt{n}}\|x\|$ for all $x \in C([0,1])$. Thus $\|A_n - A\| \leq \frac{1}{\sqrt{n}} \to 0,\ n \to \infty$.

(5) If the sequence $\{A_n : n \geq 1\}$ converges weakly (in particular, if it converges strongly or uniformly) to the operator A, then $(A_n x)(t) \to (Ax)(t),\ n \to \infty$, for all $x \in C([0,1])$ and $t \in [0,1]$. We have

$$(A_n x)(t) = t^n \int_0^1 x(s)ds \to (Ax)(t) = \begin{cases} 0, & t \in [0,1), \\ \int_0^1 x(s)ds, & t = 1, \end{cases} \quad n \to \infty.$$

Note that if $\int_0^1 x_0(s)ds \neq 0$ then $Ax_0 \notin C([0,1])$, a contradiction. Therefore the sequence $\{A_n : n \geq 1\}$ does not converge weakly, consequently does not converge strongly or uniformly.

(6) Determine an operator A to which the sequence can converge. Neither weak nor strong convergence in $L_p(\mathbb{R})$ implies convergence almost everywhere. However, if for every $x \in L_p(\mathbb{R})$ the sequence $\{A_n x : n \geq 1\}$ converges almost everywhere, it is natural to consider the limit as a "candidate" for Ax. Let $x \in L_p(\mathbb{R})$. Note that

$$(A_n x)(t) = e^{-(t-n)^2}x(t) \to 0,\ n \to \infty \ (\mathrm{mod}\, m).$$

Now let us study in which sense (if any) the sequence $\{A_n : n \geq 1\}$ converges to the operator $A = 0$.

For all $x \in L_p(\mathbb{R})$ we have $\|A_n x - Ax\|^p = \int_{\mathbb{R}} e^{-p(t-n)^2}|x(t)|^p dt \to 0$, $n \to \infty$, by Lebesgue's dominated convergence theorem (check the conditions of the theorem yourself!). Thus $A_n \xrightarrow{s} A$, hence $A_n \xrightarrow{w} A$.

Now we will show that $A_n \not\rightrightarrows A$. Set $x_n = \chi_{[n,n+1]}$, $n \geq 1$. Since $\|x_n\| = 1$, we have

$$\|A_n - A\| \geq \|A_n x_n - Ax_n\| = \left(\int_n^{n+1} e^{-p(t-n)^2}dt\right)^{1/p} = \left(\int_0^1 e^{-pt^2}dt\right)^{1/p}.$$

Therefore $\|A_n - A\| \not\to 0,\ n \to \infty$.

(7) If the sequence $\{A_n : n \geq 1\}$ converges to A at least weakly, then for every $x \in L_p(\mathbb{R})$ and for all $s, t \in \mathbb{R}$, $s < t$, we have $\int_s^t (A_n x)(u)du \to \int_s^t (Ax)(u)du$ (for $1 < p < +\infty$ it follows from the criterion of weak convergence in $L_p(\mathbb{R})$ (problem 5.7), while for $p = 1$ it is true because $f(x) = \int_s^t x(u)du$ belongs to $(L_1(R))^*$). Since $(A_n x)(u) = 0$, $u \in [s, t]$, for $n > t$, then $\int_s^t (A_n x)(u)du = 0$, $n > t$. Thus $\int_s^t (A_n x)(u)du \to 0$, $n \to \infty$. Hence $\int_s^t (Ax)(u)du = 0$ for all $s < t$, so $Ax = 0$. Therefore $A = 0$ is the unique "candidate" for a limit. Now let us study in which sense (if any) the sequence $\{A_n : n \geq 1\}$ converges to the operator A.

For all $x \in L_p(\mathbb{R})$ we have

$$\|A_n x - Ax\|^p = \int_{\mathbb{R}} |(A_n x)(t)|^p dt = \int_n^{+\infty} |x(t-n)|^p dt = \int_0^{+\infty} |x(t)|^p dt,$$

$n \geq 1$. If x is not a zero function on $[0, +\infty)$, then $A_n x \not\to Ax$, $n \to \infty$. Therefore $A_n \overset{s}{\not\to} A$, hence, $A_n \not\rightrightarrows A$.

Let $1 < p < +\infty$. We shall prove that $A_n \overset{w}{\to} A$. Apply the criterion of weak convergence in $L_p(\mathbb{R})$. For every $x \in L_p(\mathbb{R})$ we have (a) $\|A_n x\|^p = \int_n^{\infty} |x(t-n)|^p dt = \int_0^{+\infty} |x(t)|^p dt$, $n \geq 1$, hence the sequence $\{\|A_n x\| \mid n \geq 1\}$ is bounded. Condition (b) has already been verified. Hence $A_n x \overset{w}{\to} 0$.

It remains to deal with $p = 1$. Set $g(x) = \int_{\mathbb{R}} x(t)dt$. Then $g \in (L_1(\mathbb{R}))^*$ and $g(A_n x) = \int_n^{+\infty} x(t-n)dt = \int_0^{+\infty} x(u)du \not\to 0$, $n \to \infty$, for example, for $x(t) = x_0(t) = e^{-|t|}$, $t \in \mathbb{R}$. Therefore by the definition of weak convergence $A_n x_0 \overset{w}{\not\to} 0$, which implies that $A_n \overset{w}{\not\to} A$.

7.2 Under what conditions on the number $\alpha \in \mathbb{C}$ does the sequence of operators $A_n : l_p \to l_p$, $1 \leq p \leq +\infty$, defined by the formula

$$A_n x = (\alpha^n x_1, \alpha^{n-1} x_2, \ldots, \alpha x_n, 0, 0, \ldots), \; x \in l_p,$$

converge uniformly, weakly, strongly?

Solution Let the sequence $\{A_n : n \geq 1\}$ converge weakly. Then for any $x \in l_p$ the sequence $\{A_n x : n \geq 1\}$ converges weakly in l_p, and hence converges coordinatewise. In particular, the sequence of numbers $\{\alpha^n x_1 : n \geq 1\}$ must converge. If

$x_1 \neq 0$, it follows that the sequence $\{\alpha^n : n \geq 1\}$ converges. Hence $|\alpha| < 1$ or $\alpha = 1$. Thus for the sequence of operators $\{A_n : n \geq 1\}$ to converge in any sense, the condition $\alpha \in B(0, 1) \cup \{1\}$ is necessary for any $1 \leq p \leq +\infty$. Let us determine whether this condition is sufficient for uniform, strong or weak convergence. Note that the limit operator is $A = I$ for $\alpha = 1$ and $A = 0$ for $|\alpha| < 1$.

Let $1 \leq p < +\infty$. If $\alpha = 1$, then similarly to the solution of problem 7.1, item (2), we have $A_n \not\rightrightarrows A$.

If $|\alpha| < 1$, then for every $x \in l_p$ and for every $\varepsilon > 0$ there exists $m \in \mathbb{N}$ such that $\sum\limits_{k=m+1}^{\infty} |x_k|^p < \frac{\varepsilon}{2}$. For $n \geq m$ we have

$$\|A_n x - Ax\|_p^p = |\alpha|^{np}|x_1|^p + |\alpha|^{(n-1)p}|x_2|^p + \ldots + |\alpha|^p|x_n|^p \leq$$

$$\leq |\alpha|^{(n+1-m)p} \sum_{k=1}^{m} |x_k|^p + \sum_{k=m+1}^{n} |x_k|^p \leq |\alpha|^{(n+1-m)p} \|x\|_p^p + \frac{\varepsilon}{2} < \varepsilon,$$

when n is such that $|\alpha|^{(n+1-m)p} \|x\|_p^p < \frac{\varepsilon}{2}$. Therefore $A_n \xrightarrow{s} A$. However, if $x^{(n)} = e_n$, $n \geq 1$, then $\|A_n x^{(n)} - Ax^{(n)}\|_p = |\alpha|$. Therefore $\|A_n - A\| \geq |\alpha|$, $n \geq 1$, hence $A_n \rightrightarrows A$ only when $\alpha = 0$ (in this case $A_n = 0$, $n \geq 1$).

Let $p = +\infty$. If $\alpha = 1$, then for $x = (1, 1, \ldots)$ we have $\|A_n x - Ax\| = 1$, $n \geq 1$, hence $A_n \overset{s}{\nrightarrow} A$. We will also show that $A_n \overset{w}{\nrightarrow} A$. Indeed, by the Hahn–Banach Theorem the functional $f(x) = \lim\limits_{n\to\infty} x_n$, $x \in c$, can be extended to the functional F on l_∞. For $x = (1, 1, \ldots)$ we have $F(A_n x) = 0$, $n \geq 1$, and $F(Ax) = 1$.

If $|\alpha| < 1$, then for $x = (1, 1, \ldots)$ we have $\|A_n x - Ax\| = |\alpha|$, $n \geq 1$. Hence $A_n \xrightarrow{s} A$ only for $\alpha = 0$. Now we will show that $A_n \xrightarrow{w} A = 0$. Indeed, let $x \in l_\infty$ be an arbitrary fixed element. It is necessary to prove that $A_n x \xrightarrow{w} 0$ in l_∞, i.e. $F(A_n x) \to 0$, $n \to \infty$, for every $F \in l_\infty^*$. Since $A_n x \in c_0$, $n \geq 1$, and $F \restriction c_0 \in c_0^*$, the condition we have to prove is equivalent to $A_n x \xrightarrow{w} 0$ in c_0. In accordance with problem 5.8, item (1) the convergence in c_0 is equivalent to coordinate-wise convergence and boundedness of norms. For the sequence $\{A_n x : n \geq 1\}$ both these conditions are fulfilled, therefore $A_n x \xrightarrow{w} 0$ in c_0 and in l_∞.

Thus if $1 \leq p < +\infty$, then the sequence converges uniformly only for $\alpha = 0$, and converges strongly and weakly for $|\alpha| < 1$ and for $\alpha = 1$; if $p = +\infty$, then the sequence converges uniformly and strongly only for $\alpha = 0$, and converges weakly for $|\alpha| < 1$.

7.3 Let X be a Banach space, and the sequences $\{x, x_n : n \geq 1\} \subset X$ and $\{A, A_n : n \geq 1\} \subset \mathcal{L}(X)$ be such that $x_n \to x$, $A_n \xrightarrow{w} A$, $n \to \infty$. Prove that $A_n x_n \xrightarrow{w} Ax$, $n \to \infty$.

Solution For each $z \in X$ the sequence $\{A_n z : n \geq 1\}$ is bounded, because it converges weakly. Hence by the Banach–Steinhaus theorem the sequence $\{\|A_n\| : n \geq 1\}$ is bounded. For arbitrary functional $f \in X^*$ we have

$$|f(A_n x_n) - f(Ax)| \leq |f(A_n x_n) - f(A_n x)| + |f(A_n x) - f(Ax)| \leq$$
$$\leq \|f\| \cdot \|A_n\| \cdot \|x_n - x\| + |f(A_n x) - f(Ax)| \to 0, \ n \to \infty.$$

Problems to Solve

7.4° Prove the following statements:

(1) if the limit exists, then it is unique for any operator convergence;
(2) uniform convergence implies strong convergence;
(3) strong convergence implies weak convergence;
(4) the converse statements to the statements (2) and (3) are not true;
(5) in a finite-dimensional space uniform, strong and weak convergences are equivalent.

7.5° What does the uniform, strong and weak operator convergence look like for operators from $\mathcal{L}(X, \mathbb{K})$, where X is an NLS, i.e., for functionals?

7.6° Let X_1 and X_2 be normed linear spaces, $\{A_n : n \geq 1\} \subset \mathcal{L}(X_1, X_2)$, $A \in \mathcal{L}(X_1, X_2)$. Prove that for the convergence $A_n \overset{s}{\to} A$, $n \to \infty$, it is sufficient, and if X_1 is a Banach space, then it is also necessary, that the following conditions are fulfilled:

(a) the sequence $\{\|A_n\| : n \geq 1\}$ is bounded;
(b) there exists a total set $M \subset X_1$ such that $A_n x \to Ax$, $n \to \infty$, in X_2 for all $x \in M$.

7.7° Formulate and prove theorems, analogous to problem 7.6°, for weak operator convergence.

7.8° Examine the sequence of operators $A_n : l_p \to l_p$, $n \geq 1$, for uniform, strong and weak convergence.

(1) $A_n x = (\underbrace{0, \ldots, 0}_{n}, x_1, x_2, \ldots)$, $1 \leq p < +\infty$;
(2) $A_n x = (x_{n+1}, x_{n+2}, \ldots)$, $1 \leq p \leq +\infty$;
(3) $A_n x = (\underbrace{0, \ldots, 0}_{n-1}, x_n, 0, 0, \ldots)$, $1 \leq p < +\infty$;
(4) $A_n x = (\underbrace{0, \ldots, 0}_{n-1}, x_n, x_{n+1}, \ldots)$, $1 \leq p < +\infty$;
(5) $A_n x = (x_n, x_{n-1}, x_2, \ldots x_1, 0, 0, \ldots)$, $1 \leq p \leq +\infty$;
(6) $A_n x = (\frac{x_1}{n}, \frac{x_2}{n-1}, \ldots, \frac{x_n}{1}, 0, 0, \ldots)$, $1 \leq p < +\infty$;

(7) $A_n x = (\underbrace{0, \dots, 0}_{n-1}, \sqrt{n}\, x_{2n}, 0, 0, \dots)$, $1 \le p \le +\infty$;
(8) $A_n x = \left(\sum\limits_{m=1}^{n} x_m, 0, 0, \dots \right)$, $1 \le p \le +\infty$;
(9) $A_n x = (\frac{1+n}{1+n} x_1, \frac{1+n}{1+2n} x_2, \frac{1+n}{1+3n} x_3, \dots, \frac{1+n}{1+n^2} x_n, 0, 0, \dots)$, $1 \le p \le +\infty$;
(10) $A_n x = (\sqrt[n]{n} x_1, \sqrt[n]{n} x_2, \dots, \sqrt[n]{n} x_n, \dots)$, $1 \le p \le +\infty$;
(11) $A_n x = (\sqrt[n]{n} x_1, \sqrt[n]{n} x_2, \dots, \sqrt[n]{n} x_n, 0, 0, \dots)$, $1 \le p < +\infty$;
(12) $A_n x = (\frac{x_n}{2n}, \frac{x_{n+1}}{2n+1}, \dots, \frac{x_{2n}}{3n}, 0, 0, \dots)$, $1 \le p \le +\infty$;
(13) $A_n x = (\frac{1+n}{1+n} x_1, \frac{1+n}{1+2n} x_2, \frac{1+n}{1+3n} x_3, \dots, \frac{1+n}{1+kn} x_k, \dots)$, $1 \le p \le +\infty$;
(14) $A_n x = (\frac{1+n}{1+n} x_1, \frac{1+n}{1+2n} x_2, \frac{1+n}{1+3n} x_3, \dots, \frac{1+n}{1+kn} x_k, 0, 0, \dots)$, $1 \le p \le +\infty$, where $k \in \mathbb{N}$ is fixed.

7.9* Examine the sequence of operators from items (3), (6) and (11) of problem 7.8° for uniform, strong or weak convergence in the case of $p = +\infty$.

7.10 Find the necessary and sufficient conditions for the sequence of complex numbers $\{\alpha_n : n \ge 1\}$, under which the sequence of operators $A_n : l_p \to l_p$, $1 \le p \le +\infty$, $n \ge 1$, converges uniformly, strongly, weakly, if

(1) $A_n x = (\alpha_n x_1, 0, 0, \dots)$;
(2) $A_n x = (\underbrace{0, 0, \dots, 0}_{n-1}, \alpha_n x_1, 0, 0, \dots)$;
(3) $A_n x = (\alpha_1 x_1, \alpha_2 x_2, \dots, \alpha_n x_n, 0, 0, \dots)$;
(4) $A_n x = (\alpha_n x_1, \alpha_n x_2, \dots, \alpha_n x_n, 0, 0, \dots)$;
(5) $A_n x = (\underbrace{0, 0, \dots, 0}_{n-1}, \alpha_1 x_1, \alpha_2 x_2, \dots, \alpha_n x_n, 0, 0, \dots)$;
(6) $A_n x = (\underbrace{0, 0, \dots, 0}_{n-1}, \alpha_n x_1, \alpha_{n-1} x_2, \dots, \alpha_1 x_n, 0, 0, \dots)$;
(7) $A_n x = (\alpha_n x_n, \alpha_{n-1} x_{n+1}, \dots, \alpha_1 x_{2n-1}, 0, 0, \dots)$;
(8) $A_n x = (\alpha_n x_1, \alpha_{n+1} x_2, \dots)$;
(9) $A_n x = (\alpha_n x_1, \alpha_{n+1} x_2, \dots, \alpha_{2n-1} x_n, 0, 0, \dots)$;
(10)* $A_n x = (\alpha_n x_1, \alpha_{n-1} x_2, \dots, \alpha_1 x_n, 0, 0, \dots)$.

7.11° Examine the sequence of operators $A_n : C([0,1]) \to C([0,1])$ for uniform, strong, and weak convergence.

(1) $(A_n x)(t) = t^n (1-t) x(t)$;
(2) $(A_n x)(t) = t^n x(t)$;
(3) $(A_n x)(t) = e^{-nt} x(t)$;
(4) $(A_n x)(t) = x(t) \cdot \sin nt$;
(5) $(A_n x)(t) = \int\limits_0^1 (t^n + s^n) x(s) ds$;
(6) $(A_n x)(t) = x(t) \cdot \sin \frac{t}{n}$;
(7) $(A_n x)(t) = \int\limits_0^1 \frac{(t+s)^n}{3} x(s) ds$;
(8) $(A_n x)(t) = \int\limits_0^1 \frac{(t+s)^n}{3^n} x(s) ds$;
(9) $(A_n x)(t) = \int\limits_0^t \left(\sum\limits_{k=0}^{n} \frac{(-1)^k s^{2k+1}}{(2k+1)!} \right) x(s) ds$;
(10) $(A_n x)(t) = x\left(t^{1+\frac{1}{n}} \right)$;
(11) $(A_n x)(t) = (t^{10n} - t^{2n}) x(t)$;

(12) $(A_n x)(t) = t \int_0^1 x\left(\sin^n \frac{\pi}{2} s\right) ds;$

(13) $(A_n x)(t) = t \int_0^1 x(s) \sin^n \frac{\pi}{2} s ds;$

(14) $(A_n x)(t) = \int_0^1 t \frac{1+(2s)^n}{1+(2s)^{2n}} x(s) ds;$

(15) $(A_n x)(t) = \frac{\ln(1+nt)}{1+nt} x(t);$

(16) $(A_n x)(t) = \frac{\ln(1+t)}{1+nt} x(t);$

(17) $(A_n x)(t) = \int_0^1 t^2 \frac{\ln(1+ns)}{1+ns} x(s) ds;$

(18) $(A_n x)(t) = \int_0^{\frac{t^n}{2^n}} x(s) ds;$

(19) $(A_n x)(t) = \int_0^{t^n} x(s) ds;$

(20) $(A_n x)(t) = (1-t)^n \int_0^{t^n} x(s) ds;$

(21) $(A_n x)(t) = n \int_t^{t+\frac{1}{n}} x(s) ds$, where $x(s) = x(1)$ for $s > 1$.

7.12° Examine the sequence of operators $A_n : L_p(T) \to L_p(T)$ for uniform, strong, and weak convergence.

(1) $T = \mathbb{R}$, $(A_n x)(t) = x(t+n)$, $1 \le p < +\infty$;

(2) $T = \mathbb{R}$, $(A_n x)(t) = \begin{cases} x(t), & t \ge n, \\ 0, & t < n, \end{cases}$ $1 \le p < +\infty$;

(3) $T = \mathbb{R}$, $(A_n x)(t) = \frac{1}{1+|t-n|} x(t)$, $1 \le p < +\infty$;

(4) $T = [0, 1]$, $(A_n x)(t) = t^n \int_0^1 s^n x(s) ds$, $1 \le p < +\infty$;

(5) $T = [0, 1]$, $(A_n x)(t) = x(t) \cos nt$, $p = 2$;

(6) $T = [0, 1]$, $(A_n x)(t) = x(t) \cos e^n t$, $p = 2$;

(7) $T = [0, 1]$, $(A_n x)(t) = x(t) \cos ne^t$, $p = 2$;

(8)* $T = [0, 1]$, $(A_n x)(t) = x(t) \cos e^{nt}$, $p = 2$;

(9) $T = \mathbb{R}$, $(A_n x)(t) = x(t) \cdot \chi_{[\ln n, 2 \ln n]}(t)$, $1 \le p < +\infty$;

(10) $T = [0, 1]$, $(A_n x)(t) = \sqrt[n]{t} x(t)$, $p = 1$;

(11) $T = [0, 1]$, $(A_n x)(t) = \int_0^1 t^2 (s^n - s^{2n}) x(s) ds$, $1 \le p < +\infty$.

7.13° Let $\{p_n : n \ge 1\} \subset C([a, b])$ be a fixed sequence,

$$(A_n x)(t) = p_n(t) x(t),\ t \in [a, b],\ x \in C([a, b]).$$

Under what conditions on the sequence $\{p_n : n \ge 1\}$ the sequence of operators $\{A_n : n \ge 1\}$ converges (1) uniformly; (2) strongly; (3) weakly? Determine the form of the limit operator.

7.14 Let X_1, X_2 be Banach spaces.

(1) Prove that the space $\mathcal{L}(X_1, X_2)$ is complete with respect to the strong convergence of operators, i.e., every sequence $\{A_n : n \ge 1\} \subset \mathcal{L}(X_1, X_2)$, for which $\{A_n x : n \ge 1\}$ is a Cauchy sequence of elements from X_2 for each $x \in X_1$, converges strongly to some operator $A \in \mathcal{L}(X_1, X_2)$.

(2) Under what conditions on the Banach spaces X_1 and X_2 is the space $\mathcal{L}(X_1, X_2)$ complete with respect to weak convergence?

7.15 In $L_p(\mathbb{R})$, $1 \leq p < +\infty$, define the shift operator

$$(A_s x)(t) = x(t+s), \ t \in \mathbb{R}, \ x \in L_p(\mathbb{R}), \ s \in \mathbb{R}.$$

Let $s_n \to s$, $n \to \infty$, and $s_n \neq s$, $n \geq 1$. Prove that $A_{s_n} \xrightarrow{s} A_s$, but $A_{s_n} \not\rightrightarrows A_s$.

7.16 Let X be the space $C^1([0,1])$ with the norm $\|x\| = \max\limits_{t \in [0,1]} |x(t)|$, $x \in C^1([0,1])$. Define the operator $A_n : X \to C([0,1])$ by the formula $(A_n x)(t) = n\big(x(t+\frac{1}{n}) - x(t)\big)$, $t \in [0,1]$, $x \in X$, $n \geq 1$ (if $t > 1$, put $x(t) = x(1) + x'_-(1)(t-1)$.) Prove that

(1) for every $x \in X$ the sequence $\{A_n x : n \geq 1\}$ converges in norm in $C([0,1])$;
(2) the sequence $\{\|A_n\| : n \geq 1\}$ is not bounded;
(3) the space $\mathcal{L}(X, C([0,1]))$ is not complete with respect to the strong operator topology.

Do these statements contradict the Uniform Boundedness Principle?

7.17* Denote by $\widetilde{C}$ the space of all 2π-periodic continuous functions on $\mathbb{R}$. It can be identified with a subspace in $C([-\pi,\pi])$, which consists of functions x such that $x(-\pi) = x(\pi)$. For any $n \geq 1$ and $t \in [-\pi,\pi]$ put

$$(S_n x)(t) = \frac{1}{\pi} \int_{-\pi}^{\pi} D_n(t-s) x(s) ds,$$

where $D_n(t) = \frac{1}{2} + \sum\limits_{k=1}^{n} \cos kt = \frac{\sin \frac{2n+1}{2} t}{2 \sin \frac{t}{2}}$ is the Dirichlet kernel. Prove that

(1) $S_n \xrightarrow{s} I$, $S_n \not\rightrightarrows I$ in $L_2([-\pi,\pi])$;
(2) $\int\limits_{-\pi}^{\pi} |D_n(t)| dt > A \ln n$, $n \geq 1$, for some constant $A > 0$;
(3) there exist functions $x \in \widetilde{C}$ ($x \in L_1([-\pi,\pi])$), for which the Fourier series with respect to the trigonometric system does not converge to x in $\widetilde{C}$ (in $L_1([-\pi,\pi])$, respectively);
(4) for any $t_0 \in [-\pi,\pi]$ there exists a function $x \in \widetilde{C}$, for which the Fourier series at the point t_0 does not converge to $x(t_0)$;
(5) there exist functions $x \in C([-\pi,\pi])$ and $y \in L_1([-\pi,\pi])$ such that the series

$$\frac{a_0 \alpha_0}{2} + \sum_{n=1}^{\infty} (a_n \alpha_n + b_n \beta_n)$$

diverges, where $\{a_0, a_n, b_n \; : \; n \geq 1\}$, $\{\alpha_0, \alpha_n, \beta_n \; : \; n \geq 1\}$ are the Fourier coefficients of the functions x and y, respectively.

7.18* Let X_1 be a Banach space, X_2 be an NLS and the mapping $B : X_1 \times X_1 \to X_2$ be linear and continuous in each variable. Prove that B is joint continuous.

7.19 Let X be an NLS which consists of algebraic polynomials of one variable with a norm $\|x\| = \int\limits_0^1 |x(t)|dt$, and $B(x, y) = \int\limits_0^1 x(t)y(t)dt$. Prove that the functional B is linear and continuous in each variable, but is not continuous on $X \times X$.

7.20° Let X be a Banach space and $\{A, B, A_n, B_n : n \geq 1\} \subset \mathcal{L}(X)$. Prove that

(1) $A_n \overset{s}{\to} A, \; B_n \overset{s}{\to} B \; \Rightarrow \; A_n B_n \overset{s}{\to} AB;$
(2) $A_n \rightrightarrows A, \; B_n \rightrightarrows B \; \Rightarrow \; A_n B_n \rightrightarrows AB;$
(3) $A_n \overset{w}{\to} A, \; B_n \overset{w}{\to} B \; \not\Rightarrow \; A_n B_n \overset{w}{\to} AB.$

7.21° Let X be a Banach space, $\{x, x_n \; : \; n \geq 1\} \subset X$ and $\{A, A_n \; : \; n \geq 1\} \subset \mathcal{L}(X)$. Prove that

(1) $A_n \overset{s}{\to} A, \; x_n \to x \; \Rightarrow \; A_n x_n \to Ax;$
(2) $A_n \rightrightarrows A, \; x_n \overset{w}{\to} x \; \Rightarrow \; A_n x_n \overset{w}{\to} Ax;$
(3) $A_n \overset{s}{\to} A, \; x_n \overset{w}{\to} x \; \not\Rightarrow \; A_n x_n \overset{w}{\to} Ax.$

7.22° Let H be a Hilbert space, $\{A_n : n \geq 1\} \subset \mathcal{L}(H)$, $\sup\limits_{n \geq 1} \|A_n\| < +\infty$, and

$$\exists L \subset H, \; \overline{L} = H, \; \forall x, y \in L \; : \; \exists \lim_{n \to \infty} (A_n x, y).$$

Prove that there exists $A \in \mathcal{L}(H)$ such that $A_n \overset{w}{\to} A$.

7.23* Let H be a separable Hilbert space. Prove that any closed ball in $\mathcal{L}(H)$ is a weakly compact set (i.e., every bounded sequence in $\mathcal{L}(H)$ contains a weakly convergent subsequence).

7.24 Let X be a Banach space, $A \in \mathcal{L}(X)$, $\{\lambda_k : k \geq 0\} \subset \mathbb{R}$, $\varphi(t) = \sum\limits_{n=1}^{\infty} \lambda_n t^n$, $t \in \mathbb{R}$, be the sum of a power series which converges on $\mathbb{R}$. Prove that the sequence $\{\varphi_n(A) : n \geq 0\}$, where $\varphi_n(A) = \sum\limits_{k=0}^{n} \lambda^k A^k$, $n \geq 0$, $A^0 = I$, converges in $\mathcal{L}(X)$ to some operator $\varphi(A) \in \mathcal{L}(X)$.

Prove that under the condition $\lambda_k \geq 0$, $k \geq 0$, the inequality $\|\varphi(A)\| \leq \varphi(\|A\|)$ holds.

7.25 Prove that for a bounded operator $A \in \mathcal{L}(X)$, where X is a Banach space, the operators

$$e^A = \sum_{k=0}^{\infty} \frac{A^k}{k!}, \quad \sin A = \sum_{k=0}^{\infty} \frac{(-1)^k A^{2k+1}}{(2k+1)!}, \quad \cos A = \sum_{k=0}^{\infty} \frac{(-1)^k A^{2k}}{(2k)!},$$

are well-defined, and e^A, $\sin A$, $\cos A \in \mathcal{L}(X)$.

7.26 Let X be a Banach space, $A \in \mathcal{L}(X)$. Prove that the series $\sum_{n=0}^{\infty} A^n$ converges in $\mathcal{L}(X)$ if and only if for some $k \in \mathbb{N}$ the inequality $\|A^k\| < 1$ holds.

7.27 Prove that every linear bounded operator in a Hilbert space is a strong limit of a sequence of bounded finite-dimensional operators if and only if the space is separable.

7.28* Denote by T_n the subspace of the space $\widetilde{C}$ (see problem 7.17*) which consists of trigonometric polynomials of degree at most n. An operator $U \in \mathcal{L}(\widetilde{C})$ is called a polynomial (trigonometric) operator of degree n, if $Ux \in T_n$ for all $x \in \widetilde{C}$ and $Ux = x$ for all $x \in T_n$. Prove the following statements:

(1) The following operators are polynomial: the operator S_n from problem 7.17*; an operator that maps each function to its interpolation (trigonometric) polynomial, constructed for a fixed set of nodes.
(2) **(Zigmund–Marcinkiewicz–Berman)** If U is a polynomial operator of degree n, then $\frac{1}{2\pi} \int_{-\pi}^{\pi} (Ux^\tau)(s - \tau)d\tau = (S_n x)(s)$, $s \in \mathbb{R}$, $x \in \widetilde{C}$, where $x^\tau(t) = x(t + \tau)$, $t \in \mathbb{R}$, $\tau \in \mathbb{R}$, and for each $s \in \mathbb{R}$ the integrand is a continuous function of τ.
(3) If U is a polynomial operator of degree n, then $\|U\| \geq \|S_n\| > A \ln n$, $n \geq 1$, for some constant $A > 0$.
(4) The sequence $\{U_n : n \geq 1\}$ of polynomial operators, where U_n is an operator of degree n, does not converge strongly in $\widetilde{C}$.

7.29 A linear operator $A : C([a, b]) \to C([a, b])$, where $C([a, b])$ is a real space, is called nonnegative if for all $x \in C([a, b])$ such that $x(t) \geq 0$ for all $t \in [a, b]$, we have $(Ax)(t) \geq 0$ for all $t \in [a, b]$. Prove that

(1) The nonnegative operator is continuous, and $\|A\| = \|Ae_0\|$, where $e_0(t) = 1$, $t \in [a, b]$.
(2) **(Korovkin)** If $\{A_n : n \geq 1\}$ is a sequence of nonnegative operators, then $A_n \xrightarrow{s} I$ in $C([a, b])$ if and only if the convergence $A_n e_k \to e_k$, $n \to \infty$, holds for $e_k(t) = t^k$, $t \in [a, b]$, $k = 0, 1, 2$.

Is the analogous criterion of weak convergence of nonnegative operators valid?

7.30 Let $X = L_p([0, 1])$ or $X = C([0, 1])$, the operator $K_n : X \to X$ assigns to each function $x \in X$ the Kantorovich polynomial

$$(K_n x)(t) = \sum_{k=0}^{n} \binom{n}{k} x_{nk} t^k (1 - t)^{n-k}, \ t \in [0, 1],$$

where $x_{nk} = (n+1) \int\limits_{\frac{k}{n+1}}^{\frac{k+1}{n+1}} x(t)dt,\ 0 \le k \le n,\ n \ge 1$. Prove that

(1) $K_n \xrightarrow{s} I$ in $C([0, 1])$;
(2) $K_n \xrightarrow{s} I$ in $L_p([0, 1]),\ 1 \le p < +\infty$.

Is this statement true for $p = +\infty$?

Chapter 8
Inverse Operators

Theoretical Background

Let X, Y be normed linear spaces, and $A \in \mathcal{L}(X, Y)$ be an operator with the range $R(A) \subset Y$ such that $\operatorname{Ker} A = \{0\}$. Then one can define the inverse mapping A^{-1} : $R(A) \to X$, $A^{-1}(Ax) = x$, $x \in X$. This mapping is a linear operator, which is called the *algebraic inverse* to the operator A.

Let $A \in \mathcal{L}(X, Y)$. The operator A is called *continuously invertible*, if the following conditions are fulfilled: 1) $\operatorname{Ker} A = \{0\}$; 2) $R(A) = Y$; 3) the algebraic inverse $A^{-1} \in \mathcal{L}(Y, X)$. In this case, A^{-1} is called the *continuous inverse* to the operator A.

Theorem 8.1 (Banach) *If X, Y are Banach spaces and the operator $A \in \mathcal{L}(X, Y)$ is bijective, then the operator A is continuously invertible, i.e., $A^{-1} \in \mathcal{L}(Y, X)$.*

Theorem 8.2 *Let X, Y be normed linear spaces, $A \in \mathcal{L}(X, Y)$ and $R(A) = Y$. The operator A is continuously invertible if and only if*

$$\exists\, m > 0\ \forall x \in X\ :\ \|Ax\| \geq m\|x\|.$$

Note that the stated condition is equivalent to $\inf\limits_{\|x\|=1} \|Ax\| > 0$.

Theorem 8.3 (Criterion of Continuous Invertibility) *Let X, Y be normed linear spaces. The operator $A \in \mathcal{L}(X, Y)$ is continuously invertible if and only if*

$$\exists\, B \in \mathcal{L}(Y, X)\ \forall x \in X\ :\ BAx = x,\ \forall y \in Y\ :\ ABy = y.$$

Theorem 8.4 *Let X be a Banach space, $A \in \mathcal{L}(X)$, $\|A\| < 1$. Then the operator $I - A$ is continuously invertible and $(I - A)^{-1} = I + \sum\limits_{n=1}^{\infty} A^n$, where the series converges in operator norm.*

V. Brayman et al., *Functional Analysis and Operator Theory*, Problem Books in Mathematics, https://doi.org/10.1007/978-3-031-56427-7_8

Examples of Problems with Solutions

8.1 Let $X = l_p$, $1 \le p \le +\infty$, an operator $A \in \mathcal{L}(l_p)$ is defined by the formula $Ax = (\alpha_1 x_1, \alpha_2 x_2, \ldots)$, $x \in l_p$, where $\{\alpha_n : n \ge 1\} \subset \mathbb{K}$ is a fixed sequence such that $\sup\limits_{n\ge1} |\alpha_n| < +\infty$. Prove that

(1) $\operatorname{Ker} A = \{0\} \Leftrightarrow \alpha_n \ne 0$ for all $n \ge 1$;
(2) A is continuously invertible $\Leftrightarrow \inf\limits_{n\ge1} |\alpha_n| > 0$.

Solution The first statement follows from the fact that $Ax = 0$ if and only if $\alpha_n x_n = 0$, $n \ge 1$. Let us prove the second statement. Assume that $m = \inf\limits_{n\ge1} |\alpha_n| > 0$. Consider the operator that acts according to the formula $Bx = (\frac{x_1}{\alpha_1}, \frac{x_2}{\alpha_2}, \ldots)$, $x \in l_p$. It is clear that $B \in \mathcal{L}(l_p)$ and $\|B\| \le \frac{1}{m}$. Obviously, $BA = AB = I$. It follows from the criterion of continuous invertibility (Theorem 8.3) that A is continuously invertible and $A^{-1} = B$. Now assume that A is continuously invertible. Then it follows from Theorem 8.2 that $0 < \inf\limits_{\|x\|=1} \|Ax\| \le \inf\limits_{n\ge1} \|Ae_n\| = \inf\limits_{n\ge1} |\alpha_n|$.

8.2 Let $X = L_2([a, b])$, $p \in L_\infty([a, b])$, $(Ax)(t) = p(t)x(t)$. Prove that

(1) $\operatorname{Ker} A = \{0\} \Leftrightarrow |p(t)| > 0 \pmod{m}$;
(2) A is continuously invertible $\Leftrightarrow p^{-1} \in L_\infty([a, b])$.

Solution The reasonings are quite similar to the solution of problem 8.1 (note that the role of operator B is played by the operator of multiplication by the function p^{-1}). We will focus on the case $p^{-1} \notin L_\infty([a, b])$ and prove that A is not continuously invertible. Put

$$T_\varepsilon = \left\{t \in [a, b] \,\middle|\, |p^{-1}(t)| \ge \tfrac{1}{\varepsilon}\right\} = \left\{t \in [a, b] \,\middle|\, |p(t)| \le \varepsilon\right\},\ \varepsilon > 0.$$

The condition $p^{-1} \notin L_\infty([a, b])$ is equivalent to the condition $m(T_\varepsilon) > 0$ for all $\varepsilon > 0$. Introduce the functions $x_\varepsilon = (m(T_\varepsilon))^{-\frac{1}{2}} \chi_{T_\varepsilon} \in L_2([a, b])$. It is clear that $\|x_\varepsilon\| = 1$ and $\|Ax_\varepsilon\| \le \varepsilon \|x_\varepsilon\| = \varepsilon$. Then $\inf\limits_{\|x\|=1} \|Ax\| \le \inf\limits_{\varepsilon>0} \|Ax_\varepsilon\| \le \inf\limits_{\varepsilon>0} \varepsilon = 0$, and it follows from Theorem 8.2 that A is not continuously invertible.

Problems to Solve

8.3° Prove that the operator A^{-1}, which is the inverse to the linear operator A, is also linear.

8.4° Let $X = \mathbb{C}^m$ endowed with any norm. Prove that for a linear operator $A : X \to X$ the following statements are equivalent:

(1) $\operatorname{Ker} A = \{0\}$.

(2) The operator A has an algebraic inverse.
(3) The operator A has a continuous inverse.
(4) The matrix which represents the operator A is nondegenerate.
(5) The equation $Ax = y$ has a solution for every $y \in X$.

8.5° Let $A \in \mathcal{L}(l_2)$ be defined by the formula $Ax = (x_1, \frac{x_2}{2}, \ldots, \frac{x_n}{n}, \ldots)$, $x \in l_2$. Prove that the range $R(A)$ is not a closed set in l_2.

8.6° Let X be a Banach space, $A \in \mathcal{L}(X)$, $\|A\| < 1$. Prove that the operator $I + A$ is continuously invertible, and find $(I + A)^{-1}$.

8.7° Let X be an NLS, $A, B \in \mathcal{L}(X)$ are continuously invertible. Prove that the operator AB is continuously invertible and $(AB)^{-1} = B^{-1}A^{-1}$.

8.8°

(1) Let X be a Banach space, $A, B \in \mathcal{L}(X)$, A is continuously invertible with continuous inverse operator A^{-1}, $\|B\| < \|A^{-1}\|^{-1}$. Prove that $A + B$ is continuously invertible, and find $(A + B)^{-1}$.
(2) Prove that the set of continuously invertible operators is open in $\mathcal{L}(X)$.

8.9 Let X be an NLS. Prove that the operator $A \in \mathcal{L}(X)$ is continuously invertible if and only if A^2 is continuously invertible.

8.10° Denote by X_0 and X_1 the space $\{x \in C^1([0, 1]) \mid x(0) = 0\}$ endowed with norms $\|x\|_0 = \max\limits_{t\in[0,1]} |x(t)|$ and $\|x\|_1 = \max\limits_{t\in[0,1]} |x(t)| + \max\limits_{t\in[0,1]} |x'(t)|$, respectively. Let $A_i : C([0, 1]) \to X_i$, $(A_i x)(t) = \int\limits_0^t x(s)ds$, $i = 0, 1$. Find A_i^{-1}. Prove that A_1^{-1} is continuous, and A_0^{-1} is not bounded.

8.11 Let X_1 be the space from problem 8.10°. Prove that each of the following operators $A : X_1 \to C([0, 1])$ is continuously invertible, and find A^{-1}.

(1) $(Ax)(t) = 2x'(t) - 3x(t)$;
(2) $(Ax)(t) = x'(t) - tx(t)$;
(3) $(Ax)(t) = x'(t) + e^t x(t)$;
(4) $(Ax)(t) = x'(t) + 2t^2 x(t)$;
(5) $(Ax)(t) = e^t x'(t) - 2tx(t)$;
(6) $(Ax)(t) = (t + 1)x'(t) + tx(t)$;
(7) $(Ax)(t) = x'(t) + \alpha(t)x(t)$, $\alpha \in C([0, 1])$.

8.12° Let $K(t, s) = f(t)g(s)$, where $f, g \in C([0, 1])$ and $\int\limits_0^1 f(t)g(t)dt \neq 1$. Prove that the operator $A : C([0, 1]) \to C([0, 1])$, where

$$(Ax)(t) = x(t) - \int_0^1 K(t, s)x(s)ds, \ t \in [0, 1],$$

is continuously invertible. Find A^{-1}.

8.13 Let $K(t,s) = f(t)g(s)$, where $f, g \in C([0,1])$. Prove that the operator $A : C([0,1]) \to C([0,1])$, where $(Ax)(t) = x(t) - \int\limits_0^t K(t,s)x(s)ds,\ t \in [0,1]$, is continuously invertible. Find A^{-1}.

8.14° In the space $C([0,1])$ the operators A and B are defined by the formulas $(Ax)(t) = (t+1)x(t),\ (Bx)(t) = x(t^2),\ t \in [0,1]$. Find $(AB)^{-1}$ and $(BA)^{-1}$.

8.15° Let X be an NLS, $A,\ BA \in \mathcal{L}(X)$ be continuously invertible. Prove that the operator B is also continuously invertible.

8.16 Let X be an NLS, $A, B \in \mathcal{L}(X)$.

(1) Let $(I - AB)^{-1} \in \mathcal{L}(X)$. Prove that $(I - BA)^{-1} \in \mathcal{L}(X)$.
(2) Prove that the property $(AB)^{-1} \in \mathcal{L}(X)$ does not imply that $(BA)^{-1}$ exists.

8.17 Let X be an NLS, $A, B \in \mathcal{L}(X)$.

(1) Let the operators AB and BA have algebraic inverses. Does it imply that the operators A and B have algebraic inverses?
(2) Let the operators AB and BA be continuously invertible. Does it imply that the operators A and B are continuously invertible?

8.18° Let X be an NLS. Prove that if for an operator $A \in \mathcal{L}(X)$ there exists a sequence $\{x_n : n \geq 1\} \subset X$ such that $\|x_n\| = 1$ and $Ax_n \to 0,\ n \to \infty$, then A is not continuously invertible.

8.19° Prove that the operator $A : C([0,1]) \to C([0,1])$, defined by the formula $(Ax)(t) = \alpha(t)x(t),\ \alpha \in C([0,1])$, is continuously invertible if and only if $\alpha(t) \neq 0,\ t \in [0,1]$.

8.20 Let the operators $A_i : C([0,1]) \to C^i([0,1]),\ i = 0, 1$, be defined by the formula $(A_i x)(t) = \int\limits_0^t e^{t-s}x(s)ds$. For each of the cases $i = 0, 1$ answer the following questions: (1) Does A_i^{-1} exist? (2) Is A_i^{-1} continuous on its domain? (3) Is A_i continuously invertible?

8.21° Let $A, B : C([-1,1]) \to C([-1,1])$, where $(Ax)(t) = x(t^2),\ (Bx)(t) = x(t^3),\ t \in [-1,1]$. Prove that A^{-1} does not exist, while B is continuously invertible. Find B^{-1}.

8.22 Let $X_2 = \{x \in C^2([0,1]) : x(0) = x(1) = 0\}$ with a norm $\|x\|_2 = \sum\limits_{k=0}^{2} \max\limits_{t\in[0,1]} |x^{(k)}(t)|$, The operator $A : X_2 \to C([0,1])$ is defined by the formula $(Ax)(t) = x''(t) - 4x(t),\ t \in [0,1]$. Find A^{-1} and prove that A is continuously invertible.

8.23 Let $X_3 = \{x \in C^3([0,1]) : x(0) = x'(0) = x''(0) = 0\}$, $\|x\|_3 = \sum_{k=0}^{3} \max_{t\in[0,1]} |x^{(k)}(t)|$, the operator $A : X_3 \to C([0,1])$ is defined by the formula $(Ax)(t) = x'''(t) - x''(t),\ t \in [0,1]$. Find A^{-1} and prove that A is continuously invertible.

8.24° Is the operator $A : l_2 \to l_2$ continuously invertible if

(1) $Ax = (x_1, x_3, x_2, x_4, x_5, \ldots)$;
(2) $Ax = (x_1 + x_2, x_2 - x_3, x_3, x_4, \ldots)$;
(3) $Ax = (x_1 + x_2, x_3, 2x_1 + 2x_2, x_4, \ldots)$;
(4) $Ax = (0, x_1, x_2, \ldots)$;
(5) $Ax = (x_2, x_3, x_4, x_5, \ldots)$;
(6) $Ax = (x_1 - x_2, x_2, x_3, \ldots)$;
(7) $Ax = (x_1 + x_2 + x_3, x_1 - x_2 + x_3, x_1 + x_2 - x_3, x_4, x_5, \ldots)$?

Find the operator A^{-1}, if it exists.

8.25° Let $A : C([0,1]) \to C([0,1]),\ (Ax)(t) = \int_0^1 (ts + t^2s^2)x(s)ds$. Prove that the operator A^{-1} does not exist.

8.26° Let $A : C([-1,1]) \to C([-1,1]),\ (Ax)(t) = \frac{1}{2}(x(t) + x(-t))$. Does A^{-1} exist?

8.27 Let X be a Banach space with respect to the norms $\|\cdot\|_1$ and $\|\cdot\|_2$. Assume that $\|\cdot\|_2$ can be estimated by $\|\cdot\|_1$, i.e., there exists $c > 0$ such that $\|x\|_2 \le c\|x\|_1,\ x \in X$. Prove that the norms $\|\cdot\|_1$ and $\|\cdot\|_2$ are equivalent. Is the completeness of the space in both norms essential?

8.28 What conditions on the function $\alpha \in C([a,b])$ imply that

(1) $\|x\|_\alpha = \max_{a\le t\le b} \alpha(t)|x(t)|$ is a norm in $C([a,b])$?
(2) Convergence in the norm $\|x\|_\alpha$ is equivalent to the uniform convergence in $C([a,b])$?
(3) The space $C([a,b])$ with the norm $\|x\|_\alpha$ is complete?

8.29 Let $\mu(T) < +\infty$, M be a subspace of $L_2(T,\mu)$, and $M \subset L_\infty(T,\mu)$. Prove that

(1) $\exists C > 0\ \forall x \in M : \|x\|_\infty \le C\|x\|_2$;
(2)* $\dim M < +\infty$.

8.30* Let $\{e_n : n \ge 1\}$ be a Schauder basis in a Banach space X with a norm $\|\cdot\|$ (see problem 1.89). For $x = \sum_{i=1}^{\infty} \alpha_i(x)e_i$ we put $P_kx = \sum_{i=1}^{k} \alpha_i(x)e_i,\ k \ge 1$, and $|||x||| = \sup_{k\ge1} \|P_kx\|,\ x \in X$. Prove that

(1) $(X, |||\cdot|||)$ is a Banach space;

(2) The norms $|||\cdot|||$ and $\|\cdot\|$ are equivalent.
(3) $\alpha_i \in X^*$, $i \geq 1$, and $P_k \in \mathcal{L}(X)$, $k \geq 1$, and there exists $C > 0$ such that $\|P_k\| \leq C$ for all $k \geq 1$;
(4) $\exists C > 0\ \forall m > k\ \forall \lambda_1, \ldots, \lambda_m \in \mathbb{K}: \left\|\sum_{i=1}^{k} \lambda_i e_i\right\| \leq C \left\|\sum_{i=1}^{m} \lambda_i e_i\right\|.$

8.31 Let $\{e_n : n \geq 1\}$ be a Schauder basis in a Banach space X.

(1) Prove that a sequence $\{x_n = \sum_{i=1}^{\infty} \alpha_i(x_n)e_i : n \geq 1\} \subset X$ converges to $x = \sum_{i=1}^{\infty} \alpha_i(x)e_i$ in X if and only if the following conditions are fulfilled:
   (a) $\forall i \geq 1 : \alpha_i(x_n) \to \alpha_i(x),\ n \to \infty$;
   (b) $\sup_{n\geq 1} \left\|\sum_{i=k+1}^{\infty} \alpha_i(x_n)e_i\right\| \to 0,\ k \to \infty.$
(2) Derive from item (1) the convergence criterion in l_p, $1 \leq p < +\infty$.

8.32 Let $\{e_n : n \geq 1\}$ be a Schauder basis in a Banach space X, $\|e_n\| = 1,\ n \geq 1$, $\sup_{n\geq 1} \|P_n\| = C < +\infty$, where $P_n x = \sum_{i=1}^{n} \alpha_i(x)e_i$ for $x = \sum_{i=1}^{\infty} \alpha_i(x)e_i$. Prove that any sequence $\{v_n : n \geq 1\} \subset X$ such that $\sum_{n\geq 1} \|e_n - v_n\| < \frac{1}{2C}$ is also a Schauder basis in X.

Chapter 9
Classes of Linear Operators in Hilbert Space

Theoretical Background

Let H be a Hilbert space.

A function $b : H \times H \to \mathbb{K}$ is called a *bilinear form*, if it is linear with respect to the first variable and antilinear with respect to the second variable (in the case of $\mathbb{K} = \mathbb{R}$ it is linear with respect to each variable):

$$\forall x_1, x_2, y \in H\ \forall \alpha, \beta \in \mathbb{K} : b(\alpha x_1 + \beta x_2, y) = \alpha b(x_1, y) + \beta b(x_2, y),$$
$$\forall x, y_1, y_2 \in H\ \forall \alpha, \beta \in \mathbb{K} : b(x, \alpha y_1 + \beta y_2) = \overline{\alpha} b(x, y_1) + \overline{\beta} b(x, y_2).$$

A bilinear form b is called *Hermitian*, if $b(x, y) = \overline{b(y, x)},\ x, y \in H$. A bilinear form b is called *bounded*, if

$\exists\, C \geq 0\ \forall x, y \in H\ :\ |b(x, y)| \leq C\|x\| \cdot \|y\|$.

The smallest of such numbers $C \geq 0$ is called the *norm of the bilinear form b* and is denoted $\|b\|$. If b is a bilinear form, then the function $b[x] = b(x, x),\ x \in H$, is called a quadratic form.

Let $A \in \mathcal{L}(H)$. The bilinear form $b_A(x, y) = (Ax, y),\ x, y \in H$, is called the bilinear form induced by the operator A.

Theorem 9.1 *For every bounded bilinear form b there exists a unique operator $A \in \mathcal{L}(H)$ such that $b = b_A$, moreover $\|b\| = \|A\|$.*

The operator $A^* \in \mathcal{L}(H)$ is called the *adjoint operator* to an operator $A \in \mathcal{L}(H)$, if $(Ax, y) = (x, A^*y)$ for all $x, y \in H$. For every operator $A \in \mathcal{L}(H)$ the adjoint operator A^* exists and is unique, moreover $\|A^*\| = \|A\|$.

The operator $A \in \mathcal{L}(H)$ is called *self-adjoint*, if $A^* = A$, i.e. $(Ax, y) = (x, Ay)$ for every $x, y \in H$. The operator $A \in \mathcal{L}(H)$ is called *normal*, if $AA^* = A^*A$. A self-adjoint operator $A \in \mathcal{L}(H)$ is called *nonnegative*, if $(Ax, x) \geq 0$ for all $x \in H$; this property is denoted $A \geq 0$. If self-adjoint operators $A, B \in \mathcal{L}(H)$ are such that $A - B \geq 0$, then we say that $A \geq B$, or $B \leq A$. A self-adjoint operator $A \in \mathcal{L}(H)$

V. Brayman et al., *Functional Analysis and Operator Theory*, Problem Books in Mathematics, https://doi.org/10.1007/978-3-031-56427-7_9

is called *bounded from below* by a number $c \in \mathbb{R}$ (*bounded from above* by a number $d \in \mathbb{R}$), if $A \geq cI$ ($A \leq dI$, respectively).

Let G be a subspace of H. The operator $P_G : H \to H$, where $P_G x = \text{pr}_G x$, is called the operator of orthogonal projection (or *orthoprojector*).

Let H_1, H_2 be Hilbert spaces. A linear operator $V : H_1 \to H_2$ is called *isometric*, if it preserves the scalar product (i.e., $(Vx, Vy)_2 = (x, y)_1,\ x, y \in H_1$). A linear operator $U : H_1 \to H_2$ is called *unitary*, if U is isometric and surjective (i.e., $R(U) = H_2$).

The notion of an adjoint operator can also be introduced for operators that act in normed linear spaces. Let X, Y be normed linear spaces, $A \in \mathcal{L}(X, Y)$. The operator $A' : Y^* \to X^*$, which is defined by the formula $(A'f)(x) = f(Ax),\ f \in Y^*,\ x \in X$, is called the adjoint operator of the operator A.

Theorem 9.2 *This definition is correct, i.e. it defines the operator* $A' \in \mathcal{L}(Y^*, X^*)$ *uniquely. Moreover* $\|A'\| = \|A\|$.

Remark Let H be a Hilbert space and $T : H \to H^*$ be an antilinear isometry which assigns to each element $a \in H$ the functional $(Ta)(x) = (x, a),\ x \in H$. Then for an operator $A \in \mathcal{L}(H)$ operators $A^* \in \mathcal{L}(H)$ and $A' \in \mathcal{L}(H^*)$ are connected by the formula $A' = TA^*T^{-1}$.

Examples of Problems with Solutions

9.1 Let H be a Hilbert space, $A \in \mathcal{L}(H)$. Prove that

(1) $A^{**} = (A^*)^* = A$; 2) $\|A^*\| = \|A\| = \sqrt{\|A^*A\|} = \sqrt{\|AA^*\|}$;
(2) if $A = A^*$, then $\|A^2\| = \|A\|^2$.

Solution

(1) For all $x, y \in H$ we have

$$(A^*x, y) = \overline{(y, A^*x)} = \overline{(Ay, x)} = (x, Ay).$$

From here we can conclude that $A^{**} = A$. Let us explain this once in detail. Since $(A^*x, y) = (x, A^{**}y)$, then $(x, A^{**}y) = (x, Ay),\ x, y \in H$. For $x = A^{**}y - Ay$ we get $\|A^{**}y - Ay\|^2 = 0$. Therefore $A^{**}y = Ay$ for all $y \in H$, i.e., $A^{**} = A$.

(2) If $A = 0$, then $A^* = 0$ and the equalities are obviously fulfilled. Consider $A \neq 0$. For all $x \in H$ we have

$$\|Ax\|^2 = (Ax, Ax) = |(A^*Ax, x)| \leq$$
$$\leq \|A^*Ax\| \cdot \|x\| \leq \|A^*A\| \cdot \|x\|^2 \leq \|A^*\| \cdot \|A\| \cdot \|x\|^2.$$

Hence $\|A\|^2 \le \|A^*A\| \le \|A^*\| \cdot \|A\|$, thus $\|A\| \le \|A^*\|$. To prove the opposite inequality, we use (1) and the already proven inequality: $\|A^*\| \le \|A^{**}\| = \|A\|$. Hence $\|A^*\| = \|A\|$ and $\|A^*A\| = \|A\|^2$. It remains to apply the last equality to the operator A^* to get $\|AA^*\| = \|A^*\|^2$.

(3) It follows immediately from (2).

9.2 Prove the following statements.

(1) If (T, μ) is a space with a σ-finite measure and A is an integral operator in the complex Hilbert space $L_2(T, \mu)$ with a kernel $K \in L_2(T \times T, \mu \times \mu)$, i.e.

$$(Ax)(t) = \int_T K(t,s)x(s)d\mu(s), \ t \in T, \ x \in L_2(T, \mu),$$

then A^* is an integral operator with a kernel $K^*(t,s) = \overline{K(s,t)}, \ (t,s) \in T \times T$.

(2) If H is a complex separable Hilbert space and the operator A is represented by the matrix $(a_{jk})_{j,k=1}^{\infty}$, then the operator A^* is represented by the matrix $(a^*_{jk})_{j,k=1}^{\infty}$, where $a^*_{jk} = \overline{a_{kj}}, \ j,k \in \mathbb{N}$.

Solution

(1) For arbitrary $x, y \in L_2(T, \mu)$ we will find an element $y^* \in L_2(T, \mu)$ such that $(Ax, y) = (x, y^*)$. Then $y^* = A^*y$. We have

$$(Ax, y) = \int_T (Ax)(t)\overline{y(t)}dt = \int_T \left(\int_T K(t,s)x(s)ds \right) \overline{y(t)}dt =$$

$$= \int_T \left(\int_T K(t,s)x(s)\overline{y(t)}dt \right) ds = \int_T x(s) \overline{\left(\int_T \overline{K(t,s)}y(t)dt \right)} ds$$

(one can change the order of integration by Fubini's theorem). From this equality it follows similarly to the solution of item (1) of problem 9.1 that $(A^*y)(s) = y^*(s) = \int_T \overline{K(t,s)}y(t)dt$. By swapping the variables t and s, we get that A^* is an integral operator with the kernel $K^*(t,s) = \overline{K(s,t)}, \ (s,t) \in T \times T$.

(2) The adjoint operator A^* is represented by some matrix $(a^*_{jk})_{j,k=1}^{\infty}$, where $a^*_{jk} = (A^*e_k, e_j), \ j,k \in \mathbb{N}$. Since

$$(A^*e_k, e_j) = (e_k, Ae_j) = \overline{(Ae_j, e_k)} = \overline{a_{kj}},$$

we have $a^*_{jk} = \overline{a_{kj}}, \ j,k \in \mathbb{N}$.

Remark 1 A kernel $K \in L_2(T \times T, \mu \times \mu)$ is called *Hermitian* if $K(t,s) = \overline{K(s,t)}$ (mod $\mu \times \mu$). It follows from problem 9.2, item (1) that the integral operator with a Hermitian kernel is self-adjoint.

Remark 2 In the real space $K^*(t,s) = K(s,t),\ (t,s) \in T \times T$, and $a^*_{jk} = a_{kj},\ j,k \in \mathbb{N}$, respectively.

9.3 Find the adjoint to the operator $A : H \to H$ if

(1) $H = l_2,\ Ax = (0, x_1, x_2, \ldots);$
(2) $H = L_2([0,1]),\ (Ax)(t) = \int\limits_0^t x(s)ds.$

Solution

(1) For any $x \in l_2$ and $y \in l_2$ we have to find an element $y^* \in l_2$ such that $(Ax, y) = (x, y^*)$. Note that

$$(Ax, y) = 0 \cdot \overline{y_1} + x_1 \cdot \overline{y_2} + x_2 \cdot \overline{y_3} + \ldots = x_1\overline{y_2} + x_2\overline{y_3} + \ldots$$

Therefore for $y^* = (y_2, y_3, \ldots)$ we have $(Ax, y) = (x, y^*)$. Hence $A^*y = (y_2, y_3, \ldots),\ y \in l_2$.

Remark For the operator $A : l_p \to l_p,\ 1 \le p < +\infty$, the adjoint operator A' is defined by the same formula, but $A' : l_q \to l_q$, where q is the conjugate index to p.

(2) For each $x \in L_2([0,1])$ and each $y \in L_2([0,1])$ we have to find an element $y^* \in L_2([0,1])$ such that $(Ax, y) = (x, y^*)$. We have

$$(Ax, y) = \int\limits_0^1 (Ax)(t)\overline{y(t)}dt = \int\limits_0^1 \left(\int\limits_0^t x(s)\overline{y(t)}ds \right) dt.$$

Now we change the order of integration and get

$$(Ax, y) = \int\limits_0^1 \left(\int\limits_s^1 x(s)\overline{y(t)}dt \right) ds = \int\limits_0^1 x(s)\overline{\left(\int\limits_s^1 y(t)dt \right)} ds,$$

therefore $y^*(s) = \int\limits_s^1 y(t)dt$. By swapping the variables t and s, we conclude that $(A^*y)(t) = \int\limits_t^1 y(s)ds,\ t \in [0,1],\ y \in L_2([0,1])$.

9.4 Let H be a Hilbert space. Prove the following statements.

(1) If $U : H \to H,\ R(U) = H$ and $(Ux, Uy) = (x, y),\ x, y \in H$, then U is a continuous linear operator (and hence, U is a unitary operator) and $\|U\| = 1$.

(2) An operator $U \in \mathcal{L}(H)$ is unitary if and only if it is continuously invertible and $U^{-1} = U^*$; in this case U^* is also a unitary operator.

Solution

(1) Let us check that U is a linear operator, i.e. $U(\alpha x) = \alpha Ux$ and $U(x+y) = Ux + Uy$ for all $x, y \in H$ and $\alpha \in \mathbb{K}$.

Using the properties of scalar product and the equality from the problem statement, we get

$$\|U(\alpha x) - \alpha Ux\|^2 = (U(\alpha x) - \alpha Ux, U(\alpha x) - \alpha Ux) =$$
$$= (U(\alpha x), U(\alpha x)) - \alpha(Ux, U(\alpha x)) - \overline{\alpha}(U(\alpha x), Ux) + \alpha\overline{\alpha}(Ux, Ux) =$$
$$= (\alpha x, \alpha x) - \alpha(x, \alpha x) - \overline{\alpha}(\alpha x, x) + \alpha\overline{\alpha}(x, x) = 0,$$

hence $U(\alpha x) - \alpha Ux = 0$. Quite similarly it can be verified that $\|U(x+y) - Ux - Uy\|^2 = 0$. Thus $U(x+y) = Ux + Uy$, so U is a linear operator.

For $x = y$ we get $(Ux, Ux) = (x, x)$, therefore $\|Ux\| = \|x\|$, $x \in H$. It follows that U is a continuous operator and $\|U\| = 1$.

(2) *Necessity.* Let U be a unitary operator, i.e. $R(U) = H$ and $\|Ux\| = \|x\|$, $x \in H$. According to the criterion of continuous invertibility, it follows that there exists $U^{-1} \in \mathcal{L}(H)$.

Since $(x, U^*y) = (Ux, y) = (Ux, UU^{-1}y) = (x, U^{-1}y)$ for all $x, y \in H$, then $U^* = U^{-1}$ (see the solution of item (1) of problem 9.1).

Now let us verify that $U^* = U^{-1}$ is a unitary operator. Indeed, we have

$$(U^{-1}x, U^{-1}y) = (UU^{-1}x, UU^{-1}y) = (x, y), \; x, y \in H,$$

and the range of the operator U^{-1} is the same as the domain of the operator U, i.e. $R(U^{-1}) = H$.

Sufficiency. Let $U^{-1} \in \mathcal{L}(H)$ and $U^{-1} = U^*$. Then $R(U) = H$, because U is bijective, and also $(Ux, Uy) = (x, U^*Uy) = (x, U^{-1}Uy) = (x, y)$, $x, y \in H$.

9.5 Let H be a Hilbert space. Prove the following statements.

(1) If $P : H \to H$ is an orthogonal projector onto a subspace $G \subset H$, then $P \in \mathcal{L}(H)$, and if $G \neq \{0\}$, then $\|P\| = 1$.
(2) Operator $P \in \mathcal{L}(H)$ is an orthogonal projector onto some subspace $G \subset H$ if and only if $P^* = P$ and $P^2 = P$. In this case $G = \{x \in H \; : \; Px = x\}$.

Solution

(1) Let us show that P is a linear operator. Let $x, y \in H$ and $\alpha, \beta \in \mathbb{K}$. By Theorem 2.3 we have

$$\alpha x + \beta y = \alpha(\mathrm{pr}_G x + \mathrm{pr}_{G^\perp} x) + \beta(\mathrm{pr}_G y + \mathrm{pr}_{G^\perp} y) =$$
$$= (\alpha \mathrm{pr}_G x + \beta \mathrm{pr}_G y) + (\alpha \mathrm{pr}_{G^\perp} x + \beta \mathrm{pr}_{G^\perp} y) =: z_1 + z_2.$$

Since $z_1 \in G,\ z_2 \in G^\perp$, then due to the uniqueness of the decomposition of a Hilbert space we have $z_1 = \mathrm{pr}_G(\alpha x + \beta y),\ z_2 = \mathrm{pr}_{G^\perp}(\alpha x + \beta y)$. The first of these equalities means that P is a linear operator.

From the equality $\|\mathrm{pr}_G x\|^2 + \|\mathrm{pr}_{G^\perp} x\|^2 = \|\mathrm{pr}_G x + \mathrm{pr}_{G^\perp} x\|^2 = \|x\|^2$ we obtain that $\|\mathrm{pr}_G x\| \le \|x\|,\ x \in H$, hence $\|Px\| \le \|x\|,\ x \in H$. Thus P is a continuous operator and $\|P\| \le 1$. If $G = \{0\}$, then $Px = 0,\ x \in H$, hence $P = 0$ and $\|P\| = 0$. If $G \ne \{0\}$, then there exists $g \in G,\ g \ne 0$. Then $Pg = g$ and $\frac{\|Pg\|}{\|g\|} = 1$, hence $\|P\| = 1$.

(2) *Necessity*. Let P be an orthogonal projector onto a subspace G. For arbitrary $x, y \in H$ there exist $x_1, y_1 \in G$, and $x_2, y_2 \in G^\perp$ such that $x = x_1 + x_2$ and $y = y_1 + y_2$. Since $y_2 \perp x_1,\ x_2 \perp y_1$, then

$$(Px, y) = (x_1, y_1 + y_2) = (x_1, y_1),\ (x, Py) = (x_1 + x_2, y_1) = (x_1, y_1).$$

Thus $(Px, y) = (x, Py)$, hence $P^* = P$.

Further, $P^2 x = Px_1 = x_1 = Px$, hence $P^2 = P$.

Sufficiency. Let $P^* = P$ and $P^2 = P$. Consider the set $G = \{x \in H : Px = x\}$. Then $G = \mathrm{Ker}(P - I)$, hence G is a subspace of H. Now let us prove that $Px = \mathrm{pr}_G x,\ x \in H$. Since $P(Px) = Px$, then $Px \in G$. Furthermore, $(x - Px) \perp G$. Indeed, for all $y \in G$ we have $(x - Px, y) = (x - Px, Py) = (Px - P^2 x, y) = 0$. Thus P is an orthogonal projector onto G.

Problems to Solve

9.6° Let H be a Hilbert space, $A, B \in \mathcal{L}(H)$. Prove the following statements:

(1) $I^* = I,\ 0^* = 0$;
(2) $(\alpha A + \beta B)^* = \overline{\alpha} A^* + \overline{\beta} B^*,\ \alpha, \beta \in \mathbb{C}$;
(3) $(AB)^* = B^* A^*$;
(4) if $A = A^*,\ B^* = B$, then $(AB)^* = AB \Leftrightarrow AB = BA$;
(5) if $A^{-1} \in \mathcal{L}(H)$ exists, then $(A^*)^{-1} \in \mathcal{L}(H)$ also exists, and $(A^*)^{-1} = (A^{-1})^*$.

9.7° Let H be a Hilbert space over the field $\mathbb{K}$, $b : H \times H \to \mathbb{K}$ be a bilinear form, $b[x] = b(x, x),\ x \in H$, be a corresponding quadratic form. Prove that

(1) if $\mathbb{K} = \mathbb{C}$, then

$$4b(x, y) = b[x + y] - b[x - y] + ib[x + iy] - ib[x - iy],\ x, y \in H$$

(polarization identity);

(2) if $\mathbb{K} = \mathbb{R}$, then the form b is symmetric (i.e., $b(x, y) = b(y, x),\ x, y \in H$) if and only if $4b(x, y) = b[x + y] - b[x - y],\ x, y \in H$.

9.8 Prove that the continuity of a bilinear form b is equivalent to each of the following conditions:

(1) the bilinear form b is bounded;
(2) there exists an operator $A \in \mathcal{L}(H)$ such that $b(x, y) = (Ax, y),\ x, y \in H$; in this case $\|b\| = \|A\|$;
(3) there exists an operator $C \in \mathcal{L}(H)$ such that $b(x, y) = (x, Cy),\ x, y \in H$; in this case $\|b\| = \|C\|$.

9.9 Let $b : H \times H \to \mathbb{C}$ be a bilinear form on a complex Hilbert space H, $b[x] = b(x, x),\ x \in H$. Prove that

(1) $b(x, y) = \frac{1}{N} \sum\limits_{k=1}^{N} b[x + e^{\frac{2\pi ik}{N}} y]e^{\frac{2\pi ik}{N}},\ x, y \in H,\ N \in \mathbb{N},\ N \geq 3$;
(2) $b(x, y) = \frac{1}{2\pi} \int\limits_{0}^{2\pi} b[x + e^{i\varphi} y]e^{i\varphi} d\varphi,\ x, y \in H.$

9.10° Prove that in a complex Hilbert space the following statements are equivalent:

(1) operator $A \in \mathcal{L}(H)$ is self-adjoint;
(2) the bilinear form b_A, induced by the operator A, is Hermitian;
(3) the quadratic form $b_A[x] = b_A(x, x),\ x \in H$, attains only real values.

Prove that in a real Hilbert space the operator $A \in \mathcal{L}(H)$ is self-adjoint if and only if the bilinear form b_A is symmetric.

9.11°

(1) Let H be a complex Hilbert space, $A \in \mathcal{L}(H)$. Prove that $A = 0$ if and only if $(Ax, x) = 0$ for all $x \in H$.
(2) Provide an example of a real Hilbert space H and a nonzero operator $A \in \mathcal{L}(H)$ such that $(Ax, x) = 0$ for all $x \in H$.

9.12° Let H be a Hilbert space over the field $\mathbb{K}$ and $A, B, C, D \in \mathcal{L}(H)$. Prove that

(1) if $A^* = A$, then $A^2 \geq 0$;
(2) the condition $A^* = A$ in item (1) is essential;

(3) if $A \le B,\ B \le A$, then $A = B$;
(4) if $A \le B,\ B \le C$, then $A \le C$;
(5) if $A \le C,\ B \le D$, then $A + B \le C + D$;
(6) if $A \le B,\ \lambda \ge 0$, then $\lambda A \le \lambda B$;
(7) if $A \le B$, then $-B \le -A$;
(8) if $A^* = A,\ B^* = B$ and $AB = BA$, then $A^2 + B^2 \ge 2AB$;
(9) if $A^* = A$, then there exist $m, M \in \mathbb{R}$ such that $mI \le A \le MI$;
(10) if $A^* = A,\ B^* = B$, then it is not necessarily that $A \le B$ or $B \le A$;
(11) if $A^* = A,\ A \ge 0$ and $A^{-1} \in \mathcal{L}(H)$, then $A^{-1} \ge 0$;
(12) if A and B are unitary operators, then AB is also a unitary operator;
(13) if $A^* = A,\ B^* = B,\ A \ge B$, then the operators $C^*AC,\ C^*BC$ are self-adjoint and $C^*AC \ge C^*BC$;
(14) if $A^2 = A$, then $R(A) = \operatorname{Ker}(I - A)$, in particular, the set $R(A)$ is a subspace of H;
(15) if $\mathbb{K} = \mathbb{C},\ \alpha, \beta \in \mathbb{C},\ |\alpha| = |\beta| = 1$, then the operator $\alpha A + \beta A^*$ is normal;
(16) if A is an orthoprojector, then $A \ge 0$;
(17) if A is an orthoprojector, then $I - A$ is also an orthoprojector;
(18) if $\mathbb{K} = \mathbb{C},\ A = A^*,\ B = B^*$, then the operator $C = i(AB - BA)$ is self-adjoint.

9.13° Let H be a Hilbert space, $A \in \mathcal{L}(H)$. Prove that

(1) A is isometric $\Leftrightarrow$ $\|Ax\| = \|x\|,\ x \in H$;
(2) if H is a finite-dimensional space, then A is unitary if and only if A is isometric;
(3) A is isometric $\Leftrightarrow$ $A^*A = I$;
(4) A is unitary $\Leftrightarrow$ $A^*A = I$ and $AA^* = I$;
(5) if H is an infinite-dimensional space, there exists A which is isometric but not unitary.

9.14° Let a linear operator A in Hilbert space $\mathbb{C}^m$ be represented by a matrix $\mathcal{A} = (a_{jk})_{j,k=1}^m$, let $\mathcal{A}^* = \overline{\mathcal{A}}^T$ be the transposed complex conjugate matrix, and $\mathcal{I}$ be the identity matrix. Prove that

(1) A is self-adjoint $\Leftrightarrow$ $\mathcal{A}^* = \mathcal{A}$;
(2) A is unitary $\Leftrightarrow$ A is isometric $\Leftrightarrow$ $\mathcal{A}^*\mathcal{A} = \mathcal{I}$;
(3) A is nonnegative $\Leftrightarrow$ $\mathcal{A}^* = \mathcal{A}$ and all roots of the equation $\det(\mathcal{A} - \lambda\mathcal{I}) = 0$ are nonnegative;
(4) if operator A in Hilbert space $\mathbb{R}^m$ is represented by matrix $\mathcal{A}$ and $\mathcal{A}^* = \mathcal{A}^T$ is the transposed matrix, then statements (1)–(3) are true, but the condition $(Ax, x) \ge 0,\ x \in \mathbb{R}^m$, does not imply that operator A is self-adjoint, because it is not necessarily $\mathcal{A}^* = \mathcal{A}$.

9.15° Find the adjoint operator of an operator $A : l_2 \to l_2$.

(1) $Ax = (\alpha_1 x_1, \alpha_2 x_2, \ldots)$;
(2) $Ax = (x_2, x_3, \ldots)$;
(3) $Ax = (x_1, \ldots, x_j, 0, 0, \ldots)$;

(4) $Ax = (\underbrace{0, \dots, 0}_{j-1}, x_1, 0, 0, \dots);$
(5) $Ax = (\underbrace{x_1, \dots, x_1}_{j}, 0, 0, \dots);$
(6) $Ax = (x_2, x_1, x_4, x_3, x_6, x_5, \dots);$
(7) $Ax = (\alpha_j x_j, \alpha_{j+1} x_{j+1}, \dots);$
(8) $Ax = (0, 0, \alpha_1 x_1, \alpha_2 x_2, \dots);$
(9) $Ax = (2x_1 + 3x_2, 3x_1 + 5x_2, x_3, 0, 0, \dots);$
(10) $Ax = (x_1, 4x_2 + 7x_3, 7x_2 + 2x_3, 0, x_5, 0, 0, \dots);$
(11) $Ax = (3x_1 - 2x_2, x_2, x_3, x_4, \dots),$

where $j \in \mathbb{N}$ is fixed, and $\{\alpha_n : n \geq 1\} \subset \mathbb{C}$ is a fixed bounded sequence.

9.16° Find the adjoint operator of an operator $A : H \to H$.

(1) $H = L_2(\mathbb{R}),\ (Ax)(t) = \chi_{[-\alpha,\alpha]}(t)x(t)$, where $\alpha > 0$;
(2) $H = L_2([0, 1]),\ (Ax)(t) = tx(t)$;
(3) $H = L_2(\mathbb{R}),\ (Ax)(t) = a(t)x(t + \tau)$, where $a \in L_\infty(\mathbb{R})$ is a given function, $\tau \in \mathbb{R}$;
(4) $H = L_2(\mathbb{R}),\ (Ax)(t) = \frac{1}{2}(x(t) + x(-t))$;
(5) $H = L_2([0, 1]),\ (Ax)(t) = \int\limits_0^1 e^{t+s}x(s)ds$;
(6) $H = L_2([0, 1]),\ (Ax)(t) = t^3 \int\limits_0^{t^2} x(s)ds$;
(7) $H = L_2([0, 1]),\ (Ax)(t) = \int\limits_0^1 t^2 s^3 x(s)ds$;
(8) $H = L_2([0, 1]),\ (Ax)(t) = \int\limits_{t^5}^{t} tsx(s)ds$;
(9) $H = L_2([-1, 1]),\ (Ax)(t) = \int\limits_{t^2}^{1} (t - 3s^2)x(s)ds$;
(10) $H = L_2([0, 1]),\ (Ax)(t) = x(t^\alpha)$, where $0 < \alpha \leq 1$;
(11) $Ax = (x, y)z$, where $y, z \in H$ are fixed elements.

9.17° Find the adjoint operator of an operator $A \in \mathcal{L}(H)$. Prove that A is a unitary operator.

(1) $H = L_2(\mathbb{R}),\ (Ax)(t) = \sqrt{7}t^3 x(t^7 - 1)$;
(2) $H = L_2([0, \infty)),\ (Ax)(t) = 2t\sqrt{t}x(t^4)$.

9.18° Let $\{\alpha_n : n \geq 1\} \subset \mathbb{C}$ be a fixed bounded sequence,

$$Ax = (\alpha_1 x_1, \alpha_2 x_2, \dots),\ x = (x_1, x_2, \dots) \in l_2.$$

Express the following statements in terms of the numbers $\alpha_n,\ n \geq 1$:

(1) A is self-adjoint;
(2) A is normal;
(3) A is unitary;
(4) A is isometric;
(5) A is an orthogonal projector;
(6) $\operatorname{Ker} A = \{0\}$.

9.19° Let $(T, \mathcal{F}, \mu)$ be a space with a σ-finite measure, A be an operator of multiplication by the function $p \in L_\infty(T, \mu)$ in the space $L_2(T, \mu)$. Prove that

(1) $A = A^* \ \Leftrightarrow$ the function p is almost everywhere real;
(2) A is unitary $\Leftrightarrow\ |p| = 1$ almost everywhere;
(3) $A \geq 0 \ \Leftrightarrow\ p \geq 0$ almost everywhere;
(4) A is an orthogonal projector $\Leftrightarrow\ p^2 = p$ almost everywhere, i.e. $p = \chi_C \pmod{\mu}$, where $C \in \mathcal{F}$ is a measurable set.

9.20° Let $s \in \mathbb{R},\ (A_s x)(t) = x(t+s),\ t \in \mathbb{R},\ x \in L_2(\mathbb{R})$ be a shift operator.

(1) Find A_s^* and A_s^{-1} and prove that A_s is a unitary operator.
(2) Let $A = \frac{1}{2}(A_{s_0} + A_{s_1}),\ s_0, s_1 \in \mathbb{R}$. Prove that $A^* = A$ if and only if $s_0 + s_1 = 0$.

9.21° Under what conditions on a subspace of a Hilbert space does an orthoprojector onto this subspace have a continuous inverse?

9.22° Let H be a separable Hilbert space and $\{e_n : n \geq 1\}$ be an orthonormal basis in H.

(1) Find the orthogonal projector onto the linear span of the set $\{e_1, \dots, e_k\},\ k \in \mathbb{N}$.
(2) Prove that the operator $Ax = \sum\limits_{n=1}^{\infty} x_{\sigma(n)} e_n,\ x = \sum\limits_{n=1}^{\infty} x_n e_n \in H$, where $\sigma : \mathbb{N} \to \mathbb{N}$ is a fixed bijection, is a unitary operator.

9.23 Provide an example of a Hilbert space H and operators $A, B \in \mathcal{L}(H)$ such that

(1) A is normal, but not self-adjoint;
(2) $A^2 = A$, but A is not an orthogonal projector;
(3) A is a unitary operator, but not an orthogonal projector;
(4) $A^* = A$, but does not satisfy either of the inequalities $A \geq 0,\ A \leq 0$;
(5) $A^2 = I$, but does not satisfy either of the inequalities $A \geq 0,\ A \leq 0$;
(6) $A^* = A$, but the range $R(A)$ is not closed in H;
(7) $A^* = A$, but the set $\{(Ax, x) \mid x \in H,\ \|x\| = 1\}$ is not closed in $\mathbb{R}$;
(8) $A \geq 0,\ B \geq 0$, but AB is not positive;
(9) A is not normal;
(10) $A \neq 0$, but $A^2 = 0$;
(11) $A = A^*,\ B = B^*,\ A^2 = B^2$, but $AB \neq BA$.

9.24 Let H be a Hilbert space over the field $\mathbb{K},\ A \in \mathcal{L}(H),\ A \geq 0$. Prove that

(1) $|(Ax, y)|^2 \leq (Ax, x)(Ay, y),\ x, y \in H$;
(2) $\|Ax\|^2 \leq \|A\|(Ax, x),\ x \in H$.

9.25 Let $A \in \mathcal{L}(H)$. Is the set $G = \{x \in H \mid (Ax, x) = 0\}$ necessarily a subspace in the Hilbert space H, if (1) $A^* = A$; (2) $A \geq 0$?

9.26 Let H be a Hilbert space, $A \in \mathcal{L}(H)$, $A^* = A$, $x \in H$, $x \notin \operatorname{Ker} A$. Prove that

(1) $A^n x \neq 0$ for all $n \in \mathbb{N}$;
(2) the sequence $\left\{\alpha_n = \frac{\|A^{n+1}x\|}{\|A^n x\|} : n \geq 1\right\}$ converges.

9.27 Let H be a Hilbert space, $A \in \mathcal{L}(H)$. Prove that

(1) $\|A\| = \sup\limits_{x,y \in S(0,1)} |(Ax, y)|$;
(2) if $A^* = A$, then $\|A\| = \sup\limits_{x \in S(0,1)} |(Ax, x)|$;
(3) if $A^* = A$, then $A = 0$ if and only if $(Ax, x) = 0$ for all $x \in H$.

9.28 Let H be a complex Hilbert space, $A \in \mathcal{L}(H)$. Prove that

(1) the operators $\operatorname{Re} A = \frac{1}{2}(A + A^*)$, $\operatorname{Im} A = \frac{1}{2i}(A - A^*)$ are self-adjoint;
(2) the operator A can be uniquely represented in the form $A = B + iC$, where B and C are self-adjoint operators;
(3) the operator A is normal $\Leftrightarrow$ $\operatorname{Re} A \cdot \operatorname{Im} A = \operatorname{Im} A \cdot \operatorname{Re} A$;
(4) the operator A is unitary if and only if the following conditions are fulfilled: (a) A is normal; (b) $(\operatorname{Re} A)^2 + (\operatorname{Im} A)^2 = I$.

9.29 Let A, B be linear operators in Hilbert space H such that $(Ax, y) = (x, By)$, $x, y \in H$. Prove that $A \in \mathcal{L}(H)$ and $B = A^*$.

9.30° Let H be a Hilbert space, $A \in \mathcal{L}(H)$. Prove that $(R(A))^\perp = \operatorname{Ker} A^*$, $(\operatorname{Ker} A)^\perp = \overline{R(A^*)}$ (here the bar denotes closure).

9.31° Let H be a Hilbert space, $A \in \mathcal{L}(H)$. Prove that $AA^* = A^*A$ if and only if $\|Ax\| = \|A^*x\|$ for all $x \in H$.

9.32 Let H be a Hilbert space, $A \in \mathcal{L}(H)$ is a normal operator. Prove that

(1) $\operatorname{Ker} A = \operatorname{Ker} A^* = (R(A))^\perp$;
(2) $\|A^2\| = \|A\|^2$;
(3) if $A^2 = 0$, then $A = 0$;
(4) if $A^{-1} \in \mathcal{L}(H)$ exists, then A^{-1} is a normal operator.

9.33* Let H be a Hilbert space, $P \in \mathcal{L}(H)$, $P^2 = P$. Prove that the following statements are equivalent:

(1) P is an orthoprojector;
(2) $P = P^*$;
(3) $PP^* = P^*P$;
(4) $\operatorname{Ker} P = (R(P))^\perp$;
(5) $(Px, x) = \|Px\|^2$, $x \in H$.
(6) $(Px, x) \geq 0$, $x \in H$.

9.34* Let H_1 and H_2 be subspaces of a Hilbert space H, $H_0 = H_1 \cap H_2$, P_j be the orthogonal projector onto H_j, $j = 0, 1, 2$. Denote $H_i \ominus H_0 = \{x \in H_i \mid x \perp H_0\}$, $i = 1, 2$. Prove the following statements:

(1) $P_1 + P_2$ is an orthogonal projector $\Leftrightarrow$ $P_1 P_2 = 0$ $\Leftrightarrow$ $H_1 \perp H_2$; in this case $P_1 + P_2$ is an orthogonal projector onto $H_1 \oplus H_2$;
(2) $P_1 P_2$ is an orthogonal projector $\Leftrightarrow$ $P_1 P_2 = P_2 P_1$ $\Leftrightarrow$
$\Leftrightarrow$ $(H_1 \ominus H_0) \perp (H_2 \ominus H_0)$; in this case $P_1 P_2$ is an orthogonal projector onto H_0;
(3) $P_1 - P_2$ is an orthogonal projector $\Leftrightarrow$ $P_1 \geq P_2$ $\Leftrightarrow$ $H_1 \supset H_2$; in this case $P_1 - P_2$ is an orthogonal projector onto $H_1 \ominus H_0$;
(4) $P_1 P_2 = P_2 \Leftrightarrow P_2 P_1 = P_2 \Leftrightarrow P_1 \geq P_2 \Leftrightarrow H_1 \supset H_2$.
For an operator $A \in \mathcal{L}(H)$ express in terms of algebraic relations between A and P_1 the following statements:
(5) H_1 is invariant with respect to A (i.e., $A(H_1) \subset H_1$);
(6) both H_1 and $H_1^{\perp}$ are invariant with respect to A.

9.35° Let H be a Hilbert space and $\{A, A_n : n \geq 1\} \subset \mathcal{L}(H)$. Prove that

(1) $A_n \rightrightarrows A \Leftrightarrow A_n^* \rightrightarrows A^*$;
(2) $A_n \xrightarrow{w} A \Leftrightarrow A_n^* \xrightarrow{w} A^*$;
(3) $A_n \xrightarrow{s} A \not\Leftrightarrow A_n^* \xrightarrow{s} A^*$.

9.36 Let H be a Hilbert space and $\{A, A_n : n \geq 1\} \subset \mathcal{L}(H)$. Prove that

(1) if $A_n = A_n^*$, $n \geq 1$, $A_n \xrightarrow{w} A$, $n \to \infty$, then $A = A^*$;
(2) if $A_n \geq 0$, $n \geq 1$, $A_n \xrightarrow{w} A$, $n \to \infty$, then $A \geq 0$;
(3) if $A_n \xrightarrow{w} A$ and $\|A_n x\| \to \|Ax\|$ for all $x \in H$, then $A_n \xrightarrow{s} A$; in particular, if A, A_n, $n \geq 1$, are isometric, then $A_n \xrightarrow{w} A \Leftrightarrow A_n \xrightarrow{s} A$;
(4) if A_n, $n \geq 1$, and A are normal, $A_n \xrightarrow{s} A$, then $A_n^* \xrightarrow{s} A^*$;
(5) if A_n are isometric for each $n \geq 1$, and $A_n \xrightarrow{w} A$, then A is not necessarily isometric;
(6) if A_n are unitary for each $n \geq 1$, and $A_n \xrightarrow{w} A$, then A is not necessarily unitary.

9.37 Let H be a Hilbert space, $\{A_n : n \geq 1\} \subset \mathcal{L}(H)$ are such that $A_n \geq 0$, $n \geq 1$, and also there exists $C > 0$ such that $\|A_n\| \leq C$, $n \geq 1$. Prove that

$$A_n \xrightarrow{s} 0 \Leftrightarrow A_n \xrightarrow{w} 0 \Leftrightarrow (A_n x, x) \to 0 \text{ for all } x \in H.$$

9.38* Let H be a Hilbert space, $\{A_n : n \geq 1\} \subset \mathcal{L}(H)$ be a bounded sequence of self-adjoint operators such that $A_{n+1} \geq A_n$ for all $n \geq 1$. Prove that the sequence $\{A_n : n \geq 1\}$ converges strongly.

9.39 Let H be a Hilbert space, P_n be orthogonal projectors onto subspaces H_n, $n \geq 1$. Prove that

(1) if $P_{n+1} \geq P_n$, $n \geq 1$, then $P_n \xrightarrow{s} P$, where P is an orthogonal projector onto $\overline{\text{span}(\bigcup_{n=1}^{\infty} H_n)}$;
(2) if $P_{n+1} \leq P_n$, $n \geq 1$, then $P_n \xrightarrow{s} P$, where P is an orthogonal projector onto $\bigcap_{n=1}^{\infty} H_n$;
(3) if $H_k \perp H_j$, $k \neq j$, $k, j \geq 1$, then the series $\sum_{n=1}^{\infty} P_n$ converges strongly to the orthogonal projector onto $\overline{\text{span}(\bigcup_{n=1}^{\infty} H_n)}$;
(4) if $P_n \xrightarrow{s} P$, then P is an orthogonal projector;
(5) if the sequence $\{P_n : n \geq 1\}$ is monotone and $P_n \rightrightarrows P$, $n \to \infty$, then there exists $n_0 \in \mathbb{N}$ such that $P_n = P$, $n \geq n_0$. Is the monotonicity essential in this statement?
(6*) if $P_n \xrightarrow{w} P$, then P is not necessarily an orthoprojector;
(7) if $P_n \xrightarrow{w} P$ and P is an orthoprojector, then $P_n \xrightarrow{s} P$.

9.40* Let H be a Hilbert space over the field $\mathbb{K}$, and $A \in \mathcal{L}(H)$ be such that $AB = BA$ for all $B \in \mathcal{L}(H)$. Prove that $A = \gamma I$ for some $\gamma \in \mathbb{K}$.

9.41 Let H be a Hilbert space, $A \in \mathcal{L}(H)$ be a self-adjoint operator, $\lambda \in \mathbb{C}$. Prove that the operator $A - \lambda I$ is continuously invertible if and only if there exists $c > 0$ such that $\|(A - \lambda I)x\| \geq c\|x\|$, $x \in H$.

9.42 Let H be a Hilbert space, $A \in \mathcal{L}(H)$. Prove that

(1) if $A \geq 0$, then $A^{-1} \in \mathcal{L}(H)$ exists if and only if there exists $\lambda > 0$ such that $A \geq \lambda I$;
(2) there exists $(I + AA^*)^{-1} \in \mathcal{L}(H)$;
(3) if $A^* = A$, $m = \inf_{x \in S(0,1)} (Ax, x)$, $M = \sup_{x \in S(0,1)} (Ax, x)$, then the operator $A - \lambda I$ is continuously invertible for all $\lambda \in \mathbb{R} \setminus [m, M]$ and is not continuously invertible for $\lambda \in \{m, M\}$;
(4) if $A^* = A$, $\lambda = \alpha + \beta i \in \mathbb{C}$, then

$$\|(A - \lambda I)x\| \geq |\beta| \cdot \|x\|, \ x \in H;$$

(5) if $A^* = A$, $\lambda \in \mathbb{C} \setminus \mathbb{R}$, then there exists $(A - \lambda I)^{-1} \in \mathcal{L}(H)$.

9.43 Let H be a Hilbert space, $A \in \mathcal{L}(H)$. Prove that operator A is continuously invertible if and only if there exists $c > 0$ such that $AA^* \geq cI$ and $A^*A \geq cI$.

9.44 Let H be a Hilbert space and $A \in \mathcal{L}(H)$ be a self-adjoint operator. Prove that

(1) A^n is a self-adjoint operator for every $n \geq 1$;
(2) $\|A^n\| = \|A\|^n,\ n \geq 1$;
(3) if $A \geq 0$, then $A^n \geq 0$ for every $n \geq 1$.

9.45 Let H be a Hilbert space and $A \in \mathcal{L}(H)$ be a self-adjoint operator. Prove that the series $\sum\limits_{n=0}^{\infty} A^n$ converges in $\mathcal{L}(H)$ if and only if $\|A\| < 1$.

9.46° Let A be an operator in l_2, which is represented by a Jacobi matrix, i.e., a matrix of the form

$$\mathcal{A} = \begin{pmatrix} \alpha_1 & \beta_1 & 0 & 0 & 0 & \dots \\ \gamma_1 & \alpha_2 & \beta_2 & 0 & 0 & \dots \\ 0 & \gamma_2 & \alpha_3 & \beta_3 & 0 & \dots \\ \dots & \dots & \dots & \dots & \dots & \dots \end{pmatrix}, \text{ where } \{\alpha_j, \beta_j, \gamma_j : j \geq 1\} \subset \mathbb{K}.$$

Prove that the operator A is bounded if and only if

$$\sup_{k \geq 1}(|\alpha_k|^2 + |\beta_k|^2 + |\gamma_k|^2) < +\infty.$$

Under what conditions on $\mathcal{A}$ is the operator A self-adjoint? Consider the cases $\mathbb{K} = \mathbb{R}$ and $\mathbb{K} = \mathbb{C}$.

9.47 Prove that

(1) if $\{P_n : n \geq 0\}$ are Legendre polynomials (see item (2) of problem (2.38)), then

$$(n+1)P_{n+1}(t) - (2n+1)tP_n(t) + nP_{n-1}(t) = 0,\ t \in [-1, 1],\ n \geq 1;$$

(2) the operator $(Ax)(t) = tx(t),\ t \in [-1, 1],\ x \in L_2([-1, 1])$, is represented by a Jacobi matrix in the basis which consists of Legendre polynomials.

9.48 Let $\{e_n : n \geq 1\}$ be an orthonormal basis in a Hilbert space H. Prove that a matrix $(a_{jk})_{j,k=1}^{\infty},\ a_{jk} \in \mathbb{C},\ j \geq 1,\ k \geq 1$, represents a continuous linear operator in H if and only if the following conditions are fulfilled:

(a) the series $\sum\limits_{k=1}^{\infty} a_{jk}x_k$ converges for any $j \in \mathbb{N}$ and $x = \sum\limits_{k=1}^{\infty} x_k e_k \in H$;

(b) $\exists C > 0\ \forall x \in H : \sum\limits_{j=1}^{\infty} \left| \sum\limits_{k=1}^{\infty} a_{jk}x_k \right|^2 \leq C\|x\|^2$.

9.49 Prove that any of the following two conditions is necessary and sufficient for the matrix $(a_{jk})_{j,k=1}^{\infty}$ to represent an operator $A \in \mathcal{L}(H)$ in a given orthonormal basis:

(1) $\exists C > 0 \, \forall m, n \in \mathbb{N} \, \forall x_1, \ldots, x_n, y_1, \ldots, y_m \in \mathbb{C}$:

$$\left| \sum_{j=1}^{n} \sum_{k=1}^{m} a_{jk} x_j \overline{y_k} \right| \leq C \left(\sum_{j=1}^{n} |x_j|^2 \right)^{\frac{1}{2}} \left(\sum_{k=1}^{m} |y_k|^2 \right)^{\frac{1}{2}};$$

(2) $\exists C \geq 0 \, \forall m, n \in \mathbb{N} \, \forall x_1, \ldots, x_m \in \mathbb{C}$:

$$\sum_{j=1}^{n} \left| \sum_{k=1}^{m} a_{jk} x_k \right|^2 \leq C \sum_{k=1}^{m} |x_k|^2.$$

Prove that condition (2) is fulfilled in each of the following cases:

(a) $\sum\limits_{j=1}^{\infty} \sum\limits_{k=1}^{\infty} |a_{jk}|^2 < +\infty$; (b) $a_{jk} = \delta_{jk} a_j, \ \sup\limits_{j \geq 1} |a_j| < +\infty$.

9.50° Let $(a_{jk})_{j,k=1}^{\infty}$, $(b_{jk})_{j,k=1}^{\infty}$ be the matrices which represent operators $A, B \in \mathcal{L}(H)$ in an orthonormal basis $\{e_n : n \geq 1\}$. Find the matrices that represent the operators $A + B$, AB.

9.51° Let the operator $A \in \mathcal{L}(H)$ be represented by the matrix $(a_{jk})_{j,k=1}^{\infty}$ in some orthonormal basis of the complex Hilbert space H. Prove that

(1) A is self-adjoint $\Leftrightarrow$ the matrix $(a_{jk})_{j,k=1}^{\infty}$ is Hermitian;
(2) $A \geq 0$ $\Leftrightarrow$ the matrix $(a_{jk})_{j,k=1}^{\infty}$ is Hermitian and for every $n \in \mathbb{N}$ the matrix $(a_{jk})_{j,k=1}^{n}$ is nonnegatively definite;
(3) A is an orthogonal projector $\Leftrightarrow$ the matrix $(a_{jk})_{j,k=1}^{\infty}$ is Hermitian, and $a_{jk} = \sum\limits_{i=1}^{\infty} a_{ji} a_{ik}$, $j, k \in \mathbb{N}$;
(4) A is isometric $\Leftrightarrow$ $\{(a_{1k}, a_{2k}, \ldots, a_{jk}, \ldots) : k \geq 1\}$ is an orthonormal system in l_2.

Find the necessary and sufficient conditions for the matrix $(a_{jk})_{j,k=1}^{\infty}$ to represent a unitary operator in H.

9.52*

(1) Let $S \in \mathcal{L}(H)$ be a bounded operator in a Hilbert space H such that $\|S\| \leq 1$. Put $T_1 = 0$, $T_n = \frac{1}{2}(S + T_{n-1}^2)$, $n \geq 2$. Prove that

(a) for every $n \geq 2$ the operators T_n and $T_n - T_{n-1}$ are polynomials of S with nonnegative coefficients;
(b) $0 \leq T_{n-1} \leq T_n$ and $\|T_n\| \leq 1$ for every $n \geq 2$;

(c) there exists $T \in \mathcal{L}(H)$ such that $T_n \xrightarrow{s} T$, moreover $T \geq 0$, $\|T\| \leq 1$, $T_n^2 \xrightarrow{s} T^2$ and $(I - T)^2 = I - S$;
(d) T commutes with all operators that commute with S.

(2) Let $A \in \mathcal{L}(H)$ be a nonnegative operator in a Hilbert space H. Prove that there exists $B \in \mathcal{L}(H)$ such that $B \geq 0$ and $B^2 = A$ (the operator B is called the *square root* of the operator A and is denoted $B = \sqrt{A}$ or $B = A^{\frac{1}{2}}$), moreover B commutes with all operators that commute with A.
(3) Prove the uniqueness of the square root B of the nonnegative operator A.

9.53 For an operator $A \in \mathcal{L}(H)$ prove that $A \geq 0$ and find $\sqrt{A}$.

(1) $H = l_2,\ Ax = (\alpha_1 x_1, \alpha_2 x_2, \ldots)$, where $\alpha_k \geq 0,\ k \geq 1$;
(2) $H = l_2,\ Ax = (x_1, \ldots, x_j, 0, 0, \ldots)$, where $j \in \mathbb{N}$ is fixed,
(3) $H = l_2,\ Ax = (2x_1 + 3x_2, 3x_1 + 5x_2, x_3, 0, 0, \ldots)$;
(4) $H = L_2(\mathbb{R}),\ (Ax)(t) = \chi_{[-\alpha,\alpha]}(t)x(t)$, where $\alpha > 0$;
(5) $H = L_2([0, 1]),\ (Ax)(t) = tx(t)$;
(6) $H = L_2(\mathbb{R}),\ (Ax)(t) = \frac{1}{2}(x(t) + x(-t))$;
(7) $H = L_2([0, 1]),\ (Ax)(t) = \int_0^1 e^{t+\tau}x(\tau)d\tau$.

9.54* Prove that there does not exist an operator $B \in \mathcal{L}(l_2)$ such that $B^2 = A$ if

(1) $Ax = (x_2, x_3, \ldots),\ x \in l_2$;
(2) $Ax = (0, x_1, x_2, \ldots),\ x \in l_2$.

9.55 Let H be a Hilbert space, $A, B, C, D \in \mathcal{L}(H)$ are nonnegative operators that pairwise commute. Prove that

(1) $AB \geq 0$;
(2) if $A \leq B$, then $A^2 \leq B^2$;
(3) if $A \leq B$, then $AC \leq BC$;
(4) if $A \leq B$ and $C \leq D$, then $AC \leq BD$.

9.56 Let A be a nonnegative operator in a Hilbert space H. Prove that $\left\|\sqrt{A}\right\| = \sqrt{\|A\|}$.

9.57° Let A be a nonnegative operator in a Hilbert space H. Prove that the following statements are equivalent:

(1) $R(A)$ is dense in H;
(2) $\operatorname{Ker} A = \{0\}$;
(3) $(Ax, x) > 0$ for every $x \in H,\ x \neq 0$.

9.58 Let H be a Hilbert space, $A, B \in \mathcal{L}(H)$. Is it correct that

(1) if $AB = BA,\ A \geq 0,\ AB \geq 0$, then $B \geq 0$;
(2) if $A^2 = A$, then $A = 0$ or $A = I$;
(3) if $\sqrt{A}B = B\sqrt{A}$, then $AB = BA$?

9.59 Let H be a Hilbert space, $A \in \mathcal{L}(H)$. Prove that there exists a unique nonnegative operator $B \in \mathcal{L}(H)$ such that $\|Ax\| = \|Bx\|,\ x \in H$.

9.60 Let H be a Hilbert space, $A \in \mathcal{L}(H)$. Prove that

(1) if $0 \le A \le I$, then $A \le \sqrt{A} \le I$;
(2) if $I \le A$, then $I \le \sqrt{A} \le A$.

9.61* Let H be a Hilbert space, $A, B \in \mathcal{L}(H)$, $I \le A \le B$. Prove that the operators A, B are continuously invertible and $B^{-1} \le A^{-1}$.

9.62 Let H be a Hilbert space, $A \in \mathcal{L}(H)$, $|A| = \sqrt{A^*A}$. Prove that

(1) $|A| = 0 \Leftrightarrow A = 0$;
(2) $A \ge 0 \Leftrightarrow |A| = A$;
(3) the norms of the operators $|A|$ and A are equal;
(4) $\operatorname{Ker}|A| = \operatorname{Ker} A$, $\overline{R(|A|)} = \overline{R(A^*)}$.

9.63 Let $A \in \mathcal{L}(H)$ be a self-adjoint operator, $|A| = \sqrt{A^2}$. We put $A^+ = \frac{1}{2}(|A| + A)$, $A^- = \frac{1}{2}(|A| - A)$. Prove that

(1) $A^+A^- = A^-A^+ = 0$;
(2) $A^- = 0 \Leftrightarrow A = A^+ = |A|$;
(3) $A^+ = 0 \Leftrightarrow A = -A^- = -|A|$.

9.64 Let H be a Hilbert space and operator $A \in \mathcal{L}(H)$ be continuously invertible. Prove that

(1) operator $|A|$ is continuously invertible;
(2) operator $U = A|A|^{-1}$ is unitary.

Thus there is a representation $A = U|A|$, which is called the polar decomposition of A (in the case of $H = \mathbb{C}$ this decomposition is reduced to the representation of a complex number a in the form $a = re^{i\varphi}$).

9.65 Let H be a Hilbert space. An operator $U \in \mathcal{L}(H)$ is called *partially isometric* if $\|Ux\| = \|x\|,\ x \in (\operatorname{Ker} U)^{\perp}$.

(1) Prove that the operator U is partially isometric if and only if U^*U is an orthogonal projector on $H_1 = (\operatorname{Ker} U)^{\perp}$.
(2) Let the operator U be partially isometric. Prove that the set $H_2 = R(U)$ is closed, UU^* is an orthoprojector on H_2 and the operator U^* is also partially isometric.

9.66 Let H be a Hilbert space, $A \in \mathcal{L}(H)$. Representation $A = UB$, where operator $B \in \mathcal{L}(H)$ is nonnegative, operator $U \in \mathcal{L}(H)$ is partially isometric and $\operatorname{Ker} U = \operatorname{Ker} B$, is called the *polar decomposition* of A (compare with problem 9.64). Prove that the polar decomposition exists and is unique, moreover $B = |A|$.

9.67 Let $A = U|A|$ be the polar decomposition of an operator $A \in \mathcal{L}(H)$. Prove that $|A^*| = U|A|U^*,\ A^* = U^*|A^*|,\ A = |A^*|U,\ A^* = |A|U^*$.

9.68 Find the polar decomposition of an operator $A \in \mathcal{L}(l_2)$.

(1) $Ax = (\alpha_1 x_1, \alpha_2 x_2, \ldots)$, where $\{\alpha_n : n \ge 1\} \subset \mathbb{C}$ are given numbers;
(2) $Ax = (x_2, x_3, \ldots)$;
(3) $Ax = (x_1, \ldots, x_j, 0, 0, \ldots)$, where $j \in \mathbb{N}$ is fixed.

9.69 Let $A = U|A|$ be the polar decomposition of an operator $A \in \mathcal{L}(H)$.

(1) Under what conditions on A the operator U will be (a) isometric; (b) unitary?
(2) Let operator A be normal. Prove that the operators $|A|$ and U commute.
(3) Let $C \in \mathcal{L}(H)$ be a unitary operator such that $AC = CA$. Prove that $|A|$ and U commute with C.
(4) Is the statement of item (3) correct if C is not necessarily a unitary operator?

9.70 Let H be a complex Hilbert space, $A \in \mathcal{L}(H)$. Prove that

(1) if A is self-adjoint, then the operator $U = (A + \lambda I)(A + \overline{\lambda} I)^{-1},\ \lambda \in \mathbb{C}\backslash\mathbb{R}$, is unitary;
(2) if $(A - iI)^{-1} \in \mathcal{L}(H)$ exists and the operator $U = (A + iI)(A - iI)^{-1}$ is unitary, then A is self-adjoint;
(3) if the operator $U \in \mathcal{L}(H)$ is unitary and $(U - I)^{-1} \in \mathcal{L}(H)$ exists, then the operator $A = i(U + I)(U - I)^{-1}$ is self-adjoint.

9.71° Prove that

(1) if (T, μ) is a space with a finite measure μ and A is an integral operator in $L_p(T, \mu),\ 1 < p < +\infty$, with a kernel $K \in L_\infty(T \times T, \mu \times \mu)$, i.e.

$$(Ax)(t) = \int_T K(t, s)x(s)d\mu(s),\ t \in T,\ x \in L_p(T, \mu),$$

then A' is an integral operator in $L_q(T, \mu)$ with the kernel $\widetilde{K}(t, s) = K(s, t),\ (t, s) \in T \times T$;
(2) if $A \in \mathcal{L}(l_p),\ 1 < p < +\infty$, is represented by the matrix $(a_{jk})_{j,k=1}^\infty$, then $A' \in \mathcal{L}(l_q)$ is represented by the matrix $(\widetilde{a}_{jk})_{j,k=1}^\infty$, where $\widetilde{a}_{jk} = a_{kj},\ j, k \in \mathbb{N}$.

9.72° Find the adjoint operators to operators A from problem 9.15°, items (1)–(11), if $A : l_p \to l_p,\ 1 < p < +\infty$.

9.73 Find the adjoint operator to the operator A defined by the formula $Ax = (x_2, x_3, \ldots)$, if
(1) $A : c_0 \to c_0$; (2) $A : l_1 \to l_2$; (3) $A : l_1 \to c_0$.

9.74 Find the adjoint operator to an operator A if

(1) $A : l_1 \to l_2$ is the embedding operator: $Ax = x,\ x \in l_1$;

(2) $A : C([0, 2]) \to C([0, 1])$ is the restriction operator: $(Ax)(t) = x(t),\ t \in [0, 1],\ x \in C([0, 2])$;
(3) $A : L_p([0, 2]) \to L_p([0, 1]),\ 1 \le p < +\infty$, is the restriction operator;
(4) $A : C([0, 1]) \to C([0, 1]),\ (Ax)(t) = a(t)x(t)$, where $a \in C([0, 1])$ is a fixed function;
(5) $A : C([0, 1]) \to C([0, 1]),\ (Ax)(t) = \int_0^t x(s)ds$;
(6) X, Y are normed linear spaces, $f \in X^*$ and $y \in Y$ are fixed, $Ax = f(x)y,\ x \in X$.

9.75° Let X_1, X_2, X_3 be normed linear spaces over the field $\mathbb{K}$, $A \in \mathcal{L}(X_1, X_2)$. Prove that

(1) if $B \in \mathcal{L}(X_1, X_2),\ \alpha, \beta \in \mathbb{K}$, then $(\alpha A + \beta B)' = \alpha A' + \beta B'$;
(2) if $B \in \mathcal{L}(X_2, X_3)$, then $(AB)' = B'A'$;
(3) the restriction of the operator $A'' \in \mathcal{L}(X_1^{**}, X_2^{**})$ to X_1 coincides with A;
(4) if the spaces X_1 and X_2 are reflexive, then $A'' = A$;
(5) if A is continuously invertible, then A' is also continuously invertible and $(A')^{-1} = (A^{-1})'$.

9.76 Let X_1, X_2 be Banach spaces over the field $\mathbb{K}$, $A \in \mathcal{L}(X_1, X_2)$. Prove that

(1) $\operatorname{Ker} A = \{x \in X_1 \mid \forall f \in R(A') : f(x) = 0\}$;
(2) the set $\overline{R(A)} = \{y \in X_2 \mid \forall f \in \operatorname{Ker} A' : f(y) = 0\}$;
(3) the operator A is continuously invertible if and only if

$$\exists c > 0\ \forall x \in X_1\ \forall f \in X_2^* : \|Ax\|_2 \ge c\|x\|_1,\ \|A'f\| \ge c\|f\|;$$

(4) the operator A is continuously invertible if and only if the operator A' is continuously invertible.

Chapter 10
Compact Sets and Operators

Theoretical Background

A set M in a metric space (X, ρ) is called *compact* if every open cover of it contains a finite subcover. A set is called *precompact* if its closure is compact.

Bolzano-Weierstrass Criterion A set M in a metric space (X, ρ) is precompact if and only if every sequence of elements from M contains a convergent subsequence (the limit does not necessarily belong to M).

Hausdorff Criterion For the precompactness of a set M in a metric space (X, ρ) it is necessary, and in the case of a complete space it is sufficient that for any given $\varepsilon > 0$ there exists a finite ε-net for this set, i.e. a finite set $M_\varepsilon \subset X$ such that
$\forall x \in M \; \exists y \in M_\varepsilon \; : \; \rho(x, y) < \varepsilon.$

In a finite-dimensional NLS a set is precompact if and only if it is bounded.

In an infinite-dimensional NLS a ball is not a precompact set.

Theorem 10.1 (Ascoli–Arzelà) *A set M in the space $C([a, b])$ is precompact if and only if the following conditions are fulfilled:*

(1) $\exists C > 0 \; \forall x \in M \; \forall t \in [a, b] \; : \; |x(t)| \le C$ *(uniform boundedness);*
(2) $\forall \varepsilon > 0 \; \exists \delta > 0 \; \forall x \in M \; \forall t_1, t_2 \in [a, b], \; |t_1 - t_2| < \delta \; :$

$$|x(t_1) - x(t_2)| < \varepsilon \text{ (equicontinuity).}$$

Criteria of precompactness in spaces $l_p, \; 1 \le p < +\infty, \; c_0$ and c are given in problem 10.16°, in spaces $C^k([a, b]), \; k \in \mathbb{N}$, they are given in problem 10.26, and in spaces $L_p([a, b]), \; 1 \le p < +\infty$ they are given in problems 10.81 and 10.82.

Let X_1 and X_2 be normed linear spaces. A linear operator $A : X_1 \to X_2$ is called *compact*, if it maps every bounded set in X_1 into a precompact set in X_2. The set of

V. Brayman et al., *Functional Analysis and Operator Theory*, Problem Books in Mathematics, https://doi.org/10.1007/978-3-031-56427-7_10

all compact operators acting from X_1 to X_2 is denoted by $S_\infty(X_1, X_2)$ (or $S_\infty(X)$ in the case of $X_1 = X_2 = X$).

An operator A is compact if and only if the image of the unit ball $A(B(0, 1))$ is a precompact set in X_2.

Theorem 10.2 (Properties of Compact Operators) *Let X_1, X_2, X_3 be normed linear spaces. Then*

(1) $S_\infty(X_1, X_2) \subset \mathcal{L}(X_1, X_2)$;
(2) $S_\infty(X_1, X_2)$ *is a linear set;*
(3) *if X_2 is a Banach space, $\{A_n : n \geq 1\} \subset S_\infty(X_1, X_2)$ and $A_n \rightrightarrows A$, then $A \in S_\infty(X_1, X_2)$;*
(4) *if $A \in S_\infty(X_1, X_2)$, $B \in \mathcal{L}(X_2, X_3)$, $C \in \mathcal{L}(X_3, X_1)$, then $BA \in S_\infty(X_1, X_3)$, $AC \in S_\infty(X_3, X_2)$;*
(5) *a compact operator transforms any weakly convergent sequence in X_1 into a strongly convergent sequence in X_2.*

A linear set $J \subset \mathcal{L}(X)$ is called a *two-sided ideal* in the algebra $\mathcal{L}(X)$ if $\{AB, BA\} \subset J$ for all $A \in J$ and $B \in \mathcal{L}(X)$.

The statements (1)–(3) of Theorem 10.2 imply that if X_2 is a Banach space, then $S_\infty(X_1, X_2)$ is a subspace in $\mathcal{L}(X_1, X_2)$. The statements (1), (2), (4) imply that $S_\infty(X)$ is a two-sided ideal in the algebra $\mathcal{L}(X)$.

An operator $A \in \mathcal{L}(X_1, X_2)$, for which $\dim(R(A)) < +\infty$, is called *finite-dimensional*. The set of finite-dimensional operators is denoted by $S_0(X_1, X_2)$. Every finite-dimensional operator is compact, and $S_0(X)$ is a two-sided ideal in $\mathcal{L}(X)$ (see problem 10.34°).

Examples of Problems with Solutions

10.1 Determine whether the following sets are precompact (compact) in the respective spaces:

(1) $M = \{e_n : n \geq 1\}$ in l_p, $1 \leq p < +\infty$;
(2) $M = \{x_n(t) = e^{-nt},\ t \in [0, 1] : n \geq 1\}$ in $C([0, 1])$, in $L_p([0, 1])$, $1 \leq p < +\infty$;
(3) $M = \{x_\alpha(t) = \cos \alpha t, t \in [0, \pi] \,|\, \alpha \in [0, 2]\}$ in $C([0, \pi])$, in $L_p([0, \pi])$, $1 \leq p < +\infty$;
(4) $M = \{x_n(t) = \cos nt,\ t \in [0, \pi] : n \geq 1\}$ in $C([0, \pi])$, in $L_p([0, \pi])$, $1 \leq p < +\infty$;
(5) M is a bounded set in $C([0, 1])$, which consists of polynomials of degree at most n, where n is fixed;
(6) $M = \left\{ x \in C^1([0, 1]) \,\middle|\, \int\limits_0^1 |x(t)|dt \leq 1,\ |x'(t)| \leq 3,\ t \in [0, 1] \right\}$ in $C([0, 1])$.

Solution To determine whether a set is precompact, usually the Bolzano-Weierstrass criterion is used.

(1) *Method I.* Since $\|e_n - e_m\|_p = 2^{1/p} > 1,\ n \neq m$, any subsequence of the sequence $\{e_n : n \geq 1\}$ is not a Cauchy sequence, and therefore it is not convergent. Hence the set M is not precompact, and therefore is not compact.
Method II. If M is precompact, then the sequence $\{e_n : n \geq 1\}$ has a convergent subsequence $\{e_{n_k} : k \geq 1\}$. It converges to 0 coordinate-wise as well as the whole sequence. But $\|e_{n_k}\| = 1,\ k \geq 1$, so this subsequence does not converge to 0, therefore it diverges. We got a contradiction.

(2) Suppose that the given set is precompact in $C([0, 1])$. Then there exists a subsequence $\{x_{n_k} : k \geq 1\}$, which converges in $C([0, 1])$ and consequently converges pointwise to some function $y \in C([0, 1])$. But

$$x_{n_k}(t) = e^{-n_k t} \to y(t) = \begin{cases} 1, & t = 0, \\ 0, & t \in (0, 1], \end{cases}, \quad k \to \infty.$$

We got that y is a discontinuous function, a contradiction. Thus M is not precompact, so it is not compact in $C([0, 1])$.

Since $x_n \to 0,\ n \to \infty$, in $L_p([0, 1]),\ 1 \leq p < +\infty$ (which is easy to verify by applying Lebesgue's dominated convergence theorem or directly calculating the integral), then every subsequence of the sequence $\{x_n : n \geq 1\}$ converges. Therefore M is a precompact set. Nevertheless M is not compact, because $0 \notin M$.

(3) Let us verify that M is a compact set in $C([0, \pi])$. Consider a sequence $\{x_{\alpha_n} : n \geq 1\} \subset M$. The sequence $\{\alpha_n : n \geq 1\} \subset [0, 2]$ is bounded, therefore by Bolzano-Weierstrass theorem it contains a subsequence $\{\alpha_{n_k} : k \geq 1\}$ which converges to some real number $\alpha_0 \in [0, 2]$. We will show that $x_{\alpha_{n_k}} \to x_{\alpha_0},\ k \to \infty$, in $C([0, \pi])$. Indeed,

$$\max_{t \in [0,\pi]} |x_{\alpha_{n_k}} - x_{\alpha_0}(t)| = \max_{t \in [0,\pi]} |\cos \alpha_{n_k} t - \cos \alpha_0 t| \leq$$
$$\leq \max_{t \in [0,\pi]} |\alpha_{n_k} t - \alpha_0 t| = \pi |\alpha_{n_k} - \alpha_0| \to 0,\ k \to \infty$$

(we used the inequality $|\cos u - \cos v| \leq |u - v|,\ u, v \in \mathbb{R}$, which follows from Lagrange's theorem). Thus M is compact, and therefore precompact set in $C([0, \pi])$.

Remark The precompactness of M in $C([0, \pi])$ simply follows from the Ascoli–Arzelà theorem, but reasonings, similar to the solution above, are required to verify that M is a closed set.

The set M is also compact in $L_p([0, \pi]),\ 1 \leq p < +\infty$. It follows from the compactness of M in $C([0, \pi])$, because the convergence in $C([0, \pi])$ implies convergence in $L_p([0, \pi])$.

(4) The set M is not precompact in $L_p([0, \pi])$, $1 \le p < +\infty$. Indeed, since $\int\limits_0^\tau x_n(t)dt \to 0$, $n \to \infty$, $\tau \in (0, \pi]$ (in particular, $x_n \xrightarrow{w} 0$, $n \to \infty$, in $L_p([0, \pi])$ for $1 < p < \infty$), then any subsequence of the sequence $\{x_n : n \ge 1\}$ may converge only to 0. Thus there is no convergent subsequence because of

$$\|x_n\|_p^p = \int\limits_0^\pi |\cos nt|^p dt = \frac{1}{n}\int\limits_0^{\pi n} |\cos u|^p du = \int\limits_0^\pi |\cos u|^p du > 0, \ n \ge 1.$$

It follows that M is not precompact in $C([0, \pi])$ either.

Remark The fact that the set M is not precompact in $C([0, \pi])$ can be also deduced from the Ascoli–Arzelà theorem. Indeed, M is not equicontinuous: for $\varepsilon = 1$ and for any $\delta > 0$ one can take $n > \frac{\pi}{\delta}$, $t_1 = 0$ and $t_2 = \frac{\pi}{n}$. Then $|t_1 - t_2| < \delta$ but for $x_n \in M$ we have $|x_n(t_1) - x_n(t_2)| = 2 > \varepsilon$.

(5) Let P_n be the space of polynomials of degree at most n with the uniform norm. Since M is a bounded set in the finite-dimensional space P_n, it is precompact in this space, and therefore also in $C([0, 1])$. The set M is compact if it is closed, and is not compact if it is not closed.

(6) We will show that the set M is precompact in $C([0, 1])$. Let us verify that conditions of the Ascoli–Arzelà theorem are fulfilled. Equicontinuity: for all $x \in M$ and $0 \le t_1 < t_2 \le 1$ by Lagrange's theorem there exists $\theta \in [t_1, t_2]$ such that $|x(t_1) - x(t_2)| = |x'(\theta)| \cdot |t_1 - t_2| \le 3|t_1 - t_2|$. Uniform boundedness: for all $x \in M$ and $s, t \in [0, 1]$ we have

$$|x(s)| \le |x(t)| + |x(s) - x(t)| \le |x(t)| + 3|s - t| \le |x(t)| + 3.$$

Integrate this inequality and obtain that

$$\forall x \in M \ \forall s \in [0, 1] \ : \ |x(s)| \le \int\limits_0^1 |x(t)|dt + 3 \le 4.$$

The set M is not compact because it is not closed. Indeed, let $x_0(t) = |t - \frac{1}{2}|$ and $x_n(t) = \sqrt{(t - \frac{1}{2})^2 + \frac{1}{n}}$, $t \in [0, 1]$, $n \ge 1$. Then $\{x_n : n \ge 1\} \subset M$, $x_n \to x_0$, $n \to \infty$, in $C([0, 1])$, but $x_0 \notin M$.

10.2 Let $H = L_2([a, b])$, $\{p_k \mid 1 \le k \le n\}$ and $\{q_k \mid 1 \le k \le n\}$ be finite sets of functions from $L_2([a, b])$. Prove that the operator defined by the formula

$$(Ax)(t) = \int\limits_a^b \left(\sum_{k=1}^n p_k(t)q_k(s) \right) x(s)ds$$

is compact in H.

Solution Put $f_k(x) = \int\limits_a^b q_k(s)x(s)ds,\ 1 \le k \le n$. Then $f_k \in H^*$ and $Ax(t) = \sum\limits_{k=1}^{n} f_k(x)p_k(t)$. Therefore operator A is linear and bounded. Since $R(A) \subset \text{span}(\{p_k \mid 1 \le k \le n\})$, A is a finite-dimensional operator. Therefore A is compact.

10.3 Let $\{a_n : n \ge 1\}$ be a bounded sequence of complex numbers. Consider the operator $A : l_2 \to l_2$, $Ax = (a_1x_1, a_2x_2, \ldots)$. Prove that A is compact if and only if $\lim\limits_{n\to\infty} a_n = 0$.

Solution *Sufficiency.* Let $\lim\limits_{n\to\infty} a_n = 0$. We put

$$A_n x = (a_1x_1, a_2x_2, \ldots, a_nx_n, 0, 0, \ldots),\ n \ge 1.$$

Operators A_n are compact because they are finite-dimensional. Moreover (see problem 6.10°) $\|A - A_n\| = \sup\limits_{k\ge n+1} |a_k| \to 0,\ n \to \infty$. Hence A is the uniform limit of the sequence of compact operators $\{A_n : n \ge 1\}$. Therefore $A \in S_\infty(l_2)$ by item (3) of Theorem 10.2.

Necessity. Method I. Suppose that $A \in S_\infty(l_2)$. Consider the standard basis in l_2: $e_k = (\underbrace{0, \ldots, 0}_{k-1}, 1, 0, 0, \ldots),\ k \ge 1$. Since $e_k \overset{w}{\to} 0$, then by item (5) of Theorem 10.2 we have $Ae_k \to 0$, because A is compact. Therefore $|a_k| = \|Ae_k\| \to 0,\ k \to \infty$.

Method II. Assume that $A \in S_\infty(l_2)$, but a_n does not converge to 0. Then for some $\varepsilon > 0$ there exists a subsequence $\{\alpha_{n_k} : k \ge 1\}$ such that $|a_{n_k}| \ge \varepsilon,\ k \ge 1$. Since $A \in S_\infty(l_2)$ and $M = \{e_{n_k} : k \ge 1\}$ is a bounded set, then the set $A(M) = \{a_{n_k}e_{n_k} : k \ge 1\}$ is precompact. But this is not the case, because none of its subsequences converge (there is a coordinate-wise convergence to 0 but $\|a_{n_k}e_{n_k}\| = |a_{n_k}| \ge \varepsilon,\ k \ge 1$). We got a contradiction.

10.4 Let $K \in C([a,b]^2)$ and $(Ax)(t) = \int\limits_a^b K(t,s)x(s)ds,\ t \in [a,b],\ x \in C([a,b])$. Prove that $A \in S_\infty(C([a,b]))$.

Solution By problem 6.15 we have $A \in \mathcal{L}(C([a,b]))$.

Method I. Let $\overline{B}(0,1)$ be the unit ball in $C([a,b])$. We will prove that $A(\overline{B}(0,1))$ is a precompact set in $C([a,b])$. Let us verify that conditions of the Ascoli–Arzelà theorem are fulfilled. The uniform boundedness follows from the fact that $A(\overline{B}(0,1)) \subset \overline{B}(0, \|A\|)$. Now we will establish that functions from $A(\overline{B}(0,1))$ are equicontinuous. Since $K \in C([a,b]^2)$, then by Cantor's theorem K is uniformly continuous. Therefore

$$\forall\, \varepsilon > 0\, \exists\, \delta > 0\, \forall\, (t_1,s_1), (t_2,s_2) \in [a,b]^2,\ \rho((t_1,s_1),(t_2,s_2)) < \delta : \quad |K(t_1,s_1) - K(t_2,s_2)| < \tfrac{\varepsilon}{b-a}$$

(here ρ is the Euclidean distance in $\mathbb{R}^2$). Then for all $x \in \overline{B}(0, 1)$ and for all $t_1, t_2 \in [a, b]$, for which $|t_1 - t_2| < \delta$, we have

$$|(Ax)(t_1) - (Ax)(t_2)| \le \int_a^b |K(t_1, s) - K(t_2, s)| \cdot |x(s)|ds \le$$

$$\le \frac{\varepsilon}{b-a} \int_a^b |x(s)|ds \le \varepsilon.$$

By the Ascoli–Arzelà theorem, $A(\overline{B}(0, 1))$ is a precompact set in $C([a, b])$.

Method II. By the Stone–Weierstrass theorem, there exists a sequence of polynomials $\{K_n : n \ge 1\}$ such that $K_n \rightrightarrows K$, $n \to \infty$, on $[a, b]^2$. Let A_n be integral operators with kernels K_n, $n \ge 1$. In accordance with problem 6.15 we have

$$\|A_n - A\| = \max_{t\in[a,b]} \int_a^b |K_n(t, s) - K(t, s)|ds \le$$

$$\le (b - a) \max_{t,s\in[a,b]} |K_n(t, s) - K(t, s)| \to 0, \ n \to \infty,$$

hence $A_n \rightrightarrows A$. Since the kernel K_n can be represented as

$$K_n(t, s) = \sum_{k=0}^{m_n} t^k p_{k,n}(s), \ (t, s) \in [a, b]^2,$$

where $p_{k,n}$ are polynomials, $0 \le k \le m_n$, $m_n \in \mathbb{N}$, then the operators A_n are finite-dimensional, and therefore compact. By item (3) of Theorem 10.2, we obtain that $A \in S_\infty(C([a, b]))$.

Problems to Solve

10.5° Let M be a set in a Banach space such that for every $\varepsilon > 0$ there exists a precompact ε-net for the set M. Prove that the set M is precompact.

10.6° Prove that every subset of a precompact set is precompact.

10.7° Prove that if at least one (open or closed) ball in an NLS is precompact, then all open and closed balls in this space are precompact.

10.8 Let M be a precompact set in a metric space (X, ρ). A sequence $\{x_n : n \geq 1\} \subset X$ is such that $\rho(x_n, M) \to 0,\ n \to \infty$. Prove that the set $K = \{x_n : n \geq 1\}$ is precompact.

10.9° Let $M = \{x_n : n \geq 1\}$ be a precompact set in an NLS X, $x_0 \in X$. Prove that $x_n \to x_0,\ n \to \infty$, in X, if at least one of the following conditions is fulfilled:

(1) x_0 is the only limit point of the set M;
(2) $x_n \xrightarrow{w} x_0,\ n \to \infty$.

10.10 Let $\{M_n : n \geq 1\}$ be a sequence of compact sets in a metric space. Prove that if $\bigcap\limits_{k=1}^{n} M_k \neq \varnothing,\ n \geq 1$, then $\bigcap\limits_{k=1}^{\infty} M_k \neq \varnothing$.

10.11 Let M be a compact set in a metric space (X, ρ). Prove that for any $\varepsilon > 0$ the set M can be represented as $M = \bigcup\limits_{k=1}^{n} M_k$, where M_k are closed sets for which

$$\operatorname{diam} M_k = \sup_{x,y \in M_k} \rho(x, y) \leq \varepsilon,\ 1 \leq k \leq n.$$

10.12 Let X be an NLS, $M \subset X$, $N \subset X$. Is it correct that the set $M + N$:

(1) is compact, if M and N are compact;
(2) is precompact, if M and N are precompact;
(3) is closed, if M is compact and N is closed;
(4) is closed, if M and N are closed?

10.13° Let (X, ρ) be a metric space, $M \subset X$. A function $f : M \to \mathbb{R}$ is called *lower semicontinuous* (*upper semicontinuous*) on the set M, if

$$\forall x_0 \in M\ \forall \varepsilon > 0\ \exists \delta > 0\ \forall x \in B(x_0, \delta) \cap M\ :\ f(x) > f(x_0) - \varepsilon$$

(or $f(x) < f(x_0) + \varepsilon$, respectively). Prove that

(1) if M is a compact set, then every lower (upper) semicontinuous function on M attains its minimum (maximum) value on M;
(2) the converse statement to item (1) is also correct;
(3) if every continuous function on M is bounded, then M is a compact set.

10.14 Let M be a set in a Banach space such that every real continuous function on M is uniformly continuous. Does it follow that M is a compact set?

10.15 Let $\{L_n : n \geq 1\}$ be a sequence of finite-dimensional subspaces of a Banach space X such that $L_n \subset L_{n+1},\ n \geq 1$, and the set $\bigcup\limits_{n=1}^{\infty} L_n$ is dense in X. Prove that a bounded set M is precompact in X if and only if $\sup\limits_{x \in M} \rho(x, L_n) \to 0,\ n \to \infty$.

10.16° Prove that a bounded set M is precompact

(1) in the space $l_p,\ 1 \le p < +\infty$, if and only if

$$\forall \varepsilon > 0\ \exists N = N(\varepsilon) \in \mathbb{N}\ \forall x \in M\ :\ \sum_{k=N+1}^{\infty} |x_k|^p < \varepsilon;$$

(2) in the space c_0 if and only if

$$\forall \varepsilon > 0\ \exists N = N(\varepsilon) \in \mathbb{N}\ \forall x \in M\ \forall k \ge N\ :\ |x_k| < \varepsilon;$$

(3) in the space c if and only if

$$\forall \varepsilon > 0\ \exists N = N(\varepsilon) \in \mathbb{N}\ \forall x \in M\ \forall k \ge N\ :\ |x_k - \lim_{n\to\infty} x_n| < \varepsilon.$$

10.17° Formulate and prove the criterion of precompactness in the separable Hilbert space H with an orthonormal basis $\{e_n : n \ge 1\}$.

10.18 Prove that the following sets are precompact in $C([a, b])$ (here $k_1,\ k_2$ are fixed nonnegative constants):

(1) $\left\{\int\limits_a^t x(s)ds,\ t \in [a, b]\ \middle|\ x \in M\right\}$, where M is a bounded set in $C([a, b])$;
(2) $\{x \in C([a, b]) \mid |x(a)| \le k_1,\ |x(t_1) - x(t_2)| \le k_2|t_1 - t_2|,\ t_1, t_2 \in [a, b]\}$;
(3) $\{x \in C^1([a, b]) \mid |x(t_0)| \le k_1,\ |x'(t)| \le k_2,\ t \in [a, b]\}$, where $t_0 \in [a, b]$ is fixed;
(4) $\left\{\int\limits_a^t K(t, s)x(s)du,\ t \in [a, b]\ \middle|\ x \in B(0, 1)\right\}$, where $K \in C([a, b]^2)$;
(5) $\left\{x \in C^1([a, b])\ \middle|\ |x(a)| \le k_1,\ \int\limits_a^b |x'(t)|^2 dt \le k_2\right\}$;
(6) $\left\{x \in C^1([a, b])\ \middle|\ \int\limits_a^b (|x(t)|^p + |x'(t)|^p)dt \le k_1\right\}$, where $1 < p < +\infty$;
(7) $\{x \in C^2([a, b]) \mid |x(t)| \le k_1, |x''(t)| \le k_2,\ t \in [a, b]\}$;
(8) $\{x \in C^1([a, b]) \mid |x(t)| \le k_1, t \in [a, b]$;

$$|x'(t_1) - x'(t_2)| \le k_2|t_1 - t_2|,\ t_1, t_2 \in [a, b]\};$$

(9) an equicontinuous set of continuous functions, whose absolute values are bounded by a fixed constant at some point $t_0 \in [a, b]$;
(10) a bounded set of polynomials of degree n.

Which of these sets are compact in $C([a, b])$?

10.19 Which of the following sets are precompact (compact) in $C([0, 1])$:

(1) $\{t^n \mid n \geq 1\}$;
(2) $\{\sin n\sqrt{t} \mid n \geq 1\}$;
(3)* $\{\sin(t + n) \mid n \geq 1\}$;
(4) $\{\sin \alpha t \mid \alpha \in \mathbb{R}\}$;
(5) $\{\sin \alpha t \mid \alpha \in [1, 3]\}$;
(6) $\{\arctan \alpha t \mid \alpha \in \mathbb{R}\}$;
(7) $\{e^{t-\alpha} \mid \alpha \in \mathbb{R}\}$;
(8) $\{e^{t-\alpha} \mid \alpha \geq 0\}$;
(9) $\left\{\sum\limits_{n=1}^{\infty} \frac{a_n}{n^3} e^{-nt} \mid a_n \in (-1, 1), n \geq 1\right\}$;
(10) $\left\{\sum\limits_{n=1}^{\infty} \frac{a_n}{t+n^2} \mid a_n \in (-1, 1), n \geq 1\right\}$?

10.20 Under what conditions on a set $A \subset \mathbb{R}$ are the following sets precompact (compact) in $C([0, 1])$:

(1) $\{\sin \alpha t \mid \alpha \in A\}$;
(2) $\{\arctan \alpha t \mid \alpha \in A\}$;
(3) $\{e^{t-\alpha} \mid \alpha \in A\}$;
(4) $\{\frac{1}{1+|t-\alpha|} \mid \alpha \in A\}$?

10.21° Are the following sets precompact in $L_2([0, 1])$:

(1) $\{t^n \mid n \geq 1\}$;
(2) $\{\sin \pi n t \mid n \geq 1\}$;
(3) $\{\sin \alpha t \mid \alpha \in \mathbb{R}\}$;
(4) $\{\arctan \alpha t \mid \alpha \in \mathbb{R}\}$?

10.22° Prove the compactness of the following sets in l_2 :

(1) $\left\{x \in l_2 \mid \forall n \geq 1 \ : \ |x_n| \leq \frac{1}{n}\right\}$;
(2) $\left\{x \in l_2 \mid \sum\limits_{n=1}^{\infty} n^2 |x_n|^2 \leq 1\right\}$.

10.23° Let $a_n > 0, \ n \in \mathbb{N}$. Formulate conditions on $\{a_n : n \geq 1\}$ under which the following sets are compact in l_2 :

(1) $\left\{x \in l_2 \mid \forall n \geq 1 \ : \ |x_n| \leq \frac{1}{a_n}\right\}$;
(2) $\left\{x \in l_2 \mid \sum\limits_{n=1}^{\infty} a_n |x_n|^2 \leq 1\right\}$.

10.24° Prove that every set which is precompact in $C^1([a, b])$ is also precompact in $C([a, b])$. Is it true that every subset of $C^1([a, b])$ which is precompact in $C([a, b])$ is also precompact in $C^1([a, b])$?

10.25° Prove that the ball $\overline{B}(0, 1)$ of the space $C^1([a, b])$ is precompact, but not a closed set in $C([a, b])$.

10.26 Formulate and prove the criterion of precompactness in the spaces $C^k([a, b]), \ k \in \mathbb{N}$.

10.27 (Ascoli–Arzelà) Let (K, ρ) be a compact metric space. Prove that a set M is precompact in the space $C(K)$ if and only if it is uniformly bounded and equicontinuous (i.e.,

$$\forall \varepsilon > 0 \ \exists \delta > 0 \ \forall t_1, t_2 \in K, \ \rho(t_1, t_2) < \delta, \ \forall x \in M \ : \ |x(t_1) - x(t_2)| < \varepsilon).$$

10.28 Let X_1, X_2 be normed linear spaces and X_1 is a subset of X_2. Find the necessary and sufficient conditions for every precompact set in X_1 to be precompact in X_2.

10.29° Is it true that every precompact set in an NLS X_1 is precompact in an NLS X_2, if

(1) $X_1 = L_2([a, b])$, $X_2 = L_1([a, b])$;
(2) $X_1 = C([a, b])$, $X_2 = L_2([a, b])$;
(3) $X_1 = l_1$, $X_2 = l_\infty$;
(4) $X_1 = l_1$, $X_2 = l_2$?

10.30 Let T be some set, $B(T)$ be the linear space of all bounded functions from T to $\mathbb{R}$, and $\|x\|_\infty = \sup\limits_{t\in T} |x(t)|$, $x \in B(T)$. Prove the following statements.

(1) $B(T)$ with the norm $\|\cdot\|_\infty$ is a Banach space.
(2) A set M is precompact in the space $B(T)$ if and only if it is bounded and for every $\varepsilon > 0$ there exists a finite partition $T = \bigcup\limits_{k=1}^{n} T_k$ such that each function from M on each subset T_k varies at most by ε, i.e.,

$$\forall\, 1 \le k \le n\ \forall x \in M\ \forall t_1, t_2 \in T_k\ :\ |x(t_1) - x(t_2)| \le \varepsilon.$$

10.31 Let T be some set, Y be a compact subset of an NLS X, $\|\cdot\|$ be a norm in X, $B(T, Y)$ be the linear space of all mappings from T to Y, and $\|x\|_\infty = \sup\limits_{t\in T} \|x(t)\|$, $x \in B(T, Y)$. Prove that $B(T, Y)$ with the norm $\|\cdot\|_\infty$ is a Banach space. Formulate and prove the criterion of precompactness in this space.

10.32° Let X_1, X_2 be normed linear spaces. Prove that a linear operator $A : X_1 \to X_2$ is compact if and only if for every bounded sequence $\{x_n : n \ge 1\} \subset X_1$ the sequence $\{Ax_n : n \ge 1\} \subset X_2$ contains a convergent subsequence.

10.33° Prove that in an infinite-dimensional NLS

(1) the identity operator is not compact;
(2) every compact operator is not continuously invertible.

10.34° Let X_1, X_2, X_3 be the normed linear spaces. Prove that

(1) $S_0(X_1, X_2) \subset S_\infty(X_1, X_2)$;
(2) $S_0(X_1, X_2)$ is a linear set in $\mathcal{L}(X_1, X_2)$;
(3) if $A \in S_0(X_1, X_2)$, $B \in \mathcal{L}(X_2, X_3)$ and $C \in \mathcal{L}(X_3, X_1)$, then $BA \in S_0(X_1, X_3)$, $AC \in S_0(X_3, X_2)$;
(4) if $A \in S_0(X_1, X_2)$, then $A' \in S_0(X_2^*, X_1^*)$, moreover $\dim R(A) = \dim R(A')$.

10.35° Let X_1, X_2 be normed linear spaces, $A : X_1 \to X_2$ be a linear operator. Is it correct that $A \in S_\infty(X_1, X_2)$, if
(1) $\dim X_1 < +\infty$; (2) $\dim X_2 < +\infty$?

10.36° Which of the following operators in l_2 are compact:

(1) $Ax = (0, x_1, x_2, \ldots, x_n, \ldots)$;
(2) $Ax = (0, x_1, \frac{x_2}{2}, \ldots, \frac{x_n}{n}, \ldots)$;
(3) $Ax = (x_2, x_3, \ldots, x_{100}, 0, 0, \ldots)$;
(4) $Ax = (x_3, \frac{x_4}{2}, \ldots, \frac{x_n}{n-2}, \ldots)$?

10.37° Find the necessary and sufficient condition on a bounded sequence $\{\alpha_n : n \geq 1\} \subset \mathbb{C}$ under which the operator $Ax = (\alpha_1 x_1, \alpha_2 x_2, \ldots)$, $x \in X$, is compact in the space X, if
(1) $X = l_p$, $1 \leq p < +\infty$; (2) $X = l_\infty$; (3) $X = c$; 4) $X = c_0$.

10.38° Find the necessary and sufficient condition for the compactness of the operator $A : l_2 \to l_2$, which is defined by a Jacobi matrix

$$\mathcal{A} = \begin{pmatrix} \alpha_1 & \beta_1 & 0 & 0 & 0 & \ldots \\ \gamma_1 & \alpha_2 & \beta_2 & 0 & 0 & \ldots \\ 0 & \gamma_2 & \alpha_3 & \beta_3 & 0 & \ldots \\ \ldots & \ldots & \ldots & \ldots & \ldots & \ldots \end{pmatrix}, \text{ where } \{\alpha_j, \beta_j, \gamma_j : j \geq 1\} \subset \mathbb{K}.$$

10.39° Is it necessary that $A \in S_\infty(X_1, X_2)$, if

(1) X_1 is an NLS, X_2 is a Banach space, $\{A_n : n \geq 1\} \subset S_\infty(X_1, X_2)$ and $A_n \xrightarrow{s} A$;
(2) X_1, X_2 are normed linear spaces, $\{A_n : n \geq 1\} \subset S_\infty(X_1, X_2)$ and $A_n \rightrightarrows A$?

10.40° Let $K \in C([a, b]^2)$ and

$$(Ax)(t) = \int_a^t K(t, s)x(s)ds, \ t \in [a, b], \ x \in C([a, b]).$$

Prove that $A \in S_\infty(C([a, b]))$.

10.41° Which of the following operators $A : C([0, 1]) \to C([0, 1])$ are compact:

(1) $(Ax)(t) = x(0) + t^2 x(1)$;
(2) $(Ax)(t) = \int_0^t s^3 x(s)ds$;
(3) $(Ax)(t) = \int_0^1 e^{its} x(s)ds$;
(4) $(Ax)(t) = tx(t)$;
(5) $(Ax)(t) = x(t^3)$;
(6) $(Ax)(t) = 3x(t) + \int_0^1 e^{ts} x(s)ds$;
(7) $(Ax)(t) = \frac{1}{2}(x(t) + x(1-t))$;
(8) $(Ax)(t) = \int_0^1 x(s^2)ds$?

10.42° Prove that operator $A : X_1 \to X_2$, which is defined by the formula

$$(Ax)(t) = \int_a^b K(t,s)x(s)ds,\ t \in [a,b],\ x \in X_1,$$

is compact, if

(1) $X_1 = L_2([a,b]),\ X_2 = C([a,b]),\ K \in C([a,b]^2)$;
(2) $X_1 = X_2 = L_2([a,b]),\ K \in L_2([a,b]^2)$.

10.43° Which of the operators $A : L_2([0,1]) \to L_2([0,1])$ are compact:

(1) $(Ax)(t) = \int_0^1 tsx(s)ds$;

(2) $(Ax)(t) = p(t)\int_0^1 q(s)x(s)ds$, where $p, q \in L_2([0,1])$;

(3) $(Ax)(t) = x(t) + \int_0^1 (ts^2 + s)x(s)ds$;

(4) $(Ax)(t) = \int_0^t x(s)ds$;

(5) $(Ax)(t) = \int_0^t tsx(s)ds$?

10.44° Under what conditions on a function a is the operator $(Ax)(t) = a(t)x(t)$ compact

(1) in the space $C([0,1])$, if $a \in C([0,1])$?
(2) in the space $L_2([0,1])$, if $a \in L_\infty([0,1])$?

10.45 Prove that the operator of embedding an NLS X_1 into an NLS X_2, i.e., the operator $A : X_1 \to X_2$, defined as $Ax = x,\ x \in X_1$, is compact if

(1) $X_1 = C^1([a,b]),\ X_2 = C([a,b])$;
(2) $X_1 = C^{k+1}([a,b]),\ X_2 = C^k([a,b]),\ k \in \mathbb{N}$;
(3) $X_1 = C^1([a,b]),\ X_2 = L_p([a,b]),\ 1 \le p \le +\infty$.

10.46 Is the operator of embedding an NLS X_1 into an NLS X_2 compact if

(1) $X_1 = l_1,\ X_2 = l_2$;
(2) $X_1 = l_1,\ X_2 = c_0$;
(3) $X_1 = C([0,1]),\ X_2 = L_2([0,1])$;
(4) $X_1 = L_2([0,1]),\ X_2 = L_1([0,1])$?

10.47° Under what conditions on the sequence $\alpha = \{\alpha_n : n \ge 1\} \subset [1, +\infty)$ is the operator of embedding the space $l_{2,\alpha}$ (see problem 2.6°) into c_0 compact?

10.48 Is the operator $A : X_1 \to X_2,\ (Ax)(t) = x'(t),\ t \in [0, 1]$, compact if

(1) $X_1 = C^1([0, 1]),\ X_2 = C([0, 1])$;
(2) $X_1 = C^2([0, 1]),\ X_2 = C^1([0, 1])$;
(3) $X_1 = C^2([0, 1]),\ X_2 = C([0, 1])$?

10.49° Prove that the operator of orthogonal projection in a Hilbert space is compact if and only if this operator is finite-dimensional.

10.50 Provide an example of an operator $A \in \mathcal{L}(l_2)$ such that

(1) $A \notin S_\infty(l_2)$, but $A^2 \in S_\infty(l_2)$;
(2) $A^k \notin S_\infty(l_2),\ 1 \le k \le m-1$, but $A^m \in S_\infty(l_2)$, where $m \ge 3$ is fixed.

10.51° Let H be a Hilbert space, $A, B \in \mathcal{L}(H)$. Is it true that

(1) if $A + B \in S_\infty(H)$, then $A, B \in S_\infty$;
(2) if $A + B, A - B \in S_\infty(H)$, then $A, B \in S_\infty$;
(3) if $A + B \in S_\infty(H),\ R(A) \perp R(B)$, then $A, B \in S_\infty$?

10.52 Let H be a Hilbert space, $A : H \to H$ be a linear operator. Is it necessary that $A \in \mathcal{L}(H)$, if

(1) $AB \in \mathcal{L}(H)$ for all $B \in S_\infty(H)$;
(2) $BA \in \mathcal{L}(H)$ for all $B \in S_\infty(H)$?

10.53 Prove that the range of every compact operator is a separable set.

10.54° Let $c_0, c_1, \ldots, c_n \in \mathbb{C}$. Can a compact operator A in an infinite-dimensional NLS satisfy the equation $\sum_{k=0}^{n} c_k A^k = 0$? (Here $A^0 = I$.)

10.55 Can a compact operator A in an infinite-dimensional Banach space X have an algebraic inverse, which is defined (1) on X; (2) on $R(A)$?

10.56° Let X be an NLS, $A \in \mathcal{L}(X)$ and there exists $c > 0$ such that $\|Ax\| \ge c\|x\|$ for all $x \in X$. Under what conditions on X can the operator A be compact?

10.57° Let $X_1,\ X_2$ be normed linear spaces, $\dim X_1 = +\infty$ and $A \in S_\infty(X_1, X_2)$. Prove that there exists a sequence $\{x_n : n \ge 1\} \subset X_1$ such that $\|x_n\| = 1$ and $Ax_n \to 0,\ n \to \infty$.

10.58

(1) A set M is called *nowhere dense* in a metric space if its closure has no interior points. Prove that in an infinite-dimensional NLS X every precompact set is nowhere dense.
(2) A subset in a metric space is called a *set of the first category*, if it is a countable union of nowhere dense sets. Show that if $X_1,\ X_2$ are normed linear spaces, $\dim X_2 = +\infty$ and $A \in S_\infty(X_1, X_2)$, then $R(A)$ is a set of the first category in X_2.

10.59 Show that $C^1([a,b])$ is a set of the first category in a Banach space X, if (1) $X = C([a,b])$; (2) $X = L_p([a,b])$, $1 \le p \le +\infty$.

10.60* Let X_1 be an NLS, X_2 be a Banach space, and $A \in S_\infty(X_1, X_2)$ be such that the set $R(A)$ is closed in X_2. Show that $\dim R(A) < +\infty$.

10.61 Show that in a Hilbert space the subspace that lies in the range of a compact operator is finite-dimensional.

10.62 Let M be a subspace of $C([a,b])$ such that $M \subset C^1([a,b])$. Show that M is finite-dimensional.

10.63 Let H be a Hilbert space and $A \in \mathcal{L}(H)$. Prove that the operator A is compact if and only if there exists a sequence of finite-dimensional linear operators which converges uniformly to A.

10.64° Let H be a Hilbert space, $A \in \mathcal{L}(H)$. Prove that operators A, A^*, AA^*, and A^*A are simultaneously compact or not compact.

10.65 Let H be a Hilbert space, $A \in \mathcal{L}(H)$. Prove that

(1) if $A^2 \in S_\infty(H)$ and A is normal, then $A \in S_\infty(H)$;
(2) if $A^n \in S_\infty(H)$ for some $n > 1$ and A is normal, then $A \in S_\infty(H)$;
(3) if $A \in S_\infty(H)$ and $A \ge 0$, then $\sqrt{A} \in S_\infty(H)$ (see the definition of the operator $\sqrt{A}$ in problem 9.52*).

10.66° Let X be a Banach space, and the operator $A \in \mathcal{L}(X)$ be such that

$$\forall \varepsilon > 0 \ \exists B_\varepsilon \in S_0(X) \ \exists C_\varepsilon \in \mathcal{L}(X), \ \|C_\varepsilon\| < \varepsilon \ : \ A = B_\varepsilon + C_\varepsilon.$$

Prove that $A \in S_\infty(X)$.

10.67° Let $\{e_n : n \ge 1\}$ be an orthonormal basis in a Hilbert space H, $A \in \mathcal{L}(H)$ and

$$\sup \left\{ \|Ax\| \,\middle|\, x \perp e_k, 1 \le k \le n; \|x\| = 1 \right\} \to 0, \ n \to \infty.$$

Prove that A is a compact operator.

10.68 Let X_1 be a reflexive Banach space (for example, a Hilbert space), X_2 be an NLS, and $A \in S_\infty(X_1, X_2)$. Prove that the set $A(\overline{B}(0,1))$ is compact in X_2.

10.69

(1) Let H be a Hilbert space, $A \in S_\infty(H)$. Prove that there exists $h \in H \backslash \{0\}$ such that $\|Ah\| = \|A\| \cdot \|h\|$.
(2) Provide an example of a Banach space X and an operator $A \in S_\infty(X)$ for which the statement of item (1) is false.

10.70

(1) Let X_1 be a reflexive Banach space, X_2 be an NLS. Prove that $A \in \mathcal{L}(X_1, X_2)$ is compact if and only if it transforms every weakly convergent sequence into a strongly convergent one.
(2) Provide an example of Banach spaces X_1, X_2 and an operator $A \in \mathcal{L}(X_1, X_2)$, which transforms every weakly convergent sequence into a strongly convergent one, but is not compact.

10.71 Let $1 < p < +\infty$ and $A \in \mathcal{L}\big(L_p([a,b]), C([a,b])\big)$. Prove that $A \in S_\infty\big(L_p([a,b]), L_p([a,b])\big)$.

10.72 Let H be a Hilbert space, $A \in \mathcal{L}(H)$ and $A \geq 0$. Prove that the operator A is compact if and only if for every sequence $\{x_n : n \geq 1\} \subset H$ which converges weakly to zero the relation $(Ax_n, x_n) \to 0,\ n \to \infty$, holds.

10.73 Let H be a Hilbert space, $A \in \mathcal{L}(H),\ B \in S_\infty(H)$. Prove that $A \in S_\infty(H)$ in each of the following cases: (1) $0 \leq A \leq B$; (2) $A^*A \leq B^*B$.

10.74* **(Schauder)** Let $A \in \mathcal{L}(X_1, X_2)$, where X_1 is an NLS and X_2 is a Banach space. Prove that $A \in S_\infty(X_1, X_2)$ if and only if $A' \in S_\infty(X_2^*, X_1^*)$.

10.75 Let X be a reflexive Banach space. Prove that
(1) $S_\infty(X, l_1) = \mathcal{L}(X, l_1)$; (2) $S_\infty(c_0, X) = \mathcal{L}(c_0, X)$.

10.76

(1) Let X, Y be Banach spaces, $\{A, A_n : n \geq 1\} \subset \mathcal{L}(X, Y),\ A_n \xrightarrow{s} A,\ n \to \infty$, and M be a precompact set in the space X. Prove that $\{A_n x : n \geq 1\}$ converges to Ax uniformly on M.
(2) Let X be a Banach space, $\{f, f_n : n \geq 1\} \subset X^*,\ f_n \xrightarrow{w-*} f,\ n \to \infty$, and M be a precompact set in the space X. Prove that $\{f_n(x) : n \geq 1\}$ converges to $f(x)$ uniformly on M.

10.77*

(1) Let X be a separable Banach space, $\{x_n : n \geq 1\}$ be a countable sequence, which is dense in X, B be the ball $\overline{B}(0,1)$ in X^*. Put

$$\rho(f, g) = \sum_{n=1}^{\infty} \frac{1}{2^n} \cdot \frac{|f(x_n) - g(x_n)|}{1 + |f(x_n) - g(x_n)|}, \quad f, g \in B.$$

Prove that (B, ρ) is a compact metric space, and the convergence in (B, ρ) is equivalent to weak-$*$ convergence.
(2) (**Gelfand**). Prove that for a set M in a Banach space X to be compact it is necessary, and in the case of a separable space X it is also sufficient, that for each sequence $\{f_n : n \geq 1\} \subset X^*$ which converges weakly-$*$ to zero, the convergence $f_n(x) \to 0,\ n \to \infty$, is uniform for $x \in M$.

10.78 Let $A \in \mathcal{L}(X_1, X_2)$, where X_1 is an NLS and X_2 is a separable Banach space. Prove that $A \in S_\infty(X_1, X_2)$ if and only if A' transforms every weak-$*$ convergent sequence in X_2^* into a strongly convergent sequence in X_1^*.

10.79 (Mazur) Let X be a Banach space, $\{A_n : n \geq 1\} \subset S_\infty(X)$, and $A_n \xrightarrow{s} I,\ n \to \infty$. Prove that a set $M \subset X$ is precompact if and only if the following conditions are fulfilled:

(a) M is a bounded set;
(b) $A_n x \to x,\ n \to \infty$, uniformly in $x \in M$.
Verify that condition (b) can be replaced by condition
(b') $\forall\, \varepsilon > 0\ \exists\, n \in \mathbb{N}\ \forall\, x \in M\ :\ \|A_n x - x\| < \varepsilon$,
and condition $\{A_n : n \geq 1\} \subset S_\infty(X)$ can be replaced by condition

$$\exists\, k \in \mathbb{N}\ :\ \{A_n^k : n \geq 1\} \subset S_\infty(X).$$

10.80 Let $1 \leq p < +\infty,\ x \in L_p([a, b]),\ h > 0$. The function

$$(A_h x)(t) = x_h(t) = \frac{1}{2h} \int_{t-h}^{t+h} x(s)ds,\ t \in [a, b],$$

where $x(t) = 0,\ t \in \mathbb{R} \backslash [a, b]$, is called the *Steklov mean function*. Prove that

(1) $x_h \in C([a, b])$ and $|x_h(t)| \leq \frac{\|x\|_p}{(2h)^{\frac{1}{p}}},\ t \in [a, b]$;
(2) $\|x_h\|_p \leq \|x\|_p$ for $p = 1$;
(3) $\|x_h\|_p \leq \|x\|_p$ for $1 < p < +\infty$;
(4) $\|A_h x - x\|_p \to 0,\ h \to 0+$;
(5) $A_h \in S_\infty\big(L_p([a, b]), C([a, b])\big),\ 1 < p < +\infty$;
(6) $A_h^2 \in S_\infty\big(L_1([a, b]), C([a, b])\big)$;
(7) $A_h \in S_\infty\big(L_p([a, b])\big),\ 1 < p < +\infty$, and $A_h^2 \in S_\infty\big(L_1([a, b])\big)$.

10.81 (Kolmogorov, $p > 1$; Tulaikov, $p = 1$) Prove that a set M is precompact in the space $L_p([a, b]),\ 1 \leq p < +\infty$, if and only if the following conditions are fulfilled:

(a) M is a bounded set;
(b) $\forall\, \varepsilon > 0\ \exists\, h > 0\ \forall\, x \in M\ :\ \|x_h - x\|_p < \varepsilon$.

Remark Condition (a) follows from condition (b), see problem 12.29.

10.82 (M. Riesz) Prove that a set M is precompact in the space $L_p([a,b])$, $1 \leq p < +\infty$, if and only if the following conditions are fulfilled:

(a) M is a bounded set;
(b) the set M is equicontinuous in p-mean i.e.,

$$\forall \varepsilon > 0\ \exists \delta > 0\ \forall h \in (0,\delta)\ \forall x \in M\ :\ \int_a^b |x(t+h) - x(t)|^p dt < \varepsilon^p$$

(here $x(t+h) = 0$, if $t + h > b$).

Remark Condition (a) follows from condition (b), see problem 12.29.

10.83° Let B_n, $n \geq 1$, be Bernstein operators:

$$(B_n x)(t) = \sum_{k=0}^{n} \binom{n}{k} x\left(\frac{k}{n}\right) t^k (1-t)^{n-k},\ t \in [0,1],\ x \in C([0,1]).$$

Is it true that $B_n \rightrightarrows I$?

10.84* Let X, Y and Z be Banach spaces, $A \in S_\infty(X, Z)$, $B \in \mathcal{L}(Y, Z)$, and $R(B) \subset R(A)$. Prove that $B \in S_\infty(Y, Z)$.

10.85 Let X_1, X_2 be normed linear spaces, $M \subset X_1$ be a linear set which is dense in X_1, and $A \in \mathcal{L}(X_1, X_2)$. Prove that if the restriction of the operator A on M belongs to $S_\infty(M, X_2)$, then $A \in S_\infty(X_1, X_2)$.

10.86 Let $W_p^1([a,b])$, $p \geq 1$, be the completion of the space $C^1([a,b])$ in the norm $\|x\|_{W_p^1} = \left(\int_a^b |x(t)|^p dt + \int_a^b |x'(t)|^p dt \right)^{\frac{1}{p}}$, $x \in C^1([a,b])$. Prove that

(1) for every function $x \in C^1([a,b])$ there exists $\theta \in [a,b]$ such that

$$x(t) = \int_\theta^t x'(s)ds + \frac{1}{b-a} \int_a^b x(s)ds,\ t \in [a,b];$$

(2) for $p \geq 1$ the inequality $\|x\|_C \leq c\|x\|_{W_p^1}$, $x \in C^1([a,b])$, holds, where $c > 0$ is a constant, and $\|\cdot\|_C$ and $\|\cdot\|_{W_p^1}$ are norms in $C([a,b])$ and $W_p^1([a,b])$, respectively;
(3) for $p \geq 1$ we have $W_p^1([a,b]) \subset C([a,b])$ and the embedding operator $A : W_p^1([a,b]) \to C([a,b])$, $Ax = x$, is continuous;
(4) for $p > 1$ the embedding operator $A : W_p^1([a,b]) \to C([a,b])$, $Ax = x$, is compact.

10.87 Let $p \geq 1$. Prove that

(1) there exists a unique continuous linear operator $D : W_p^1([a, b]) \to L_p([a, b])$ such that $Dx(t) = x'(t),\ t \in [a, b]$, for all $x \in C^1([a, b])$;
(2) the operator D is not compact.

10.88 Let $(T, \mathcal{F}, \mu)$ be a space with a finite measure, $p > 1,\ \frac{1}{p} + \frac{1}{q} = 1,\ K : T \times T \to \mathbb{C}$ be a measurable function such that $\operatorname{ess\,sup}_{t\in T} \int_T |K(t, s)|^q d\mu(s) < +\infty$. Prove that the operator $A : L_p(T, \mu) \to L_p(T, \mu)$, defined by the formula $(Ax)(t) = \int_T K(t, s)x(s)d\mu(s),\ t \in T$, is compact.

10.89 Let $(T, \mathcal{F}, \mu)$ be a space with a measure, $K : T \times T \to \mathbb{R}$ be a measurable function, $c_1 = \operatorname{ess\,sup}_{s\in T} \int_T |K(t, s)|d\mu(t),\ c_2 = \operatorname{ess\,sup}_{t\in T} \int_T |K(t, s)|d\mu(s)$.

Prove that the operator $(Ax)(t) = \int_T K(t, s)x(s)d\mu(s),\ t \in T$, has the following properties:

(1) if $c_1 < +\infty$, then $A \in \mathcal{L}(L_1(T, \mu))$ and $\|A\| \leq c_1$;
(2) if $c_1 < +\infty,\ c_2 < +\infty$, then $A \in \mathcal{L}(L_p(T, \mu))$ and $\|A\| \leq c_1^{\frac{1}{p}} c_2^{\frac{1}{q}},\ 1 < p < +\infty$;
(3) under the conditions of item (2) the operator A is not necessarily compact in the space $L_p(T, \mu),\ 1 \leq p < +\infty$, even if $\mu(T) < +\infty$;
(4) if $1 < p < +\infty$, $c_2 < +\infty,\ \mu(T) < +\infty$ and

$$c_{1,n} = \operatorname*{ess\,sup}_{s\in T} \int_{B_n(s)} |K(t, s)|d\mu(t) \to 0,\ n \to \infty,$$

where $B_n(s) = \{t \in T \mid |K(t, s)| > n\},\ n \geq 1,\ s \in T$, then $A \in S_\infty(L_p(T, \mu))$.

10.90 Let $K : \mathbb{R} \to \mathbb{C}$ be a 2π-periodic Lebesgue measurable function and $K \in L_1([0, 2\pi])$. Prove that for $1 < p < +\infty$ the operator

$$(Ax)(t) = \int_0^{2\pi} K(t - s)x(s)ds,\ t \in [0, 2\pi],\ x \in L_p([0, 2\pi]),$$

is compact.

10.91 Prove the compactness of the operator $A : X \to X$, defined by the formula $(Ax)(t) = \int\limits_a^b K(t,s)x(s)ds,\ t \in [a,b],\ x \in X$, if

(1) $X = C([a,b]),\ K : [a,b]^2 \to \mathbb{R}$ is a Lebesgue measurable function such that $K(t,\cdot) \in L_1([a,b])$ for all $t \in [a,b]$, and

$$\forall\, t_0 \in [a,b] : \int\limits_a^b |K(t,s) - K(t_0,s)|ds \to 0,\ t \to t_0;$$

(2) $X = C([a,b]),\ K(t,s) = \frac{K_0(t,s)}{|t-s|^\alpha},\ (t,s) \in [a,b]^2,\ t \neq s$, where $K_0 \in C([a,b]^2),\ \alpha < 1$;
(3) $X = L_2([a,b]),\ K(t,s) = \frac{K_0(t,s)}{|t-s|^\alpha},\ (t,s) \in [a,b]^2,\ t \neq s$, where $K_0 \in L_\infty([a,b]^2),\ \alpha < 1$.

10.92 Let $u \in L_1(\mathbb{R})$. Prove that the operator

$$(Ax)(t) = \int\limits_{\mathbb{R}} u(t-s)x(s)ds,\ t \in \mathbb{R},\ x \in L_2(\mathbb{R}),$$

is linear and continuous in $L_2(\mathbb{R})$. Under what conditions on the function u is the operator A compact?

10.93 Let $u \in L_1([0,+\infty))$ and $(Ax)(t) = \int\limits_0^t u(t-s)x(s)ds,\ t \in [0,+\infty),\ x \in L_2([0,+\infty))$. Prove that $A \in \mathcal{L}(L_2([0,+\infty)))$. Under what conditions on the function u is the operator A compact?

10.94 Let $u \in L_1([0,+\infty))$ and $(Ax)(t) = \int\limits_0^{+\infty} u(t+s)x(s)ds,\ t \in [0,+\infty),\ x \in L_2([0,+\infty))$. Prove that (1) $A \in \mathcal{L}(L_2([0,+\infty)))$; (2) if $\int\limits_0^{+\infty} t|u(t)|^2dt < +\infty$, then $A \in S_\infty(L_2([0,+\infty)))$.

10.95 Let $(Ax)(t) = \int\limits_0^{+\infty} \frac{x(s)}{t+s}ds,\ t \in [0,+\infty),\ x \in L_2([0,+\infty))$. Prove that (1) $A \in \mathcal{L}(L_2([0,+\infty)))$; (2) $A \notin S_\infty(L_2([0,+\infty)))$.

10.96

(1) Prove that the operator $A : L_1([0,1]) \to L_1([0,1])$, defined by the formula $(Ax)(t) = \int\limits_0^t x(s)ds,\ t \in [0,1],\ x \in L_1([0,1])$, is compact.

(2) Provide an example of a bounded Borel function $K : [0, 1]^2 \to \mathbb{R}$ such that the operator $A : L_1([0, 1]) \to L_1([0, 1])$, defined by the formula

$$(Ax)(t) = \int_0^1 K(t, s)x(s)ds, \ t \in [0, 1], \ x \in L_1([0, 1]),$$

is not compact.

10.97 Let $C_b(\mathbb{R})$ be the linear space of continuous functions which are bounded on $\mathbb{R}$ endowed with the norm $\|x\| = \sup\limits_{t\in\mathbb{R}} |x(t)|, \ x \in C_b(\mathbb{R})$, and let a function $K \in C(\mathbb{R}^2)$ be such that $\sup\limits_{t\in\mathbb{R}} \int\limits_{\mathbb{R}} |K(t, s)|ds < \infty$. Consider the integral operator

$$(Ax)(t) = \int_{\mathbb{R}} K(t, s)x(s)ds, \ t \in \mathbb{R}, \ x \in C_b(\mathbb{R}).$$

Prove that (1) $A \in \mathcal{L}(C_b(\mathbb{R}))$; (2) it is not necessarily that $A \in S_\infty(C_b(\mathbb{R}))$.

10.98 Let $X = C([0, 1])$ or $X = L_p([0, +\infty))$, $1 < p < +\infty$. The operator A is defined by the formula $(Ax)(t) = \frac{1}{t}\int\limits_0^t x(s)ds, \ t > 0, \ x \in X$ (in the case of $X = C([0, 1])$ additionally $(Ax)(0) := x(0)$). Prove that

(1) $A \in \mathcal{L}(C([0, 1]))$;
(2) $A \notin S_\infty(C([0, 1]))$;
(3)* $A \in \mathcal{L}(L_p([0, +\infty)))$;
(4) $A \notin S_\infty(L_p([0, +\infty)))$.

Chapter 11
Spectrum of Linear Operators

Theoretical Background

Let X be a complex Banach space. A number $\lambda \in \mathbb{C}$ is called a *regular point* of an operator $A \in \mathcal{L}(X)$, if the operator $A - \lambda I$ is continuously invertible. The set of all regular points of the operator A is called the *resolvent set* of the operator A and is denoted $\rho(A)$. The set $\sigma(A) = \mathbb{C} \backslash \rho(A)$ is called the *spectrum* of the operator A.

An *eigenvalue* of the operator $A \in \mathcal{L}(X)$ is a number $\lambda \in \mathbb{C}$, for which there exists $x \in X$, $x \neq 0$, such that $Ax = \lambda x$. Elements $x \in X$, that satisfy this equation, are called *eigenvectors* of the operator A, which correspond to the eigenvalue λ. The set of eigenvalues of the operator $A \in \mathcal{L}(X)$ is called the *point spectrum* of the operator A and is denoted $\sigma_p(A)$. Thus

$$\sigma_p(A) = \big\{\lambda \in \sigma(A) \mid \operatorname{Ker}(A - \lambda I) \neq \{0\}\big\}.$$

The *continuous spectrum* of the operator A is defined as the set

$$\sigma_c(A) = \big\{\lambda \in \sigma(A) \mid \operatorname{Ker}(A - \lambda I) = \{0\} \text{ and } \overline{R(A - \lambda I)} = X\big\}.$$

The *residual spectrum* of the operator A is defined as the set

$$\sigma_r(A) = \big\{\lambda \in \sigma(A) \mid \operatorname{Ker}(A - \lambda I) = \{0\} \text{ and } \overline{R(A - \lambda I)} \neq X\big\}.$$

Clearly, $\sigma(A) = \sigma_p(A) \cup \sigma_c(A) \cup \sigma_r(A)$ and the sets $\sigma_p(A)$, $\sigma_c(A)$, $\sigma_r(A)$ are pairwise disjoint.

Theorem 11.1 *For every $A \in \mathcal{L}(X)$ the spectrum $\sigma(A)$ is a closed bounded nonempty subset of $\mathbb{C}$.*

The operator $R_z(A) = (A - zI)^{-1}$, $z \in \rho(A)$, is called the *resolvent* of the operator A.

V. Brayman et al., *Functional Analysis and Operator Theory*, Problem Books in Mathematics, https://doi.org/10.1007/978-3-031-56427-7_11

The resolvent satisfies the *Hilbert's identity*

$\forall z_1, z_2 \in \rho(A) \ : \ R_{z_1}(A) - R_{z_2}(A) = (z_1 - z_2) R_{z_1}(A) R_{z_2}(A)$.

The operator-valued function $F(z) = R_z(A),\ z \in \rho(A)$, is analytic, moreover $F'(z) = F^2(z)$.

The spectral radius of the operator $A \in \mathcal{L}(X)$ is the number $r(A) = \max\limits_{z \in \sigma(A)} |z|$, i.e., $r(A)$ is the radius of the smallest circle with the center at the origin that contains $\sigma(A)$. It is easy to show that $r(A) \le \|A\|$ and for $|z| > r(A)$ an equality $R_z(A) = -\sum\limits_{n=0}^{\infty} z^{-(n+1)} A^n$ holds, where the series converges uniformly.

Theorem 11.2 $r(A) = \lim\limits_{n\to\infty} \sqrt[n]{\|A^n\|}$.

The following theorems provide a description of the spectra of self-adjoint and unitary operators.

Theorem 11.3 *Let H be a Hilbert space, $A = A^* \in \mathcal{L}(H)$. Then $\sigma(A) \subset [-\|A\|, \|A\|]$ and at least one of the ends of the segment $[-\|A\|, \|A\|]$ belongs to $\sigma(A)$. In particular, $r(A) = \|A\|$.*

Theorem 11.4 *Let H be a Hilbert space, $A \in \mathcal{L}(H)$ be a unitary operator. Then $\sigma(A) \subset \{\lambda \in \mathbb{C} \mid |\lambda| = 1\}$.*

If A is a normal (for example, self-adjoint or unitary) operator in a Hilbert space H, then $\sigma_r(A) = \varnothing$ (see problem 11.21).

Examples of Problems with Solutions

11.1 Let $1 \le p < +\infty$. The operator $A \in \mathcal{L}(l_p)$ is defined by the formula $Ax = (\alpha_1 x_1, \alpha_2 x_2, \ldots),\ x \in l_p$, where $(\alpha_1, \alpha_2, \ldots) \in l_\infty$. Prove that

(1) $\sigma_p(A) = \{\alpha_n : n \ge 1\}$; (2) $\sigma(A) = \overline{\{\alpha_n : n \ge 1\}}$; (3) $\sigma_r(A) = \varnothing$.

Solution

(1) It follows from the fact that $(A - \lambda)x = 0$ if and only if $(\alpha_n - \lambda)x_n = 0$ for every $n \ge 1$.

(2) Put $m = m(\lambda) = \inf\limits_{n\ge1} |\alpha_n - \lambda|$. It is clear that $\lambda \in \overline{\{\alpha_n : n \ge 1\}}$ if and only if $m = 0$. If the operator $A - \lambda I$ is continuously invertible, then

$$m = \inf_{n\ge1} \|(A - \lambda I)e_n\| \ge \inf_{\|x\|=1} \|(A - \lambda I)x\| > 0$$

by Theorem 8.2. Conversely, let $m > 0$. Consider the operator

$$By = \left(\tfrac{y_1}{\alpha_1-\lambda}, \tfrac{y_2}{\alpha_2-\lambda}, \ldots\right),\ y \in l_p.$$

It is clear that $B \in \mathcal{L}(l_p)$, $\|B\| \leq \frac{1}{m}$ and $B(A - \lambda I) = (A - \lambda I)B = I$. By Theorem 8.3 the operator $A - \lambda I$ is continuously invertible and $R_\lambda(A) = (A - \lambda I)^{-1} = B$.

(3) If $\lambda \notin \sigma_p(A)$, then the set $R(A - \lambda I)$ contains all sequences with finitely many nonzero coordinates, because

$$(y_1, \dots, y_n, 0, \dots) = (A - \lambda I)\left(\tfrac{y_1}{\alpha_1 - \lambda}, \dots, \tfrac{y_n}{\alpha_n - \lambda}, 0, \dots\right)$$

for all $n \in \mathbb{N}$ and $y_1, \dots, y_n \in \mathbb{C}$. Therefore $\overline{R(A - \lambda I)} = l_p$ and $\lambda \notin \sigma_r(A)$.

11.2 Let H be a Hilbert space, $A \in \mathcal{L}(H)$. Prove that

(1) $\sigma(A^*) = \{\overline{\lambda} \mid \lambda \in \sigma(A)\}$;
(2) $\{\overline{\lambda} \mid \lambda \in \sigma_r(A)\} \subset \sigma_p(A^*) \subset \{\overline{\lambda} \mid \lambda \in \sigma_p(A) \cup \sigma_r(A)\}$.

Solution

(1) It follows from item (5) of problem 9.6° and item (1) of problem 9.1 that the operator $A - \lambda I$ is continuously invertible if and only if the operator $(A - \lambda I)^* = A^* - \overline{\lambda} I$ is continuously invertible. Therefore $\rho(A^*) = \{\overline{\lambda} \mid \lambda \in \rho(A)\}$, hence $\sigma(A^*) = \mathbb{C} \setminus \rho(A^*) = \mathbb{C} \setminus \{\overline{\lambda} \mid \lambda \in \rho(A)\} = \{\overline{\lambda} \mid \lambda \in \sigma(A)\}$.
(2) Let $\lambda \in \sigma_r(A)$. Then $\overline{R(A - \lambda I)} \neq H$, hence there exists $y \in H$, $y \neq 0$, such that $((A - \lambda I)x, y) = (x, A^*y - \overline{\lambda} y) = 0$ for all $x \in H$. Hence $A^*y = \overline{\lambda} y$, therefore $\overline{\lambda} \in \sigma_p(A^*)$. Conversely, if $\overline{\lambda} \in \sigma_p(A^*)$, then there exists $y \in H$, $y \neq 0$, such that $A^*y = \overline{\lambda} y$. Hence $((A - \lambda I)x, y) = (x, A^*y - \overline{\lambda} y) = 0$ for all $x \in H$. Therefore $\overline{R(A - \lambda I)} \neq H$, so $\lambda \in \sigma_p(A) \cup \sigma_r(A)$.

Remark The spectra of the operator $A \in \mathcal{L}(X)$ and the adjoint operator $A' \in \mathcal{L}(X^*)$, where X is a Banach space, are related as follows: $\sigma(A') = \sigma(A)$,

$$\sigma_r(A) \subset \sigma_p(A') \subset (\sigma_p(A) \cup \sigma_r(A)) \text{ and } \sigma_p(A) \subset (\sigma_p(A') \cup \sigma_r(A'))$$

(see problem 11.27).

11.3 Let $X = L_2([a, b])$, $(Ax)(t) = tx(t)$, $t \in [a, b]$. Find $\sigma(A)$, $\sigma_p(A)$, $\sigma_c(A)$, and $\sigma_r(A)$.

Solution Since the operator A is self-adjoint, then $\sigma_r(A) = \varnothing$. Let us find $\sigma_p(A)$. If $Ax = \lambda x$, then $(t - \lambda)x(t) = 0 \,(\mathrm{mod}\, m)$, and consequently $x(t) = 0 \,(\mathrm{mod}\, m)$, because $t - \lambda \neq 0$ for $t \neq \lambda$. Hence $\sigma_p(A) = \varnothing$. Now let us find $\sigma(A) = \sigma_c(A)$. If $\lambda \notin [a, b]$, then the function $(t - \lambda)^{-1}$ is bounded on $[a, b]$ and the operator $(Bx)(t) = (t - \lambda)^{-1}x(t)$ is continuous on X. Clearly, $B(A - \lambda I) = (A - \lambda I)B = I$, hence $\lambda \in \rho(A)$ and $R_\lambda(A) = B$. If $\lambda \in [a, b]$, then the function $y(t) = 1$, $t \in [a, b]$, does not belong to the range of the operator $A - \lambda I$, because $(t - \lambda)^{-1} \notin L_2([a, b])$. It follows that $R(A - \lambda I) \neq X$ and $\lambda \in \sigma(A)$. Hence $\sigma(A) = \sigma_c(A) = [a, b]$.

11.4 Let $X = l_2$. The operators $A, B \in \mathcal{L}(l_2)$ are defined as follows:

$$Ax = (x_2, x_3, x_4, \dots), \ \ Bx = (0, x_1, x_2, \dots), \ \ x \in l_2.$$

Find $\sigma(A), \ \sigma_p(A), \ \sigma_c(A), \ \sigma_r(A)$ and do the same for the operator B.

Solution Firstly, we will find the eigenvalues of operator A. Let $Ax = \lambda x, \ x \neq 0$. Then $x_{n+1} = \lambda x_n, \ n \geq 1$, so $x = x_1(1, \lambda, \lambda^2, \dots, \lambda^n, \dots)$, where $x_1 \neq 0$. The condition $x \in l_2$ is fulfilled if and only if $|\lambda| < 1$. Hence $\sigma_p(A) = \{\lambda \in \mathbb{C} \mid |\lambda| < 1\}$. Similarly, if $Bx = \lambda x, \ x \neq 0$, then $0 = \lambda x_1$ and $x_n = \lambda x_{n+1}, \ n \geq 1$, whence $x = 0$. Therefore $\sigma_p(B) = \varnothing$.

Since the set $\sigma(A)$ is closed, $\overline{\sigma_p(A)} \subset \sigma(A)$. On the other hand, since $r(A) \leq \|A\|$ and $\|A\| = 1$, then $\sigma(A) \subset \{\lambda \in \mathbb{C} \mid |\lambda| \leq 1\}$. Hence $\sigma(A) = \{\lambda \in \mathbb{C} \mid |\lambda| \leq 1\}$. Since $A^* = B$ (see item (1) of problem 9.3), we deduce from item (1) of problem 11.2 that

$$\sigma(B) = \{\lambda \mid \overline{\lambda} \in \sigma(A)\} = \{\lambda \in \mathbb{C} \mid |\lambda| \leq 1\}.$$

Finally, in view of item (2) of problem 11.2, the following inclusions hold:

$$\{\overline{\lambda} \mid \lambda \in \sigma_r(A)\} \subset \sigma_p(B) \subset \{\overline{\lambda} \mid \lambda \in \sigma_p(A) \cup \sigma_r(A)\},$$

$$\{\overline{\lambda} \mid \lambda \in \sigma_r(B)\} \subset \sigma_p(A) \subset \{\overline{\lambda} \mid \lambda \in \sigma_p(B) \cup \sigma_r(B)\}.$$

Since $\sigma_p(B) = \varnothing$, then $\sigma_r(A) = \varnothing$ and $\{\overline{\lambda} \mid \lambda \in \sigma_r(B)\} = \sigma_p(A)$, whence $\sigma_r(B) = \{\lambda \in \mathbb{C} \mid |\lambda| < 1\}$. Therefore $\sigma_c(A) = \sigma(A) \setminus (\sigma_p(A) \cup \sigma_r(A)) = \{\lambda \in \mathbb{C} \mid |\lambda| = 1\}$, and similarly $\sigma_c(B) = \{\lambda \in \mathbb{C} \mid |\lambda| = 1\}$.

11.5 Let $X = L_2(\mathbb{R}), \ (Ax)(t) = x(t+1), \ t \in \mathbb{R}$. Find $\sigma(A), \ \sigma_p(A), \ \sigma_c(A)$, and $\sigma_r(A)$.

Solution Since the operator A is unitary, it follows from Theorem 11.4 that $\sigma(A) \subset \{\lambda \mid |\lambda| = 1\}$ and $\sigma_r(A) = \varnothing$. Let us find the eigenvalues of the operator A. Assume that $Ax = \lambda x$, where $x \in L_2(\mathbb{R}), \ x \neq 0$. Then $x(t+1) = \lambda x(t) \pmod{m}$, whence

$$x(t) = \sum_{k \in \mathbb{Z}} \chi_{[k,k+1)}(t) x(t) = \sum_{k \in \mathbb{Z}} \lambda^k \chi_{[k,k+1)}(t) x(t-k) \pmod{m}.$$

It follows that

$$\|x\|_2^2 = \sum_{k \in \mathbb{Z}} \int_k^{k+1} |x(t-k)|^2 \, dt = \sum_{k \in \mathbb{Z}} \int_0^1 |x(t)|^2 \, dt.$$

The latter expression can only be equal to 0 or $+\infty$, a contradiction. Therefore $\sigma_p(A) = \varnothing$.

Now let us show that $\sigma(A) = \{\lambda \mid |\lambda| = 1\}$. Indeed, for each $\lambda \in \mathbb{C}$, $|\lambda| = 1$, define a function $x_{\lambda,n} \in L_2(\mathbb{R})$ as follows:

$$x_{\lambda,n}(t) = \sum_{k=1}^{n} \lambda^k \chi_{[k,k+1)}(t), \; n \geq 1.$$

Then $\|x_{\lambda,n}\| = \sqrt{n}$ and $\big((A - \lambda I)x_{\lambda,n}\big)(t) = \lambda\chi_{[0,1)}(t) - \lambda^{n+1}\chi_{[n,n+1)}(t)$, whence $\|(A - \lambda I)x_{\lambda,n}\| = \sqrt{2}$. Thus there does not exist $m > 0$ such that $\|(A - \lambda I)x\| \geq m\|x\|$, $x \in L_2(\mathbb{R})$. According to Theorem 8.2, the operator $A - \lambda I$ is not continuously invertible, so $\lambda \in \sigma(A)$. Therefore $\sigma(A) = \sigma_c(A) = \{\lambda \in \mathbb{C} \mid |\lambda| = 1\}$.

Problems to Solve

11.6°

(1) Prove that the spectrum of a linear operator A in a finite-dimensional normed space X coincides with the set of eigenvalues of the matrix that represents the operator A in some basis.
(2) Give an example of an operator $A \in \mathcal{L}(\mathbb{C}^m)$, $m \geq 2$, such that $r(A) = 0$, but $A \neq 0$.

11.7° Let $A \in \mathcal{L}(X)$. Prove that for $|\lambda| > \|A\|$ the inequality $\|R_\lambda(A)\| \leq (|\lambda| - \|A\|)^{-1}$ holds.

11.8° Let X be a Banach space, $A \in \mathcal{L}(X)$, $\lambda \in \rho(A)$. Prove that A and $R_\lambda(A)$ commute.

11.9° The operator $A \in \mathcal{L}(l_\infty)$ is defined by the formula $Ax = (\alpha_1 x_1, \alpha_2 x_2, \ldots)$, $x \in l_\infty$, where $(\alpha_1, \alpha_2, \ldots) \in l_\infty$. Find $\sigma(A)$, $\sigma_p(A)$, $\sigma_c(A)$, $\sigma_r(A)$, the resolvent and the spectral radius of the operator A.

11.10° Find the spectrum, the set of eigenvalues, the spectral radius and the resolvent of an operator $A : l_p \to l_p$, $1 \leq p \leq +\infty$.

(1) $Ax = (x_1 + x_2, x_2, x_3, \ldots)$;
(2) $Ax = (x_1 + 2x_2, 3x_1 + 4x_2, x_3, x_4, \ldots)$;
(3) $Ax = (-2x_3, x_1, 4x_2, x_4, x_5, \ldots)$;
(4) $Ax = (x_1, \ldots, x_k, 0, 0, \ldots)$, where $k \in \mathbb{N}$ is fixed;
(5) $Ax = (0, x_1, x_2, 0, 0, \ldots)$;
(6) $Ax = (x_1 + 2x_2, 2x_1 + 4x_2, -x_3, -x_4, \ldots)$;
(7) $Ax = (x_1 + 2x_2, 2x_1 + x_2, x_3, x_4, \ldots)$;
(8) $Ax = (3x_1 + 5x_2, -2x_1 + x_2, 0, 0, \ldots)$;
(9) $Ax = (\alpha x_1 + \beta x_2, \gamma x_1 + \delta x_2, x_3, x_4, \ldots)$;
(10) $Ax = (3x_1 + x_3, 2x_2, 5x_3 - x_1, x_4, x_5, \ldots)$.

11.11° Find the norm, the spectral radius, the set of eigenvalues and the spectrum of an operator $A : l_2 \to l_2$.

(1) $Ax = (0, x_1, \frac{x_2}{2}, \dots, \frac{x_n}{n}, \dots)$;
(2) $Ax = (x_2, x_3, \dots, x_{100}, 0, 0, \dots)$;
(3) $Ax = (2x_2, \frac{3}{2}x_3, \dots, \frac{n+1}{n}x_{n+1}, \dots)$;
(4) $Ax = (4x_2, \frac{9}{4}x_3, \dots, \frac{(n+1)^2}{n^2}x_{n+1}, \dots)$.

11.12 Find $\sigma(A)$, $\sigma_p(A)$, $\sigma_c(A)$, $\sigma_r(A)$ and $R_\lambda(A)$.

(1) for the operator $A : C([a, b]) \to C([a, b])$, defined by the formula $(Ax)(t) = (t + 1)x(t)$, $t \in [a, b]$;
(2) for the operator $A : L_2([a, b]) \to L_2([a, b])$, defined by the formula $(Ax)(t) = e^t x(t)$, $t \in [a, b]$;
(3) for the operator $A : C([0, 2\pi]) \to C([0, 2\pi])$, defined by the formula $(Ax)(t) = \sin t\, x(t)$, $t \in [0, 2\pi]$;
(4) for the operator $A : L_2([0, 1]) \to L_2([0, 1])$, defined by the formula $(Ax)(t) = \chi_{[0,\frac{1}{2}]}(t)x(t)$, $t \in [0, 1]$;
(5) for the operator $A : C([0, 1]) \to C([0, 1])$, defined by the formula $(Ax)(t) = a(t)x(t)$, $t \in [0, 1]$, $a \in C([0, 1])$;
(6) for the operator $A : L_2([0, 1]) \to L_2([0, 1])$, defined by the formula $(Ax)(t) = a(t)x(t)$, $t \in [0, 1]$, $a \in C([0, 1])$.

11.13 Let $X = L_p(T, \mu)$, $1 \le p < +\infty$, where μ is a σ-finite measure on T, $a \in L_\infty(T, \mu)$, $(Ax)(t) = a(t)x(t)$, $t \in T$, $x \in X$. Prove that

$$\sigma(A) = \left\{\lambda \in \mathbb{C} \,\middle|\, \forall \varepsilon > 0 : \mu\big(\{t \in T \mid |a(t) - \lambda| < \varepsilon\}\big) > 0\right\}$$

(this set is called the *essential range* of the function a). Find the set of eigenvalues of the operator A.

11.14° Let X be a Banach space, an operator $A \in \mathcal{L}(X)$ has a continuous inverse A^{-1}. Prove that

(1) $\sigma(A^{-1}) = \{\frac{1}{\lambda} \mid \lambda \in \sigma(A)\}$;
(2) A and A^{-1} have the same eigenvectors.

11.15* Let X be a Banach space, $A \in \mathcal{L}(X)$. Prove that

(1) $\sigma(A^2) = \{\lambda^2 \mid \lambda \in \sigma(A)\}$;
(2) $\sigma_p(A^2) = \{\lambda^2 \mid \lambda \in \sigma_p(A)\}$;
(3) if P is a polynomial, then $\sigma(P(A)) = \{P(\lambda) \mid \lambda \in \sigma(A)\}$;
(4) if P is a polynomial, then $\sigma_p(P(A)) = \{P(\lambda) \mid \lambda \in \sigma_p(A)\}$.

Remark If $P(z) = \sum\limits_{k=0}^{n} c_k z^k$, $n \in \mathbb{N} \cup \{0\}$, $c_0, c_1, \dots, c_n \in \mathbb{C}$, then by definition $P(A) = \sum\limits_{k=0}^{n} c_k A^k$, where $A^0 = I$.

11.16 Let X be a Banach space, $A \in \mathcal{L}(X)$ and there exists a nonzero polynomial P such that $P(A) = 0$. Prove that

(1) $\sigma(A) \subset \{\lambda \in \mathbb{C} \mid P(\lambda) = 0\}$;
(2) if P is a nonzero polynomial of the smallest possible degree such that $P(A) = 0$, then $\sigma(A) = \sigma_p(A) = \{\lambda \in \mathbb{C} \mid P(\lambda) = 0\}$.

11.17° Let X be a Banach space, $A \in \mathcal{L}(X)$ and $A^2 = A$. Prove that $\sigma(A) \subset \{0, 1\}$. Provide examples of operators A, for which $A^2 = A$ and
(1) $\sigma(A) = \{0\}$; (2) $\sigma(A) = \{1\}$; (3) $\sigma(A) = \{0, 1\}$.

11.18 Let H be a Hilbert space, $A \in \mathcal{L}(H)$ and $A^* = A$. Prove that if $\sigma(A) \subset \{0, 1\}$, then A is an orthogonal projector.

11.19° Let $X = C([-1, 1])$. Find the spectra of the operators

$$(Ax)(t) = \tfrac{x(t)+x(-t)}{2}, \ (Bx)(t) = \tfrac{x(t)-x(-t)}{2}, \ t \in [-1, 1], \ x \in X.$$

11.20° Let $X = C([-1, 1])$. Find the eigenvalues, the eigenvectors, and the spectrum of the operator $A \in \mathcal{L}(X)$, which is defined by the formula $(Ax)(t) = x(-t), \ t \in [-1, 1], \ x \in X$.

11.21 Let H be a Hilbert space, $A \in \mathcal{L}(H)$ be a normal operator (for example, self-adjoint or unitary). Prove that
(1) $\sigma_r(A) = \varnothing$; (2) $r(A) = \|A\|$.

11.22° Let $X = L_2([-1, 1])$, $a, b \in \mathbb{C}$. Find the eigenvalues, the spectrum, and the resolvent of the operator $A \in \mathcal{L}(X)$, which is defined by the formula $(Ax)(t) = ax(t) + bx(-t), \ t \in [-1, 1], \ x \in X$.

11.23 Prove that any nonempty compact subset of $\mathbb{C}$ is the spectrum of some continuous linear operator.

11.24° Let X be a Banach space, $A \in \mathcal{L}(X)$ be an operator such that $\|Ax\| = \|x\|, \ x \in X$, and $0 \in \rho(A)$. Prove that

$$\sigma(A) \subset \big\{\lambda \in \mathbb{C} \,\big|\, |\lambda| = 1\big\}.$$

11.25° Let $X = C([0, 1])$ and $(Ax)(t) = x(t^2), \ t \in [0, 1]$. Prove that $\sigma(A) \subset \big\{\lambda \in \mathbb{C} \,\big|\, |\lambda| = 1\big\}$.

11.26° Let H be a Hilbert space, P be an orthogonal projector in H, $P \neq 0$ and $P \neq I$. Prove that $\sigma(P) = \sigma_p(P) = \{0, 1\}$, moreover

$$R_\lambda(P) = -\tfrac{1}{\lambda}(I - P) - \tfrac{1}{\lambda - 1}P, \ \lambda \in \mathbb{C} \setminus \{0, 1\}.$$

11.27 Let X be a Banach space, $A \in \mathcal{L}(X)$. Prove that
(1) $\sigma(A') = \sigma(A)$; (2) $\sigma_r(A) \subset \sigma_p(A')$; (3) $\sigma_p(A) \subset (\sigma_p(A') \cup \sigma_r(A'))$; (4) $\sigma_p(A') \subset (\sigma_p(A) \cup \sigma_r(A))$.

Remark The inclusion $\sigma_r(A') \subset \sigma_p(A)$ is not always true, see problem 11.28, item (2).

11.28 Let $\frac{1}{p} + \frac{1}{q} = 1$, $Ax = (x_2, x_3, \ldots)$, $x \in l_p$, $Bx = (0, x_1, x_2, \ldots)$, $x \in l_q$. Find $\sigma(A)$, $\sigma_p(A)$, $\sigma_c(A)$, $\sigma_r(A)$ and do the same for the operator B, if (1) $1 < p < +\infty$; (2)* $p = 1$; (3) $p = +\infty$.

11.29 Find $\sigma(A)$, $\sigma_p(A)$, $\sigma_c(A)$, $\sigma_r(A)$, the resolvent, and the spectral radius of the operator $A : l_p \to l_p$, $1 < p < +\infty$, if

(1) $Ax = (x_3, x_4, \ldots)$;
(2) $Ax = (0, 0, x_1, x_2, \ldots)$.

11.30° Let y, z be fixed elements of a Hilbert space H. The operator $A \in \mathcal{L}(H)$ is defined by the formula $Ax = (x, y)z$. Find $\sigma_p(A)$, $\sigma(A)$, $R_\lambda(A)$, A^n for all $n \geq 1$, and $r(A)$.

11.31° An operator $A \in \mathcal{L}(X)$, where X is a Banach space, is called *quasinilpotent* if $\lim\limits_{n\to\infty} \|A^n\|^{\frac{1}{n}} = 0$. Prove that A is quasinilpotent if and only if $\sigma(A) = \{0\}$.

11.32° Consider the operator $A \in \mathcal{L}(L_2([0, 1]))$, which is defined by the formula $(Ax)(t) = \int\limits_0^t x(s)ds$, $t \in [0, 1]$.

(1) find A^n for all $n \geq 1$, $r(A)$ and $\sigma(A)$;
(2) find $R_\lambda(A)$;
(3) find $A + A^*$ and prove that it is a projector onto a one-dimensional subspace, which consists of constant functions.

11.33 Find the spectrum of the Volterra integral operator in $L_2([0, 1])$ with a continuous kernel K, i.e., the operator $(Ax)(t) = \int\limits_0^t K(t, s)x(s)ds$, $t \in [0, 1]$, $x \in L_2([0, 1])$, where $K \in C(\{(t, s) \mid 0 \leq s \leq t \leq 1\})$.

11.34 Find the spectrum of the operator $A \in \mathcal{L}(L_2([-1, 1]))$, which is defined by the formula $(Ax)(t) = \int\limits_{-t}^t x(s)ds$, $t \in [-1, 1]$.

11.35 Let X_1, X_2 be Banach spaces, $A_1 \in \mathcal{L}(X_1)$ and $A_2 \in \mathcal{L}(X_2)$. Operators A_1 and A_2 are called *similar*, if there exists a continuously invertible operator $B : X_2 \to X_1$ such that $A_2 = B^{-1}A_1B$. Let the operators A_1 and A_2 be similar. Prove that

(1) $\sigma(A_1) = \sigma(A_2)$;
(2) $\sigma_p(A_1) = \sigma_p(A_2)$;
(3) $\sigma_c(A_1) = \sigma_c(A_2)$ and $\sigma_r(A_1) = \sigma_r(A_2)$.

11.36 Let X be a Banach space, $A \in \mathcal{L}(X)$. Prove that the sets $\sigma(A)$, $\sigma_p(A)$, $\sigma_c(A)$, $\sigma_r(A)$ remain the same, if the norm in X is replaced with any equivalent norm.

11.37 Let $s \in \mathbb{R}\backslash\{0\}$ be a fixed number. The operator A_s in $L_2(\mathbb{R})$ is defined by the formula $(A_s x)(t) = x(t+s),\ t \in \mathbb{R}$. Prove that $\sigma_p(A_s) = \sigma_r(A_s) = \varnothing$, and $\sigma(A_s) = \sigma_c(A_s) = \{\lambda \in \mathbb{C} \mid |\lambda| = 1\}$.

11.38 Let X be a Banach space, $A, B \in \mathcal{L}(X)$ and $AB = BA$. Prove that

(1) $r(AB) \le r(A)r(B)$;
(2)* $r(A+B) \le r(A) + r(B)$.

Are these statements correct if A and B do not commute?

11.39° Let $X = C([0, \frac{1}{2}])$ and $(Ax)(t) = tx(t^2),\ t \in [0, \frac{1}{2}],\ x \in X$. Prove that $r(A) = 0$.

11.40° Let X be a Banach space, $A \in \mathcal{L}(X)$. Prove that if for $\lambda \in \mathbb{C}$ there exists a sequence $\{x_n : n \ge 1\} \subset X$, such that $\|x_n\| = 1$ for $n \ge 1$ and

$$\lim_{n\to\infty} (Ax_n - \lambda x_n) = 0,$$

then $\lambda \in \sigma(A)$.

11.41 Let H be a Hilbert space, $A \in \mathcal{L}(H),\ A = A^*$. Prove that

(1) $\lambda \in \rho(A) \Leftrightarrow \exists c > 0\ \forall x \in H : \|(A - \lambda I)x\| \ge c\|x\|$;
(2) $\lambda \in \sigma(A) \Leftrightarrow \exists \{x_n : n \ge 1\} \subset H,\ \|x_n\| = 1$:

$$\lim_{n\to\infty} (Ax_n - \lambda x_n) = 0.$$

11.42 Let H be a Hilbert space, $A \in \mathcal{L}(H),\ A = A^*$. Prove that if $R(A - \lambda I) = H$, then $\lambda \in \rho(A)$.

11.43 The operator A in $C([0, 1])$ is defined by the formula

$$(Ax)(t) = \frac{1}{t}\int_0^t x(s)ds,\ 0 < t \le 1,\ (Ax)(0) = x(0).$$

Prove that $r(A) = 1$.

11.44° Let H be a Hilbert space, $A \in \mathcal{L}(H),\ A = A^*,\ \lambda \in \rho(A) \cap \mathbb{R}$. Prove that the resolvent $(A - \lambda I)^{-1}$ is a self-adjoint operator.

11.45° Let X be a Banach space, $A, B \in \mathcal{L}(X)$. Prove that for all $\lambda \in \rho(A) \cap \rho(B)$ the following identity holds:

$$(A - \lambda I)^{-1} - (B - \lambda I)^{-1} = (A - \lambda I)^{-1}(B - A)(B - \lambda I)^{-1}.$$

11.46 Prove that $r(AB) = r(BA)$ for all $A, B \in \mathcal{L}(X)$.

11.47* Let X be a Banach space, $A, B \in \mathcal{L}(X)$. Prove that $\sigma(AB)\setminus\{0\} = \sigma(BA)\setminus\{0\}$ and for all $\lambda \in \rho(AB)\setminus\{0\}$ the identity

$$A(BA - \lambda I)^{-1}B - \lambda(AB - \lambda I)^{-1} = I$$

holds.

11.48 Let X be a Banach space, $A, B \in \mathcal{L}(X)$ and $AB - BA = cI$ for some $c \in \mathbb{C}$. Prove that $c = 0$.

Chapter 12
Spectral Theory of Compact Operators

Theoretical Background

Theorem 12.1 *Let A be a compact operator in an infinite-dimensional Banach space. Then*

(1) $0 \in \sigma(A)$;
(2) $\sigma(A)$ *is a countable set, the only possible limit point of which is 0;*
(3) *any nonzero point of the spectrum is an eigenvalue of finite multiplicity, i.e., if* $\lambda \in \sigma(A)\backslash\{0\}$, *then* $0 < \dim \operatorname{Ker}(A - \lambda I) < +\infty$.

The description of compact self-adjoint operators is given by the following theorem.

Theorem 12.2 (Hilbert–Schmidt) *Let H be a Hilbert space, A be a compact self-adjoint operator in H. Then there exists an orthonormal system $\{\varphi_n : 1 \leq n \leq N\}$, $N \leq +\infty$, and numbers $\{\lambda_n : 1 \leq n \leq N\} \subset \mathbb{R}$ such that*

$$A = \sum_{n=1}^{N} \lambda_n(\cdot, \varphi_n)\varphi_n.$$

In this case, $\{\lambda_n : n \geq 1\}\backslash\{0\} = \sigma_p(A)\backslash\{0\} = \sigma(A)\backslash\{0\}$, φ_n is an eigenvector which corresponds to the eigenvalue λ_n, and also if $N = +\infty$, then $\lambda_n \to 0$, $n \to \infty$, and the series converges in operator norm.

The series (or a finite sum if $N < +\infty$) in Theorem 12.2 is called the *Schmidt expansion*.

In Theorem 12.2 it can be assumed that $\lambda_n \neq 0$, $n \geq 1$. Then

$$x = \sum_{n=1}^{N}(x, \varphi_n)\varphi_n + Px, \ x \in H,$$

where P is the orthogonal projector onto $\operatorname{Ker} A$.

V. Brayman et al., *Functional Analysis and Operator Theory*, Problem Books in Mathematics, https://doi.org/10.1007/978-3-031-56427-7_12

An operator $A \in \mathcal{L}(H)$, where H is a separable Hilbert space, is called a *Hilbert–Schmidt operator* if for some orthonormal basis $\{\varphi_n : n \geq 1\}$ in H the series $\sum\limits_{n=1}^{\infty} \|A\varphi_n\|^2$ converges. In this case, the number $\|A\|_2 = \left(\sum\limits_{n=1}^{\infty} \|A\varphi_n\|^2\right)^{\frac{1}{2}}$ is called the *Hilbert–Schmidt norm* (or *absolute norm*). The convergence of the mentioned series and the number $\|A\|_2$ do not depend on the choice of the basis, hence the definition is correct. The set of Hilbert–Schmidt operators is denoted by $S_2(H)$. Each Hilbert–Schmidt operator is compact.

An integral operator

$$(Ax)(t) = \int\limits_a^b K(t,s)x(s)ds, \ t \in [a,b], \ x \in L_2([a,b]),$$

is a Hilbert–Schmidt operator if and only if $K \in L_2([a,b]^2)$. In this case, $\|A\|_2 = \left(\int\limits_a^b \int\limits_a^b |K(t,s)|^2 dt ds\right)^{\frac{1}{2}}$ (see problem 12.41).

Examples of Problems with Solutions

12.1 Find the spectrum and the eigenvalues of an operator A in $C([0,1])$, which is defined by the formula

(1) $(Ax)(t) = \int\limits_0^1 e^{2t-s}x(s)ds$; (2) $(Ax)(t) = \int\limits_0^1 (t^2 + ts)x(s)ds$;

(3) $(Ax)(t) = 3x(t) + \int\limits_0^{\pi} (\cos t - 2\cos s)x(s)ds$.

Solution

(1) The operator A is compact, because it is finite-dimensional. Therefore each nonzero point of its spectrum is an eigenvalue. Let $x \neq 0$ be a solution of the equation $Ax = \lambda x$ for some $\lambda \neq 0$. Then $e^{2t} \int\limits_0^1 e^{-s}x(s)ds = \lambda x(t)$. Hence $x(t) = ce^{2t}$, where $c \neq 0$, because $x \neq 0$. We substitute the expression for x into the equation and get $e^{2t} \int\limits_0^1 e^{-s}ce^{2s}\,ds = \lambda c e^{2t}$. Therefore $\int\limits_0^1 e^s\,ds = \lambda$, whence $\lambda = e-1$. So, $\lambda = e-1$ is the unique nonzero eigenvalue of the operator A. It corresponds to the eigenfunction ce^{2t}, where $c \in \mathbb{C}\backslash\{0\}$ is arbitrary. It follows from Theorem 12.1 that $\sigma(A) = \{0, e-1\}$.

Notice that $\lambda = 0$ is also an eigenvalue of the operator A, because there exists a nonzero function x such that $\int_0^1 e^{-s}x(s)ds = 0$. For example, we can substitute $x(t) = 1 + \alpha e^t$ into the equation and determine the coefficient α.

(2) The operator A is compact because it is finite-dimensional. Therefore all nonzero points of its spectrum are eigenvalues. Let $\lambda \neq 0$, $x \neq 0$ and $Ax = \lambda x$, i.e. $t^2 \int_0^1 x(s)ds + t \int_0^1 sx(s)ds = \lambda x(t)$. Left-hand side of the equation contains a linear combination of functions t^2 and t, hence $x(t) = c_1t^2 + c_2t$, where at least one of the constants c_1, c_2 is nonzero, because $x \neq 0$. Substitute the expression for x into the equation: $t^2 \int_0^1 (c_1s^2 + c_2s)ds + t \int_0^1 s(c_1s^2 + c_2s)ds = \lambda(c_1t^2 + c_2t)$. We equate the coefficients of t^2 and of t separately and get the system of equations

$$\begin{cases} \frac{1}{3}c_1 + \frac{1}{2}c_2 = \lambda c_1, \\ \frac{1}{4}c_1 + \frac{1}{3}c_2 = \lambda c_2, \end{cases} \quad \text{or} \quad \begin{cases} (\frac{1}{3} - \lambda)c_1 + \frac{1}{2}c_2 = 0, \\ \frac{1}{4}c_1 + (\frac{1}{3} - \lambda)c_2 = 0. \end{cases}$$

A homogeneous system of linear equations has nontrivial solutions if and only if its determinant equals 0. Hence $(\frac{1}{3} - \lambda)^2 - \frac{1}{8} = 0$, The roots are $\lambda = \frac{1}{3} \pm \frac{1}{2\sqrt{2}}$. Solving the system for these values of λ, we find that $c_2 = \pm\frac{1}{\sqrt{2}}c_1$. So nonzero eigenvalues are $\lambda_1 = \frac{1}{3} - \frac{1}{2\sqrt{2}}$ and $\lambda_2 = \frac{1}{3} + \frac{1}{2\sqrt{2}}$. The corresponding eigenfunctions are $x_1(t) = c\big(t^2 - \frac{1}{\sqrt{2}}t\big)$ and $x_2(t) = c\big(t^2 + \frac{1}{\sqrt{2}}t\big)$, respectively, where $c \in \mathbb{C}\backslash\{0\}$ is arbitrary.

Notice that $\lambda = 0$ is also an eigenvalue of the operator A, because there exists a nonzero function x such that $\int_0^1 x(s)ds = 0$ and $\int_0^1 sx(s)ds = 0$. For example, we can substitute $x(t) = t^2 + \alpha t + \beta$ into these equations and determine the coefficients α and β.

Therefore $\sigma(A) = \sigma_p(A) = \big\{0, \frac{1}{3} \pm \frac{1}{2\sqrt{2}}\big\}$.

(3) Let $(Bx)(t) = \int_0^\pi (\cos t - 2\cos s)x(s)ds$, $x \in C([0, 1])$. The operator B is compact because it is finite-dimensional. Since $A = 3I + B$ and the identity operator I in an infinite-dimensional space is not compact, the operator A is not compact. It follows from the equality $A - \lambda I = B - (\lambda - 3)I$ that $\sigma(A) = \{3 + \lambda \mid \lambda \in \sigma(B)\}$ and $\sigma_p(A) = \{3 + \lambda \mid \lambda \in \sigma_p(B)\}$.

Let us find the eigenvalues of the operator B. Let $\lambda \neq 0$, $x \neq 0$ and $Bx = \lambda x$, i.e. $\cos t \int_0^\pi x(s)ds - 2\int_0^\pi \cos s\, x(s)ds = \lambda x(t)$. Left-hand side of the equation contains a linear combination of functions $\cos t$ and 1, hence

$x(t) = c_1 \cos t + c_2$, where at least one of the constants c_1, c_2 is nonzero, because $x \neq 0$. Substitute the expression for x into the equation:

$$\cos t \int_0^\pi (c_1 \cos s + c_2) ds - 2 \int_0^\pi \cos s \, (c_1 \cos s + c_2) ds = \lambda(c_1 \cos t + c_2).$$

We equate the coefficients of $\cos t$ and of 1 and get the system of equations

$$\begin{cases} \pi c_2 = \lambda c_1, \\ -\pi c_2 = \lambda c_2. \end{cases} \text{ or } \begin{cases} \lambda c_1 - \pi c_2 = 0, \\ \pi c_1 + \lambda c_2 = 0. \end{cases}$$

The determinant $\lambda^2 + \pi^2$ of the system of linear equations equals 0 for $\lambda = \pm \pi i$. Solving the system for these values of λ, we find that $c_2 = \pm c_1 i$. Therefore nonzero eigenvalues of the operator B are $\lambda_1 = -\pi i$ and $\lambda_2 = \pi i$. The corresponding eigenfunctions are $x_1(t) = c\big(\cos t - i\big)$ and $x_2(t) = c\big(\cos t + i\big)$, where $c \in \mathbb{C} \backslash \{0\}$ is arbitrary. Notice that $\lambda = 0$ is also an eigenvalue of the operator B, because there exists a nonzero function x such that $\int_0^\pi x(s) ds = 0$ and $\int_0^1 \cos s \, x(s) ds = 0$. For example, one can take $x(t) = \sin 2t$. Therefore $\sigma_p(B) = \{0, \pm \pi i\}$. Hence $\sigma(B) = \{0, \pm \pi i\}$, because the operator B is compact, and consequently $\sigma(A) = \sigma_p(A) = \{3, 3 \pm \pi i\}$.

12.2 Find the spectrum, the eigenvalues, the eigenfunctions, and the norm of the operator A in $L_2([0, 1])$, defined by the formula $(Ax)(t) = \int_0^1 \min(t, s) x(s) ds$.

Solution Since the kernel $K(t, s) = \min(t, s)$ belongs to $L_2([0, 1]^2)$, then A is a compact operator as a Hilbert–Schmidt operator, and since $K(t, s) = \overline{K(s, t)}$, then A is a self-adjoint operator. Let us find its nonzero eigenvalues. Let $Ax = \lambda x$, $\lambda \neq 0$, $x \neq 0$, i.e.

$$\int_0^t s x(s) ds + t \int_t^1 x(s) ds = \lambda x(t). \qquad (*)$$

For an arbitrary $x \in L_2([0, 1])$ the left-hand side of the equation is a continuous function of t, therefore the right-hand side is also continuous, whence $x \in C([0, 1])$. But if $x \in C([0, 1])$, then the left-hand side belongs to $C^1([0, 1])$, hence $x \in C^1([0, 1])$. By the similar reasonings we obtain that any eigenfunction of the operator A, which corresponds to $\lambda \neq 0$, belongs to $C^\infty([0, 1])$. Now we can differentiate the equation twice:

$$tx(t) + \int_t^1 x(s)ds - tx(t) = \lambda x'(t), \qquad (**)$$

$$-x(t) = \lambda x''(t).$$

It follows from the equations (∗), (∗∗) that $x(0) = 0$ and $x'(1) = 0$. Hence the eigenfunction of the operator A, which corresponds to the eigenvalue $\lambda \neq 0$, is a solution of the boundary value problem

$$\lambda x''(t) + x(t) = 0, \ x(0) = 0, \ x'(1) = 0.$$

For $\lambda < 0$ the general solution of the equation $\lambda x''(t) + x(t) = 0$ has the form $x(t) = c_1 e^{zt} + c_2 e^{-zt}$, where $z = \sqrt{-\frac{1}{\lambda}}$. Since $x(0) = x'(1) = 0$, it is not difficult to find that $c_1 = c_2 = 0$, hence there are no nontrivial solutions for $\lambda < 0$. For $\lambda > 0$ the general solution has the form $x(t) = c_1 \cos zt + c_2 \sin zt$, where $z = \sqrt{\frac{1}{\lambda}}$. Since $x(0) = x'(1) = 0$, we obtain that $c_1 = 0$ and $c_2 \cos z = 0$. Therefore the nontrivial solution exists when $z_n = \sqrt{\frac{1}{\lambda_n}} = \frac{\pi}{2} + \pi n, \ n \geq 0$. Hence the operator A has eigenvalues $\lambda_n = \pi^{-2}(n + \frac{1}{2})^{-2}, \ n \geq 0$. The corresponding eigenfunctions are $x_n(t) = \sin(n + \frac{1}{2})\pi t, \ n \geq 0$. These functions are orthogonal because they correspond to different eigenvalues of the self-adjoint operator. After normalization, we obtain the orthonormal sequence $\varphi_n(t) = \sqrt{2} \sin(n + \frac{1}{2})\pi t, \ n \geq 0$.

For $\lambda = 0$ we can't make a conclusion that $x \in C^\infty([0, 1])$. Nevertheless it is possible to differentiate equation (∗) due to the following Theorem ([21], Ch.9, Sec. 31, Theorem 8): if $x \in L_1([a, b])$ and $y(t) = \int_a^t x(s)ds, \ t \in [a, b]$, then there exists a derivative y' almost everywhere on $[a, b]$ and $y'(t) = x(t) \pmod{m}$. We differentiate (∗) twice and obtain that $x(t) = 0$. Thus the equation $Ax = 0$ has only trivial solution, so $\lambda = 0$ is not an eigenvalue of the operator A.

Hence $\sigma_p(A) = \left\{\pi^{-2}(n + \frac{1}{2})^{-2} \mid n \geq 0\right\}$ and $\sigma(A) = \{0\} \cup \sigma_p(A)$. Since the operator A is self-adjoint, then by Theorem 11.3 we have $\|A\| = r(A) = \sup\{\lambda_n : n \geq 0\} = \frac{4}{\pi^2}$.

Remark By the Hilbert–Schmidt theorem

$$(Ax)(t) = \sum_{n=0}^{\infty} \lambda_n (x, \varphi_n)\varphi_n(t) =$$

$$= \sum_{n=0}^{\infty} \frac{2}{\pi^2 (n + \frac{1}{2})^2} \int_0^1 x(s) \sin(n + \tfrac{1}{2})\pi s \, ds \cdot \sin(n + \tfrac{1}{2})\pi t.$$

This is the Schmidt expansion. It converges in $L_2([0,1])$ uniformly for $x \in \overline{B}(0,1) \subset L_2([0,1])$. Changing the order of integration and summation, we get

$$(Ax)(t) = \int_0^1 \left(\frac{2}{\pi^2} \sum_{n=0}^{\infty} \frac{\sin(n+\frac{1}{2})\pi t \cdot \sin(n+\frac{1}{2})\pi s}{(n+\frac{1}{2})^2} \right) x(s)ds.$$

For justification one can apply the Lebesgue's dominated convergence theorem or use continuity of scalar product and the fact that the series in parentheses converges uniformly in $s \in [0,1]$ for each $t \in [0,1]$, so it converges in $L_2([0,1])$. In partcular, it follows that

$$\min(t,s) = \frac{2}{\pi^2} \sum_{n=0}^{\infty} \frac{\sin(n+\frac{1}{2})\pi t \cdot \sin(n+\frac{1}{2})\pi s}{(n+\frac{1}{2})^2}, \quad t,s \in [0,1].$$

Problems to Solve

12.3° Which of the following sets can be the spectrum of a compact operator in l_2: (1) $[0,1]$, (2) $\{0,1\}$, (3) $\{0\}$, (4) $\{1\}$, (5) $\{\frac{1}{n} : n \geq 1\}$, (6) $\{0\} \cup \{\frac{1}{n} : n \geq 1\}$, (7) $\{0\} \cup \{\frac{i}{n} : n \geq 1\}$?

12.4° Determine whether the operator $(Ax)(t) = \int_0^1 ts^3 x(s)ds$ is compact, and find its spectrum: (1) in the space $L_2([0,1])$; (2) in the space $C([0,1])$.

12.5° Let $u, v \in C([a,b])$. Determine whether the operator

$$(Ax)(t) = \int_a^b u(t)v(s)x(s)ds$$

is compact, and find its spectrum: (1) in the space $L_2([a,b])$; (2) in the space $C([a,b])$.

12.6° Find the spectrum of an operator A in $C([0,1])$.

(1) $(Ax)(t) = \int_0^1 e^{t-s} x(s)ds$;

(2) $(Ax)(t) = t^3 \int_0^1 s^2 x(s)ds$;

(3) $(Ax)(t) = t \int_0^1 \sin s \, x(s)ds$;

(4) $(Ax)(t) = \sin t \cdot x(\frac{\pi}{6})$;

(5) $(Ax)(t) = \int_0^1 s^\alpha x(s)ds$, $\alpha \geq 0$;

(6) $(Ax)(t) = e^t \int_0^1 sx(s)ds$;

(7) $(Ax)(t) = 2x(0) + tx(1)$;

(8) $(Ax)(t) = t \int_0^1 \frac{x(s)}{1+s^2} ds$.

12.7° Find the spectrum of an operator A in $L_2([0, 1])$.

(1) $(Ax)(t) = t^{\alpha} \int\limits_0^1 x(s)ds, \ \alpha > -\frac{1}{2}$;

(2) $(Ax)(t) = e^{2t} \int\limits_0^1 sx(s)ds$;

(3) $(Ax)(t) = t^5 \int\limits_0^1 sx(s)ds$;

(4) $(Ax)(t) = \int\limits_0^1 e^{t-3s}x(s)ds$;

(5) $(Ax)(t) = t^2 \int\limits_0^1 s^2x(s)ds$;

(6) $(Ax)(t) = t \int\limits_0^1 \frac{x(s)}{1+s}ds$;

(7) $(Ax)(t) = t^{\alpha} \int\limits_0^1 s^{\beta}x(s)ds, \ \alpha, \beta \geq 0$;

(8) $(Ax)(t) = \chi_{[\frac{1}{3}, \frac{2}{3}]}(t) \cdot \int\limits_0^1 e^s x(s)ds$.

12.8° Find the spectrum of an operator $A \in \mathcal{L}(X)$.

(1) $(Ax)(t) = \int\limits_{-1}^1 (3t^2s - 2ts^3)x(s)ds, \ X = C([-1, 1])$;

(2) $(Ax)(t) = \int\limits_0^1 (4t^4s^3 + t)x(s)ds, \ X = L_p([0, 1]), \ 1 \leq p \leq +\infty$;

(3) $(Ax)(t) = \int\limits_0^1 (8t^3 + 5s^4)x(s)ds, \ X = C([0, 1])$;

(4) $(Ax)(t) = \int\limits_0^1 (s - 3t^2s^3)x(s)ds, \ X = L_p([0, 1]), \ 1 \leq p \leq +\infty$.

12.9° Determine whether an operator A is compact in $L_2([a, b])$, find its spectrum and norm.

(1) $(Ax)(t) = \int\limits_0^{2\pi} (\cos t \cos s + 1)x(s)ds, \ [a, b] = [0, 2\pi]$;

(2) $(Ax)(t) = \int\limits_0^{2\pi} \sin(t - s)x(s)ds, \ [a, b] = [0, 2\pi]$;

(3) $(Ax)(t) = \int\limits_0^{\pi} \cos(t + s)x(s)ds, \ [a, b] = [0, \pi]$;

(4) $(Ax)(t) = 2x(t) + \int\limits_{-1}^1 (ts - 2t^4s^4)x(s)ds, \ [a, b] = [-1, 1]$;

(5) $(Ax)(t) = \int\limits_{0}^{1}(3ts + 4t^2s^2)x(s)ds,\ [a, b] = [0, 1];$

(6) $(Ax)(t) = \int\limits_{-1}^{1}(t^2s - ts^2)x(s)ds,\ [a, b] = [-1, 1];$

(7) $(Ax)(t) = 3ix(t) + \int\limits_{0}^{\pi}2(\cos t - \cos s)x(s)ds,\ [a, b] = [0, \pi];$

(8) $(Ax)(t) = \int\limits_{0}^{1}(t^4 + s^4)x(s)ds,\ [a, b] = [0, 1];$

(9) $(Ax)(t) = x(t) + \int\limits_{-1}^{1}(ts + 3t^2s^2)x(s)ds,\ [a, b] = [-1, 1];$

(10) $(Ax)(t) = -x(t) + \int\limits_{-\pi}^{\pi}\cos(t - s)x(s)ds,\ [a, b] = [-\pi, \pi].$

12.10° Determine whether an operator A in $l_p,\ 1 \le p \le +\infty$, is compact, find its spectrum. Also find the norm of the operator in case $p = 2$.

(1) $Ax = (x_1 + 3x_2, 3x_1 + x_2, 2x_3, 5x_4, 5x_5, \ldots);$
(2) $Ax = (ix_2, -ix_1, 2ix_4, -2ix_3, x_5, 0, 0, \ldots);$
(3) $Ax = (x_1 + x_3, x_2 + 2x_4, x_1 + x_3, 2x_2 + x_4, 0, 0, \ldots);$
(4) $Ax = (x_2 + x_3, x_1 + x_3, x_1 + x_2, 2x_4, 2x_5, \ldots).$

12.11 Find the spectrum, the spectral radius, and the norm of the operator A in l_2, defined by the formula $Ax = (2x_1 + \sqrt{2}x_2, 3x_2, 0, \ldots)$.

12.12 Let X be a Banach space, $A \in S_\infty(X)$ and $\lambda \ne 0$. Prove that $R(A - \lambda I)$ is a closed subspace in X.

12.13° Provide an example of a compact operator in l_2, for which

(1) 0 is not an eigenvalue;
(2) 0 is an eigenvalue of finite multiplicity;
(3) 0 is an eigenvalue of infinite multiplicity.

12.14° Let A be a compact self-adjoint operator in an infinite-dimensional Hilbert space. Prove that if A has finitely many eigenvalues, then 0 is an eigenvalue of A of infinite multiplicity.

12.15° Let $\{\varphi_n : n \ge 1\}$ be an orthonormal basis in the Hilbert space $H,\ A \in \mathcal{L}(H)$ and $A\varphi_n = \lambda_n\varphi_n,\ n \ge 1$, where $\{\lambda_n : n \ge 1\}$ is a bounded sequence of numbers. Prove that $Ax = \sum\limits_{n=1}^{\infty}\lambda_n(x, \varphi_n)\varphi_n$ for all $x \in H$.

12.16° Let $A \in \mathcal{L}(L_2([0, \pi])),\ (A\varphi_n)(t) = \frac{1}{n^2}\varphi_n(t),\ t \in [0, \pi]$, where $\varphi_n(t) = \sin nt,\ n \ge 1$. Prove that A is a Hilbert–Schmidt integral operator. Find its kernel K.

12.17° Let H be a Hilbert space, $\{\varphi_n : n \geq 1\}$ be an orthonormal sequence in H, $\{\lambda_n : n \geq 1\} \subset \mathbb{C}\backslash\{0\}$ be a bounded sequence, $Ax = \sum_{n=1}^{\infty} \lambda_n (x, \varphi_n)\varphi_n$, $x \in H$. Prove that

(1) $A \in \mathcal{L}(H)$;
(2) $\sigma_p(A)\backslash\{0\} = \{\lambda_n : n \geq 1\}$;
(3) $0 \notin \sigma_p(A)$ if and only if $\{\varphi_n : n \geq 1\}$ is a basis in H.

Under what conditions on the sequence $\{\lambda_n : n \geq 1\}$ is the operator A self-adjoint? nonnegative? normal? compact? a Hilbert–Schmidt operator?

12.18 Let A be a self-adjoint compact operator in a Hilbert space H, $A = \sum_{n=1}^{\infty} \lambda_n (\cdot, \varphi_n)\varphi_n$ be its Schmidt expansion. Prove that

(1) $\forall k \geq 1 : A^k x = \sum_{n=1}^{\infty} \lambda_n^k (x, \varphi_n)\varphi_n$, $x \in H$;
(2) $\forall \lambda \in \mathbb{C}\backslash\{0, \lambda_n, n \geq 1\}$:

$$R_\lambda(A)y = \sum_{n=1}^{\infty} \frac{(y, \varphi_n)}{\lambda_n - \lambda}\varphi_n - \frac{1}{\lambda}Py = \frac{1}{\lambda}\left(\sum_{n=1}^{\infty} \frac{\lambda_n (y, \varphi_n)}{\lambda_n - \lambda}\varphi_n - y\right), \quad y \in H,$$

where P is the orthogonal projector onto Ker A.

12.19 Let H be a separable Hilbert space, A be a compact self-adjoint operator in H, $\{\varphi_n : n \geq 1\}$ be an orthonormal basis in H, such that $A\varphi_n = \lambda_n\varphi_n$, $n \geq 1$. For fixed $\lambda \in \mathbb{C}\backslash\{0\}$ and $y \in H$ consider the equation $(A - \lambda I)x = y$. Prove that

(1) if $\lambda_n \neq \lambda$ for all $n \geq 1$, then the solution is unique and has the form

$$x = \frac{1}{\lambda}\left(\sum_{n=1}^{\infty} \frac{\lambda_n}{\lambda_n - \lambda}(y, \varphi_n)\varphi_n - y\right);$$

(2) if $J = \{n \geq 1 \,|\, \lambda_n = \lambda\} \neq \varnothing$ (the set J is finite, because the operator A is compact), then the solution exists if and only if $(y, \varphi_n) = 0$ for all $n \in J$. In this case, the general solution has the form

$$x = \frac{1}{\lambda}\left(\sum_{n\in\mathbb{N}\backslash J} \frac{\lambda_n}{\lambda_n - \lambda}(y, \varphi_n)\varphi_n - y\right) + \sum_{n\in J} C_n\varphi_n,$$

where C_n, $n \in J$, are arbitrary constants.

12.20 Let A be a nonnegative compact operator in a Hilbert space H. Prove that there exists a unique nonnegative compact operator B such that $B^2 = A$. In this case, if $A = \sum_{n=1}^{\infty} \lambda_n(\cdot, \varphi_n)\varphi_n$ is the Schmidt expansion, then $B = \sum_{n=1}^{\infty} \sqrt{\lambda_n}(\cdot, \varphi_n)\varphi_n$.

Remark The operator B is called *square root* of the operator A. The square root of the operator A is denoted as $B = \sqrt{A}$ or $B = A^{\frac{1}{2}}$. In problem 9.52* the existence and uniqueness of the square root of an arbitrary nonnegative operator A is proven by another method without assumption that the operator A is compact.

12.21° Let $K(t, s) = \sum_{n=1}^{\infty} \frac{1}{n^2} \sin n\pi t \sin n\pi s,\ t, s \in [-1, 1]$. Consider the operator A in $L_2([-1, 1])$, defined by the formula

$$(Ax)(t) = \int_{-1}^{1} K(t, s)x(s)ds,\ t \in [-1, 1].$$

(1) Prove that A is a self-adjoint Hilbert–Schmidt operator (in particular, A is a compact operator).
(2) Find $\sigma(A)$, $r(A)$ and $\|A\|$, as well as an orthonormal sequence of eigenfunctions of the operator A.
(3) Find the Schmidt expansion of A.

12.22 Let K be an even 2π-periodic real-valued function that belongs to $L_2([-\pi, \pi])$, $K(t) = \sum_{n\in\mathbb{Z}} c_n e^{int},\ t \in \mathbb{R}$ be its Fourier series. Consider the operator A in $L_2([-\pi, \pi])$, defined by the formula

$$(Ax)(t) = \int_{-\pi}^{\pi} K(t - s)x(s)ds,\ t \in [-\pi, \pi].$$

(1) Prove that A is a self-adjoint Hilbert–Schmidt operator and the functions $\varphi_n(t) = \frac{1}{\sqrt{2\pi}} e^{int},\ t \in [-\pi, \pi],\ n \in \mathbb{Z}$, which form an orthonormal basis in the space $L_2([-\pi, \pi])$, are the eigenfunctions of the operator A.
(2) Find $\sigma_p(A)$, $\sigma(A)$, $r(A)$ and $\|A\|$.
(3) Find the Schmidt expansion of A.

12.23 Let K be an even 2π-periodic real-valued function that belongs to $L_2([-\pi, \pi])$, $K(t) = \frac{a_0}{2} + \sum_{n=1}^{\infty} a_n \cos nt,\ t \in \mathbb{R}$ be its Fourier series. Consider the operator A in $L_2([-\pi, \pi])$, defined by the formula

$$(Ax)(t) = \int_{-\pi}^{\pi} K(t + s)x(s)ds,\ t \in [-\pi, \pi].$$

(1) Prove that A is a self-adjoint Hilbert–Schmidt operator with the following eigenvalues and orthonormal eigenfunctions: $\lambda_0 = \pi a_0$, $\varphi_0(t) = \frac{1}{\sqrt{2\pi}}$; $\lambda_n^{(1)} = \pi a_n$, $\varphi_n^{(1)}(t) = \frac{\cos nt}{\sqrt{\pi}}$; $\lambda_n^{(2)} = -\pi a_n$, $\varphi_n^{(2)}(t) = \frac{\sin nt}{\sqrt{\pi}}$, $n \geq 1$.
(2) Find $\sigma(A)$, $r(A)$ and $\|A\|$.
(3) Find the Schmidt expansion of A.

12.24° Let A be the integral operator in $L_2([-\pi, \pi])$ with a kernel $K(t, s)$. Find the norm and the spectrum of A and the eigenfunctions that correspond to nonzero eigenvalues.

(1) $K(t, s) = \sum_{n=1}^{N} \lambda_n \cos n(t - s)$, $N \in \mathbb{N}$, $\lambda_1, \ldots, \lambda_N \in \mathbb{C} \setminus \{0\}$;
(2) $K(t, s) = \cos^4(t - s)$;
(3) $K(t, s) = \sum_{n=1}^{\infty} \frac{1}{n^2} \cos nt \cos ns$;
(4) $K(t, s) = \sum_{n=1}^{\infty} \frac{1}{n^2} \sin(2n + 1)t \sin(2n + 1)s$;
(5) $K(t, s) = \sum_{n=0}^{\infty} \frac{1}{(2n+1)^2} \cos(2n + 1)t \cos(2n + 1)s$;
(6) $K(t, s) = \sum_{n=0}^{\infty} \frac{1}{(2n+1)^2} \cos(2n + 1)(t - s)$.

12.25 Find the norm, the spectrum, the eigenvalues and the eigenfunctions of an operator A in $L_2([a, b])$, defined by the formula $(Ax)(t) = \int_a^b K(t, s)x(s)ds$, $t \in [a, b]$, in the following cases:

(1) $K(t, s) = \begin{cases} t(1 - s), & t \leq s, \\ s(1 - t), & s \leq t, \end{cases} \quad a = 0, b = 1;$

(2) $K(t, s) = \begin{cases} \sin t \cos s, & t \leq s, \\ \cos t \sin s, & s \leq t, \end{cases} \quad a = 0, b = \pi;$

(3) $K(t, s) = \begin{cases} \cos t \sin s, & t \leq s, \\ \sin t \cos s, & s \leq t, \end{cases} \quad a = 0, b = \pi;$

(4) $K(t, s) = \begin{cases} (t + 1)s, & t \leq s, \\ t(s + 1), & s \leq t, \end{cases} \quad a = 0, b = 1;$

(5) $K(t, s) = \begin{cases} \sin t \sin(1 - s), & t \leq s, \\ \sin(1 - t) \sin s, & s \leq t, \end{cases} \quad a = 0, b = 1;$

(6) $K(t, s) = \begin{cases} (1 + t)(1 - s), & t \leq s, \\ (1 - t)(1 + s), & s \leq t, \end{cases} \quad a = -1, b = 1.$

12.26* Find the norm of the operator A in $L_2([0, 1])$, defined by the formula $(Ax)(t) = \int\limits_0^t x(s)ds$.

12.27 Let H be a Hilbert space, $A \in \mathcal{L}(H)$ and $A^* = A$. Prove that if $\sigma(A)$ is a countable set which has no nonzero limit points, and every nonzero point of its spectrum is an eigenvalue of finite multiplicity, then the operator A is compact.

12.28* (Nikolsky) Prove that Theorem 12.1 remains valid for the operator $A \in \mathcal{L}(X)$, where X is a Banach space, if the condition $A \in S_\infty(X)$ is replaced with the condition $A^k \in S_\infty(X)$ for some $k \in \mathbb{N}$.

12.29 (Sudakov)

(1) Let X be a Banach space, $M \subset X$, $A \in \mathcal{L}(X)$, $1 \notin \sigma_p(A)$, $A^k \in S_\infty(X)$ for some $k \in \mathbb{N}$, and there exists $C > 0$ such that $\|Ax - x\| < C$ for all $x \in M$. Prove that the set M is bounded.
(2) Prove that in the compactness criteria in $L_p([a, b])$ by Kolmogorov and M. Riesz (see problems 10.81 and 10.82) condition (a) follows from condition (b).

12.30° Let H be a separable Hilbert space, $\{\varphi_n : n \geq 1\}$ and $\{\psi_n : n \geq 1\}$ be orthonormal sequences in H, $\{s_n : n \geq 1\}$ be a sequence of numbers such that $\lim\limits_{n\to\infty} s_n = 0$. For any $x \in H$ put $A_n x = \sum\limits_{k=1}^{n} s_k(x, \varphi_k)\psi_k$, $n \geq 1$, and $Ax = \sum\limits_{n=1}^{\infty} s_n(x, \varphi_n)\psi_n$. Prove that

(1) $A_n \rightrightarrows A$, $n \to \infty$;
(2) A is a compact operator in H;
(3) A is a Hilbert–Schmidt operator if and only if $\sum\limits_{n=1}^{\infty} |s_n|^2 < \infty$.

12.31° Let A be the operator from problem 12.30°. Find A^*, AA^*, A^*A.

12.32 (Canonical form of a compact operator) Prove that any compact operator A in a Hilbert space H can be represented as

$$A = \sum_{n=1}^{N} s_n(\cdot, \varphi_n)\psi_n,$$

where $N \leq +\infty$, $\{\varphi_n : 1 \leq n \leq N\}$ and $\{\psi_n : 1 \leq n \leq N\}$ are orthonormal sequences and $\{s_n : 1 \leq n \leq N\}$ are positive eigenvalues of the operator $|A| = \sqrt{A^*A}$ (square root is defined in problem 12.20). In this case, if $N = +\infty$, then the series converges in operator norm. The numbers s_n, $1 \leq n \leq N$, are called the *singular values* of the operator A.

12.33* Let H be a Hilbert space, $A \in S_\infty(H)$, $\|A^*x\| \leq \|Ax\|$, $x \in H$. Prove that $A^*A = AA^*$.

12.34 (Hilbert) Let H be a Hilbert space, $A \in S_\infty(H)$, $A^* = A$, $\{\lambda_n^+ : 1 \le n \le N^+\}$, $\{\lambda_n^- : 1 \le n \le N^-\}$, where $N^+, N^- \le +\infty$, are the sequences of positive and negative eigenvalues of the operator A, numbered counting their multiplicities, and $\lambda_1^- \le \lambda_2^- \le \ldots < 0 < \ldots \le \lambda_2^+ \le \lambda_1^+$ (one or both of the sequences $\{\lambda_n^+\}$, $\{\lambda_n^-\}$ may be empty). Also, let $\varphi_n^\pm$ be the unit eigenvectors corresponding to the eigenvalues $\lambda_n^\pm$. Prove that

$\lambda_1^+ = \max\{(Ax, x) \mid \|x\| = 1\} = (A\varphi_1^+, \varphi_1^+)$,
$\lambda_n^+ = \max\{(Ax, x) \mid \|x\| = 1,\ (x, \varphi_k^+) = 0,\ 1 \le k \le n-1\} =$
$= (A\varphi_n^+, \varphi_n^+),\ 2 \le n \le N^+$.

Remark For the numbers λ_n^-, similar formulas are true with max replaced with min, because $\{-\lambda_n^- : n \ge 1\}$ are the positive eigenvalues of the operator $-A$.

12.35 (Courant–Fisher; Min-max theorem) Let $\mathcal{L}_n$ be the family of all n-dimensional subspaces of H, $n \ge 1$. In notations from problem 12.34, prove that for all $1 \le n \le N^+$

$$(1)\ \lambda_n^+ = \min_{L \in \mathcal{L}_{n-1}} \max_{x \in L^\perp,\, \|x\|=1} (Ax, x);\quad (2)\ \lambda_n^+ = \max_{L \in \mathcal{L}_n} \min_{x \in L,\, \|x\|=1} (Ax, x).$$

12.36* (Cauchy interlacing theorem) Let H be a Hilbert space, A be a compact self-adjoint operator in H, $G \subset H$ be a subspace of codimension 1 (i.e., $\dim G^\perp = 1$), P be the orthogonal projector onto G. Prove that the eigenvalues $\lambda_n^\pm$ of the operator A and the eigenvalues $\mu_n^\pm$ of the operator PAP, numbered in the same way as in problem 12.34, satisfy the following inequalities:

$\lambda_1^+ \ge \mu_1^+ \ge \lambda_2^+ \ge \mu_2^+ \ge \ldots > 0$, if A has positive eigenvalues,
$\lambda_1^- \le \mu_1^- \le \lambda_2^- \le \mu_2^- \le \ldots < 0$, if A has negative eigenvalues.

12.37 Let H be a Hilbert space, $A \in S_\infty(H)$. Denote by $s_n(A)$, $1 \le n \le N_A$, $N_A \le +\infty$, the singular values of the operator A (see problem 12.32), written in the order of decreasing taking into account their multiplicities. In case $N_A < +\infty$ put $s_n(A) = 0$, $n > N_A$. Prove that

(1) $s_n(A^*) = s_n(A)$, $n \ge 1$;
(2) $s_n(A) = \min\limits_{L:\, \dim L \le n-1} \max\limits_{x \in L^\perp,\, \|x\|=1} \|Ax\|$, $1 \le n \le \dim H$, in particular, $s_1(A) = \|A\|$;
(3) if $B \in \mathcal{L}(H)$, then
$$s_n(BA) \le \|B\|\, s_n(A),\ s_n(AB) \le \|B\|\, s_n(A),\ n \ge 1;$$
(4) if $B \in S_\infty(H)$, $A \ge B \ge 0$, then $s_n(A) \ge s_n(B)$, $n \ge 1$;
(5) $s_n(A) = \min \|A - T\|$, where the minimum is taken over all operators $T \in S_0(H)$, for which $\operatorname{rank} T = \dim R(T) \le n-1$, $n \ge 1$;
(6) if $B \in S_\infty(H)$, then $s_{m+n-1}(A + B) \le s_m(A) + s_n(B)$, $n, m \ge 1$;
(7) if $B \in S_\infty(H)$, then $|s_n(A) - s_n(B)| \le \|A - B\|$, $n \ge 1$.

12.38 Let H be a separable Hilbert space, $A \in S_\infty(H)$, $\{s_n(A) : n \geq 1\}$ be a sequence of singular values of the operator A, numbered with respect to their multiplicities and, possibly, supplemented by zeros. Prove that A is a Hilbert–Schmidt operator if and only if $\sum\limits_{n=1}^{\infty} s_n^2(A) < +\infty$. Moreover $\|A\|_2 = \left(\sum\limits_{n=1}^{\infty} s_n^2(A)\right)^{1/2}$.

12.39 Let H be a separable Hilbert space, $A \in S_\infty(H)$, $\{s_n(A) : n \geq 1\}$ be a sequence of singular values of the operator A. It is said that the operator A belongs to the Schatten class $S_p(H)$, where $1 \leq p < +\infty$, if $\|A\|_p := \left(\sum\limits_{n=1}^{\infty} s_n^p(A)\right)^{1/p} < +\infty$. Prove that

(1) $A \in S_p(H)$ if and only if $A^* \in S_p(H)$;
(2) $\|A\|_p \geq \|A\|$, $A \in S_p(H)$;
(3) if $A \in S_p(H)$ and $B \in \mathcal{L}(H)$, then $AB, BA \in S_p(H)$, and

$$\|AB\|_p \leq \|A\|_p \cdot \|B\|, \quad \|BA\|_p \leq \|A\|_p \cdot \|B\|;$$

(4) $\|A\|_p = \max \left(\sum\limits_{n=1}^{\infty} |(Ae_n, g_n)|^p\right)^{1/p}$, $A \in S_\infty(H)$, where the maximum is taken over all pairs of orthonormal bases $\{e_n : n \geq 1\}$, $\{g_n : n \geq 1\}$;
(5) if $A, B \in S_p(H)$, then $A + B \in S_p(H)$ and

$$\|A + B\|_p \leq \|A\|_p + \|B\|_p;$$

(6) $S_p(H)$ endowed with the norm $\|\cdot\|_p$ is a Banach space;
(7) the set of finite-dimensional operators $S_0(H)$ is dense in $S_p(H)$.

Remark In view of problem 12.38, the definition of the Schatten class $S_p(H)$ for $p = 2$ aligns with the definition of the set of Hilbert–Schmidt operators $S_2(H)$.

12.40 Let H be a separable Hilbert space. An operator A that belongs to the Schatten class $S_1(H)$ is called *nuclear* (or *trace-class operator*), and the number $\|A\|_1$ is called the *nuclear norm*. Prove the following statements:

(1) If $A \in S_1(H)$ and $\{e_n : n \geq 1\}$ is a complete orthonormal basis in H, then the series $\sum\limits_{n=1}^{\infty} (Ae_n, e_n)$ converges absolutely and its sum does not depend on the choice of basis $\{e_n : n \geq 1\}$. This sum is denoted by $\operatorname{tr} A$ and is called the *trace of the operator* A.
(2) If $A \in S_1(H)$ and $A \geq 0$, then $\operatorname{tr} A = \sum\limits_{n=1}^{\infty} \lambda_n$, where λ_n, $n \geq 1$, are the eigenvalues of the operator A, numbered with respect to their multiplicities.

Remark Lidskii's theorem states that the equality $\operatorname{tr} A = \sum\limits_{n=1}^{\infty} \lambda_n$ is valid for all $A \in S_1(H)$. In particular, if an operator A has no eigenvalues, then $\operatorname{tr} A = 0$.

(3) If $A, B \in S_2(H)$, then $AB, BA \in S_1(H)$,

$$\|AB\|_1 \le \|A\|_2 \cdot \|B\|_2, \quad \|BA\|_1 \le \|A\|_2 \cdot \|B\|_2.$$

(4) If $A, B \in S_2(H)$, then $\operatorname{tr} AB = \operatorname{tr} BA$.
(5) For every $A \in S_1(H)$ there exist $B, C \in S_2(H)$ such that $A = BC$.
(6) The norm $\|A\|_2$ is defined by the scalar product $(A, B) := \operatorname{tr} B^* A$.
(7) $S_2(H)$ is a separable Hilbert space with an orthonormal basis, which consists of one-dimensional operators $A_{m,n} = (\cdot, e_n)g_m$, $m, n \in \mathbb{N}$, where $\{e_n : n \ge 1\}$ and $\{g_n : n \ge 1\}$ are orthonormal bases in H.

12.41 Let $H = L_2(T, \mu)$ be a separable Hilbert space. Prove that $A \in S_2(H)$ if and only if A is an integral operator of the form $(Ax)(t) = \int\limits_T K(t, s)x(s)d\mu(s)$, $t \in T$, $x \in H$, where $K \in L_2(T \times T, \mu \times \mu)$. Moreover $\|A\|_2^2 = \int\limits_T \int\limits_T |K(t, s)|^2 d\mu(t)d\mu(s)$.

12.42 Let H_1, H_2 be separable Hilbert spaces, $\{e_n : n \ge 1\}, \{f_n : n \ge 1\}$ be orthonormal bases in H_1 and H_2, respectively. An operator $A \in \mathcal{L}(H_1, H_2)$ is called a Hilbert–Schmidt operator if its Hilbert–Schmidt norm $\|A\|_2 = \left(\sum\limits_{j,k=1}^{\infty} |(Ae_k, f_j)|^2\right)^{\frac{1}{2}}$ is finite. Prove that

(1) Hilbert–Schmidt norm does not depend on the choice of orthonormal bases in spaces H_1 and H_2;
(2) functions $\sinh t$, e^{ikt}, $t \in [-\pi, \pi]$, $k \in \mathbb{Z}$, form a complete orthogonal system in the space $W_2^1([-\pi, \pi])$ (see problems 2.32°, 2.33*);
(3) the embedding operator of the space $W_2^1([-\pi, \pi])$ into $L_2([-\pi, \pi])$ is a Hilbert–Schmidt operator.

12.43° Let $\alpha = (\alpha_1, \alpha_2, \ldots)$, $\alpha_k \ge 1$ for all $k \ge 1$, $l_{2,\alpha}$ be the Hilbert space of all sequences $x = (x_1, x_2, \ldots)$, which satisfy the condition

$$\|x\|_\alpha = \left(\sum_{n=1}^{\infty} \alpha_n x_n^2\right)^{\frac{1}{2}} < \infty,$$

endowed with the corresponding scalar product. Prove the following statements:

(1) $l_{2,\alpha} \subset l_2$, and the norm of the embedding operator $A : l_{2,\alpha} \to l_2$, $Ax = x$, does not exceed one;
(2) the operator A is compact if and only if $\lim\limits_{n\to\infty} \alpha_n = \infty$;
(3) $A : l_{2,\alpha} \to l_2$ is a Hilbert–Schmidt operator if and only if $\sum\limits_{n=1}^{\infty} \frac{1}{\alpha_n} < \infty$.

Chapter 13
Integral Equations

Theoretical Background

Fredholm integral equation of the second kind is the equation of the form

$$x(t) = \lambda \int_a^b K(t, s)x(s)ds + y(t), \ t \in [a, b],$$

where K and y are known functions, x is the function to find, λ is a parameter. We will restrict ourselves to considering cases where $x, y \in C([a, b])$ and $K \in C([a, b]^2)$ or $x, y \in L_2([a, b])$ and $K \in L_2([a, b]^2)$.

Volterra integral equation of the second kind is the equation of the form

$$x(t) = \lambda \int_a^t K(t, s)x(s)ds + y(t), \ t \in [a, b].$$

We will assume that $x, y \in C([a, b])$ and $K \in C(\triangle_{a,b})$ or $x, y \in L_2([a, b])$ and $K \in L_2(\triangle_{a,b})$, where $\triangle_{a,b} := \{(t, s) \in [a, b]^2 \mid s \leq t\}$.

Fredholm integral equations of the first kind and *Volterra integral equations of the first kind* have the form $\int_a^b K(t, s)x(s)ds = y(t), \ t \in [a, b]$, and $\int_a^t K(t, s)x(s)ds = y(t), \ t \in [a, b]$, respectively.

Fredholm integral equations of the second kind can be solved by the method of successive approximations for small $|\lambda|$. More precisely, let $|\lambda| \cdot r(A) < 1$,

V. Brayman et al., *Functional Analysis and Operator Theory*, Problem Books in Mathematics, https://doi.org/10.1007/978-3-031-56427-7_13

where $r(A)$ is the spectral radius of the integral operator with kernel K. Then the successive approximations of the form

$$x_0(t) = 0,\ x_n(t) = \lambda \int_a^b K(t,s)x_{n-1}(s)ds + y(t),\ n \geq 1,$$

converge in the norm of the space $L_2([a,b])$ or $C([a,b])$, respectively, to the unique solution of the equation

$$x(t) = \lambda \int_a^b K(t,s)x(s)ds + y(t).$$

Note that $r(A) \leq \|A\| = \max_{t\in[a,b]} \int_a^b |K(t,s)|ds$ for the equation in $C([a,b])$ and $r(A) \leq \|A\| \leq \left(\int_a^b \int_a^b |K(t,s)|^2 dtds\right)^{\frac{1}{2}}$ for the equation in $L_2([a,b])$.

Volterra integral equations of the second kind can be solved by the method of successive approximations for any λ.

The kernels

$$K_1(t,s) = K(t,s),\ K_n(t,s) = \int_s^t K(t,\tau)K_{n-1}(\tau,s)d\tau,\ n \geq 2,\ a \leq s \leq t \leq b,$$

are called the *iterated kernels* of the Volterra equation, and the expression

$$R_\lambda(t,s) = \sum_{n=1}^{\infty} \lambda^{n-1} K_n(t,s),\ a \leq s \leq t \leq b,$$

is called *the resolvent* of this equation. The series in the definition of the resolvent converges for every $\lambda \in \mathbb{C}$ in $C(\Delta_{a,b})$ if $K \in C(\Delta_{a,b})$ and in $L_2(\Delta_{a,b})$ if $K \in L_2(\Delta_{a,b})$. The solution of the Volterra integral equation of the second kind is given by the formula $x(t) = \lambda \int_a^t R_\lambda(t,s)y(s)ds + y(t)$.

The kernels

$$K_1(t,s) = K(t,s),\ K_n(t,s) = \int_a^b K(t,\tau)K_{n-1}(\tau,s)d\tau,$$

$$n \geq 2,\ (t,s) \in [a,b]^2,$$

are called *iterated kernels* of the Fredholm equation, and the expression

$$R_\lambda(t,s) = \sum_{n=1}^{\infty} \lambda^{n-1} K_n(t,s), \ \ (t,s) \in [a,b]^2,$$

is called the *the resolvent* of this equation. The series in the definition of the resolvent converges for sufficiently small in modulus values of λ in $C([a,b]^2)$ when $K \in C([a,b]^2)$ and in $L_2([a,b]^2)$ if $K \in L_2([a,b]^2)$. The solution of the Fredholm integral equation of the second kind for these values of λ is given by the formula $x(t) = \lambda \int\limits_a^b R_\lambda(t,s)y(s)ds + y(t)$.

The Fredholm equation of the second kind with a degenerate kernel, i.e., a kernel of the form $K(t,s) = \sum\limits_{k=1}^{n} f_k(t)g_k(s)$, can be solved by determining coefficients c_k in the expression $x(t) = \sum\limits_{k=1}^{n} c_k f_k(t) + y(t)$. The substitution of this expression into the integral equation reduces the equation to a linear algebraic system of equations.

If the integral operator $(Ax)(t) = \int\limits_a^b K(t,s)x(s)ds$ is compact in $L_2([a,b])$, in particular, if $K \in L_2([a,b]^2)$, the following result is true for Fredholm equation of the second kind.

Theorem 13.1 (Fredholm Alternative) *Let H be a Hilbert space, $A \in S_\infty(H)$, $\lambda \in \mathbb{C}$. Then*

(1) either the equation $x - \lambda Ax = y$ has a unique solution in H for all $y \in H$, or the homogeneous equation $x - \lambda Ax = 0$ has a nonzero solution;
(2) the homogeneous equation $x - \lambda Ax = 0$ and the homogeneous adjoint equation $f - \overline{\lambda} A^* f = 0$ have the same finite number of linearly independent solutions in H;
(3) the equation $x - \lambda Ax = y$ has solutions in H for those and only for those $y \in H$, which are orthogonal to all solutions of the homogeneous adjoint equation.

Remark 1 The first case in (1) occurs if $\lambda = 0$ or $\frac{1}{\lambda} \in \rho(A)$, and the second case occurs if $\frac{1}{\lambda} \in \sigma_p(A)$. The *characteristic numbers* of the integral equation $x(t) = \lambda \int\limits_a^b K(t,s)x(s)ds + y(t)$ are the values of λ, for which the corresponding homogeneous equation has a nonzero solution. The characteristic numbers of the integral equation are the reciprocals of the nonzero eigenvalues of the integral operator with the same kernel.

Remark 2 Sometimes just the statement (1) of Theorem 13.1 is referred to as *Fredholm alternative.*

Let us state the Fredholm alternative for the linear equation $x - \lambda Ax = y$ with a compact operator A in a Banach space X.

Theorem 13.2 (Fredholm Alternative; see, for Example, [8], Ch. 9, Sec. 2) *Let X be a Banach space, $A \in S_\infty(X)$. For $\lambda \in \mathbb{C}$ consider the couple of the equations*

$$x - \lambda Ax = y, \ (1) \qquad f - \lambda A' f = g, \ (2)$$

where $x, y \in X$, $f, g \in X^*$. Exactly one of the following two cases holds true:

Case 1. Equation (1) has a solution for all elements $y \in X$, equation (2) has a solution for all functionals $g \in X^*$, and the homogeneous equations $x - \lambda Ax = 0$ and $f - \lambda A' f = 0$ have only zero solutions.

Case 2. The homogeneous equations $x - \lambda Ax = 0$ and $f - \overline{\lambda} A' f = 0$ have the same number of linearly independent solutions $x_1, \ldots, x_n$ and $f_1, \ldots, f_n$, respectively, equation (1) has solutions for those and only those elements y, for which $f_k(y) = 0$, $1 \le k \le n$, and equation (2) has solutions for those and only those functionals g, for which $g(x_k) = 0$, $1 \le k \le n$.

Theorem 13.3 (The Bilinear Expansion of the Hermitian Kernel) *Let $K \in L_2([a,b]^2)$ be a Hermitian kernel (i.e., $K(t,s) = \overline{K(s,t)}$ (mod $m \times m$)), A be an integral operator in $L_2([a,b])$ with kernel K, $\{\lambda_n : n \ge 1\}$ be all nonzero eigenvalues of the operator A, numbered counting their multiplicities, $\{\varphi_n : n \ge 1\}$ be an orthonormal system in H, composed of the corresponding eigenfunctions of the operator A, i.e., $A\varphi_n = \lambda_n \varphi_n$, $n \ge 1$. Then*

$$K(t,s) = \sum_{n=1}^{\infty} \lambda_n \varphi_n(t) \overline{\varphi_n(s)},$$

where the series converges in $L_2([a,b]^2)$.

Remark The converse statement is also correct: if $\{\varphi_n : n \ge 1\}$ is an orthonormal system in $L_2([a,b])$, $\lambda_n \in \mathbb{C}\backslash\{0\}$, $n \ge 1$, and $K(t,s) = \sum_{n=1}^{\infty} \lambda_n \varphi_n(t) \overline{\varphi_n(s)}$, where the series converges in $L_2([a,b]^2)$, then the integral operator A with kernel K has nonzero eigenvalues λ_n (with corresponding multiplicities), and the eigenfunctions φ_n, $n \ge 1$ (see item (2) of problem 12.17°).

Under the conditions of Theorem 13.3 the Fredholm integral equation $x = \lambda Ax + y$ has characteristic numbers $\mu_n = \frac{1}{\lambda_n}$, $n \ge 1$, and the solution of the equation can be written as follows:

(1) Let $\lambda \in \mathbb{C}$ be such that $\lambda \ne \mu_n$, $n \ge 1$. Then for every $y \in L_2([a,b])$ the solution is unique and has the form $x = \lambda \sum_{n=1}^{\infty} \frac{(y,\varphi_n)}{\mu_n - \lambda} \varphi_n + y$, where $(\cdot,\cdot)$ is the scalar product in $L_2([a,b])$.
(2) Let $\lambda \in \mathbb{C}$ be such that $J = \{n \in \mathbb{N} : \lambda = \mu_n\} \ne \varnothing$. Then the integral equation has a solution for those and only those $y \in L_2([a,b])$, for which

$(y, \varphi_n) = 0$ for all $n \in J$. In this case, the general solution has the form $x = \lambda \sum\limits_{n\in\mathbb{N}\setminus J} \frac{(y,\varphi_n)}{\mu_n-\lambda}\varphi_n + \sum\limits_{n\in J} c_n\varphi_n + y$, where $c_n,\ n \in J$, are arbitrary constants.

Examples of Problems with Solutions

13.1 Construct the resolvent of the integral equation

$$x(t) = 4\int\limits_0^t e^{t-s}x(s)ds + e^{3t},\ t \in [0, T],$$

and apply it to solve the equation.

Solution We have $K_1(t, s) = e^{t-s}$. Let us calculate the iterated kernels:

$$K_2(t, s) = \int\limits_s^t e^{t-\tau}e^{\tau-s}d\tau = (t-s)e^{t-s},$$

$$K_3(t, s) = \int\limits_s^t e^{t-\tau}(\tau-s)e^{\tau-s}d\tau = e^{t-s}\int\limits_s^t(\tau-s)d\tau = \frac{(t-s)^2}{2!}e^{t-s}.$$

We can hypothesize that $K_n(t, s) = \frac{(t-s)^{n-1}}{(n-1)!}e^{t-s},\ n \geq 1$. We will verify this by induction. The base of induction is already there, we make the step:
$K_{n+1}(t, s) = \int\limits_s^t e^{t-\tau}\frac{(\tau-s)^{n-1}}{(n-1)!}e^{\tau-s}d\tau = e^{t-s}\int\limits_s^t\frac{(\tau-s)^{n-1}}{(n-1)!}d\tau = \frac{(t-s)^n}{n!}e^{t-s}.$

Thus the hypothesis is verified. Now we find the resolvent:

$$R_\lambda(t, s) = \sum_{n=1}^{\infty}\lambda^{n-1}\frac{(t-s)^{n-1}}{(n-1)!}e^{t-s} = e^{(\lambda+1)(t-s)},\ 0 \leq s \leq t < \infty.$$

For $\lambda = 4$ and $y(t) = e^{3t}$ we get $x(t) = y(t) + \lambda\int\limits_0^t R_\lambda(t, s)y(s)ds =$

$$= e^{3t} + 4\int\limits_0^t e^{5(t-s)}e^{3s}ds = e^{3t} + 4e^{5t}\int\limits_0^t e^{-2s}ds =$$

$$= e^{3t} - 2e^{5t}(e^{-2t} - 1) = 2e^{5t} - e^{3t}.$$

13.2 Solve in the space $L_2([0, 1])$ the integral equation with a degenerate kernel

$$x(t) = \lambda \int_0^1 (28t^5s^2 - 15t^3s)x(s)ds + \frac{1}{\sqrt[3]{t}}, \quad \lambda \in \mathbb{C}.$$

Solution Since the kernel of the equation is a linear combination of the functions t^5 and t^3, we look for the solution in the form $x(t) = c_1t^5 + c_2t^3 + \frac{1}{\sqrt[3]{t}}$. Substitute this expression into the equation:

$$c_1t^5 + c_2t^3 + \frac{1}{\sqrt[3]{t}} = \lambda \int_0^1 (28t^5s^2 - 15t^3s)\big(c_1s^5 + c_2s^3 + \frac{1}{\sqrt[3]{s}}\big)ds + \frac{1}{\sqrt[3]{t}},$$

$$c_1t^5 + c_2t^3 = \lambda t^5\big(\frac{7}{2}c_1 + \frac{14}{3}c_2 + \frac{21}{2}\big) - \lambda t^3\big(\frac{15}{7}c_1 + 3c_2 + 9\big).$$

Now we equate the coefficients of t^5 and t^3 and obtain the system of equations

$$\begin{cases} \big(\frac{7\lambda}{2} - 1\big)c_1 + \frac{14\lambda}{3}c_2 = -\frac{21\lambda}{2}, \\ \frac{15\lambda}{7}c_1 + (3\lambda + 1)c_2 = -9\lambda. \end{cases}$$

The determinant of this system is $\triangle = \frac{(7\lambda-2)(3\lambda+1)}{2} - 10\lambda^2 = \frac{\lambda^2+\lambda-2}{2}$. It equals 0 at $\lambda = 1$ and $\lambda = -2$.

For $\lambda = 1$ we obtain the system $\begin{cases} \frac{5}{2}c_1 + \frac{14}{3}c_2 = -\frac{21}{2}, \\ \frac{15}{7}c_1 + 4c_2 = -9, \end{cases}$ which has an infinite number of solutions, because the equations are proportional. The solutions of the integral equation are the functions $x(t) = c_1t^5 + \big(-\frac{9}{4} - \frac{15}{28}c_1\big)t^3 + \frac{1}{\sqrt[3]{t}}$, where $c_1 \in \mathbb{C}$ is arbitrary. For $\lambda = -2$ we have an inconsistent system

$$\begin{cases} -8c_1 - \frac{28}{3}c_2 = 21, \\ -\frac{30}{7}c_1 - 5c_2 = 18. \end{cases}$$

Therefore the integral equation has no solutions.

If $\lambda \neq 1$ and $\lambda \neq -2$, then the system has a unique solution $c_1 = \frac{21\lambda}{\lambda+2}$, $c_2 = -\frac{18\lambda}{\lambda+2}$. Therefore the integral equation has a unique solution

$$x(t) = \frac{21\lambda}{\lambda + 2}t^5 - \frac{18\lambda}{\lambda + 2}t^3 + \frac{1}{\sqrt[3]{t}}.$$

13.3 Find all values of the parameters p, q, r, for which the integral equation $x(t) = \lambda \int\limits_{-1}^{1} (1 + 2t + 3st)x(s)ds + pt^2 + qt + r,\ t \in [-1, 1]$, has a solution in the space $L_2([-1, 1])$ for every $\lambda \in \mathbb{C}$.

Solution Apply the Fredholm alternative. Consider the homogeneous adjoint equation $f(t) = \overline{\lambda} \int\limits_{-1}^{1} (1 + 2s + 3ts) f(s)ds$ and solve it as an equation with a degenerate kernel. We look for a solution in the form $f(t) = c_1 + c_2 t$. Substitute this expression into the equation, we get

$$c_1 + c_2 t = \overline{\lambda} \int\limits_{-1}^{1} (1 + 2s + 3ts)(c_1 + c_2 s)ds,$$

$$c_1 + c_2 t = \overline{\lambda}\big(2c_1 + \frac{4}{3}c_2\big) + \overline{\lambda} t \cdot 2c_2.$$

Now we equate the coefficients at t and constant terms and obtain the system of equations

$$\begin{cases} \big(2\overline{\lambda} - 1\big)c_1 + \frac{4}{3}\overline{\lambda}c_2 = 0, \\ \big(2\overline{\lambda} - 1\big)c_2 = 0. \end{cases}$$

This system has a unique solution for $\lambda \neq \frac{1}{2}$. If $\lambda = \frac{1}{2}$, then $c_2 = 0$ and c_1 is arbitrary, so the homogeneous adjoint equation has solutions of the form $f(t) = c_1 f_1(t)$, where $f_1(t) = 1,\ t \in [-1, 1]$.

In accordance with the Fredholm alternative, the initial equation always has a solution for $\lambda \neq \frac{1}{2}$, and for $\lambda = \frac{1}{2}$ it has solutions if and only if $(y, f_1) = 0$, i.e., $\int\limits_{-1}^{1} (pt^2 + qt + r)dt = \frac{2p}{3} + 2r = 0$. Thus the condition is satisfied for all p, q, r such that $p + 3r = 0$.

13.4 Find the characteristic numbers and corresponding normalized eigenfunctions and solve the integral equation with a symmetric kernel

$$x(t) = \lambda \int\limits_{0}^{2\pi} \big(\cos(t - s) + 2\sin 2t \sin 2s\big)x(s)ds + \cos t,\ t \in [0, 2\pi].$$

Solution We have $K(t, s) = \cos(t - s) + 2\sin 2t \sin 2s =$

$$= \pi\left(\frac{\cos t}{\sqrt{\pi}} \cdot \frac{\cos s}{\sqrt{\pi}} + \frac{\sin t}{\sqrt{\pi}} \cdot \frac{\sin s}{\sqrt{\pi}}\right) + 2\pi\left(\frac{\sin 2t}{\sqrt{\pi}} \cdot \frac{\sin 2s}{\sqrt{\pi}}\right).$$

Note that $\frac{\cos t}{\sqrt{\pi}}, \frac{\sin t}{\sqrt{\pi}}, \frac{\sin 2t}{\sqrt{\pi}}$ is an orthonormal system in $L_2([0, 2\pi])$. Then it follows from remark after the Theorem 13.3 that the nonzero eigenvalues of the integral operator with kernel K are $\lambda_1 = \lambda_2 = \pi$ and $\lambda_3 = 2\pi$, while the corresponding normalized eigenfunctions are $\varphi_1(t) = \frac{\cos t}{\sqrt{\pi}}$, $\varphi_2(t) = \frac{\sin t}{\sqrt{\pi}}$ and $\varphi_3(t) = \frac{\sin 2t}{\sqrt{\pi}}$. Then the characteristic numbers of the integral equation are $\mu_1 = \mu_2 = \frac{1}{\pi}$ and $\mu_3 = \frac{1}{2\pi}$.

For the function $y(t) = \cos t$ we calculate $(y, \varphi_1) = \int\limits_0^{2\pi} \cos t \frac{\cos t}{\sqrt{\pi}} dt = \sqrt{\pi}$, $(y, \varphi_2) = \int\limits_0^{2\pi} \cos t \frac{\sin t}{\sqrt{\pi}} dt = 0$, $(y, \varphi_3) = \int\limits_0^{2\pi} \cos t \frac{\sin 2t}{\sqrt{\pi}} dt = 0$.

If $\lambda \neq \frac{1}{\pi}$ and $\lambda \neq \frac{1}{2\pi}$, then

$$x(t) = \lambda \left(\frac{(y, \varphi_1)}{\frac{1}{\pi} - \lambda} \varphi_1(t) + \frac{(y, \varphi_2)}{\frac{1}{\pi} - \lambda} \varphi_2(t) + \frac{(y, \varphi_3)}{\frac{1}{2\pi} - \lambda} \varphi_3(t) \right) + y(t) =$$

$$= \frac{\lambda\pi \cos t}{1 - \lambda\pi} + 0 + 0 + \cos t = \frac{\cos t}{1 - \lambda\pi}.$$

For $\lambda = \mu_1 = \mu_2 = \frac{1}{\pi}$ there are no solutions, because $(y, \varphi_1) \neq 0$. For $\lambda = \mu_3 = \frac{1}{2\pi}$ and $(y, \varphi_3) = 0$, the solution has the form

$$x(t) = \lambda \left(\frac{(y, \varphi_1)}{\frac{1}{\pi} - \lambda} \varphi_1(t) + \frac{(y, \varphi_2)}{\frac{1}{\pi} - \lambda} \varphi_2(t) \right) + c_3\varphi_3(t) + y(t) =$$

$$= \frac{\lambda\pi \cos t}{1 - \lambda\pi} + 0 + C \sin 2t + \cos t = \frac{\cos t}{1 - \lambda\pi} + C \sin 2t,$$

where $C \in \mathbb{C}$ is an arbitrary constant.

Problems to Solve

13.5° Solve the integral equations by the method of successive approximations, taking $x_0(t) = 0$:

(1) $x(t) = \int\limits_0^t \frac{x(s)}{1+s^2} ds + \arctan t, \ t \in [0, T], \ T > 0$;

(2) $x(t) = \int\limits_0^t e^{-(t-s)} x(s) ds + 1, \ t \in [0, 1]$;

(3) $x(t) = \int\limits_0^t (t - s) x(s) ds + 1, \ t \in [0, 1]$;

(4) $x(t) = \int_0^1 e^{-t-s}x(s)ds + 1,\ t \in [0, 1]$;

(5) $x(t) = \int_0^1 \frac{x(s)}{1+s^2}ds + \arctan t,\ t \in [0, 1]$;

(6) $x(t) = \int_0^1 \frac{te^{ts}}{2(e^t-1)}x(s)ds + \frac{1}{2},\ t \in [0, 1]$;

(7) $x(t) = \int_0^1 sx(s)ds + 1,\ t \in [0, 1]$;

(8) $x(t) = \int_1^2 \frac{t}{2s}x(s)ds + 1,\ t \in [1, 2]$.

13.6° Construct the first two successive approximations of the integral equation, taking $x_0(t) = 0$:

(1) $x(t) = \int_0^1 \frac{x(s)}{(1+s+t)^2}ds + 1,\ t \in [0, 1]$;

(2) $x(t) = \int_0^1 \frac{1}{3}(ts + s^2)x(s)ds + t,\ t \in [0, 1]$;

(3) $x(t) = \int_0^{\frac{\pi}{4}} \frac{1}{10} \arctan(ts)x(s)ds + 1,\ t \in [0, \frac{\pi}{4}]$.

13.7° Construct the resolvent of an integral equation

$$x(t) = \lambda \int_0^t K(t, s)x(s)ds + y(t),\ t \in [0, T],$$

and solve the equation.

(1) $K(t, s) = e^{t^2-s^2},\ \lambda = 2,\ y(t) = e^{t^2+2t}$;
(2) $K(t, s) = e^{t-s},\ \lambda = 2,\ y(t) = \sin t$;
(3) $K(t, s) = \frac{1+t^2}{1+s^2},\ \lambda = 1,\ y(t) = 1 + t^2$;
(4) $K(t, s) = 3^{t-s},\ \lambda = -1,\ y(t) = t3^t$;
(5) $K(t, s) = \frac{2+\cos t}{2+\cos s},\ \lambda = 1,\ y(t) = e^t \sin t$;
(6) $K(t, s) = \frac{\cosh t}{\cosh s},\ \lambda = -1,\ y(t) = t \cosh t$;
(7) $K(t, s) = e^{s-t},\ \lambda = 1,\ y(t) = te^{\frac{t^2}{2}}$.

13.8 Construct the resolvent of an integral equation

$$x(t) = \lambda \int_a^b K(t, s)x(s)ds + y(t),\ t \in [a, b],$$

and solve the equation.

(1) $K(t,s) = e^{t+s}$, $[a,b] = [0,1]$, $\lambda = \frac{1}{e^2-1}$, $y(t) = \sin \pi t$;
(2) $K(t,s) = \sin t \cos s$, $[a,b] = [0, \frac{\pi}{2}]$, $\lambda = -1$, $y(t) = \cos t$;
(3) $K(t,s) = te^s$, $[a,b] = [-1,1]$, $\lambda = \frac{e}{3}$, $y(t) = (3t^2+1)e^{-t}$;
(4) $K(t,s) = t^2s^2$, $[a,b] = [-1,1]$, $\lambda = 2$, $y(t) = e^t$;
(5) $K(t,s) = (1+t)(1-s)$, $[a,b] = [-1,0]$, $\lambda = 1$, $y(t) = \pi \cos \pi t$;
(6) $K(t,s) = ts + t^2s^2$, $[a,b] = [-1,1]$, $\lambda = 1$, $y(t) = 3(t+1)$;
(7) $K(t,s) = 1 + (2t-1)(2s-1)$, $[a,b] = [0,1]$, $\lambda = \frac{1}{2}$, $y(t) = 3t^2$;
(8) $K(t,s) = \sin t \cos s + \cos 2t \sin 2s$, $[a,b] = [0, 2\pi]$, $\lambda = -\frac{1}{\pi}$, $y(t) = \cos t + \sin t$;
(9) $K(t,s) = \sin t \sin s + \cos 2t \cos 2s$, $[a,b] = [0, 2\pi]$, $\lambda = \frac{1}{2\pi}$, $y(t) = \cos 2t$;
(10) $K(t,s) = \cos t \cos s + 3 \sin 2t \sin 2s$, $[a,b] = [0, 2\pi]$, $\lambda = \frac{1}{6\pi}$, $y(t) = 5 \cos t$;
(11) $K(t,s) = 2 + \cos t \cos s$, $[a,b] = [0, 2\pi]$, $\lambda = -\frac{1}{8\pi}$, $y(t) = 2t$;
(12) $K(t,s) = e^t \cos s$, $[a,b] = [0, \pi]$, $\lambda = \frac{1}{e^\pi+1}$, $y(t) = \frac{t}{4}$.

13.9 Let R_λ be the resolvent of an integral equation

$$x(t) = \lambda \int_a^b K(t,s)x(s)ds + y(t), \ t \in [a,b],$$

with a kernel $K \in C([a,b]^2)$, $M = \max_{a \le t,s \le b} |K(t,s)| > 0$. Show that for λ such that $|\lambda| < (M(b-a))^{-1}$ and for all $t, s \in [a,b]$ the function R_λ satisfies the following relations:

(1) $R_\lambda(t,s) = \lambda \int_a^b K(t,u)R_\lambda(u,s)du + K(t,s)$;
(2) $R_\lambda(t,s) = \lambda \int_a^b K(u,s)R_\lambda(t,u)du + K(t,s)$;
(3) $\frac{\partial R_\lambda(t,s)}{\partial \lambda} = \int_a^b R_\lambda(t,u)R_\lambda(u,s)du$.

13.10° Solve an integral equation

$$x(t) = \lambda \int_a^b K(t,s)x(s)ds + y(t), \ t \in [a,b], \ \lambda \in \mathbb{C}$$

with a degenerate kernel.

(1) $K(t,s) = 1$, $y(t) = 1$, $[a,b] = [-1,1]$;
(2) $K(t,s) = ts + t^2s^2$, $y(t) = t^2 + t^4$, $[a,b] = [-1,1]$;
(3) $K(t,s) = \sin(t-2s)$, $y(t) = \cos 2t$, $[a,b] = [0, \pi]$;

(4) $K(t,s) = \sqrt[3]{t} + \sqrt[3]{s},\ y(t) = 1 - 6t^2,\ [a,b] = [-1,1];$
(5) $K(t,s) = t^4 + 5t^3 s,\ y(t) = t^2 - t^4,\ [a,b] = [-1,1];$
(6) $K(t,s) = 2ts^3 + 5t^2 s^2,\ y(t) = 7t^4 + 3,\ [a,b] = [-1,1];$
(7) $K(t,s) = t^2 - ts,\ y(t) = t^2 + t,\ [a,b] = [-1,1];$
(8) $K(t,s) = \sin(2t+s),\ y(t) = \pi - 2t,\ [a,b] = [0,\pi];$
(9) $K(t,s) = \cos(2t+s),\ y(t) = \sin t,\ [a,b] = [0,\pi];$
(10) $K(t,s) = \sin(3t+s),\ y(t) = \cos t,\ [a,b] = [0,\pi];$
(11) $K(t,s) = \sin s + s\cos t,\ y(t) = 1 - \frac{2t}{\pi},\ [a,b] = [0,\pi];$
(12) $K(t,s) = \cos^2(t-s),\ y(t) = 1 + \cos 4t,\ [a,b] = [0,\pi];$
(13) $K(t,s) = \cos t\cos s + \cos 2t\cos 2s,\ y(t) = \cos 3t,\ [a,b] = [0,2\pi];$
(14) $K(t,s) = \cos t\cos s + 2\sin 2t\sin 2s,\ y(t) = \cos t,\ [a,b] = [0,2\pi];$
(15) $K(t,s) = \sin t\sin s + 3\cos 2t\cos 2s,\ y(t) = \sin t,\ [a,b] = [0,2\pi];$

13.11 Solve an integral equation

$$x(t) = \int_a^b K(t,s)x(s)ds + y(t),\ t \in [a,b],$$

with a degenerate kernel.

(1) $K(t,s) = \sum_{k=1}^{100} \frac{\sin kt \sin ks}{\sqrt{k}},\ y(t) = \sin\sqrt{2}t,\ [a,b] = [0,\pi];$
(2) $K(t,s) = s^2 + t^2,\ y(t) = 1,\ [a,b] = [0,1];$
(3) $K(t,s) = \cos(t-s),\ y(t) = \sin t,\ [a,b] = [0,\frac{\pi}{2}];$
(4) $K(t,s) = \begin{cases} 4ts, & 0 \le s \le 1, \\ 3t^2, & 1 < s \le 2, \end{cases}\ y(t) = 1,\ [a,b] = [0,2];$
(5) $K(t,s) = \begin{cases} 3t^2 - 2s, & -1 \le s \le 0, \\ 5ts^3, & 0 < s \le 1, \end{cases}\ y(t) = t^2 + 1,$
$[a,b] = [-1,1].$

13.12° Determine for which $\lambda \in \mathbb{C}$ an integral equation

$$x(t) = \lambda \int_a^b K(t,s)x(s)ds + y(t),\ t \in [a,b],$$

has a solution in the space $L_2([a,b])$.

(1) $K(t,s) = e^{t-s},\ y(t) = 1,\ t \in [a,b];$
(2) $K(t,s) = ts^2 e^{-s^4},\ y(t) = t,\ t \in [a,b].$

13.13° For which $y \in L_2([-1, 1])$ does an equation

$$x(t) = \lambda \int_{-1}^{1} K(t, s)x(s)ds + y(t)$$

have a solution in $L_2([-1, 1])$ for arbitrary $\lambda \in \mathbb{C}$, if

(1) $K(t, s) = ts^3$, (2) $K(t, s) = \frac{s}{t+2}$, (3) $K(t, s) = \frac{t+1}{\sqrt[3]{s}}$?

For which λ is the solution unique?

13.14° Find all values of parameters p, q, r, for which an integral equation

$$x(t) = \lambda \int_{a}^{b} K(t, s)x(s)ds + y(t), \ t \in [a, b],$$

has a solution in the space $L_2([a, b])$ for every $\lambda \in \mathbb{C}$.

(1) $K(t, s) = ts^5 + t^4s^2$, $y(t) = pt^2 + qt + r$, $[a, b] = [-1, 1]$;
(2) $K(t, s) = 1 + t^3s$, $y(t) = pt^2 + qt + r$, $[a, b] = [-1, 1]$;
(3) $K(t, s) = t^2 + ts^2$, $y(t) = pt + q$, $[a, b] = [-1, 1]$;
(4) $K(t, s) = 2t^5s^3 + t^2s^2$, $y(t) = pt + q$, $[a, b] = [-1, 1]$;
(5) $K(t, s) = its + t^2s^4$, $y(t) = pt^2 + q$, $[a, b] = [-1, 1]$;
(6) $K(t, s) = 5t^2s + 3ts^2$, $y(t) = pt + qt^3$, $[a, b] = [-1, 1]$;
(7) $K(t, s) = \sqrt[3]{t} - \sqrt[3]{s}$, $y(t) = pt^2 + qt + r$, $[a, b] = [-1, 1]$;
(8) $K(t, s) = 3t + ts - 5t^2s^2$, $y(t) = pt^5$, $[a, b] = [-1, 1]$;
(9) $K(t, s) = 2t^5s + 3it^2s^2$, $y(t) = pt^3 + qt$, $[a, b] = [-1, 1]$;
(10) $K(t, s) = s \sin t + \cos s$, $y(t) = pt + q$, $[a, b] = [-\frac{\pi}{2}, \frac{\pi}{2}]$;
(11) $K(t, s) = \cos(t + s)$, $y(t) = p \sin t + q$, $[a, b] = [0, \pi]$;
(12) $K(t, s) = t \cos s + \sin t \sin s$, $y(t) = p + q \cos t + r \sin t$, $[a, b] = [-\pi, \pi]$;
(13) $K(t, s) = t \sin s + \cos t$, $y(t) = pt + q$, $[a, b] = [-\pi, \pi]$.

13.15° For which functions $y \in C([0, \pi])$ does the integral equation

$$x(t) = \int_{0}^{\pi} \sin(t - s)x(s)ds + y(t), \ t \in [0, \pi],$$

have a solution in the space $C([0, \pi])$?

13.16° Find all real values of the parameter p, for which the integral equation $x(t) = \lambda \int_0^1 (pt - s)x(s)ds + y(t)$, $t \in [0, 1]$, has a solution for all $\lambda \in \mathbb{R}$ and all $y \in L_2([0, 1])$.

13.17° Find all values of λ, for which an integral equation

$$x(t) = \lambda \int_a^b K(t,s)x(s)ds + y(t), \ t \in [a,b],$$

has a unique solution for every $y \in C([a,b])$.

(1) $K(t,s) = \frac{1}{2} + \cos^2(t+s), \ [a,b] = [0, 2\pi]$;
(2) $K(t,s) = \sin(t-s), \ [a,b] = [0, 2\pi]$;
(3) $K(t,s) = 45t^2s^2 - 2, \ [a,b] = [0,1]$.

13.18 Find all values of λ, for which an integral equation

$$x(t) = \lambda \int_0^1 K(t,s)x(s)ds + y(t), \ t \in [0,1],$$

has a solution for every $y \in L_p([0,1])$.

(1) $K(t,s) = \left(\frac{t}{s}\right)^{\frac{2}{5}} + \left(\frac{s}{t}\right)^{\frac{2}{5}}, \ p = 2$;
(2) $K(t,s) = \frac{s}{\sqrt{1-t}} + \frac{1-s}{\sqrt{t}}, \ p = \frac{3}{2}$;
(3) $K(t,s) = \frac{\sqrt{s}}{\sqrt[4]{t}} + \frac{\sqrt{t}}{\sqrt[4]{s}}, \ p = 3$.
(4) $K(t,s) = \frac{t^7}{\sqrt{1-s^2}}, \ p = +\infty$.

13.19 Let $f, a_1, \ldots, a_n \in C([0,T]), \ T > 0$, and $c_0, c_1, \ldots, c_{n-1} \in \mathbb{C}$. Prove that the function $x \in C^n([0,T])$, for which $x^{(k)}(0) = c_k, \ k = 0, 1, \ldots, n-1$, is a solution of the differential equation

$$x^{(n)}(t) + a_1(t)x^{(n-1)}(t) + \ldots + a_n(t)x(t) = f(t), \ t \in [0,T],$$

if and only if the function $y = x^{(n)}$ is a solution of the Volterra integral equation of the second kind $y(t) = -\int_0^t K(t,s)y(s)ds + F(t), \ t \in [0,T]$, where

$$K(t,s) = \sum_{k=1}^{n} a_k(t)\frac{(t-s)^{k-1}}{(k-1)!}, \quad F(t) = f(t) - \sum_{k=1}^{n}\left(\sum_{j=0}^{k-1} c_{n-k+j}\frac{t^j}{j!}\right)a_k(t).$$

13.20° Find the eigenvalues and corresponding normalized eigenfunctions of an integral equation $x(t) = \lambda \int_a^b K(t,s)x(s)ds, \ t \in [a,b]$.

(1) $K(t,s) = ts, \ [a,b] = [1,2]$;
(2) $K(t,s) = \sum_{n=1}^{\infty} \frac{\sin nt \sin ns}{n^2+n}, \ [a,b] = [0,\pi]$;
(3) $K(t,s) = 5ts - 2t - 2s + 1, \ [a,b] = [0,1]$;

(4) $K(t, s) = 6i(t - s)$, $[a, b] = [0, 1]$;
(5) $K(t, s) = \sin(t + s)$, $[a, b] = [-\pi, \pi]$;
(6) $K(t, s) = \frac{1}{1-\alpha e^{i(t-s)}}$, $0 < \alpha < 1$, $[a, b] = [-\pi, \pi]$.

13.21 Find the eigenvalues and corresponding normalized eigenfunctions of an integral equation with a symmetric kernel

$$x(t) = \lambda \int_a^b K(t, s)x(s)ds + y(t), \ t \in [a, b],$$

and solve the equation for each $\lambda \in \mathbb{C}$.

(1) $K(t, s) = \sin t \sin s$, $[a, b] = [0, 2\pi]$, $y(t) = \cos t + \sin t$;
(2) $K(t, s) = \cos^2(t - s)$, $[a, b] = [-\pi, \pi]$, $y(t) = \sin 2t$;
(3) $K(t, s) = \begin{cases} t(1 - s), & t \le s, \\ s(1 - t), & s \le t, \end{cases}$ $[a, b] = [0, 1]$, $y(t) = \frac{1}{4} \sin 2\pi t + \frac{1}{9} \sin 3\pi t$;
(4) $K(t, s) = \begin{cases} (t + 1)s, & t \le s, \\ t(s + 1), & s \le t, \end{cases}$ $[a, b] = [0, 1]$, $y(t) = \sin \pi t + \pi \cos \pi t$;
(5) $K(t, s) = \begin{cases} (t + 1)(1 - s), & t \le s, \\ (1 - t)(s + 1), & s \le t, \end{cases}$ $[a, b] = [-1, 1]$, $y(t) = 1$;
(6) $K(t, s) = \begin{cases} (e^t - e^{-t})(e^s + e^{2-s}), & t \le s, \\ (e^s - e^{-s})(e^t + e^{2-t}), & s \le t, \end{cases}$ $[a, b] = [0, 1]$, $y(t) = t$;
(7) $K(t, s) = \begin{cases} \sin t \cos s, & t \le s, \\ \sin s \cos t, & s \le t, \end{cases}$ $[a, b] = [0, \frac{\pi}{2}]$, $y(t) = \cos t$;
(8) $K(t, s) = \begin{cases} \sin t \cos s, & t \le s, \\ \sin s \cos t, & s \le t, \end{cases}$ $[a, b] = [0, \pi]$, $y(t) = \sin \frac{3}{2}t$;
(9) $K(t, s) = \begin{cases} \cos t \sin s, & t \le s, \\ \cos s \sin t, & s \le t, \end{cases}$ $[a, b] = [0, \pi]$, $y(t) = t$;
(10) $K(t, s) = \begin{cases} \sin t \sin(1 - s), & t \le s, \\ \sin(1 - t) \sin s, & s \le t, \end{cases}$ $[a, b] = [0, 1]$, $y(t) = \cos \pi t$;
(11) $K(t, s) = \min(t, s)$, $[a, b] = [0, 1]$, $y(t) = \sin \pi t$.

13.22 Transform an integral equation $\mu x(t) = \int_a^b K(t, s)x(s)ds + y(t)$, $t \in [a, b]$, into a differential one and solve it:

(1) $y(t) = \frac{t}{2}$, $K(t, s) = \begin{cases} \frac{\pi^2}{32} t(2 - s), & t \le s, \\ \frac{\pi^2}{32} s(2 - t), & s < t, \end{cases}$ $[a, b] = [0, 2]$, $\mu = 1$;

(2) $y(t) = te^t$, $K(t,s) = \begin{cases} -\frac{1}{\sinh 1}\sinh t \sinh(s-1), & t \le s, \\ -\frac{1}{\sinh 1}\sinh s \sinh(t-1), & s < t, \end{cases}$ $[a,b] = [0,1]$, $\mu = 1$;

(3) $y(t) = \sin^3 \pi t$, $K(t,s) = \begin{cases} (s-1)t, & 0 \le t \le s, \\ (t-1)s, & s < t \le 1, \end{cases}$ $[a,b] = [0,1]$, $\mu = 0$;

(4) $y(t) = 2t^2 - 3t + \frac{5}{2}$, $K(t,s) = |t-s| + t^2$, $[a,b] = [1,2]$, $\mu = 0$.

13.23 By differentiation, transform a Fredholm integral equation of the first kind $\int_a^b K(t,s)x(s)ds = y(t)$, $t \in [a,b]$, into an equation of the second kind:

(1) $K(t,s) = \frac{1}{|t-s|+4}$, $[a,b] = [0,1]$;
(2) $K(t,s) = \sin|t-s|$, $[a,b] = [1,2]$;
(3) $K(t,s) = \frac{e^{i|t-s|}}{1+|t-s|}$, $[a,b] = [0,1]$;
(4) $K(t,s) = t + s + 2\,\mathrm{sign}(t-s)$, $[a,b] = [0,1]$.

13.24 By differentiation, transform a Volterra integral equation of the first kind $\int_a^t K(t,s)x(s)ds = y(t)$, $t \in [a,b]$, into an equation of the second kind:

(1) $K(t,s) = t - s + 1$, $y(t) = t$, $[a,b] = [0,1]$;
(2) $K(t,s) = e^{ts}$, $y(t) = e^t - 1$, $[a,b] = [0,1]$;
(3) $K(t,s) = \sin(t-s)$, $y(t) = \cos t - 1$, $[a,b] = [0,\pi]$.

13.25 By change of variable and differentiation, transform a Volterra integral equation of the first kind

$$\int_a^{\varphi(t)} K(t,s)x(s)ds = y(t),\ t \in [a,b],$$

into an equation of the second kind:

(1) $K(t,s) = 1 - t^2 + s$, $y(t) = t^4$, $\varphi(t) = t^2$, $[a,b] = [0,1]$;
(2) $K(t,s) = t + s$, $y(t) = t^4 + \frac{t^6}{2}$, $\varphi(t) = t^3$, $[a,b] = [1,2]$;
(3) $K(t,s) = 1 + t + s$, $y(t) = \frac{t^2-1}{2} + \frac{t\sqrt{t}-1}{3}$, $\varphi(t) = \sqrt{t}$, $[a,b] = [0,1]$.

13.26 Let K be a 2π-periodic function such that $K(t) = |t|$ for $t \in [-\pi,\pi]$. Find the eigenvalues and normalized eigenfunctions of an integral equation

$$x(t) = \lambda \int_{-\pi}^{\pi} K(t+s)x(s)ds + y(t),\ t \in [-\pi,\pi],$$

and solve the equation for each $\lambda \in \mathbb{C}$.

(1) $y(t) = 1 + 2\sin t - \cos t$;
(2) $y(t) = \cos 2t + \sin 3t$;
(3) $y(t) = \operatorname{sign} t$;
(4) $y(t) = t^2$.

13.27 Let K be a function from problem 13.26. Find the characteristic numbers and normalized eigenfunctions of an integral equation

$$x(t) = \lambda \int_{-\pi}^{\pi} K(t-s)x(s)ds + y(t),\ t \in [-\pi, \pi],$$

and solve the equation for each $\lambda \in \mathbb{C}$.

(1) $y(t) = 1 + \sin t - 2\cos 2t$;
(2) $y(t) = \sin t + \sin 2t + \sin 3t$;
(3) $y(t) = \operatorname{sign} t$;
(4) $y(t) = t^2$.

13.28 Let $K \in C(\mathbb{R})$ be a 2π-periodic function such that $\int_{-\pi}^{\pi} K(t)e^{ikt}dt \neq 0$ for all $k \in \mathbb{Z}$. Prove that $\mu_k = \left(\int_{-\pi}^{\pi} K(t)e^{ikt}dt\right)^{-1}$ and $\varphi_k(t) = \frac{1}{\sqrt{2\pi}}e^{-ikt}$, $k \in \mathbb{Z}$, are the characteristic numbers and corresponding normalized eigenfunctions of the integral equation

$$x(t) = \lambda \int_{-\pi}^{\pi} K(t-s)x(s)ds.$$

13.29 Let $\{\mu_n : n \geq 1\}$ be a sequence of characteristic numbers and $\{\varphi_n : n \geq 1\}$ be a corresponding orthonormal sequence of eigenfunctions of the integral equation $x(t) = \lambda \int_a^b K(t,s)x(s)ds,\ t \in [a,b]$, with Hermitian kernel $K \in C([a,b]^2)$. Prove that

(1) $\sum_{n=1}^{\infty} \frac{|\varphi_n(t)|^2}{\mu_n^2} = \int_a^b |K(t,s)|^2 ds,\ t \in [a,b]$;
(2) $\sum_{n=1}^{\infty} \frac{1}{\mu_n^2} = \int_{[a,b]^2} |K(t,s)|^2 dtds$;
(3) $(Ax, x) = \sum_{n=1}^{\infty} \frac{|(x,\varphi_n)|^2}{\mu_n},\ x \in L_2([a,b])$, where A is the integral operator with kernel K.

13.30

(1) Find all $\lambda \in \mathbb{R}$, for which the homogeneous integral equation $x(t) = \lambda \int_{\mathbb{R}} e^{-|t-s|}x(s)ds,\ t \in \mathbb{R}$, has a nonzero solution $x \in L_\infty(\mathbb{R})$.

(2) Prove that for all $\lambda < \frac{1}{2}$ and $y \in L_\infty(\mathbb{R})$ the integral equation $x(t) = \lambda \int_{\mathbb{R}} e^{-|t-s|} x(s)ds + y(t),\ t \in \mathbb{R}$, has a unique solution in the space $L_\infty(\mathbb{R})$, which is expressed by the formula

$$x(t) = \frac{\lambda}{\sqrt{1-2\lambda}} \int_{\mathbb{R}} e^{-\sqrt{1-2\lambda}|t-s|} y(s)ds + y(t),\ t \in \mathbb{R}.$$

13.31 Let $y \in C^1([0, T]),\ T > 0,\ y(0) = 0$ and $0 < \alpha < 1$. Prove that the function $x(t) = \frac{\sin \alpha\pi}{\pi} \int_0^t \frac{y'(s)}{(t-s)^{1-\alpha}} ds$ satisfies the *Abel equation*

$$\int_0^t \frac{x(s)}{(t-s)^\alpha} ds = y(t),\ t \in [0, T].$$

Chapter 14
Generalized Functions

Theoretical Background

Let $f : \mathbb{R} \to \mathbb{C}$ be a continuous function. The *support* of the function f is the closure of the set $\{x \mid f(x) \neq 0\}$. The support is denoted as $\operatorname{supp} f$. A function f is called *a function with bounded support* if its support $\operatorname{supp} f$ is a bounded set, i.e., there exists $C > 0$ such that $f(x) = 0$ for all x, for which $|x| > C$.

The space of test functions $\mathcal{D}(\mathbb{R})$ is the set of all infinitely differentiable functions from $\mathbb{R}$ to $\mathbb{C}$ with bounded support. A sequence $\{\varphi_n\} \subset \mathcal{D}(\mathbb{R})$ is said to converge to the function $\varphi \in \mathcal{D}(\mathbb{R})$, if

(a) $\exists\, R > 0\ \forall\, n \geq 1\ :\ \operatorname{supp} \varphi_n \subset [-R, R]$,
(b) $\forall\, k \geq 0\ :\ \varphi_n^{(k)}(x) \rightrightarrows \varphi^{(k)}(x)$ on $\mathbb{R}$, $n \to \infty$.

Notation: $\varphi_n \xrightarrow{\mathcal{D}(\mathbb{R})} \varphi$.

The following example of a test function is known as the ε-"cap":

$$\omega_\varepsilon(x) = \begin{cases} C_\varepsilon \exp\{-\frac{\varepsilon^2}{\varepsilon^2 - x^2}\}, & \text{if } |x| < \varepsilon, \\ 0, & \text{if } |x| \geq \varepsilon, \end{cases}$$

where $\varepsilon > 0$ and the constant C_ε is chosen in such a way that $\int\limits_{\mathbb{R}} \omega_\varepsilon(x)dx = 1$.

For every $1 \leq p < +\infty$ the set $\mathcal{D}(\mathbb{R})$ is a dense subset of $L_p(\mathbb{R})$, see item (3) of problem 14.19.

Denote by $\mathcal{D}'(\mathbb{R})$ the set of all continuous linear functionals on the space of test functions $\mathcal{D}(\mathbb{R})$. A functional $f \in \mathcal{D}'(\mathbb{R})$ is called a *generalized function* or *distribution*. The value of a functional f on a test function φ is denoted by $f(\varphi)$ or $\langle f, \varphi \rangle$. To indicate the argument of the test functions, sometimes instead of f and $\langle f, \varphi \rangle$ we write $f(x)$ and $\langle f(x), \varphi(x) \rangle$.

V. Brayman et al., *Functional Analysis and Operator Theory*, Problem Books in Mathematics, https://doi.org/10.1007/978-3-031-56427-7_14

A sequence $\{f_n : n \geq 1\} \subset \mathcal{D}'(\mathbb{R})$ is said to converge to $f \in \mathcal{D}'(\mathbb{R})$, if $f_n(\varphi) \to f(\varphi)$, $n \to \infty$, for every $\varphi \in \mathcal{D}(\mathbb{R})$. In particular, a series of generalized functions $u_1 + u_2 + \ldots + u_n + \ldots$ is said to converge in $\mathcal{D}'(\mathbb{R})$ to the generalized function f, if for every $\varphi \in \mathcal{D}(\mathbb{R})$ the series $\sum\limits_{n=1}^{\infty} \langle u_n, \varphi \rangle$ converges to the number $\langle f, \varphi \rangle$.

The space $\mathcal{D}'(\mathbb{R})$ is complete. It means that any sequence $\{f_n : n \geq 1\} \subset \mathcal{D}'(\mathbb{R})$, for which there exists a limit $\lim\limits_{n\to\infty} \langle f_n, \varphi \rangle$ for every $\varphi \in \mathcal{D}(\mathbb{R})$, is convergent in the space $\mathcal{D}'(\mathbb{R})$ to some $f \in \mathcal{D}'(\mathbb{R})$.

It is said that a generalized function f equals zero on an open set $G \subset \mathbb{R}$, if $f(\varphi) = 0$ for all $\varphi \in \mathcal{D}(\mathbb{R})$ with $\operatorname{supp} \varphi \subset G$. The *support of a generalized function* f is the set of all points such that f does not equal zero in any neighborhood of them. The support of f is denoted by $\operatorname{supp} f$, it is a closed subset of the real line.

A Lebesgue measurable function $f : \mathbb{R} \to \mathbb{C}$ is called locally integrable if $f \in L_1([a, b])$ for all $a, b \in \mathbb{R}$, $a < b$. The set of such functions is denoted by $L_1^{loc}(\mathbb{R})$.

A *regular generalized function* from $\mathcal{D}'(\mathbb{R})$ is a functional of the form

$$\langle f, \varphi \rangle = \int\limits_{\mathbb{R}} f(x)\varphi(x)dx,$$

where $f \in L_1^{loc}(\mathbb{R})$. The following statement is true.

Lemma 14.1 (Du Bois-Reymond) *Let G be an open set in $\mathbb{R}$, and the function $f : G \to \mathbb{C}$ be locally integrable on G. If for any $\varphi \in \mathcal{D}(\mathbb{R})$ with support in G the condition $\int\limits_{\mathbb{R}} f(x)\varphi(x)dx = 0$ is fulfilled, then $f(x) = 0$ almost everywhere in G.*

By this lemma, there is a one-to-one correspondence between the space $L_1^{loc}(\mathbb{R})$ and the space of regular generalized functions. Therefore a function $f \in L_1^{loc}(\mathbb{R})$ can be identified with the respective regular generalized function. For example, we will write

$$\langle 1, \varphi \rangle = \int\limits_{\mathbb{R}} 1 \cdot \varphi(x)dx = \int\limits_{\mathbb{R}} \varphi(x)dx, \quad \varphi \in \mathcal{D}(\mathbb{R}).$$

If $f \in C(\mathbb{R})$, then $\operatorname{supp} f$ as a support of a generalized function coincides with $\operatorname{supp} f$ as a support of a classical function.

Each generalized function that is not regular is called a *singular generalized function*. For example, the *Dirac delta function* δ which is defined as $\langle \delta, \varphi \rangle = \varphi(0)$, $\varphi \in \mathcal{D}(\mathbb{R})$, is singular. The shifted delta functions δ_a, $a \in \mathbb{R}$, which is defined as $\langle \delta_a, \varphi \rangle = \varphi(a)$, $\varphi \in \mathcal{D}(\mathbb{R})$, is also singular. In particlar, $\delta_0 = \delta$.

The product of the function $\alpha(x) \in C^{\infty}(\mathbb{R})$ and the generalized function $f \in \mathcal{D}'(\mathbb{R})$ is the generalized function αf, which is defined as $\langle \alpha f, \varphi \rangle = \langle f, \alpha\varphi \rangle$, $\varphi \in$

$\mathcal{D}(\mathbb{R})$. Let $f \in \mathcal{D}'(\mathbb{R})$, $a, b \in \mathbb{R}$, $a \neq 0$. The generalized function $f(ax+b)$ is defined by the formula

$$\langle f(ax+b), \varphi(x)\rangle = \frac{1}{|a|}\left\langle f(t), \varphi\left(\frac{t-b}{a}\right)\right\rangle, \quad \varphi \in \mathcal{D}(\mathbb{R}).$$

For $a = 1$ we have $\langle f(x+b), \varphi(x)\rangle = \langle f(t), \varphi(t-b)\rangle$, it is the shift of the generalized function f by $(-b)$. In particular,

$$\langle \delta(x-x_0), \varphi(x)\rangle = \langle \delta(t), \varphi(t+x_0)\rangle = \varphi(x_0) = \langle \delta_{x_0}, \varphi\rangle,$$

hence $\delta(x-x_0) = \delta_{x_0}$ is the shifted delta-function.

The *derivative of a generalized function* f is the generalized function f', which is defined by the formula $\langle f', \varphi\rangle = -\langle f, \varphi'\rangle$, $\varphi \in \mathcal{D}(\mathbb{R})$.

Every generalized function f has a derivative $f^{(m)}$ of any order m. This derivative is a generalized function, which acts according to the formula $\langle f^{(m)}, \varphi\rangle = (-1)^m \langle f, \varphi^{(m)}\rangle$.

Let f be a locally integrable function on $\mathbb{R}$, the classical derivative of which is a piecewise continuous function on $\mathbb{R}$. The regular generalized function, generated by this derivative, is denoted by $\{f'(x)\}$ in contrast to the generalized derivative $(Df)(x) = f'(x)$. If f has isolated jump discontinuities at points $\{x_n : n \geq 1\}$, and classical derivative $\{f'\}$ is piecewise continuous on R, then the generalized derivative equals

$$f'(x) = \{f'(x)\} + \sum_{n=1}^{\infty}(f(x_n+) - f(x_n-))\delta_{x_n}.$$

The operation of differentiation is continuous in $\mathcal{D}'(\mathbb{R})$, i.e., if $f_n \xrightarrow{\mathcal{D}'(\mathbb{R})} f$ then $f_n' \xrightarrow{\mathcal{D}'(\mathbb{R})} f'$, $n \to \infty$.

If the generalized function f satisfies the equation $f' = 0$, then $f = \text{const}$, i.e., there exists $C \in \mathbb{C}$ such that $\langle f, \varphi\rangle = \langle C, \varphi\rangle = C \int_{\mathbb{R}} \varphi(x)dx$ for all $\varphi \in \mathcal{D}(\mathbb{R})$.

Let $a_k \in C^{\infty}(\mathbb{R})$, $0 \leq k \leq m$. Consider differential operator

$$Ly = \sum_{k=0}^{m} a_k(x) y^{(k)}, \quad y \in \mathcal{D}'(\mathbb{R}).$$

Let $f \in \mathcal{D}'(\mathbb{R})$. A *generalized solution* of the equation $Ly = f$ is a generalized function $y \in \mathcal{D}'(\mathbb{R})$, which satisfies the equation in the space $\mathcal{D}'(\mathbb{R})$, i.e., $\left\langle \sum_{k=0}^{m} a_k y^{(k)}, \varphi\right\rangle = \left\langle y, \sum_{k=0}^{m} (-1)^k (a_k\varphi)^{(k)}\right\rangle = \langle f, \varphi\rangle$ for all $\varphi \in \mathcal{D}(\mathbb{R})$. Any solution of the equation can be obtained as the sum of its partial solution and of general solution of the homogeneous equation $Ly = 0$.

Let $f, g : \mathbb{R} \to \mathbb{C}$, $f, g \in L_1(\mathbb{R})$. The function

$$h(x) = \int\limits_{\mathbb{R}} f(t)g(x-t)dt, \ x \in \mathbb{R},$$

is called the *convolution* of functions f and g and is denoted by $f*g$. It also belongs to $L_1(\mathbb{R})$.

Let $f \in \mathcal{D}'(\mathbb{R})$, $\varphi \in \mathcal{D}(\mathbb{R})$. Their *convolution* $f * g$ is the function

$$(f*g)(x) = \big\langle f(y), \varphi(x-y)\big\rangle, \ x \in \mathbb{R}.$$

In the right-hand side of this formula the generalized function $f(y)$ acts on the test function $\psi(y) = \varphi(x-y)$, $y \in \mathbb{R}$, and x is a parameter. If $f \in L_1^{loc}(\mathbb{R})$, then

$$(f*g)(x) = \int\limits_{\mathbb{R}} f(y)\varphi(x-y)dy, \ x \in \mathbb{R}.$$

Consider the differential operator with constant coefficients

$$L = D^n + a_1 D^{n-1} + \ldots + a_{n-1}D + a_n I, \ n \geq 1, \ a_1, \ldots, a_n \in \mathbb{C}.$$

The generalized function $\mathcal{E}$ is called the *fundamental solution* of the operator L, if $L\mathcal{E} = \delta_0$. Let z be a solution to the Cauchy problem

$$Lz = 0, \ z(0) = z'(0) = \ldots = z^{(n-2)}(0) = 0, \ z^{(n-1)}(0) = 1.$$

Then the fundamental solution of the operator L is the locally integrable function $\mathcal{E}(x) = \theta(x)z(x)$, $x \in \mathbb{R}$, where $\theta(x) = \chi_{[0,+\infty)}(x)$ is the Heaviside function. For $\varphi \in \mathcal{D}(\mathbb{R})$, the linear inhomogeneous differential equation $Lu = \varphi$, $u \in C^\infty(\mathbb{R})$, has a partial solution $u(x) = (\mathcal{E} * \varphi)(x)$, $x \in \mathbb{R}$. The generalization of these statements to the case of a differential operator with nonconstant coefficients is given in problem 14.59.

The Schwartz space of test functions $\mathcal{S}(\mathbb{R})$, or the space of rapidly decreasing functions, is the set of all functions $\varphi : \mathbb{R} \to \mathbb{C}$ such that

$$\varphi \in C^\infty(\mathbb{R}) \text{ and } \varphi^{(k)}(x) = o\Big(\frac{1}{|x|^m}\Big), \ x \to \pm\infty, \text{ for all } k, m \in \mathbb{N} \cup \{0\}.$$

A sequence $\{\varphi_n\} \subset \mathcal{S}(\mathbb{R})$ is said to converge to a function $\varphi \in \mathcal{S}(\mathbb{R})$, if for all $k, m \geq 0$ the sequence $\{x^m \varphi_n^{(k)}(x), \ x \in \mathbb{R} : n \geq 1\}$ converges to the function $x^m \varphi^{(k)}(x)$, $x \in \mathbb{R}$, uniformly on $\mathbb{R}$, as $n \to \infty$. Notation: $\varphi_n \xrightarrow{\mathcal{S}(\mathbb{R})} \varphi$.

For every $1 \le p < +\infty$ the set $\mathcal{S}(\mathbb{R})$ is a dense subset of $L_p(\mathbb{R})$, see problem 14.64°.

Denote by $\mathcal{S}'(\mathbb{R})$ the set of all continuous linear functionals on the space of test functions $\mathcal{S}(\mathbb{R})$. These functionals are called *tempered distributions*. The notation $\langle f, \varphi\rangle$ and the convergence of sequences are introduced in the space $\mathcal{S}'(\mathbb{R})$ similarly to the space $\mathcal{D}'(\mathbb{R})$. The operations of differentiation and linear change of argument are also defined in $\mathcal{S}'(\mathbb{R})$ similarly to $\mathcal{D}'(\mathbb{R})$.

For $\varphi \in L_1(\mathbb{R})$, and in particular for $\varphi \in \mathcal{S}(\mathbb{R})$, the *Fourier transform* F is defined as

$$(F[\varphi])(y) = \frac{1}{\sqrt{2\pi}} \int\limits_{\mathbb{R}} e^{ixy} \varphi(x)dx, \ \ y \in \mathbb{R}.$$

Fourier transform $\mathcal{S}(\mathbb{R}) \ni \varphi \mapsto F[\varphi] \in \mathcal{S}(\mathbb{R})$ is a continuous linear bijection on $\mathcal{S}(\mathbb{R})$. Its inverse map is the inverse Fourier transform F^{-1}, where

$$(F^{-1}[\varphi])(y) = \frac{1}{\sqrt{2\pi}} \int\limits_{\mathbb{R}} e^{-ixy} \varphi(x)dx, \ \ y \in \mathbb{R}, \ \ \varphi \in \mathcal{S}(\mathbb{R}).$$

For every $\varphi \in \mathcal{S}(\mathbb{R})$ we have $\|F[\varphi]\|_{L_2(\mathbb{R})} = \|\varphi\|_{L_2(\mathbb{R})}$. The extension by continuity of the operator F from $\mathcal{S}(\mathbb{R})$ to $L_2(\mathbb{R})$ is a unitary operator in $L_2(\mathbb{R})$. This extension $F[f]$, $f \in L_2(\mathbb{R})$, is called the Fourier transform of the function $f \in L_2(\mathbb{R})$. It can be defined as follows:

$$(F[f])(y) = \lim_{A\to+\infty} \frac{1}{\sqrt{2\pi}} \int\limits_{-A}^{A} e^{ixy} f(x)dx, \ \ y \in \mathbb{R},$$

where the limit is taken in the sense of convergence in $L_2(\mathbb{R})$. For $f \in L_2(\mathbb{R}) \cap L_1(\mathbb{R})$ this limit equals almost everywhere to the Fourier transform $F[f]$, defined above for the function from $L_1(\mathbb{R})$.

Let $f \in \mathcal{S}'(\mathbb{R})$. The Fourier transform $F[f]$ is the generalized function from $\mathcal{S}'(\mathbb{R})$, which acts by the formula $\langle F[f], \varphi\rangle = \langle f, F[\varphi]\rangle$ for all $\varphi \in \mathcal{S}(\mathbb{R})$. The operator F in $\mathcal{S}'(\mathbb{R})$ is a continuous linear bijection. The operator $F^{-1}[f] = F[f(-x)]$, $f \in \mathcal{S}'(\mathbb{R})$, is the inverse for the operator F on $\mathcal{S}'(\mathbb{R})$, this operator F^{-1} is called the inverse Fourier transform.

Most of the definitions given above can be extended without substantial changes to functions defined on $\mathbb{R}^m$. We will give only a few definitions for such functions.

Let $f \in \mathcal{D}'(\mathbb{R}^m)$, A be a nondegenerate matrix of size $m \times m$, $b \in \mathbb{R}^m$. The generalized function $f(Ay + b) \in \mathcal{D}'(\mathbb{R}^m)$ is defined by the formula

$$\big\langle f(Ay+b), \varphi(y)\big\rangle = \left\langle f(x), \frac{\varphi(A^{-1}(x-b))}{|\det A|}\right\rangle, \ \ \varphi \in \mathcal{D}(\mathbb{R}^m.)$$

In particular, if $A = aI$, $a \neq 0$, then $\langle f(ay+b), \varphi(y)\rangle = \left\langle f(x), \frac{\varphi(\frac{1}{a})}{a^m}\right\rangle$, $\varphi \in \mathcal{D}(\mathbb{R}^m)$.

Let $\alpha = (\alpha_1, \ldots, \alpha_m)$ be a multi-index, $\alpha_k \geq 0$, $1 \leq k \leq m$. For $f \in \mathcal{D}'(\mathbb{R}^m)$ the derivative $D^\alpha f$ is the generalized function that acts by the formula

$$\langle D^\alpha f, \varphi\rangle = (-1)^{|\alpha|}\langle f, D^\alpha \varphi\rangle, \quad \varphi \in \mathcal{D}(\mathbb{R}^m).$$

Here $|\alpha| = \alpha_1 + \ldots + \alpha_m$.

Fourier transform in the space $\mathcal{S}(\mathbb{R}^m)$ is given by the formula

$$(F[\varphi])(y) = \frac{1}{(2\pi)^{\frac{m}{2}}} \int\limits_{\mathbb{R}^m} e^{i(x,y)}\varphi(x)dx, \quad \varphi \in \mathcal{S}(\mathbb{R}^m).$$

Examples of Problems with Solutions

14.1 Let φ be a nonzero function from $\mathcal{D}(\mathbb{R})$. Does the sequence $\varphi_n(x) = \frac{1}{n}\varphi(\frac{x}{n})$, $x \in \mathbb{R}$, $n \geq 1$, converge in $\mathcal{D}(\mathbb{R})$?

Solution We will show that the condition (a) from the definition of convergence in $\mathcal{D}(\mathbb{R})$ is violated. Indeed, suppose that there exists $R > 0$ such that $\operatorname{supp}\varphi_n \subset [-R, R]$ for all $n \geq 1$. Then $\varphi(\frac{x}{n}) = 0$ for all $x \in (-\infty, -R) \cup (R, +\infty)$ and $n \geq 1$. Hence $\varphi(t) = 0$ for $t \in \bigcup\limits_{n=1}^{\infty}((-\infty, -\frac{R}{n}) \cup (\frac{R}{n}, +\infty)) = \mathbb{R} \setminus \{0\}$. But φ is continuous at zero, therefore $\varphi(0) = 0$ as well, i.e., $\varphi(t) \equiv 0$. Since φ is a nonzero function, we got a contradiction. Thus the condition (a) fails, so the sequence diverges.

14.2 Let $n \in \mathbb{N}$, $\xi \in \mathcal{D}(\mathbb{R})$ be such that $\xi(x) = 1$ in some neighborhood of zero, and $\varphi \in \mathcal{D}(\mathbb{R})$. Prove that the function

$$\psi(x) = \frac{1}{x^n}\left(\varphi(x) - \xi(x)\sum_{k=0}^{n-1}\frac{\varphi^{(k)}(0)}{k!}x^k\right), \quad x \in \mathbb{R}\setminus\{0\},$$

which is defined by continuity at $x = 0$, belongs to the space $\mathcal{D}(\mathbb{R})$.

Remark The function $\xi \in \mathcal{D}(\mathbb{R})$, for which $\xi(x) = 1$ in some neighborhood of zero, exists, see problem 14.18°.

Solution Assume that $\xi(x) = 1$ for $|x| < \delta$. Then for $x \neq 0$, $|x| < \delta$, we have

$$\psi(x) = \frac{1}{x^n}\left(\varphi(x) - \sum_{k=0}^{n-1}\frac{\varphi^{(k)}(0)}{k!}x^k\right) = \frac{1}{(n-1)!x^n}\int\limits_0^x \varphi^{(n)}(t)(x-t)^{n-1}dt$$

according to Taylor's formula with integral remainder. Make substitution $z = \frac{t}{x}$:

$$\psi(x) = \frac{1}{(n-1)!} \int_0^1 \varphi^{(n)}(xz)(1-z)^{n-1} dz.$$

It follows that ψ can be extended by continuity to zero, if we put

$$\psi(0) = \frac{1}{(n-1)!} \int_0^1 \varphi^{(n)}(0)(1-z)^{n-1} dz = \frac{\varphi^{(n)}(0)}{n!}.$$

Then ψ becomes infinitely differentiable for $|x| < \delta$, because

$$\psi^{(k)}(x) = \frac{1}{(n-1)!} \int_0^1 \varphi^{(n+k)}(xz) z^k (1-z)^{n-1} dz, \ k \geq 1.$$

The infinite differentiability of ψ at points $x \neq 0$ follows from the fact that $\varphi \in C^\infty(\mathbb{R})$, $\xi \in C^\infty(\mathbb{R})$. Since φ and ξ have bounded supports, the function ψ also has a bounded support. Therefore $\psi \in \mathcal{D}(\mathbb{R})$.

14.3 Using the definition, prove that the following functionals are the generalized functions: (1) $f(\varphi) = \int\limits_{\mathbb{R}} x\varphi(x)dx$, (2) $g(\varphi) = \sum\limits_{n=1}^{\infty} \varphi^{(n)}(n)$, $\varphi \in \mathcal{D}(\mathbb{R})$. Are they regular? Find supports of these generalized functions.

Solution

(1) Since $\varphi \in C(\mathbb{R})$ and $\operatorname{supp} \varphi \subset [-a, a]$ for some $a > 0$, then $\int\limits_{\mathbb{R}} x\varphi(x)dx = \int\limits_{-a}^{a} x\varphi(x)dx$, it is a Riemann integral over a finite interval. Thus the functional $f(\varphi)$ is correctly defined.

Let $\varphi_1, \varphi_2 \in \mathcal{D}(\mathbb{R})$ and $\lambda_1, \lambda_2 \in \mathbb{C}$. Then there exists $a > 0$ such that the supports of both functions φ_1 and φ_2 are subsets of $[-a, a]$. Hence

$$f(\lambda_1\varphi_1 + \lambda_2\varphi_2) = \int_{-a}^{a} x(\lambda_1\varphi_1 + \lambda_2\varphi_2)dx =$$

$$= \lambda_1 \int_{-a}^{a} x\varphi_1 dx + \lambda_2 \int_{-a}^{a} x\varphi_2 dx = \lambda_1 f(\varphi_1) + \lambda_2 f(\varphi_2).$$

Therefore f is a linear functional on $\mathcal{D}(\mathbb{R})$. Now we shall prove its continuity. Let $\varphi_n \xrightarrow{\mathcal{D}(\mathbb{R})} \varphi$, $n \to \infty$. Then (a) there exists $a > 0$ such that $\operatorname{supp}\varphi \subset [-a, a]$ and $\operatorname{supp}\varphi_n \subset [-a, a]$ for all $n \geq 1$; (b) $\varphi_n \rightrightarrows \varphi$ on $\mathbb{R}$, $n \to \infty$. Due to (b) we can pass to the limit under the Riemann integral sign, thus $f(\varphi_n) = \int\limits_{-a}^{a} \varphi_n(x)dx \to \int\limits_{-a}^{a} \varphi(x)dx = f(\varphi), n \to \infty$. Therefore f is a continuous linear functional on $\mathcal{D}(\mathbb{R})$, i.e., $f \in \mathcal{D}'(\mathbb{R})$.

Notice that $f(\varphi) = \int\limits_{\mathbb{R}} h(x)\varphi(x)dx$, $\varphi \in \mathcal{D}(\mathbb{R})$, where $h(x) = x$, $x \in \mathbb{R}$. Since $h \in C(\mathbb{R})$, we have $h \in L_1^{loc}(\mathbb{R})$. Thus f is a regular generalized function and the support of f coincides with the support of h. Consequently,

$$\operatorname{supp} f = \operatorname{supp} h = \overline{\{x \in \mathbb{R} \,|\, h(x) \neq 0\}} = \overline{\mathbb{R}\backslash\{0\}} = \mathbb{R}.$$

(2) Let $\varphi \in \mathcal{D}(\mathbb{R})$. Then there exists $k \in \mathbb{N}$ such that $\varphi(x) = 0$ for $|x| > k$. Hence the value of $g(\varphi)$ equals to the finite sum $\sum\limits_{n=1}^{k} \varphi^{(n)}(n)$, which is correctly defined due to $\varphi \in C^{\infty}(\mathbb{R})$.

Let us check that g is linear. If $\varphi_1, \varphi_2 \in \mathcal{D}(\mathbb{R})$ and $\lambda_1, \lambda_2 \in \mathbb{C}$, then there exists $k \in \mathbb{N}$ such that $\varphi_1(x) = 0$ and $\varphi_2(x) = 0$ for $|x| > k$. Then

$$g(\lambda_1\varphi_1 + \lambda_2\varphi_2) = \sum_{n=1}^{k} (\lambda_1\varphi_1 + \lambda_2\varphi_2)^{(n)}(n) =$$

$$= \lambda_1 \sum_{n=1}^{k} \varphi_1^{(n)}(n) + \lambda_2 \sum_{n=1}^{k} \varphi_2^{(n)}(n) = \lambda_1 g(\varphi_1) + \lambda_2 g(\varphi_2),$$

therefore g is a linear functional on $\mathcal{D}(\mathbb{R})$.

Let us check that g is continuous. Let $\varphi_m \xrightarrow{\mathcal{D}(\mathbb{R})} \varphi$. Then there exists $k \in \mathbb{N}$ such that $\operatorname{supp}\varphi \subset [-k, k]$ and $\operatorname{supp}\varphi_m \subset [-k, k]$ for all $m \geq 1$. Moreover, for each $n \in \mathbb{N}$ we have $\varphi_m^{(n)} \rightrightarrows \varphi^{(n)}$ on $\mathbb{R}$, $m \to \infty$, whence $\varphi_m^{(n)}(n) \to \varphi^{(n)}(n)$, $m \to \infty$. Therefore

$$g(\varphi_m) = \sum_{n=1}^{k} \varphi_m^{(n)}(n) \to \sum_{n=1}^{k} \varphi^{(n)}(n) = g(\varphi),\ m \to \infty.$$

Hence g is continuous so $g \in \mathcal{D}'(\mathbb{R})$.

We will show by contradiction that the function g is singular. Assume that g is regular. Then

$$\exists h \in L_1^{loc}(\mathbb{R})\ \forall \varphi \in \mathcal{D}(\mathbb{R})\ :\ g(\varphi) = \int_{\mathbb{R}} h(x)\varphi(x)dx.$$

If $\operatorname{supp}\varphi \subset (-\infty, 1)$, then $g(\varphi) = \sum_{n=1}^{\infty} \varphi^{(n)}(n) = 0$. Therefore by the Du Bois-Reymond lemma (Lemma 14.1) $h(x) = 0$ almost everywhere on $(-\infty, 1)$. Similarly, if $\operatorname{supp}\varphi \subset (k, k+1)$ for some $k \in \mathbb{N}$, then $\varphi^{(n)}(n) = 0$ for all $n \in \mathbb{N}$, $g(\varphi) = 0$ and by the Du Bois-Reymond lemma $h(x) = 0$ almost everywhere on $(k, k+1)$. Hence $h(x) = 0$ almost everywhere on $\mathbb{R}\backslash\mathbb{N} = (-\infty, -1) \cup \bigcup_{k=1}^{\infty} (k, k+1)$. But $\mathbb{N}$ has Lebesgue measure zero, hence $h(x) = 0$ almost everywhere on $\mathbb{R}$ and $g(\varphi) \equiv 0$. However, this is not the case. Indeed, consider the ε-“cap” ω_ε for $\varepsilon = \frac{1}{2}$. Choose x_0 such that $\omega_\varepsilon'(x_0) \neq 0$. Then for $\varphi_1(x) = \omega_\varepsilon(x - 1 + x_0)$ we have $\varphi_1'(1) = \omega_\varepsilon'(x_0) \neq 0$. Since $\operatorname{supp}\varphi_1 \subset (0, 2)$, it follows that $\varphi_1^{(n)}(n) = 0$ for $n \geq 2$. Therefore $g(\varphi_1) = \varphi_1'(1) \neq 0$. We got a contradiction, so g is a singular generalized function.

Let us find $\operatorname{supp} g$. Points of the open set $\mathbb{R}\backslash\mathbb{N}$ do not belong to the support of g, because if $\operatorname{supp}\varphi \subset \mathbb{R}\backslash\mathbb{N}$, then $g(\varphi) = 0$. On the other hand, $\mathbb{N} \subset \operatorname{supp} g$. Indeed, let us fix any $n \in \mathbb{N}$ and show that g does not vanish in the ε-neighborhood of the point n for $\varepsilon \in (0, 1)$. It means that $n \in \operatorname{supp} g$. Consider the δ-“cap” ω_δ for $\delta = \frac{\varepsilon}{2}$. Since ω_δ is not a polynomial, there exists a point x_0 such that $\omega_\delta^{(n)}(x_0) \neq 0$. For $\varphi_n(x) = \omega_\delta(x - n + x_0)$ we have $\operatorname{supp}\varphi_n \subset (n-2\delta, n+2\delta) = (n-\varepsilon, n+\varepsilon) \subset (n-1, n+1)$. Therefore $g(\varphi_n) = \varphi_n^{(n)}(n) = \omega_\delta^{(n)}(x_0) \neq 0$. Hence $\operatorname{supp} g = \mathbb{N}$.

14.4 Prove that the functional $\mathcal{P}\frac{1}{x}$, which is defined by the formula

$$(\mathcal{P}\frac{1}{x})(\varphi) = \text{v. p.} \int_{\mathbb{R}} \frac{\varphi(x)}{x} dx := \lim_{\varepsilon \to 0+} \int_{\mathbb{R}\backslash(-\varepsilon,\varepsilon)} \frac{\varphi(x)}{x} dx,\ \ \varphi \in \mathcal{D}(\mathbb{R}),$$

is a singular generalized function.

Solution Firstly, we will establish that the limit exists. For $0 < \varepsilon < 1$ and $\varphi \in \mathcal{D}(\mathbb{R})$ we have

$$\int_{\mathbb{R}\backslash(-\varepsilon,\varepsilon)} \frac{\varphi(x)}{x} dx = \int_{\mathbb{R}\backslash(-1,1)} \frac{\varphi(x)}{x} dx + \int_{[-1,1]\backslash(-\varepsilon,\varepsilon)} \frac{\varphi(x) - \varphi(0)}{x} dx.$$

There exists $\lim\limits_{x\to 0} \frac{\varphi(x)-\varphi(0)}{x} = \varphi'(0)$, hence $\int\limits_{-1}^{1} \frac{\varphi(x)-\varphi(0)}{x} dx$ exists as a Riemann integral. Therefore

$$\lim_{\varepsilon\to 0+} \int\limits_{\mathbb{R}\setminus(-\varepsilon,\varepsilon)} \frac{\varphi(x)}{x} dx = \int\limits_{\mathbb{R}\setminus(-1,1)} \frac{\varphi(x)}{x} dx + \int\limits_{-1}^{1} \frac{\varphi(x)-\varphi(0)}{x} dx,$$

so the functional $\mathcal{P}\frac{1}{x}$ is correctly defined. The linearity of $\mathcal{P}\frac{1}{x}$ follows immediately from its representation. To prove the continuity of $\mathcal{P}\frac{1}{x}$, we estimate $|\langle \mathcal{P}\frac{1}{x}, \varphi\rangle|$, $\varphi \in \mathcal{D}(\mathbb{R})$, under the condition $\operatorname{supp}\varphi \subset [-a, a]$, where $a > 1$. We have

$$|\langle \mathcal{P}\frac{1}{x}, \varphi\rangle| \le \int\limits_{[-a,a]\setminus(-1,1)} |\varphi(x)| dx + \int\limits_{-1}^{1} \frac{|\varphi(x)-\varphi(0)|}{|x|} dx.$$

Further, for $x \ne 0$ by the Lagrange theorem

$$\left|\frac{\varphi(x)-\varphi(0)}{x}\right| = |\varphi'(\theta_x)| \le \max_{x\in\mathbb{R}} |\varphi'(x)|,$$

where θ_x is an intermediate point between 0 and x. Hence

$$|\langle \mathcal{P}\frac{1}{x}, \varphi\rangle| \le 2a \cdot \max_{x\in\mathbb{R}} |\varphi(x)| + 2\max_{x\in\mathbb{R}} |\varphi'(x)|.$$

It is enough to prove the continuity of the functional at zero. Let $\varphi_n \xrightarrow{\mathcal{D}(\mathbb{R})} 0$, $n \to \infty$. Then there exists $a > 1$ such that $\operatorname{supp}\varphi_n \subset [-a, a]$ for all $n \ge 1$. Since $\varphi_n \rightrightarrows 0$ and $\varphi_n' \rightrightarrows 0$ on $\mathbb{R}$, $n \to \infty$,

$$|\langle \mathcal{P}\frac{1}{x}, \varphi_n\rangle| \le 2a \cdot \max_{x\in\mathbb{R}} |\varphi_n(x)| + 2\max_{x\in\mathbb{R}} |\varphi_n'(x)| \to 0,\ n \to \infty.$$

Therefore

$$|\langle \mathcal{P}\frac{1}{x}, \varphi_n\rangle| \to 0 = |\langle \mathcal{P}\frac{1}{x}, 0\rangle|,\ n \to \infty.$$

It means that $\mathcal{P}\frac{1}{x}$ is a continuous linear functional, i.e. $\mathcal{P}\frac{1}{x} \in \mathcal{D}'(\mathbb{R})$.

Assume that $\mathcal{P}\frac{1}{x}$ is a regular distribution function, i.e.,

$\exists\, f \in L_1^{loc}(\mathbb{R})\ \forall \varphi \in \mathcal{D}(\mathbb{R})$:

$$\langle \mathcal{P}\frac{1}{x}, \varphi\rangle = \lim_{\varepsilon\to 0+} \int\limits_{\mathbb{R}\setminus(-\varepsilon,\varepsilon)} \frac{\varphi(x)}{x}dx = \int\limits_{\mathbb{R}} f(x)\varphi(x)dx.$$

If $\operatorname{supp}\varphi \subset \mathbb{R}\setminus\{0\}$, then $\int\limits_{\mathbb{R}\setminus(-\varepsilon,\varepsilon)} \frac{\varphi(x)}{x}dx = \int\limits_{\mathbb{R}} \frac{\varphi(x)}{x}dx$ for all sufficiently small $\varepsilon > 0$. Hence $\int\limits_{\mathbb{R}} (\frac{\varphi(x)}{x} - f(x)\varphi(x))dx = 0$, and by Du Bois-Reymond lemma we get that $f(x) = \frac{1}{x}$ almost everywhere on $\mathbb{R}\setminus\{0\}$. Then $f \notin L_1^{loc}(\mathbb{R})$, a contradiction. Thus $\mathcal{P}\frac{1}{x}$ is a singular distribution function.

14.5 Prove that the series $\sum\limits_{n=1}^{\infty} a_n\delta_n$ converges in $\mathcal{D}'(\mathbb{R})$ for any $a_n \in \mathbb{C}$.

Solution For every $N \in \mathbb{N}$ the partial sum $S_N = \sum\limits_{n=1}^{N} a_n\delta_n$ belongs to the space $\mathcal{D}'(\mathbb{R})$ as a linear combination of shifted delta functions. Let $\varphi \in \mathcal{D}(\mathbb{R})$. There exists $m \in \mathbb{N}$ such that $\varphi(x) = 0$, $|x| > m$. For all $N \geq m$ we have

$$\langle S_N, \varphi\rangle = \sum_{n=1}^{N} a_n\varphi(n) = \sum_{n=1}^{m} a_n\varphi(n) \to \sum_{n=1}^{m} a_n\varphi(n) \in \mathbb{C},\ N \to \infty.$$

As a result of completeness of $\mathcal{D}'(\mathbb{R})$, there exists $S \in \mathcal{D}'(\mathbb{R})$ such that $S_N \xrightarrow{\mathcal{D}'(\mathbb{R})} S$, so the series converges to S in $\mathcal{D}'(\mathbb{R})$.

14.6 Let for $n \in \mathbb{N}$ the function $\mathcal{P}\frac{\cos nx}{x}$ be defined by the formula

$$\langle \mathcal{P}\frac{\cos nx}{x}, \varphi\rangle = \text{v. p.} \int\limits_{\mathbb{R}} \frac{\cos nx}{x}\varphi(x)dx := \lim_{\varepsilon\to 0+} \int\limits_{\mathbb{R}\setminus(-\varepsilon,\varepsilon)} \frac{\cos nx}{x}\varphi(x)dx,$$

$\varphi \in \mathcal{D}(\mathbb{R})$. Prove that $\mathcal{P}\frac{\cos nx}{x} \xrightarrow{\mathcal{D}'(\mathbb{R})} 0$, $n \to \infty$.

Solution Similar to the solution of problem 14.4, we have

$$\langle \mathcal{P}\frac{\cos nx}{x}, \varphi\rangle = \int\limits_{\mathbb{R}\setminus(-1,1)} \frac{\cos nx}{x}\varphi(x)dx + \int\limits_{-1}^{1} \frac{\cos nx}{x}\big(\varphi(x) - \varphi(0)\big)dx.$$

Let $\varphi \in \mathcal{D}(\mathbb{R})$ and $\varphi(x) = 0$ when $|x| > a$, where $a > 1$. Then

$$\langle \mathcal{P}\frac{\cos nx}{x}, \varphi\rangle = \int_{-a}^{-1} \cos nx \frac{\varphi(x)}{x} dx + \int_{-1}^{1} \cos nx \frac{\varphi(x) - \varphi(0)}{x} dx + \int_{1}^{a} \cos nx \frac{\varphi(x)}{x} dx.$$

Define the function $\frac{\varphi(x)-\varphi(0)}{x}$ by continuity at $x = 0$ as $\varphi'(0)$. Then each of the integrands is a product of $\cos nx$ and some continuous function. By Riemann's lemma each of the three integrals tends to zero as $n \to \infty$. Therefore $\langle \mathcal{P}\frac{\cos nx}{x}, \varphi\rangle \to 0 = \langle 0, \varphi\rangle$, $n \to \infty$. Thus $\mathcal{P}\frac{\cos nx}{x} \xrightarrow{\mathcal{D}'(\mathbb{R})} 0$, $n \to \infty$.

14.7 Prove that the generalized function $f = \sum\limits_{k=0}^{n-1} c_k \delta_0^{(k)}$, $c_0, \dots, c_{n-1} \in \mathbb{C}$ is the general solution in $\mathcal{D}'(\mathbb{R})$ of the equation $x^n f(x) = 0$, $n \in \mathbb{N}$.

Solution Let $\xi \in \mathcal{D}(\mathbb{R})$ be fixed, and $\xi = 1$ in some neighborhood of zero. According to problem 14.2, for any function $\varphi \in \mathcal{D}(\mathbb{R})$ there exists a function $\psi \in \mathcal{D}(\mathbb{R})$ such that $\varphi(x) = \xi(x) \sum\limits_{k=0}^{n-1} \frac{\varphi^{(k)}(0)}{k!} x^k + x^n \psi(x)$, $x \in \mathbb{R}$. Therefore

$$\langle f, \varphi\rangle = \sum_{k=0}^{n-1} \varphi^{(k)}(0) \langle f(x), \frac{\xi(x) x^k}{k!}\rangle + \langle f(x), x^n \psi(x)\rangle.$$

Since $x^n \in C^\infty(\mathbb{R})$, we conclude that $\langle f(x), x^n \psi(x)\rangle =$

$$\langle x^n f(x), \psi(x)\rangle = 0.$$

Put $c_k = (-1)^k \langle f(x), \frac{\xi(x)x^k}{k!}\rangle$, $0 \le k \le n-1$. Then

$$\langle f, \varphi\rangle = \sum_{k=0}^{n-1} c_k (-1)^k \varphi^{(k)}(0) = \left\langle \sum_{k=0}^{n-1} c_k \delta_0^{(k)}, \varphi \right\rangle, \quad \varphi \in \mathcal{D}(\mathbb{R}).$$

Now let us verify that $f = \sum\limits_{k=0}^{n-1} c_k \delta_0^{(k)}$ is a solution of the equation $x^n f(x) = 0$ for arbitrary $c_0, \dots, c_{n-1} \in \mathbb{C}$. Due to linearity, it is sufficient to show that $x^n \delta_0^{(k)} = 0$, $0 \le k \le n-1$. Indeed, we have

$$\langle x^n \delta_0^{(k)}, \varphi(x)\rangle = \langle \delta_0^{(k)}, x^n \varphi(x)\rangle = (-1)^k (x^n \varphi(x))^{(k)}|_{x=0} = 0$$

for all $\varphi \in \mathcal{D}(\mathbb{R})$ and $k < n$.

14.8 Let $f \in \mathcal{D}'(\mathbb{R})$ and $f(x+a) = f(x)$ for all $a \in \mathbb{R}$. Prove that $f = \text{const}$, i.e., there exists $C \in \mathbb{C}$ such that $\langle f, \varphi \rangle = C \int\limits_{\mathbb{R}} \varphi(x)dx$, $\varphi \in \mathcal{D}(\mathbb{R})$.

Solution To prove that $f' = 0$, we will show that $\langle f, \varphi' \rangle = 0$ for every $\varphi \in \mathcal{D}(\mathbb{R})$. Since $\langle f(t+a), \varphi(t) \rangle = \langle f(t), \varphi(t-a) \rangle$, it follows from the condition of the problem that

$$\begin{aligned}\langle f(t), \varphi(t-a) - \varphi(t) \rangle &= \langle f(t), \varphi(t-a) \rangle - \langle f(t), \varphi(t) \rangle \\ &= \langle f(t+a), \varphi(t) \rangle - \langle f(t), \varphi(t) \rangle = 0.\end{aligned}$$

Hence for all $a \neq 0$ we have $\langle f(t), \frac{\varphi(t-a)-\varphi(t)}{-a} \rangle = 0$. Put $a = -\frac{1}{n}$, $n \geq 1$, and consider the sequence

$$\left\{ \varphi_n(t) = \frac{\varphi(t+\frac{1}{n}) - \varphi(t)}{\frac{1}{n}},\ t \in \mathbb{R} : n \geq 1 \right\} \subset \mathcal{D}(\mathbb{R}).$$

The sequence converges in $\mathcal{D}(\mathbb{R})$ to the function $\varphi'(t)$. Indeed, if $\operatorname{supp} \varphi \subset [-b, b]$, then $\operatorname{supp} \varphi_n \subset [-b-1, b]$, $n \geq 1$. Moreover, for each $k \geq 0$ we have $\varphi_n^{(k)}(t) = \frac{\varphi^{(k)}(t+\frac{1}{n})-\varphi^{(k)}(t)}{\frac{1}{n}} = \varphi^{(k+1)}(\theta)$, where $\theta \in (t, t+\frac{1}{n})$ is an intermediate point, which depends on t, n and k. Then

$$\max_{t \in \mathbb{R}} |\varphi_n^{(k)}(t) - \varphi^{(k+1)}(t)| \leq \max_{t, \theta \in \mathbb{R},\ |\theta - t| \leq \frac{1}{n}} |\varphi^{(k+1)}(\theta) - \varphi^{(k+1)}(t)|,$$

and the right-hand side tends to zero as $n \to \infty$ due to the uniform continuity of the function $\varphi^{(k+1)}(t)$ which has bounded support. Therefore $\varphi_n \xrightarrow{\mathcal{D}(\mathbb{R})} \varphi'$. Then, due to the continuity of the generalized function f, we get that

$$0 = \langle f, \varphi_n \rangle \to \langle f, \varphi' \rangle,\ n \to \infty.$$

Therefore $\langle f', \varphi \rangle = -\langle f, \varphi' \rangle = 0$ for all $\varphi \in \mathcal{D}(\mathbb{R})$. Hence $f' = 0$, and the statement of the problem follows.

14.9 Prove that the equality $|\sin x|'' + |\sin x| = 2 \sum\limits_{k \in \mathbb{Z}} \delta_{k\pi}$ holds in $\mathcal{D}'(\mathbb{R})$.

Solution We have $|\sin x| = \sin x \cdot \operatorname{sign}(\sin x)$, $x \in \mathbb{R}$. The function $\operatorname{sign}(\sin x)$ is constant on the intervals $(k\pi, (k+1)\pi)$, $k \in \mathbb{Z}$, hence

$$\{|\sin x|'\} = (\sin x)' \operatorname{sign}(\sin x) = \cos x \operatorname{sign}(\sin x),\ x \neq k\pi,\ k \in \mathbb{Z}.$$

The function $|\sin x|$ is continuous, and its classical derivative is piecewise continuous, therefore the generalized derivative equals

$$|\sin x|' = \{|\sin x|'\} = \cos x \cdot \text{sign}(\sin x).$$

Furthermore, $\{|\sin x|''\} = (-\sin x)\,\text{sign}(\sin x) = -|\sin x|$. Since the function $h(x) = \cos x \cdot \text{sign}(\sin x)$ has jumps at points $k\pi$ of size

$$h(k\pi+) - h(k\pi-) = \cos k\pi \cdot (-1)^k - \cos k\pi(-1)^{k-1} = 2,$$

it follows that

$$|\sin x|'' = \{|\sin x|''\} + \sum_{k\in\mathbb{Z}} 2\delta_{k\pi} = -|\sin x| + 2\sum_{k\in\mathbb{Z}} \delta_{k\pi}.$$

14.10 Find the general solution of the equation $(x+1)y'' = 0$ in $\mathcal{D}'(\mathbb{R})$.

Solution Similarly to problem 14.7, we have $y'' = c_1\delta_{-1}$, $c_1 \in \mathbb{C}$, or $(y')' = c_1\delta_{-1}$. Denote $y' = z$, $z \in \mathcal{D}'(\mathbb{R})$. The equation $z' = c_1\delta_{-1}$ has a partial solution $z_0 = c_1\theta(x+1)$, where θ is the Heaviside function, and the homogeneous equation $z' = 0$ has a general solution $z = c_2$, $c_2 \in \mathbb{C}$. Therefore $z_{\text{gen}} = c_1\theta(x+1) + c_2$. Now we have the equation $y' = c_1\theta(x+1) + c_2$. Its partial solution is $y_0 = c_1(x+1)\theta(x+1) + c_2x$, and the homogeneous equation $y' = 0$ has a solution $y = c_3$, $c_3 \in \mathbb{C}$. Hence

$$y_{\text{gen}} = c_1(x+1)\theta(x+1) + c_2x + c_3,$$

where c_1, c_2, c_3 are arbitrary complex numbers.

14.11 Let $a \in \mathbb{R}\backslash\{0\}$, $\varphi \in \mathcal{D}(\mathbb{R})$. Find the general solution of the equation $\frac{d^2u}{dx^2} + a^2u = \varphi$.

Solution Firstly, we will find the fundamental solution of the operator $L = D^2 + a^2I$. Consider the Cauchy problem $z'' + a^2z = 0$, $z(0) = 0$, $z'(0) = 1$. From the characteristic equation $\lambda^2 + a^2 = 0$ we get $\lambda = \pm ia$. Therefore the ordinary homogeneous differential equation $Lz = 0$ has a general solution $z(x) = c_1 \sin ax + c_2 \cos ax$, where $c_1, c_2 \in \mathbb{C}$. From the condition $z(0) = 0$ we get $c_2 = 0$, then $z'(0) = c_1a = 1$, hence $c_1 = \frac{1}{a}$. Therefore the solution of the Cauchy problem is $z(x) = \frac{\sin ax}{a}$. Function $\mathcal{E}(x) = \frac{\theta(x)\sin ax}{a}$ is the fundamental solution of the operator L. Now we find the particular solution of the nonhomogeneous equation that has a form $(\mathcal{E} * \varphi)(x) = \int\limits_{\mathbb{R}} \frac{\theta(y)\sin ay}{a}\varphi(x-y)dy$. The general solution of the homogeneous equation is the sum of this particular solution and the general solution of the homogeneous equation: $u_{\text{gen}}(x) = c_1 \sin ax + c_2 \cos ax + \frac{1}{a}\int\limits_0^\infty \sin ay \cdot \varphi(x - y)dy$, $x \in \mathbb{R}$.

14.12 Prove that $\mathcal{D}(\mathbb{R}) \subset \mathcal{S}(\mathbb{R})$, and $\mathcal{D}(\mathbb{R})$ is dense in $\mathcal{S}(\mathbb{R})$.

Solution Let $\varphi \in \mathcal{D}(\mathbb{R})$. Then $\varphi \in C^\infty(\mathbb{R})$, and there exists $a > 0$ such that $\varphi(x) = 0$ for $|x| > a$. Let $m, k \geq 0$. For $|x| > a$ we have $|x|^m \varphi^{(k)}(x) = 0$, therefore $|x|^m \varphi^{(k)}(x) \to 0$, $x \to \pm\infty$. Thus $\varphi^{(k)}(x) = o(\frac{1}{|x|^m})$, $x \to \pm\infty$, and $\varphi \in \mathcal{S}(\mathbb{R})$. This implies the inclusion $\mathcal{D}(\mathbb{R}) \subset \mathcal{S}(\mathbb{R})$.

Now let $\psi \in \mathcal{S}(\mathbb{R})$. We approximate it by functions from $\mathcal{D}(\mathbb{R})$. Let $\xi \in \mathcal{D}(\mathbb{R})$ and $\xi(x) = 1$ for $|x| \leq 1$. Then $\psi_n(x) = \psi(x)\xi(\frac{x}{n}) \in \mathcal{D}(\mathbb{R})$. Let us prove that $\psi_n \xrightarrow{\mathcal{S}(\mathbb{R})} \psi$. Note that

$$\sup_{x\in\mathbb{R}} |x|^m |\psi_n(x) - \psi(x)| = \sup_{|x|>n} (|x|^m \cdot |\psi(x)| \cdot |\xi(\frac{x}{n}) - 1|) \leq$$

$$\leq (1 + \max_{y\in\mathbb{R}} |\xi(y)|) \cdot \sup_{|x|>n} (|x|^m \cdot |\psi(x)|) \to 0, \ n \to \infty.$$

Let $k \in \mathbb{N}$. It follows from the Leibniz formula that

$$\psi_n^{(k)}(x) = \psi^{(k)}(x)\xi(\frac{x}{n}) + \sum_{j=1}^{k} \binom{k}{j} \frac{1}{n^j} \xi^{(j)}(\frac{x}{n})\psi^{(k-j)}(x).$$

Thus, as above, $\sup\limits_{x\in\mathbb{R}} |x|^m |\psi^{(k)}(x)\xi(\frac{x}{n}) - \psi^{(k)}(x)| \to 0$, $n \to \infty$, and also for all $1 \leq j \leq k$ we have $\frac{1}{n^j} \sup\limits_{x\in\mathbb{R}} |x|^m \cdot |\xi^{(j)}(\frac{x}{n})\psi^{k-j}(x)| \leq \frac{\text{const}}{n^j} \to 0$, $n \to \infty$. It follows that $\sup\limits_{x\in\mathbb{R}} |x|^m |\psi_n^{(k)}(x) - \psi^{(k)}(x)| \to 0$, $n \to \infty$, for all $k \geq 1$ (for $k = 0$ the convergence has already been established above). Therefore the sequence $\{\psi_n\} \subset \mathcal{D}(\mathbb{R})$ converges to ψ in the space $\mathcal{S}(\mathbb{R})$, which proves that $\mathcal{D}(\mathbb{R})$ is dense in $\mathcal{S}(\mathbb{R})$.

14.13 Prove that the series $\sum\limits_{k\in\mathbb{Z}} e^{k^2}\delta_k$ converges in $\mathcal{D}'(\mathbb{R})$, but diverges in $\mathcal{S}'(\mathbb{R})$.

Solution For $n, m \in \mathbb{N}$ we set $S_{nm} = \sum\limits_{k=-n}^{m} e^{k^2}\delta_k$. These functions belong to $\mathcal{D}'(\mathbb{R})$ and $S'(\mathbb{R})$ as linear combinations of shifted delta functions. Let $\varphi \in \mathcal{D}(\mathbb{R})$. Since $\operatorname{supp}\varphi \subset [-p, p]$ for some $p \in \mathbb{N}$, then

$$\langle S_{nm}, \varphi\rangle = \sum_{k=-n}^{m} e^{k^2}\varphi(k) \to \sum_{k=-p}^{p} e^{k^2}\varphi(k)$$

as $m, n \to \infty$. Completeness of $\mathcal{D}'(\mathbb{R})$ implies the convergence of S_{nm} to some generalized function from $\mathcal{D}'(\mathbb{R})$, so the series converges in $\mathcal{D}'(\mathbb{R})$. Furthermore, for $e^{-x^2} \in \mathcal{S}(\mathbb{R})$ we have

$$\langle S_{nm}, e^{-x^2} \rangle = \sum_{k=-n}^{m} e^{k^2} \cdot e^{-k^2} = n + m + 1 \to +\infty, \ m, n \to +\infty.$$

Therefore S_{nm} diverges in $\mathcal{S}'(\mathbb{R})$ as $m, n \to \infty$.

14.14 Prove that for all $f \in L_2(\mathbb{R})$ the equality $(F^2[f])(x) = f(-x)$ holds for almost all $x \in \mathbb{R}$.

Solution Firstly, we will prove this equality for $\varphi \in \mathcal{S}(\mathbb{R})$. Denote $\psi(x) = \varphi(-x)$, $x \in \mathbb{R}$. We have

$$(F[\varphi])(y) = \frac{1}{\sqrt{2\pi}} \int\limits_{\mathbb{R}} e^{ixy} \varphi(x) dx = \Big| x = -t \Big| =$$

$$= \frac{1}{\sqrt{2\pi}} \int\limits_{\mathbb{R}} e^{-ity} \varphi(-t) dt = F^{-1}[\psi](y).$$

Then $F^2[\varphi] = F(F^{-1}[\psi]) = \psi$, and the equality is proven for $\mathcal{S}(\mathbb{R})$. Now let $f \in L_2(\mathbb{R})$. Since the set $\mathcal{S}(\mathbb{R})$ is dense in $L_2(\mathbb{R})$, there exists a sequence $\{\varphi_n : n \geq 1\} \subset \mathcal{S}(\mathbb{R})$, which converges to f in $L_2(\mathbb{R})$. We have

$$(F^2[\varphi_n])(y) = \varphi_n(-y), \ y \in \mathbb{R}, \ n \geq 1.$$

The operators F and F^2 are continuous in $L_2(\mathbb{R})$, hence $F^2[\varphi_n](y) \to F^2[f](y)$ and $\varphi_n(-y) \to f(-y)$, $n \to \infty$, where in both cases the convergence is in $L_2(\mathbb{R})$. It follows that $(F^2[f])(y) = f(-y)$ for almost all $y \in \mathbb{R}$.

14.15 Prove that $F[\delta_0] = \frac{1}{\sqrt{2\pi}}$ and $F[1] = \sqrt{2\pi}\,\delta_0$, where F acts in $\mathcal{S}'(\mathbb{R})$.

Solution Let $\varphi \in \mathcal{S}(\mathbb{R})$. We have

$$\langle F[\delta_0], \varphi \rangle = \langle \delta_0, F[\varphi] \rangle = \langle \delta_0(y), \frac{1}{\sqrt{2\pi}} \int\limits_{\mathbb{R}} e^{ixy} \varphi(x) dx \rangle =$$

$$= \frac{1}{\sqrt{2\pi}} \int\limits_{\mathbb{R}} \varphi(x) dx = \langle \frac{1}{\sqrt{2\pi}}, \varphi \rangle.$$

Since φ is an arbitrary test function, we conclude that $F[\delta_0]$ is a regular tempered distribution $\frac{1}{\sqrt{2\pi}} \in \mathcal{S}'(\mathbb{R})$. Therefore $F[\delta_0] = \frac{1}{\sqrt{2\pi}}$, $F^{-1}F[\delta_0] = F^{-1}[\frac{1}{\sqrt{2\pi}}]$.

It follows that $\delta_0 = \frac{1}{\sqrt{2\pi}} F^{-1}[1]$, i.e. $F^{-1}[1] = \sqrt{2\pi}\,\delta_0$. But for the tempered distribution $f(x) = 1$ we have $f(-x) = 1$, so $F^{-1}[1] = F[f(-x)] = F[1]$. Therefore $F[1] = \sqrt{2\pi}\,\delta_0$.

14.16 Let $\theta(x_1, \ldots, x_m) = \theta(x_1)\theta(x_2)\ldots\theta(x_m)$, where $\theta(x_i)$ is the Heaviside function. Prove that in the space $\mathcal{D}'(\mathbb{R}^m)$ the following equality holds:

$$\frac{\partial^m \theta}{\partial x_1 \partial x_2 \ldots \partial x_m} = \delta_0,$$

where $x = (x_1, \ldots, x_m)$.

Solution Let $\varphi \in \mathcal{D}'(\mathbb{R}^m)$. Integrating sequentially, we obtain

$$\langle \frac{\partial^m \theta}{\partial x_1 \partial x_2 \ldots \partial x_m}, \varphi \rangle = (-1)^m \langle \theta, \frac{\partial^m \varphi}{\partial x_1 \partial x_2 \ldots \partial x_m} \rangle =$$

$$= (-1)^m \int_0^\infty dx_1 \int_0^\infty dx_2 \ldots \int_0^\infty \frac{\partial^m \varphi}{\partial x_1 \partial x_2 \ldots \partial x_m} dx_m =$$

$$= (-1)^{m-1} \int_0^\infty dx_1 \ldots \int_0^\infty \frac{\partial^{m-1} \varphi(x_1, \ldots, x_{m-1}, 0)}{\partial x_1 \partial x_2 \ldots \partial x_{m-1}} dx_{m-1} =$$

$$= \ldots = -\int_0^\infty \frac{\partial \varphi(x_1, 0, \ldots, 0)}{\partial x_1} dx_1 = \varphi(0) = \langle \delta_0, \varphi \rangle.$$

Problems to Solve

14.17° Let $\varphi \in \mathcal{D}(\mathbb{R})$ be a nonzero function. Do the following sequences converge in $\mathcal{D}(\mathbb{R})$:

(1) $\{\frac{1}{n}\varphi(x),\ x \in \mathbb{R} : n \geq 1\}$;
(2) $\{\frac{1}{n}\varphi(nx),\ x \in \mathbb{R} : n \geq 1\}$;
(3) $\{\varphi(x + n),\ x \in \mathbb{R} : n \geq 1\}$?

14.18° Let $\varepsilon > 0$. For a locally integrable function $f : \mathbb{R} \to \mathbb{C}$ consider the function $f_\varepsilon : \mathbb{R} \to \mathbb{C}$, defined as follows:

$$f_\varepsilon(x) = \int_{\mathbb{R}} \omega_\varepsilon(x - y) f(y) dy,\ x \in \mathbb{R}.$$

Prove the following statements:

(1) if $f = \chi_{[-a,a]}$ and $\varepsilon < a < +\infty$, then $f_\varepsilon \in \mathcal{D}(\mathbb{R})$, moreover $0 \le f_\varepsilon(x) \le 1$ for all $x \in \mathbb{R}$, $f_\varepsilon(x) = 1$ for $|x| \le a - \varepsilon$ and $f_\varepsilon(x) = 0$ for $|x| \ge a + \varepsilon$;
(2) if $f \in C(\mathbb{R})$ and $\operatorname{supp} f \subset [-a, a]$, then $f_\varepsilon \in \mathcal{D}(\mathbb{R})$, and $f_\varepsilon(x) = 0$ for $|x| \ge a + \varepsilon$.

14.19 Let f_ε be a function defined in problem 14.18°. Prove the following statements:

(1) if $f \in L_p(\mathbb{R})$, $1 \le p < +\infty$, then for arbitrary $\varepsilon > 0$ it holds that $f_\varepsilon \in L_p(\mathbb{R}) \cap C^\infty(\mathbb{R})$, and $\|f_\varepsilon\|_p \le \|f\|_p$;
(2) if there exists $a > 0$ such that $f = 0$ almost everywhere for $|x| \ge a$, then for $f \in C(\mathbb{R})$, we have the convergence $f_\varepsilon \to f$ as $\varepsilon \to 0+$ in $C(\mathbb{R})$, and for $f \in L_p(\mathbb{R})$, $1 \le p < \infty$, we have the convergence $f_\varepsilon \to f$ in $L_p(\mathbb{R})$;
(3) the set $\mathcal{D}(\mathbb{R})$ is dense in $L_p(\mathbb{R})$, $1 \le p < +\infty$.

14.20 Let $f \in L_\infty(\mathbb{R})$, and there exists $a > 0$ such that $f = 0$ almost everywhere for $|x| \ge a$. Prove that there exists a sequence $\{g_n\} \subset \mathcal{D}(\mathbb{R})$ such that $\|g_n\|_\infty \le \|f\|_\infty$, $n \ge 1$, and $g_n \to f \pmod m$, $n \to \infty$.

14.21 The set $M \subset \mathcal{D}(\mathbb{R})$ is said to be bounded in $\mathcal{D}(\mathbb{R})$ if the supports of all functions from M belong to the same segment $[a, b]$, and

$$\forall k \in \mathbb{N} \cup \{0\}\ \exists C_k > 0\ \forall \varphi \in M\ :\ \max_{a \le x \le b} |\varphi^{(k)}(x)| \le C_k.$$

Prove that every bounded in $\mathcal{D}(\mathbb{R})$ set is relatively compact, i.e., every sequence of its elements contains a subsequence which converges in $\mathcal{D}(\mathbb{R})$.

14.22 Prove that the operations of differentiation and multiplication by a function $\alpha \in C^\infty(\mathbb{R})$ are continuous in $\mathcal{D}(\mathbb{R})$, i.e. the convergence $\varphi_n \xrightarrow{\mathcal{D}(\mathbb{R})} \varphi$ implies that (1) $\varphi_n' \xrightarrow{\mathcal{D}(\mathbb{R})} \varphi'$; (2) $\alpha\varphi_n \xrightarrow{\mathcal{D}(\mathbb{R})} \alpha\varphi$, $n \to \infty$.

14.23 Let $\xi \in \mathcal{D}(\mathbb{R})$, $\xi(x) = 1$ in a neighborhood of zero, and the function $\alpha \in C^\infty(\mathbb{R})$ has a unique zero of order 1 at the point $x = 0$. Let $\varphi \in \mathcal{D}(\mathbb{R})$. Prove that the function

$$\psi(x) = \frac{\varphi(x) - \xi(x)\varphi(0)}{\alpha(x)}, \quad x \in \mathbb{R}\setminus\{0\},$$

which is defined by continuity at $x = 0$, belongs to the space $\mathcal{D}(\mathbb{R})$.

14.24° Prove that the function $\varphi \in \mathcal{D}(\mathbb{R})$ is a derivative of some other function $\psi \in \mathcal{D}(\mathbb{R})$ if and only if it satisfies the condition $\int_{\mathbb{R}} \varphi(x)dx = 0$.

14.25° Prove that every function $\varphi \in \mathcal{D}(\mathbb{R})$ can be represented in the form

$$\varphi(x) = \eta(x) \int_{\mathbb{R}} \varphi(y)dy + \psi'(x), \quad x \in \mathbb{R},$$

where $\psi \in \mathcal{D}(\mathbb{R})$, and η is an arbitrary fixed function from $\mathcal{D}(\mathbb{R})$, which satisfies the condition $\int_{\mathbb{R}} \eta(x)dx = 1$.

14.26° Prove that the following functionals $f : \mathcal{D}(\mathbb{R}) \to \mathbb{C}$ are generalized function, i.e., $f \in \mathcal{D}'(\mathbb{R})$.

(1) $f(\varphi) = \varphi(0) + \varphi(1)$;

(2) $f(\varphi) = \varphi'(0)$;

(3) $f(\varphi) = \int_{\mathbb{R}} \varphi(x)dx$;

(4) $f(\varphi) = \int_{-1}^{1} |x|\varphi'(x)dx$;

(5) $f(\varphi) = \int_{-2}^{2} \operatorname{sign} x\varphi'(x)dx$;

(6) $f(\varphi) = \int_{\mathbb{R}} \ln|x|\varphi(x)dx$;

(7) $f(\varphi) = \int_{\mathbb{R}} e^{x}\varphi'(x)dx$;

(8) $f(\varphi) = \int_{\mathbb{R}} e^{-x}\varphi^{(n)}(x)dx,\ n \in \mathbb{N}$;

(9) $f(\varphi) = \varphi'(1) + \int_{0}^{2\pi} \sin x\varphi(x)dx$;

(10) $f(\varphi) = \varphi(0) + \int_{1}^{3} x\varphi''(x)dx$.

In which of these cases is the function regular? Find the support of f.

14.27 Prove that the following functionals are singular generalized functions from $\mathcal{D}'(\mathbb{R})$.

(1) $\langle \mathcal{P}\frac{1}{x^2}, \varphi\rangle = \text{v. p.} \int_{\mathbb{R}} \frac{\varphi(x)-\varphi(0)}{x^2}dx$;

(2) $\langle \mathcal{P}\frac{1}{x^3}, \varphi\rangle = \text{v. p.} \int_{\mathbb{R}} \frac{\varphi(x)-\varphi(0)-\varphi'(0)x}{x^3}dx$;

(3) $\langle \mathcal{P}\frac{1}{x^n}, \varphi\rangle = \text{v. p.} \int_{\mathbb{R}} \frac{\varphi(x)-\varphi(0)-\varphi'(0)x-\ldots-\frac{\varphi^{(n-1)}(0)}{(n-1)!}x^{n-1}}{x^n}dx,\ n \in \mathbb{N}$;

(4) $\langle \mathcal{P}\frac{\cos ax}{x}, \varphi\rangle = \text{v. p.} \int_{\mathbb{R}} \frac{\cos ax}{x}\varphi(x)dx,\ a > 0$,

where $\text{v. p.} \int_{\mathbb{R}} f(x)dx := \lim_{\varepsilon\to 0+} \left(\int_{-\infty}^{-\varepsilon} f(x)dx + \int_{\varepsilon}^{\infty} f(x)dx \right)$.

14.28 Let $f \in \mathcal{D}'(\mathbb{R})$ and $\langle f, \varphi\rangle \geq 0$ for every nonnegative $\varphi \in \mathcal{D}(\mathbb{R})$. Prove that there exists a measure μ, defined on the σ-algebra of Borel sets $\mathcal{B}(\mathbb{R})$, such that $\langle f, \varphi\rangle = \int\limits_{\mathbb{R}} \varphi(x)d\mu(x)$ for all $\varphi \in \mathcal{D}(\mathbb{R})$.

14.29° Prove that $\delta_a \xrightarrow{\mathcal{D}'(\mathbb{R})} 0$, $a \to +\infty$.

14.30° Prove that

(1) $\eta_\varepsilon \xrightarrow{\mathcal{D}'(\mathbb{R})} \delta_0$, $\varepsilon \to 0+$, where $\eta_\varepsilon(x) = \begin{cases} \frac{1}{2\varepsilon}, & |x| \leq \varepsilon, \\ 0, & |x| > \varepsilon; \end{cases}$

(2) $f_\varepsilon \xrightarrow{\mathcal{D}'(\mathbb{R})} \delta_0 \int\limits_{\mathbb{R}} f(x)dx$, $\varepsilon \to 0+$, where $f \in L_1(\mathbb{R})$ and $f_\varepsilon(x) := \frac{1}{\varepsilon} f(\frac{x}{\varepsilon})$.

14.31 Prove that

(1) $\frac{\varepsilon}{x^2+\varepsilon^2} \xrightarrow{\mathcal{D}'(\mathbb{R})} \pi\delta_0$, $\varepsilon \to 0+$;
(2) $\frac{1}{\sqrt{\varepsilon}} e^{-\frac{x^2}{\varepsilon}} \xrightarrow{\mathcal{D}'(\mathbb{R})} \sqrt{\pi}\delta_0$, $\varepsilon \to 0+$;
(3) $\frac{1}{x} \sin \frac{x}{\varepsilon} \chi_{[0,+\infty)}(x) \xrightarrow{\mathcal{D}'(\mathbb{R})} \frac{\pi}{2}\delta_0$, $\varepsilon \to 0+$;
(4) $\frac{1}{2x+\varepsilon} \chi_{(\varepsilon,4\varepsilon)}(x) \xrightarrow{\mathcal{D}'(\mathbb{R})} \frac{\ln 3}{2}\delta_0$, $\varepsilon \to 0+$;
(5) $t^n e^{ixt} \xrightarrow{\mathcal{D}'(\mathbb{R})} 0$, $t \to +\infty$, for every $n \geq 0$.

14.32 (Sokhotsky's formulas) Prove that

$$\frac{1}{x \pm i\varepsilon} \xrightarrow{\mathcal{D}'(\mathbb{R})} \mathcal{P}\frac{1}{x} \mp i\pi\delta_0, \ \ \varepsilon \to 0+ .$$

14.33° Prove that

(1) $\forall\, \alpha \in C^\infty(\mathbb{R})$: $\alpha\delta_0 = \alpha(0)\delta_0$;
(2) $\forall\, a \in \mathbb{R}\setminus\{0\}$: $\delta_0(ax) = \frac{1}{|a|}\delta_0(x)$.

14.34 Let $\alpha \in C^\infty(\mathbb{R})$ have a unique zero of order 1 at the point $x = 0$. Find the general solution of the equation $\alpha f = 0$ in $\mathcal{D}'(\mathbb{R})$.

14.35° Let $f \in \mathcal{D}'(\mathbb{R})$, $\operatorname{supp} f \subset [-a, a]$, $\eta \in C^\infty(\mathbb{R})$, and there exists $\varepsilon > 0$ such that $\eta(x) = 1$ for all $x \in [-a-\varepsilon, a+\varepsilon]$. Prove that $\eta f = f$.

14.36 Prove that in $\mathcal{D}'(\mathbb{R})$ the following equalities hold:

(1) $x^n \mathcal{P}\frac{1}{x} = x^{n-1}$, $n \in \mathbb{N}$;
(2) $x\mathcal{P}\frac{1}{x^2} = \mathcal{P}\frac{1}{x}$;
(3) $x^n \mathcal{P}\frac{1}{x^k} = x^{n-k}$, $n, k \in \mathbb{N}$, $n \geq k$;
(4) $x^n \mathcal{P}\frac{1}{x^k} = \mathcal{P}\frac{1}{x^{k-n}}$, $n, k \in \mathbb{N}$, $n < k$,

where $\mathcal{P}\frac{1}{x^n}$, $n \geq 1$, are generalized functions, defined in problems 14.4 and 14.27.

14.37° Find the derivatives in $\mathcal{D}'(\mathbb{R})$ of the functions

(1) $f(x) = |x|,\ x \in \mathbb{R}$;
(2) $f(x) = \operatorname{sign}(\sin x),\ x \in \mathbb{R}$.

14.38° Prove that for any $f \in \mathcal{D}'(\mathbb{R})$, $a \in \mathbb{R}$ and $n \in \mathbb{N}$ the following equality is true: $(f(x+a))^{(n)} = f^{(n)}(x+a)$.

14.39° Calculate the derivatives in $\mathcal{D}'(\mathbb{R})$ for $n \geq 1$:

(1) $\theta^{(n)}(x-a)$;
(2) $(\theta(x)e^{ax})^{(n)}$;
(3) $(\operatorname{sign} x)^{(n)}$;
(4) $|x|^{(n)}$;
(5) $(\operatorname{sign}(\cos x))^{(n)}$;
(6) $[x]^{(n)}$,

where $\theta(x) = \chi_{[0,+\infty)}(x)$ is the Heaviside function, $[x]$ is the integer part of a number x.

14.40° Find the derivatives of the first three orders in $\mathcal{D}'(\mathbb{R})$ of the following functions:

(1) $f(x) = \theta(x)\sin x$;
(2) $f(x) = \theta(x)\cos x$;
(3) $f(x) = \begin{cases} x^2, & |x| \leq 2, \\ 0, & |x| > 2; \end{cases}$
(4) $f(x) = \begin{cases} 1, & x < 0, \\ x+1, & 0 \leq x < 1, \\ x^2+1, & x \geq 1; \end{cases}$
(5) $f(x) = \begin{cases} 0, & x < -1, \\ (x+1)^2, & -1 \leq x < 0, \\ x^2+1, & x \geq 0; \end{cases}$
(6) $f(x) = \sin x\, \chi_{[-\pi,\pi]}(x)$;
(7) $f(x) = \begin{cases} x^2, & x < 1, \\ (x-2)^2, & 1 \leq x < 2, \\ 0, & x \geq 2; \end{cases}$
(8) $f(x) = |\cos x|\chi_{[0,2\pi]}(x)$.

14.41° Prove that

(1) $\frac{1}{\varepsilon^2} f'(\frac{x}{\varepsilon}) \xrightarrow{\mathcal{D}'(\mathbb{R})} \delta_0' \int\limits_{\mathbb{R}} f(x)dx$, $\varepsilon \to 0+$, where $f \in L_1(\mathbb{R}) \cap C^1(\mathbb{R})$;

(2) $\frac{x\varepsilon}{(x^2+\varepsilon^2)^2} \xrightarrow{\mathcal{D}'(\mathbb{R})} -\frac{\pi}{2}\delta_0'$, $\varepsilon \to 0+$;

(3) $\frac{x}{\varepsilon\sqrt{\varepsilon}} e^{-\frac{x^2}{\varepsilon}} \xrightarrow{\mathcal{D}'(\mathbb{R})} -\frac{\sqrt{\pi}}{2}\delta_0'$, $\varepsilon \to 0+$;

(4) $\frac{1}{x^2}\chi_{[\varepsilon,2\varepsilon)}(x) - \frac{1}{\varepsilon}\delta_\varepsilon + \frac{1}{2\varepsilon}\delta_{2\varepsilon} \xrightarrow{\mathcal{D}'(\mathbb{R})} -\ln 2\, \delta_0'$, $\varepsilon \to 0+$;

(5) $\frac{e}{\varepsilon^2} e^{-\frac{x}{\varepsilon}} \chi_{[\varepsilon,+\infty)}(x) - \frac{1}{\varepsilon}\delta_\varepsilon \xrightarrow{\mathcal{D}'(\mathbb{R})} -\delta_0'$, $\varepsilon \to 0+$.

14.42 Prove that in $\mathcal{D}'(\mathbb{R})$ the following equalities hold:

(1) $(\ln|x|)' = \mathcal{P}\frac{1}{x}$;
(2) $\left(\mathcal{P}\frac{1}{x}\right)' = -\mathcal{P}\frac{1}{x^2}$;
(3) $\left(\mathcal{P}\frac{1}{x^2}\right)' = -2\mathcal{P}\frac{1}{x^3}$;
(4) $\left(\mathcal{P}\frac{1}{x^n}\right)' = -n\mathcal{P}\frac{1}{x^{n+1}}$, $n \geq 1$,

where $\mathcal{P}\frac{1}{x^n}$, $n \geq 1$, are generalized functions defined in problems 14.4 and 14.27.

14.43 Let $\alpha \in C^\infty(\mathbb{R})$ and $f \in \mathcal{D}'(\mathbb{R})$. Prove that $(\alpha f)' = \alpha' f + \alpha f'$. Calculate $(\alpha\theta)'$, where θ is the Heaviside function.

14.44 Prove in $\mathcal{D}'(\mathbb{R})$ the following equalities

(1) $x\delta_0^{(n)} = -n\delta_0^{(n-1)}$, $n \in \mathbb{N}$;
(2) $e^x\delta_0''' = \delta_0''' - 3\delta_0'' + 3\delta_0' - \delta_0$;
(3) $x^k\delta_0^{(n)} = (-1)^k\binom{n}{k}\delta_0^{(n-k)}$, $n, k \in \mathbb{N}$, $n \geq k$;
(4) $x^k\delta_0^{(n)} = 0$, $n, k \in \mathbb{N}$, $n < k$;
(5) $\alpha\delta_0^{(n)} = \sum\limits_{k=0}^{n}(-1)^k\binom{n}{k}\alpha^{(k)}(0)\delta_0^{(n-k)}$, $\alpha \in C^\infty(\mathbb{R})$.

14.45 Prove that the generalized functions $\delta_0, \delta_0', \ldots, \delta_0^{(n)}$ are linearly independent in $\mathcal{D}'(\mathbb{R})$.

14.46° Prove in $\mathcal{D}'(\mathbb{R})$ the equality $|\cos x|'' + |\cos x| = 2\sum\limits_{k\in\mathbb{Z}}\delta_{(k+\frac{1}{2})\pi}$.

14.47 Using the expansion

$$\frac{x}{2} - \frac{x^2}{4\pi} = \frac{\pi}{6} - \frac{1}{2\pi}\sum_{k\in\mathbb{Z}\setminus\{0\}}\frac{1}{k^2}e^{ikx}, \quad x \in [0, 2\pi],$$

prove that in $\mathcal{D}'(\mathbb{R})$ the equality $\frac{1}{2\pi}\sum\limits_{k\in\mathbb{Z}}e^{ikx} = \sum\limits_{k\in\mathbb{Z}}\delta_{2\pi k}$ holds.

14.48 Prove in $\mathcal{D}'(\mathbb{R})$ the equality $\frac{2}{\pi}\sum\limits_{k=0}^{\infty}\cos(2k+1)x = \sum\limits_{k\in\mathbb{Z}}(-1)^k\delta_{\pi k}$.

14.49° Find the general solution of the equation $y'' = 0$ in $\mathcal{D}'(\mathbb{R})$.

14.50° Prove that the general solution of the equation $xu' = 1$ in $\mathcal{D}'(\mathbb{R})$ has the form $u = c_1 + c_2\theta(x) + \ln|x|$, where $c_1, c_2 \in \mathbb{C}$ are arbitrary and θ is the Heaviside function.

14.51 Find the general solutions in $\mathcal{D}'(\mathbb{R})$ of the following equations

(1) $(x-1)u = 0$;
(2) $(x^2-1)u = 0$;
(3) $xu = \mathcal{P}\frac{1}{x}$;
(4) $x^2u' = 0$;
(5) $(x+1)^2u'' = 0$;
(6) $x(x-1)u = 0$;
(7) $xu = 1$;
(8) $x^2u = 0$;
(9) $xu' = \mathcal{P}\frac{1}{x}$;
(10) $x^2u' = 1$;
(11) $(x+1)u''' = 0$.

14.52*. Find the general solution in $\mathcal{D}'(\mathbb{R})$ of the equations
(1) $\sin x \cdot y = 0$; (2) $\cos x \cdot y = 0$.

14.53 Prove that for $n > m$ the general solution in $\mathcal{D}'(\mathbb{R})$ of the equation $x^n u^{(m)} = 0$ is the generalized function

$$u = \sum_{k=0}^{m-1} a_k x^k + \sum_{k=0}^{m-1} b_k \theta(x) x^k + \sum_{k=m}^{n-1} b_k \delta_0^{(k-m)},$$

where a_k, b_k are arbitrary complex numbers.

14.54° Prove that $f_1 * f_2 = f_2 * f_1$ and $(f_1 * f_2) * f_3 = f_1 * (f_2 * f_3)$ for all $f_1, f_2, f_3 \in L_1(\mathbb{R})$.

14.55 Prove that for all $f, g \in L_1(\mathbb{R})$ and $y \in \mathbb{R}$

$$F[f * g](y) = \sqrt{2\pi}\, F[f](y)\, F[g](y).$$

14.56° Prove that

(1) $\delta_0 * \varphi = \varphi$, $\delta_a * \varphi = \varphi(x - a)$ for all $\varphi \in \mathcal{D}(\mathbb{R})$ and $a \in \mathbb{R}$.
(2) $f * \varphi \in C^\infty(\mathbb{R})$ and $(f * \varphi)^{(n)} = f * \varphi^{(n)} = f^{(n)} * \varphi$ for all $n \geq 1$, $f \in \mathcal{D}'(\mathbb{R})$ and $\varphi \in \mathcal{D}(\mathbb{R})$.
(3) $\delta_0^{(n)} * \varphi = \varphi^{(n)}$ for all $n \geq 1$ and $\varphi \in \mathcal{D}(\mathbb{R})$.

14.57° Find the fundamental solution of the operator $D + aI$, $a \in \mathbb{R}$, and the general solution of the equation $\frac{du}{dx} + au = \varphi$, $\varphi \in \mathcal{D}(\mathbb{R})$.

14.58 Find the fundamental solution of the operator L and the general solution of the equation $Lu = \varphi$, $\varphi \in \mathcal{D}(\mathbb{R})$.

(1) $L = D^2 - I$;
(2) $L = (D - aI)^n$, $a \in \mathbb{R}$, $n \geq 2$;
(3) $L = D^2 + 4D$;
(4) $L = D^2 - 2D + I$;
(5) $L = D^2 + 3D + 2I$;
(6) $L = D^2 - 4D + 5I$;
(7) $L = D^3 - I$;
(8) $L = D^3 - 3D^2 + 2D$;
(9) $L = D^4 - I$;
(10) $L = D^4 - 2D^2 + 1$.

14.59 Let $a_1, \ldots, a_n \in C^\infty(\mathbb{R})$, $t \in \mathbb{R}$, $z(x, t)$ be the solution of the linear homogeneous differential equation

$$Lz = z^{(n)} + a_1(x) z^{(n-1)} + \ldots + a_n(x) z = 0,$$

for which $z(t) = z'(t) = \ldots = z^{(n-2)}(t) = 0$, $z^{(n-1)}(t) = 1$. Prove that

(1) for any $t \in \mathbb{R}$ the function $\mathcal{E}(x, t) = \theta(x - t) z(x, t)$, $x \in \mathbb{R}$, satisfies in $\mathcal{D}'(\mathbb{R})$ the equation $L\mathcal{E} = \delta_t$ (the function $\mathcal{E}$ is called the fundamental solution of the operator L);
(2) for $\varphi \in \mathcal{D}(\mathbb{R})$ the inhomogeneous differential equation $Lu = \varphi$, $u \in C^\infty(\mathbb{R})$, has a particular solution $u(x) = \langle \mathcal{E}(x, t), \varphi(t) \rangle$, $x \in \mathbb{R}$;
(3) if all the coefficients of L are constant, then $\mathcal{E}(x, t) = \mathcal{E}(x - t, 0)$.

14.60 Find the fundamental solution $\mathcal{E}$ of the operator L.

(1) $L = D - \frac{2x}{1+x^2}$;
(2) $L = D - 2x$;
(3) $L = D + 3x^2$;
(4) $L = D - \sin x$;
(5) $L = D + \cos x$;
(6) $L = D + 4x \sin x^2$;
(7) $L = D - \frac{\arctan x}{1+x^2}$;
(8) $L = D - 4x^3 + 1$;
(9) $L = D - 2^x$;
(10) $L = D + x^2 3^{x^3+1}$.

14.61° Do the functions (1) e^{-x^2}, (2) e^{-x}, (3) xe^{-x^2}, (4) $\frac{1}{1+x^2}$, $x \in \mathbb{R}$ belong to the space $\mathcal{S}(\mathbb{R})$?

14.62°

(1) Let $\varphi \in \mathcal{S}(\mathbb{R})$ and P be a polynomial. Prove that $P\varphi \in \mathcal{S}(\mathbb{R})$.
(2) Let $\varphi \in \mathcal{S}(\mathbb{R})$ and $k \geq 1$. Prove that $\varphi^{(k)} \in \mathcal{S}(\mathbb{R})$.

14.63 Assume that the function ψ belongs to $C^\infty(\mathbb{R})$, equals zero on $(-\infty, 0)$ and is bounded together with all its derivatives. Prove that for $\sigma > 0$ the function $\psi(x)e^{-\sigma x}$, $x \in \mathbb{R}$, belongs to the space $\mathcal{S}(\mathbb{R})$.

14.64° Prove that the set $\mathcal{S}(\mathbb{R})$ is dense in $L_p(\mathbb{R})$, $1 \leq p < +\infty$.

14.65° Let φ be a nonzero function from $\mathcal{S}(\mathbb{R})$. Do the following sequences converge in $\mathcal{S}(\mathbb{R})$:

(1) $\{\frac{1}{n}\varphi(x),\ x \in \mathbb{R} : n \geq 1\}$;
(2) $\{\frac{1}{n}\varphi(nx),\ x \in \mathbb{R} : n \geq 1\}$;
(3) $\{\frac{1}{n}\varphi(\frac{x}{n}),\ x \in \mathbb{R} : n \geq 1\}$?

14.66° Prove that $\delta_0 \in \mathcal{S}'(\mathbb{R})$ and $\delta_0^{(k)} \in \mathcal{S}'(\mathbb{R})$ for all $k \geq 1$.

14.67° Prove that $e^{x^2} \in \mathcal{D}'(\mathbb{R})$, but e^{x^2} does not generate a regular tempered distribution from the space $\mathcal{S}'(\mathbb{R})$.

14.68 Prove that $\mathcal{S}'(\mathbb{R}) \subset \mathcal{D}'(\mathbb{R})$ and the embedding operator is continuous.

14.69° Let $f \in \mathcal{S}'(\mathbb{R})$ and P be a polynomial. Define the functional $\langle Pf, \varphi\rangle = \langle f, P\varphi\rangle$, $\varphi \in \mathcal{S}(\mathbb{R})$. Prove that $Pf \in \mathcal{S}'(\mathbb{R})$.

14.70 Prove that if $|a_k| \leq C(1+|k|)^m$ for all $k \in \mathbb{Z}$, where $C > 0, m \geq 0$ are some constants, then the series $\sum\limits_{k\in\mathbb{Z}} a_k\delta_k$ converges in $\mathcal{S}'(\mathbb{R})$,

14.71

(1) Let $f : \mathbb{R} \to \mathbb{C}$ be a Lebesgue measurable function for which there exists $m \geq 0$ such that $\int\limits_{\mathbb{R}} \frac{|f(x)|}{1+|x|^m}dx < \infty$. Prove that f defines a regular tempered distribution from $\mathcal{S}'(\mathbb{R})$ by the formula

$$\langle f, \varphi\rangle = \int\limits_{\mathbb{R}} f(x)\varphi(x)dx,\ \varphi \in \mathcal{S}(\mathbb{R}).$$

(2) Prove that the function $f(x) = e^x \cos e^x$, $x \in \mathbb{R}$ does not satisfy the conditions of item (1), but defines a tempered distribution from $\mathcal{S}'(\mathbb{R})$ by the formula $\langle f, \varphi\rangle = \int\limits_{-\infty}^{+\infty} e^x \cos e^x \varphi(x)dx$, $\varphi \in \mathcal{S}(\mathbb{R})$. Here the integral should be understood as an improper Riemann integral of the first kind.

14.72° Prove that for all $\varphi \in \mathcal{S}(\mathbb{R})$ the following equations hold

(1) $F[\varphi(x-a)] = e^{iay}F[\varphi],\ a \in \mathbb{R};$
(2) $F[\varphi^{(k)}] = (-iy)^k F[\varphi],\ k \geq 1;$
(3) $(F[\varphi])^{(k)} = F[(ix)^k\varphi],\ k \geq 1;$
(4) $y^k(F[\varphi])^{(l)} = i^{k+l}F[(x^l\varphi)^{(k)}],\ k \geq 1,\ l \geq 1.$

14.73 Find the Fourier transforms of the functions
(1) $e^{-\frac{x^2}{2}}$; (2) $xe^{-\frac{x^2}{2}}$; (3) $x^2e^{-\frac{x^2}{2}}$; (4) $x^3e^{-\frac{x^2}{2}}$.

14.74 Let F be the Fourier transform in $L_2(\mathbb{R})$. Prove that $\sigma_p(F) = \sigma(F) = \{1, i, -1, -i\}$.

14.75 Let $a > 0$. Find the Fourier transforms of the functions

(1) $\theta(x)e^{-ax}$;
(2) $e^{-a|x|}$;
(3) $xe^{-a|x|}$;
(4) $\frac{\sqrt{2}a}{\sqrt{\pi}(x^2+a^2)}$;
(5) $\chi_{[0,a]}(x)$;
(6) $\chi_{[-a,a]}(x)$;
(7) $e^{-\frac{x^2}{2}}\cos(ax)$;
(8) $e^{-\frac{x^2}{2}}\sin(ax)$.

14.76° Let F be the Fourier transform in $\mathcal{S}'(\mathbb{R})$. Show that for all $f \in \mathcal{S}'(\mathbb{R})$ the following equalities hold

(1) $(F[f])^{(k)} = F[(ix)^k f],\ k \geq 1;$
(2) $F[f^{(k)}] = (-iy)^k F[f],\ k \geq 1.$

14.77° Prove that

(1) $F[\delta_a] = \frac{1}{\sqrt{2\pi}}e^{iay}$, $y \in \mathbb{R}, a \in \mathbb{R}$;
(2) $F[x^k] = \sqrt{2\pi}(-i)^k\delta_0^{(k)}$, $k \geq 1$;
(3) $F[\delta_0^{(k)}] = \frac{(-iy)^k}{\sqrt{2\pi}}$, $k \geq 1$.

14.78° Prove that in $\mathcal{S}'(\mathbb{R})$ the following equalities hold

(1) $F[\frac{1}{2}(\delta_a + \delta_{-a})] = \frac{1}{\sqrt{2\pi}}\cos(ay)$, $a \in \mathbb{R}$;
(2) $F[\frac{1}{2}(\delta_a - \delta_{-a})] = \frac{1}{\sqrt{2\pi}}\sin(ay)$, $a \in \mathbb{R}$;
(3) $F[3x^2 + 2x - 1] = \sqrt{2\pi}(-3\delta_0'' + 2i\delta_0' - \delta_0)$.

14.79 Find the Fourier transforms of the following tempered distributions

(1) $\theta(x)$;
(2) $\theta(-x)$;
(3) $\operatorname{sign} x$;
(4) $|x|$;
(5) $\mathcal{P}\frac{1}{x}$;
(6) $\mathcal{P}\frac{1}{x^2}$.

14.80

(1) Operators A, B are defined by the formulae

$$(Af)(y) = \lim_{t\to+\infty} \int_0^t \cos(yx) f(x)dx, \quad (Bf)(y) = \lim_{t\to+\infty} \int_0^t \sin(yx) f(x)dx,$$

$f \in L_2([0, +\infty))$, where the limits are taken in the sense of convergence in $L_2([0, +\infty))$. Prove that $A, B \in \mathcal{L}(L_2([0, +\infty)))$.

(2) Find the spectra and the eigenvalues of operators A and B.

(3) Prove that all functions of the form $f = g \pm \sqrt{\frac{2}{\pi}} Ag$, $g \in L_2([0, +\infty))$, satisfy the equation $f = \pm\sqrt{\frac{2}{\pi}} Af$.

(4) Prove that all functions of the form $f = g \pm \sqrt{\frac{2}{\pi}} Bg$, $g \in L_2([0, +\infty))$, satisfy the equation $f = \pm\sqrt{\frac{2}{\pi}} Bf$.

14.81 Let $f \in L_1(\mathbb{R})$ and $g \in L_2(\mathbb{R})$. Prove that

(1) the formula $(f * g)(x) = \int_{\mathbb{R}} f(x - y)g(y)dy$, where the integral converges for almost all $x \in \mathbb{R}$, defines a function $f * g \in L_2(\mathbb{R})$;

(2) $F[f * g](y) = \sqrt{2\pi} F[f](y) \cdot F[g](y)$ for almost all $y \in \mathbb{R}$.

14.82 Let $f \in L_1(\mathbb{R})$. The operator $A \in \mathcal{L}(L_2(\mathbb{R}))$ is defined by the formula $Ag = f * g$ (the convolution operator with function f). Find $\sigma(A)$.

14.83 Prove that the generalized function

$$\langle f, \varphi\rangle = \int_{\{x^2+y^2<1\}} \frac{\varphi(x, y) - \varphi(0, 0)}{x^2 + y^2} dxdy + \int_{\{x^2+y^2>1\}} \frac{\varphi(x, y)}{x^2 + y^2} dxdy,$$

$\varphi \in \mathcal{D}(\mathbb{R}^2)$, satisfies the equation $(x^2 + y^2)f = 1$ in $\mathcal{D}'(\mathbb{R}^2)$.

14.84

(1) Consider the square $[0, 1]^2$ in the plane (x, y). Let the function f be equal to 1 inside this square and to 0 outside of it. Calculate f''_{xy} in $\mathcal{D}'(\mathbb{R}^2)$.

(2) Consider the square S with vertices $(-1, 0)$, $(0, -1)$, $(1, 0)$ and $(0, 1)$. Let the function f be equal to 1 inside S and to 0 outside of it. Calculate $f''_{yy} - f''_{xx}$ in $\mathcal{D}'(\mathbb{R}^2)$.

14.85 Do the functions (1) $e^{-\|x\|^2}$, (2) $e^{-\|x\|}$, (3) $\frac{1}{1+\|x\|^2}$, (4) $(x_1^2+\ldots+x_m^2)e^{-\|x\|^2}$, $x\in\mathbb{R}^m$ belong to the space $\mathcal{S}(\mathbb{R}^m)$?

14.86

(1) Let μ be a measure on $\mathcal{B}(\mathbb{R}^m)$, which attains finite values on compact sets. Prove that the equality $\alpha_\mu(\varphi) = \int\limits_{\mathbb{R}^m}\varphi(t)d\mu(t)$, $\varphi \in \mathcal{D}(\mathbb{R}^m)$, defines a generalized function $\alpha_\mu \in \mathcal{D}'(\mathbb{R}^m)$.
(2) Let μ be a measure on $\mathcal{B}(\mathbb{R}^m)$, for which there exists $k\in\mathbb{N}$ such that $(1+\|x\|)^{-k}\in L_1(\mathbb{R}^m,\mu)$ (such a measure is called a tempered measure). Prove that the formula from item (1) defines a tempered distribution $\alpha_\mu\in\mathcal{S}'(\mathbb{R}^m)$.
(3) Show that not every measure, which attains finite values on compacts, defines $\alpha_\mu\in\mathcal{S}'(\mathbb{R}^m)$.

14.87 Prove the following properties of the Fourier transform in $\mathcal{S}'(\mathbb{R}^m)$:

(1) $F[f(x-a)](y) = e^{i(y,a)}F[f](y)$, $a\in\mathbb{R}^m$;
(2) $F[f](y+b) = F[e^{i(b,x)}f(x)](y)$, $b\in\mathbb{R}^m$.

14.88 Prove in $\mathcal{S}'(\mathbb{R}^m)$ the equalities
(1) $F[\delta_0] = \frac{1}{(2\pi)^{\frac{m}{2}}}$; (2) $F[\delta_a] = \frac{e^{i(y,a)}}{(2\pi)^{\frac{m}{2}}}$, $a\in\mathbb{R}^m$; (3) $F[1] = (2\pi)^{\frac{m}{2}}\delta_0$.

14.89 Prove in $\mathcal{S}'(\mathbb{R}^m)$ the equalities

(1) $F[D^\alpha\delta_0] = \frac{(-iy)^\alpha}{(2\pi)^{\frac{m}{2}}}$;

(2) $F[x^\alpha] = (2\pi)^{\frac{m}{2}}(-1)^{|\alpha|}D^\alpha\delta_0$, where $x^\alpha = \prod\limits_{k=1}^{m} x_k^{\alpha_k}$.

Chapter 15
Answers, Hints and Solutions

Problems of Chap. 1

1.5° (1) 1; (2) 1; (3) $\frac{\pi}{\sqrt{6}}$, 1; (4) $\frac{1}{\sqrt[4]{5}}$; (5) 3; (6) 0; (7) 1; (8) $\sqrt[3]{\frac{1}{3}e^3 - \frac{3}{2}e^2 + 3e - \frac{17}{6}}$; (9) $\sqrt{\frac{3}{2} - \frac{1}{2}\cos 2}$; (10) $\frac{e^2}{4} + \frac{1}{4}$.

1.6° (1) No; (2), (3) Yes.

1.7° (1), (3), (4), (6) Yes; (2), (5) No.

1.8° (1), (2), (3) Yes.

1.9° Here and further we denote by x_0 the limit of the convergent sequence in $C([0, 1])$, $C^1([0, 1])$, $L_p([0, 1])$ and by $x^{(0)}$ the limit in l_p.

I. (1) No; (2) Yes, $x_0(t) = 0$; (3) No; (4) Yes, $x_0(t) = 1$; (5) Yes, $x_0(t) = \sin t$; (6) Yes, $x_0(t) = t$; (7) No; (8) Yes, $x_0(t) = t$; (9) Yes, $x_0(t) = t$; (10) No; (11) Yes, $x_0(t) = \varphi(t)$.

II. (1) Yes, $x_0(t) = 0$; (2) Yes, $x_0(t) = t$.

III. (1) Yes, $x^{(0)} = 0$; (2) Yes, $x^{(0)} = e_1$; (3) Yes, $x^{(0)} = 0$; (4) No; (5) Yes only for $x^{(0)} = 0$; (6) Yes only for $1 < p \le +\infty$, $x^{(0)} = (1, \frac{1}{2}, \dots, \frac{1}{n}, \dots)$; (7) Yes only for $2 < p \le +\infty$, $x^{(0)} = (1, \frac{1}{\sqrt{2}}, \dots, \frac{1}{\sqrt{n}}, \dots)$; (8) Yes only for $1 < p \le +\infty$, $x^{(0)} = 0$; (9) Yes only for $p = +\infty$, $x^{(0)} = 0$; (10)–(13) No.

IV. (1) Yes, $x_0(t) = 0$; (2) Yes, $x_0(t) = 0$; (3) Yes only for $1 \le p < 2$, $x_0(t) = 0$; (4) No; (5) Yes, $x_0(t) = 1$; (6) No.

1.14 (2) It is easy to check that the function $\|\cdot\|$ is defined correctly (i.e. the value of $\|x\|$ does not depend on the choice of polar coordinates of the point x) and satisfies the first two axioms of norm. To check the triangle inequality, we note that $A_f = \{x \in \mathbb{R}^2 \mid \|x\| \le 1\}$. Therefore for any $x, y \in \mathbb{R}^2$ if $x \ne 0$, $y \ne 0$, then

V. Brayman et al., *Functional Analysis and Operator Theory*, Problem Books in Mathematics, https://doi.org/10.1007/978-3-031-56427-7_15

$\frac{x}{\|x\|}, \frac{y}{\|y\|} \in A_f$. Since by the condition of the problem A_f is a convex set, then

$$\frac{x+y}{\|x\|+\|y\|} = \frac{\|x\|}{\|x\|+\|y\|} \cdot \frac{x}{\|x\|} + \frac{\|y\|}{\|x\|+\|y\|} \cdot \frac{y}{\|y\|} \in A_f.$$

So the norm of the element $\frac{x+y}{\|x\|+\|y\|}$ does not exceed one, whence $\|x+y\| \le \|x\| + \|y\|$. For $x = 0$ or $y = 0$ the triangle inequality is also obviously satisfied.

(3) For any norm $\|\cdot\|$ in $\mathbb{R}^2$ put $f(\varphi) = \frac{1}{\|(\cos\varphi, \sin\varphi)\|}$, $\varphi \in \mathbb{R}$. It follows from the properties of the norm that f is positive and π-periodic. For all $\rho > 0$ we have $\|(\rho\cos\varphi, \rho\sin\varphi)\| = \rho \cdot \|(\cos\varphi, \sin\varphi)\| = \frac{\rho}{f(\varphi)}$. Finally, the set A_f is convex because of $A_f = \overline{B}(0, 1)$ in the given norm.

If $1 \le p < +\infty$, then $f(\varphi) = \frac{1}{\left(|\cos\varphi|^p + |\sin\varphi|^p\right)^{\frac{1}{p}}}$;

if $p = +\infty$, then $f(\varphi) = \frac{1}{\max\{|\cos\varphi|, |\sin\varphi|\}}$.

1.15 To prove the equality, notice that for $x \in l_1$, $x \neq 0$, we have

$$\|x\|_\infty \le \|x\|_p = \|x\|_\infty \left(\sum_{n=1}^{\infty} \left(\frac{|x_n|}{\|x\|_\infty} \right)^p \right)^{\frac{1}{p}} \le$$

$$\le \|x\|_\infty \left(\sum_{n=1}^{\infty} \frac{|x_n|}{\|x\|_\infty} \right)^{\frac{1}{p}} = \|x\|_1^{\frac{1}{p}} \|x\|_\infty^{1-\frac{1}{p}} \to \|x\|_\infty, \ p \to +\infty.$$

1.16 Inclusion follows from Hölder's inequality, and equality follows from the fact that for $x \in L_\infty(T, \mu)$, $x \neq 0$, we have

$$\forall\, 0 < \varepsilon < \|x\|_\infty \ \exists\, A_\varepsilon \in \mathcal{F}, \ \mu(A_\varepsilon) > 0 \ \forall t \in A_\varepsilon \ : \ |x(t)| \ge \|x\|_\infty - \varepsilon,$$

whence $(\|x\|_\infty - \varepsilon)(\mu(A_\varepsilon))^{\frac{1}{p}} \le \|x\|_p \le \|x\|_\infty (\mu(T))^{\frac{1}{p}}$.

1.17° Consider functions $x(t) = t^{-\alpha}\chi_{(0,1]}(t)$ and $y(t) = t^{-\beta}\chi_{[1,+\infty)}(t)$, $t \in \mathbb{R}$. Select $\alpha \ge 0$ and $\beta \ge 0$ such that $x \in L_{p_1}(\mathbb{R}) \setminus L_{p_2}(\mathbb{R})$, $y \in L_{p_2}(\mathbb{R}) \setminus L_{p_1}(\mathbb{R})$.

1.18* (1) *Necessity.* Assume that for $n \ge 1$ there exist sets $A_n \in \mathcal{F}$ such that $2^n < \mu(A_n) < \infty$. Put $f_n(t) = \mu(A_n)^{-\frac{1}{p_1}} \chi_{A_n}(t)$, $n \ge 1$, and $f(t) = \sum\limits_{n=1}^{\infty} f_n(t)$. Then $\|f_n\|_{p_2} = \mu(A_n)^{\frac{p_1-p_2}{p_1 p_2}} < 2^{-\alpha n}$, where $\alpha = \frac{p_2-p_1}{p_1 p_2} > 0$. Since $\sum\limits_{n\ge 1}^{\infty} \|f_n\|_{p_2} \le \sum\limits_{n\ge 1}^{\infty} 2^{-\alpha n} < \infty$, then f is the sum of the series which converges in p_2-norm, therefore $f \in L_{p_2}(T, \mu)$. However

$$\int_T |f(t)|^{p_1} d\mu(t) = \int_T \left(\sum_{n=1}^{\infty} f_n(t) \right)^{p_1} d\mu(t) \geq$$

$$\geq \sum_{n=1}^{\infty} \int_T f_n(t)^{p_1} d\mu(t) = \sum_{n=1}^{\infty} 1 = \infty,$$

i.e. $f \notin L_{p_1}(T, \mu)$.

Sufficiency. Let $f \in L_{p_2}(T, \mu)$. It follows from Chebyshev's inequality that $\mu(\{|f(t)| \geq \frac{1}{n}\}) < \infty$, $n \geq 1$. Thus by condition of the problem $\mu(\{|f(t)| \geq \frac{1}{n}\}) \leq C$ for all $n \geq 1$, whence $\mu(\{|f(t)| > 0\}) \leq C$ by continuity of measure from below. Therefore

$$\int_T |f(t)|^{p_1} d\mu(t) \leq \mu(\{0 < |f(t)| < 1\}) +$$

$$+ \int_{\{|f(t)| \geq 1\}} |f(t)|^{p_1} d\mu(t) \leq C + \int_{\{|f(t)| \geq 1\}} |f(t)|^{p_2} d\mu(t) < \infty,$$

i.e. $f \in L_{p_1}(T, \mu)$.

(2) *Necessity.* Assume that for $n \geq 1$ there exist sets $A_n \in \mathcal{F}$ such that $0 < \mu(A_n) < \frac{1}{2^n}$. Let $f_n(t) = \mu(A_n)^{-\frac{1}{p_2}} \chi_{A_n}(t)$, $n \geq 1$, and $f(t) = \sum_{n=1}^{\infty} f_n(t)$. Quite similarly to item (1), it can be verified that $f \in L_{p_1}(T, \mu)$ but $f \notin L_{p_2}(T, \mu)$.

Sufficiency. Let $f \in L_{p_1}(T, \mu)$. It follows from Chebyshev's inequality that $\mu(\{|f(t)| \geq n\}) \to 0$, $n \to \infty$. Thus by condition of the problem there exists $N \in \mathbb{N}$ such that $\mu(\{|f(t)| \geq N\}) = 0$. Therefore

$$\int_T |f(t)|^{p_2} d\mu(t) \leq N^{p_2 - p_1} \int_T |f(t)|^{p_1} d\mu(t) < +\infty,$$

i.e. $f \in L_{p_2}(T, \mu)$.

1.19 In the case $s < +\infty$ prove that $\|x\|_r \leq \|x\|_p^{\alpha} \|x\|_s^{\beta}$ for all $x \in L_p(\mathbb{R}) \cap L_s(\mathbb{R})$, where $\alpha = \frac{r^{-1} - s^{-1}}{p^{-1} - s^{-1}}$, $\beta = \frac{p^{-1} - r^{-1}}{p^{-1} - s^{-1}}$. To do this, apply Hölder's inequality to the product $|x|^{\alpha r} \cdot |x|^{\beta r}$ and take into account the equality $\frac{1}{p/(r\alpha)} + \frac{1}{s/(r\beta)} = 1$.

1.21° (1), (2) No.

1.22 It is necessary and sufficient that $\alpha = (\alpha_1, \alpha_2, \ldots) \in l_\infty$ and $\alpha_k > 0$, $k \geq 1$. If $\alpha \notin l_\infty$, then there exists a sequence

$$\{k_1 < k_2 < \ldots < k_n < \ldots\} \subset \mathbb{N}$$

such that $\alpha_{k_n} \geq n$ for all $n \geq 1$. Consider $x = \sum_{n=1}^{\infty} \frac{1}{n} e_{k_n}$.

1.23 (1) In Hölder's inequality $\|xy\|_1 \leq \|x\|_p \|y\|_q$, $x \in L_p(T, \mu)$, $y \in L_q(T, \mu)$, equality is achieved if and only if there exist nonnegative numbers α, β such that $\alpha + \beta > 0$ and $\alpha|x|^p = \beta|y|^q \pmod{\mu}$.

(2) In the Minkowski inequality $\|x + y\|_p \leq \|x\|_p + \|y\|_p$, $x, y \in L_p(T, \mu)$, for $p > 1$ the equality is achieved if and only if there exist nonnegative numbers α, β such that $\alpha + \beta > 0$ and $\alpha x = \beta y \pmod{\mu}$, and for $p = 1$ if and only if $x\overline{y} \geq 0 \pmod{\mu}$.

1.24 Verify that both functions $f(t)\cdot\chi_{\{|t|\leq 1\}} + tf(t)\cdot\chi_{\{|t|>1\}}$ and $\chi_{\{|t|\leq 1\}} + \frac{1}{t}\cdot\chi_{\{|t|>1\}}$ belong to $L_2(\mathbb{R})$. Apply the Cauchy–Schwarz inequality to them.

1.25 Consider functions $f_1 = f \cdot \chi_{\{t\in\mathbb{R} \mid |f(t)|>1\}}$ and $f_2 = f - f_1$.

1.26 Note that for all $\varepsilon > 0$ and all $k \in \mathbb{N}$ the inequality

$$A\sigma^k \geq \int_{\mathbb{R}} |x^k f(x)|dx \geq (\sigma + \varepsilon)^k \int_{|x|\geq\sigma+\varepsilon} |f(x)|dx$$

holds. It follows that $f(x) = 0$ for $|x| > \sigma + \varepsilon$.

1.27° For $1 \leq p < \frac{1}{\alpha}$.

1.28 (1) $L = \overline{\text{span}(\{\varphi_n : n \in \mathbb{Z}\})}$, where $\varphi_n(t) = \sin \pi t \chi_{[n,n+1]}(t)$, $t \in \mathbb{R}$;

(2) $L = \overline{\text{span}(\{\psi_n : n \in \mathbb{Z}\})}$, where $\psi_n(t) = \chi_{[n+\frac{1}{2},n+1]}(t)$, $t \in \mathbb{R}$.

1.31° (1) Yes. (2) No.

1.35° Consider the sequence $x_n(t) = \left(\frac{t-a}{b-a}\right)^n$, $t \in [a, b]$, $n \geq 1$.

1.36° Yes. Use item (2) of problem 1.32.

1.37 No. Let $\alpha \in [0, 1]\backslash\mathbb{Q}$. Put $x_k(t) = e^{-k|t-\alpha|}$, $t \in [0, 1]$, $k \geq 1$. Then $\|x_k\| \to 0$, but $x_k \not\rightrightarrows 0$ on $[0, 1]$, $k \to \infty$.

1.39 All conditions are equivalent.

1.40 *Necessity.* Show that the convergence of the series $\sum_{k=1}^{\infty} \|x_k\|$ implies that the sequence of partial sums of the series $\sum_{k=1}^{\infty} x_k$ is a Cauchy sequence, so this sequence converges.

Sufficiency. Let $\{y_n : n \geq 1\} \subset X$ be a Cauchy sequence. For each $k \geq 1$ there exists N_k such that $\|y_n - y_m\| < \frac{1}{2^k}$ for all $n, m \geq N_k$. Put $n_1 = N_1$ and $n_k = \max\{n_{k-1} + 1, N_k\}$, $k \geq 2$. Then

$$\|y_{n_1}\| + \sum_{k=2}^{\infty} \|y_{n_k} - y_{n_{k-1}}\| < \|y_{n_1}\| + \sum_{k=2}^{\infty} \frac{1}{2^{k-1}} < +\infty.$$

Therefore the series $y_{n_1} + \sum\limits_{k=2}^{\infty} (y_{n_k} - y_{n_{k-1}})$ converges to some $y \in X$, i.e. $y_{n_k} \to y$, $k \to \infty$. Since $\{y_n : n \geq 1\}$ is a Cauchy sequence, it follows that $y_n \to y$, $n \to \infty$.

1.41 To prove completeness of the space $(C^n([a,b]), \|\cdot\|_n)$ use Cauchy's criterion for convergence in $C([a,b])$ and show that if $x_n^{(k)} \rightrightarrows y_k$, $n \to \infty$, on $[a,b]$, $0 \leq k \leq n$, then $y_0 \in C^n([a,b])$ and $y_0^{(k)} = y_k$, $1 \leq k \leq n$. To do this, you can apply the Newton-Leibnitz formula and the theorem on taking the limit under the integral sign in the Riemann integral. To prove incompleteness of the space $(C^n([a,b]), \|\cdot\|_{n-j})$, construct a sequence of functions $\{x_n : n \geq 1\} \subset C^{j-1}([a,b])$, which is bounded in $C([a,b])$ and converges pointwise to a discontinuous function $x : [a,b] \to \mathbb{R}$. Prove that the sequence $y_k(t) = \int\limits_a^t (t-u)^{n-j} x_k(u)du$, $t \in [a,b]$, $k \geq 1$, is a Cauchy sequence which diverges in $(C^n([a,b]), \|\cdot\|_{n-j})$.

1.43 If Q is not a compact set, then Q has a countable subset T that has no boundary points. Let M be the set of functions of the form

$$x_A(t) = \frac{\rho(t, T \setminus A)}{\rho(t, A) + \rho(t, T \setminus A)}, \text{ where } A \subset T,\ A \neq \varnothing,\ A \neq T,$$

and the distance from the point to the set is defined as in item (3) of problem 1.42. Check that $x_A \in C(Q)$ and $x_A(t) \in [0,1]$ for all $t \in Q$, moreover $x_A(t) = \chi_A(t)$ for $t \in T$. Then $M \subset C_b(Q)$ is an uncountable set and for all $x, y \in M$ if $x \neq y$, then $\|x - y\| = 1$. Deduce that each dense set in $C_b(Q)$ is uncountable.

1.44 (2) It is necessary and sufficient that the set T is finite.

1.45 No. Consider the set $\{\chi_{(c,b]}(t) \mid c \in (a,b)\}$.

1.46 No. Consider the set $\{(t-c)^\lambda \cdot \chi_{[c,b]}(t) \mid c \in (a,b)\}$.

1.47 (2) $\rho(x,y) = |\arctan x - \arctan y|$.

1.48 (1) No. Use item (2) of problem 1.23. (2) For $\alpha \geq \beta \geq 0$ we have

$$\alpha\|x\| + \beta\|y\| \geq \|\alpha x + \beta y\| = \|\alpha(x+y) + (\beta - \alpha)y\| \geq$$
$$\geq \alpha\|x+y\| - (\alpha - \beta)\|y\| = \alpha\|x\| + \beta\|y\|.$$

1.50 (1) Let $\|\cdot\|$ be any norm in X, $e_n = (0, \ldots, 0, \underbrace{1}_{n}, 0, 0, \ldots)$ and $\alpha_n = \|e_n\|^{-1}$, $n \geq 1$. The sequence $\{\alpha_n e_n : n \geq 1\}$ converges to 0 coordinate-wise, but does not converge to 0 in the norm. (2) The desired metric can be defined by the formula

$$\rho(x, y) = \sum_{n=1}^{\infty} \frac{1}{2^n} \cdot \frac{|x_n - y_n|}{1 + |x_n - y_n|}, \quad x = (x_1, x_2, \ldots), \ y = (y_1, y_2, \ldots) \in X.$$

1.51 Prove the statement for some total set in $L_1([a, b])$ first, e.g., for continuous functions.

1.55° Consider the function $\chi_{[0,+\infty)}$.

1.56 Consider the set of functions $\{\chi_{[t_0,b]} \mid t_0 \in (a, b]\}$.

1.57 Use the Stone-Weierstrass theorem.

1.58* (1) It follows from the condition of the problem that there exists a countable set $\{t_n \mid n \geq 0\} \subset K$ such that $t_n \to t_0$, $n \to \infty$. For each $n \geq 1$, choose $r_n > 0$ such that $B(t_n, r_n) \cap \{t_k \mid k \geq 1, \ k \neq n\} = \varnothing$. Then functions

$$x_n(t) = \frac{\rho(t, K \setminus B(t_n, r_n))}{\rho(t, t_n) + \rho(t, K \setminus B(t_n, r_n))}, \quad t \in K, \ n \geq 1,$$

where the distance from the point to the set is defined as in item (3) of problem 1.42, are linearly independent. (2) Verify that functions

$$\varphi_i(t) = \frac{\rho(t, K \backslash U_i)}{\sum\limits_{k=1}^{n} \rho(t, K \setminus U_k)}, \quad t \in K, \ 1 \leq i \leq n,$$

satisfy all the desired conditions. (3) Let $M = \{y_n \mid n \geq 1\}$ be a countable dense set in K. For each $m \geq 1$ there exists $n_m \in \mathbb{N}$ such that $\left\{B\left(y_k, \frac{1}{m}\right) \mid 1 \leq k \leq n_m\right\}$ is a cover of K. Let $\{\varphi_{k,m} \mid 1 \leq k \leq n_m\}$ be the partition of unity constructed for this cover. Consider the family of functions $\sum\limits_{k=1}^{n_m} (r_k + i q_k)\varphi_{k,m}$, where $r_k, q_k \in \mathbb{Q}$ and $m \in \mathbb{N}$ are arbitrary.

1.60 It is necessary and sufficient that $m(\{t \in \mathbb{R} \mid g(t) = 0\}) = 0$.

1.61° (1), (2), (5), (7)–(9), (12) No. (3), (4), (6), (10), (11) Yes. In (6) closedness follows from problem 1.34°.

1.62° (1) Yes. (2) No.

1.63 $\overline{l_1} = c_0$ in l_∞.

1.64 (1) Yes. (2), (3) No. For $1 < p \leq +\infty$ consider the sequence

$$x^{(n)} = e_1 - \frac{1}{n}(e_2 + \ldots + e_{n+1}),\ n \geq 1.$$

1.65 No. Consider the sequence $x^{(n)} = (1, \frac{1}{2}, \ldots, \frac{1}{n}, 0, \ldots), n \geq 1$.

1.68° No. Consider the set $\{e_n : n \geq 1\}$ in l_2.

1.71 To prove sufficiency, verify that for any $x, y \in X$ if $x \neq 0$, $y \neq 0$, then $\frac{x}{\varphi(x)}, \frac{y}{\varphi(y)} \in B$. Since B is a convex set we have

$$\frac{x+y}{\varphi(x)+\varphi(y)} = \frac{\varphi(x)}{\varphi(x)+\varphi(y)} \cdot \frac{x}{\varphi(x)} + \frac{\varphi(y)}{\varphi(x)+\varphi(y)} \cdot \frac{y}{\varphi(y)} \in B.$$

Therefore $\varphi\left(\frac{x+y}{\varphi(x)+\varphi(y)}\right) \leq 1$, whence $\varphi(x+y) \leq \varphi(x)+\varphi(y)$. For $x = 0$ or $y = 0$, the triangle inequality is also obviously satisfied.

1.72 Use Theorem 1.1 and problem 1.32.

1.73° Use problem 1.72.

1.74 (2) $\{e_n : n \geq 1\}$.

1.75° Use item (1) of problem 1.74.

1.76° Use problem 1.34°.

1.78 Use problem 1.4.

1.81 If $B(x_0, r) \subset M$ for some $x_0 \in M$ and $r > 0$, then due to the linearity of the set M we have $B(0, r) \subset M$ and $B(0, nr) \subset M$, $n \geq 1$, whence $M = X$.

1.82 (1) Consider the collection $\{X_A\}$ of all systems X_A of linearly independent elements in X; introduce a partial order on it by inclusion and apply Zorn's lemma (see, for example, [26], Theorem I.2 or [21], Ch. 1, Sec. 3). (3) If $\{e_n : n \geq 1\}$ is a countable Hamel basis in the Banach space X, $L_n = \text{span}(\{e_1, \ldots, e_n\})$, $n \geq 1$, then $X = \bigcup\limits_{n=1}^{\infty} L_n$. Use the Baire category theorem to show that at least one of the subspaces L_n has an interior point, then use problem 1.81. (4) Consider

$$\{x \in l_2 \,|\, x = (x_1, \ldots, x_n, 0, 0, \ldots),\ n \in \mathbb{N}\}$$

with the norm induced from l_2.

1.83 To prove necessity, notice that an NLS is nonseparable if and only if it does not have a countable ε-net for some $\varepsilon > 0$. Next, apply Zorn's lemma similarly to solution of item (1) of problem 1.82.

1.84 Show that $\{x_n : n \geq 1\}$ is a Cauchy sequence.

1.85 For a Cauchy sequence $\{x_n : n \geq 1\} \subset X$, consider the balls $B_n = \overline{B}(x_n, r_n)$, where $r_n = 2 \sup\limits_{m \geq n} \|x_m - x_n\|$, $n \geq 1$. It is obvious that $r_n \to 0$, $n \to \infty$, because $\{x_n : n \geq 1\}$ is a Cauchy sequence. If $r_{n_0} = 0$, then $x_n \to x_{n_0}$, $n \to \infty$. Otherwise $r_n > 0$ for all $n \geq 1$ and there exists a subsequence $\{n_1 < n_2 < \ldots\} \subset \mathbb{N}$ such that $r_{n_{k+1}} < \frac{1}{2} r_{n_k}$, $k \geq 1$. If $y \in B_{n_{k+1}}$, then

$$\|y - x_{n_k}\| \leq \|y - x_{n_{k+1}}\| + \|x_{n_{k+1}} - x_{n_k}\| \leq r_{n_{k+1}} + \frac{1}{2} r_{n_k} < r_{n_k},$$

so $B_{n_{k+1}} \subset B_{n_k}$. Therefore $\{B_{n_k} : k \geq 1\}$ is a sequence of closed nested balls and $r_{n_k} \to 0$, $k \to \infty$. For $\{x\} = \bigcap\limits_{k=1}^{\infty} B_{n_k}$ show that $x_n \to x$, $n \to \infty$.

1.86 (1) $X = l_2$, $M_n = \{e_k \mid k \geq n\}$, $n \geq 1$. (2) Show that the sequence $\{r_n : n \geq 1\}$ of balls' radii is nonincreasing. If $r_n \to 0$, $n \to \infty$, see problem 1.84. If there exists $n \in \mathbb{N}$ such that $r_k = r_n$ for all $k \geq n$, there is nothing to prove. Else $r_n \to \rho$, $n \to \infty$, and $r_n > \rho$, $n \geq 1$. Then balls with the same centers and radii $r_n - \rho$ are also nested.

1.89 (2) $\{e_n : n \geq 1\}$ is a basis in c_0 and l_p, $1 \leq p < +\infty$; $\{e_n : n \geq 0\}$, where $e_0 = (1, 1, \ldots, 1, \ldots)$, is a basis in c.

1.90* Firstly, we will show that if x can be represented as a sum of a series, then the representation is unique. Let $x = \sum\limits_{i=1}^{\infty} \alpha_i e_i = \sum\limits_{i=1}^{\infty} \beta_i e_i$. Apply the condition (b) for $\lambda_i = \alpha_i - \beta_i$ and direct m to infinity. We get that $\sum\limits_{i=1}^{k} (\alpha_i - \beta_i) e_i = 0$, $k \geq 1$, so $\alpha_i = \beta_i$, $i \geq 1$.

Let Y be the set of those $x \in X$ for which the representation $x = \sum\limits_{i=1}^{\infty} \alpha_i e_i$ exists. Let us verify that the set Y is closed, then condition (a) will imply that $Y = X$. Let $y_n \to y$, $n \to \infty$, where $y_n = \sum\limits_{i=1}^{\infty} \alpha_i^{(n)} e_i \in Y$, $n \geq 1$. The sequence $\{y_n : n \geq 1\}$ is a Cauchy sequence. It follows from condition (b) that for each $k \geq 1$ the sequence $\left\{\sum\limits_{i=1}^{k} \alpha_i^{(n)} e_i : n \geq 1\right\}$ is also a Cauchy sequence. Therefore for every $i \geq 1$ the sequences $\{\alpha_i^{(n)} : n \geq 1\}$ are Cauchy sequences, therefore they converge to some numbers α_i. Check that $y = \sum\limits_{i=1}^{\infty} \alpha_i e_i$.

1.91* (1) Verify that $\chi_{[\frac{l-1}{2^k}, \frac{l}{2^k}]} \in \text{span}(\{h_n : n \geq 1\})$ for all $k \in \mathbb{N}$ and $1 \leq l \leq 2^k$. Show that these characteristic functions form a total set in $L_p([0, 1])$.

(2) Use items (1), (3) and problem 1.90*.

(3) For all $x, y \in \mathbb{R}$ we have $2|x|^p \leq |x+y|^p+|x-y|^p$, because the function $|t|^p$, $t \in \mathbb{R}$, is convex. Let $f_n = \sum\limits_{i=1}^{n} \lambda_i h_i$, $f_{n+1} = f_n + \lambda_{n+1}h_{n+1}$, $[a, b]$ be the segment on which $h_{n+1} \neq 0$. Then $f_{n+1}(t) = f_n(t)$ for $t \notin [a, b)$, $f_n(t) = c =$ const for $t \in [a, b)$, $f_{n+1}(t) = c + \lambda_{n+1}$ for $t \in [a, \frac{a+b}{2})$, and $f_{n+1}(t) = c - \lambda_{n+1}$ for $t \in [\frac{a+b}{2}, b)$. Thus

$$\int\limits_a^b |f_n(t)|^p dt = |c|^p(b-a) \leq$$

$$\leq \frac{b-a}{2}(|c+\lambda_{n+1}|^p + |c-\lambda_{n+1}|^p) = \int\limits_a^b |f_{n+1}(t)|^p dt,$$

which is equivalent to the desired inequality.

1.92* Let $t_0 = 0$, $t_1 = 1$ and t_{2^k+l} be the middle of the segment $[\frac{l-1}{2^k}, \frac{l}{2^k}]$ for $k = 0, 1, 2, \ldots$ and $1 \leq l \leq 2^k$. For $n \geq 1$, the segment $[0, 1]$ is divided by the points $t_0, t_1, \ldots, t_n$ into smaller segments, and the function s_n is linear on each of them. Moreover, $s_n(t_n) \neq 0$ and $s_n(t_i) = 0$, $0 \leq i < n$. Therefore for any $\lambda_i \in \mathbb{K}$, $i \geq 0$, the functions $f_n = \sum\limits_{i=0}^{n} \lambda_i s_i$ are piecewise linear and

$$\|f_n\| = \max_{0 \leq i \leq n} |f_n(t_i)| = \max_{0 \leq i \leq n} |f_{n+1}(t_i)| \leq \max_{0 \leq i \leq n+1} |f_{n+1}(t_i)| = \|f_{n+1}\|.$$

It is easy to check that for $k \geq 0$ $\text{span}(\{s_0, s_1, \ldots, s_{2^k}\})$ consists of all functions which are linear on each of the segments $[\frac{l-1}{2^k}, \frac{l}{2^k}]$, $1 \leq l \leq 2^k$. Therefore $\overline{\text{span}(\{s_n : n\ ge 0\})} = C([0, 1])$. It remains to use problem 1.90*.

1.93 Let $\{x_n : n \geq 1\}$ be a countable dense set in A. Then

$$\left\{ \sum_{k=1}^{n} (r_k + iq_k)x_k \mid r_k, q_k \in \mathbb{Q},\ 1 \leq k \leq n;\ n \in \mathbb{N} \right\}$$

is a countable dense set in $\overline{\text{span}(A)}$, if X is a complex NLS. Similarly

$$\left\{ \sum_{k=1}^{n} r_k x_k \mid r_k \in \mathbb{Q},\ 1 \leq k \leq n;\ n \in \mathbb{N} \right\}$$

is a countable dense set in $\overline{\text{span}(A)}$, if X is a real NLS.

1.95 $1 < p < +\infty$.

1.96 (1) Use item (2) of problem 1.23.

1.97* *Necessity.* Let there exist $x \in X$ and $y_1, y_2 \in L$, $y_1 \neq y_2$, such that

$$\|x - y_1\| = \|x - y_2\| = \rho(x, L) > 0.$$

Since $\frac{y_1+y_2}{2} \in L$, then

$$\rho(x, L) \leq \|x - \frac{y_1 + y_2}{2}\| \leq \frac{1}{2}(\|x - y_1\| + \|x - y_2\|) = \rho(x, L).$$

Thus $\|x - \frac{y_1+y_2}{2}\| = \rho(x, L)$. Therefore for $x_k = \frac{x-y_k}{\|x-y_k\|}$, $k = 1, 2$, we have $\|x_1\| = \|x_2\| = \|\frac{1}{2}(x_1 + x_2)\| = 1$, however $x_1 \neq x_2$, i.e. X is not strictly normed.

Sufficiency. Assume that there exist $x_1, x_2 \in X$ such that

$$\|x_1\| = \|x_2\| = \left\|\frac{x_1 + x_2}{2}\right\| = 1$$

and $x_1 \neq x_2$. Consider a convex set $L = \big\{y_\alpha = (1 - \alpha)x_1 + \alpha x_2 \mid \alpha \in [0, 1]\big\}$. For all $\alpha \in [0, 1]$ we have $\|y_\alpha\| \leq (1 - \alpha)\|x_1\| + \alpha\|x_2\| = 1$ and

$$1 = \|y_{1/2}\| \leq \frac{1}{2}\big(\|y_\alpha\| + \|y_{1-\alpha}\|\big) \leq 1.$$

So, $\|y_\alpha\| = 1$, $\alpha \in [0, 1]$. Thus $\rho(0, L) = 1$ and y_α, $\alpha \in [0, 1]$ are elements of the best approximation in L for $x = 0$.

(1) $L = \{(x_1, 0) \mid x_1 \in \mathbb{R}\}$, $x = (0, 1)$.

(2) $L = \{(x_1, -x_1) \mid x_1 \in \mathbb{R}\}$, $x = (1, 1)$.

1.98 (2) Use proof by contradiction and item (1) of problem 1.74.

1.99* (1) Show that $f(z_1, z_2) = \frac{|z_1|^p+|z_2|^p}{2} - \left|\frac{z_1+z_2}{2}\right|^p > 0$ for all $z_1, z_2 \in \mathbb{K}$, for which $z_1 \neq z_2$, and put

$$a_c = \max\left\{f(z_1, z_2)^{-1} \;\Big|\; \frac{|z_1|^p + |z_2|^p}{2} = 1, \left|\frac{z_1 - z_2}{2}\right|^p \geq c\right\}.$$

(2) For $x_1, x_2 \in L_p(T, \mu)$ such that $\|x_1\|_p = \|x_2\|_p = 1$, consider the set

$$M = \left\{t \in T \;\Big|\; \left|\frac{x_1(t) - x_2(t)}{2}\right|^p \geq c\left(\frac{|z_1(t)|^p + |z_2(t)|^p}{2}\right)\right\},$$

where $0 < c < 1$ will be chosen later. By (1) we have

$$\int\limits_M \left|\frac{x_1(t) - x_2(t)}{2}\right|^p d\mu(t) \leq \int\limits_M \left(\frac{|x_1(t)|^p + |x_2(t)|^p}{2}\right) d\mu(t) \leq$$

$$\leq a_c \int\limits_M \left(\frac{|x_1(t)|^p + |x_2(t)|^p}{2} - \left|\frac{x_1(t) + x_2(t)}{2}\right|^p\right) d\mu(t) \leq$$

$$\leq a_c \int\limits_T \left(\frac{|x_1(t)|^p + |x_2(t)|^p}{2} - \left|\frac{x_1(t) + x_2(t)}{2}\right|^p\right) d\mu(t) = a_c\left(1 - \left\|\frac{x_1 + x_2}{2}\right\|_p^p\right)$$

and

$$\int\limits_{T\setminus M} \left|\frac{x_1(t) - x_2(t)}{2}\right|^p d\mu(t) \leq c \int\limits_{T\setminus M} \left(\frac{|x_1(t)|^p + |x_2(t)|^p}{2}\right) d\mu(t) \leq c.$$

Hence

$$\|x_1 - x_2\|_p^p = \int\limits_T |x_1(t) - x_2(t)|^p d\mu(t) \leq 2^p a_c\left(1 - \left\|\frac{x_1 + x_2}{2}\right\|_p^p\right) + 2^p c.$$

For an arbitrary $0 < \varepsilon < 1$, take $c = \frac{\varepsilon^p}{2^{p+1}}$ and determine $\delta > 0$ from the condition $1-(1-\delta)^p = \frac{\varepsilon^p}{2^{p+1}a_c}$. Then if $\|x_1\|_p = \|x_2\|_p = 1$ and $\left\|\frac{x_1+x_2}{2}\right\|_p > 1-\delta$, then $\|x_1 - x_2\|_p < \varepsilon$.

1.100* Let $\{y_n : n \geq 1\} \subset L$ be a sequence of elements for which

$$r_n = \|x - y_n\| \to \rho(x, L) = r > 0,\ n \to \infty.$$

Assume that $z_n = \frac{x-y_n}{r_n}$, $n \geq 1$ ($r_n \neq 0$, because $r_n \geq r > 0$). Then

$$z_n + z_m = \left(\frac{1}{r_n} + \frac{1}{r_m}\right)x - \left(\frac{y_n}{r_n} + \frac{y_m}{r_m}\right).$$

Since $\left(\frac{y_n}{r_n} + \frac{y_m}{r_m}\right)\left(\frac{1}{r_n} + \frac{1}{r_m}\right)^{-1} \in L$, then $\left\|x - \left(\frac{y_n}{r_n} + \frac{y_m}{r_m}\right)\left(\frac{1}{r_n} + \frac{1}{r_m}\right)^{-1}\right\| \geq r$, whence $\|z_n + z_m\| \geq \left(\frac{1}{r_n} + \frac{1}{r_m}\right)r$. Thus $\|z_n\| = \|z_m\| = 1$ and

$$\left\|\frac{z_n + z_n}{2}\right\| \geq \left(\frac{1}{2r_n} + \frac{1}{2r_m}\right)r \to 1,\ n, m \to \infty.$$

Since the space is uniformly convex, it follows that $\|z_n - z_m\| \to 0$, $n, m \to \infty$. Therefore the sequence $\{z_n : n \geq 1\}$ is a Cauchy sequence, so it converges, because the space is complete. Let $z_n \to z$, $n \to \infty$. Then $y_n = x - r_n z_n \to y = x - rz$, where $y \in L$, because L is a closed set, and $\|x - y\| = \|rz\| = r = \rho(x, L)$.

The uniqueness follows from problem 1.97* and item (1) of problem 1.98.

1.101 Use problem 1.100*.

1.102* Assume that there exist a partition $A_1, \ldots, A_n$ of the space T with $\mu(A_k) > 0$, $1 \leq k \leq n$. Prove that the functions χ_{A_k}, $1 \leq k \leq n$, are linearly independent in $L_p(T, \mu)$, thus $\dim L_p(T, \mu) \geq n$. Conversely, assume that $\sup \Phi = n_0 < \infty$. Consider the corresponding partition $\{A_1, \ldots, A_{n_0}\}$ and prove that it consists of atoms. Verify that each function $f \in L_p(T, \mu)$ is μ-almost everywhere constant on each of the elements of partition, i.e. $f = \sum\limits_{k=1}^{n_0} z_k \chi_{A_k}$. Therefore $\{\chi_{A_k} : 1 \leq k \leq n_0\}$ is a basis in $L_p(T, \mu)$.

Problems of Chap. 2

2.6° (1) $\left\{\frac{1}{\sqrt{n}} e_n : n \geq 1\right\}$; (2) $\left\{\frac{1}{n} e_n : n \geq 1\right\}$; (3) $\left\{e_n e^{\frac{n}{2}} : n \geq 1\right\}$.

2.7 It is necessary and sufficient that $\inf\limits_{k \geq 1} \alpha_k > 0$.

2.8° (1) Write the square of the norm as a scalar product and use its properties. (2) Check that for $x(t) = t$, $y(t) = 1$, $t \in [0, 1]$, parallelogram law is not true in $C([0, 1])$.

2.9° Use item (1) of problem 2.8°.

2.10 Use item (1) of problem 2.8°.

2.11 Introduce new variables $a = x - u$, $b = y - u$, $c = z - u$. We get the inequality $\|a - c\| \cdot \|b\| \leq \|a - b\| \cdot \|c\| + \|b - c\| \cdot \|a\|$. In cases $a = 0$, or $b = 0$, or $c = 0$ the inequality is obvious. For $a, b, c \neq 0$, we set $a' = \frac{a}{\|a\|^2}$, $b' = \frac{b}{\|b\|^2}$, $c' = \frac{c}{\|c\|^2}$. Note that

$$\|a' - b'\|^2 = (a' - b', a' - b') = \|a'\|^2 - (a', b') - (b', a') + \|b'\|^2 =$$
$$= \frac{1}{\|a\|^2} - \frac{(a, b) + (b, a)}{\|a\|^2 \|b\|^2} + \frac{1}{\|b\|^2} = \frac{(b - a, b - a)}{\|a\|^2 \|b\|^2} = \frac{\|a - b\|^2}{\|a\|^2 \|b\|^2},$$

i.e. $\|a'-b'\| = \frac{\|a-b\|}{\|a\|\cdot\|b\|}$. Since $\|a'-c'\| \le \|a'-b'\| + \|b'-c'\|$ (triangle inequality), then

$$\frac{\|a-c\|}{\|a\|\cdot\|c\|} \le \frac{\|a-b\|}{\|a\|\cdot\|b\|} + \frac{\|b-c\|}{\|b\|\cdot\|c\|}.$$

It remains to multiply both parts of the inequality by $\|a\|\cdot\|b\|\cdot\|c\|$.

2.13° In a real Hilbert space, the polarization identity has the form

$$(x, y) = \frac{1}{4}(\|x+y\|^2 - \|x-y\|^2).$$

2.14 Apply condition (3) to deduce that $S(nx, y) = nS(x, y)$, $S(\frac{x}{n}, y) = \frac{1}{n}S(x, y)$, $S(\frac{m}{n}x, y) = \frac{m}{n}S(x, y)$ for $m \in \mathbb{Z}, n \in \mathbb{N}, x, y \in X$. For any $\lambda \in \mathbb{C}$ there exists a sequence $\{\frac{m_k}{n_k} + i\frac{p_k}{q_k} : k \ge 1\}$ which converges to λ, where $m_k, p_k \in \mathbb{Z}, n_k, q_k \in \mathbb{N}$. Therefore it follows from conditions (4) and (5) that $S(\lambda x, y) = \lambda S(x, y)$. The space is not necessarily a Hilbert space. For example, the space $C([0, 1])$ with the scalar product $(x, y) = \int_0^1 x(t)\overline{y(t)}dt$ is not complete.

2.15* Use the statement of problem 2.14. Properties (1) and (5) follow from the properties of a norm. Let us verify property (2):

$$S(y, x) = \tfrac{1}{4}(\|y+x\|^2 - \|y-x\|^2 + i\|y+ix\|^2 - i\|y-ix\|^2) =$$
$$= \tfrac{1}{4}(\|x+y\|^2 - \|x-y\|^2 + i\|x-iy\|^2 - i\|x+iy\|^2) = \overline{S(x, y)}$$

(because $\|y+ix\| = \|i(x-iy)\| = \|x-iy\|$ and similarly $\|y-ix\| = \|x+iy\|$). Property (4) can be verified in a similar way. To check property (3), note that

$$S(x, y) = \sum_{k=0}^{3} \frac{i^k}{4}\|x + i^k y\|^2.$$

Therefore due to the parallelogram law we have

$$S(x_1, y) + S(x_2, y) = \sum_{k=0}^{3} \frac{i^k}{4}(\|x_1 + i^k y\|^2 + \|x_2 + i^k y\|^2) =$$

$$= \sum_{k=0}^{3} \frac{i^k}{8}\left(\|x_1 + x_2 + 2i^k y\|^2 + \|x_1 - x_2\|^2\right) = \sum_{k=0}^{3} \frac{i^k}{8}\|x_1 + x_2 + 2i^k y\|^2$$

(we took into account that $\sum\limits_{k=0}^{3} i^k = 1 + i - 1 - i = 0$). The last expression depends only on $x_1 + x_2$ and y, therefore

$$S(x_1, y) + S(x_2, y) = S(x_1 + x_2, y) + S(0, y) = S(x_1 + x_2, y).$$

Remark 1 Similarly, in the real NLS if the norm satisfies the parallelogram law, then $(x, y) = \frac{1}{4}(\|x + y\|^2 - \|x - y\|^2)$ is the scalar product which generates this norm.

Remark 2 It follows from the statements of problems 2.8° and 2.15* that the norm in NLS is generated by a scalar product if and only if it satisfies the parallelogram law (the Jordan-von Neumann theorem).

2.16 It is necessary and sufficient that $\operatorname*{ess\,inf}\limits_{t\in[a,b]} p(t) > 0$, i.e. there exists $\varepsilon > 0$ such that $p \geq \varepsilon \pmod m$.

2.17 Use the fact that $\left\| \sum\limits_{k=n}^{m} x_k \right\|^2 = \sum\limits_{k=n}^{m} \|x_k\|^2$, $1 \leq n \leq m$.

2.18 *Sufficiency.* Let $x_1, \dots, x_n$ be linearly dependent. Then there exist $\alpha_1, \dots, \alpha_n \in \mathbb{K}$, not all zero, for which $\sum\limits_{k=1}^{n} \alpha_k x_k = 0$. It follows that $\sum\limits_{k=1}^{n} \alpha_k (x_k, x_j) = 0$ for $j = 1, \dots, n$. Therefore the rows of the Gram determinant are linearly dependent and the determinant is equal to zero.

Necessity. Let the Gram determinant be zero. Then its rows are linearly dependent, so there exist $\alpha_1, \dots, \alpha_n \in \mathbb{K}$, not all zero, for which $\sum\limits_{k=1}^{n} \alpha_k (x_k, x_j) = \left(\sum\limits_{k=1}^{n} \alpha_k x_k, x_j \right) = 0$ for $j = 1, \dots, n$. Therefore for $y = \sum\limits_{k=1}^{n} \alpha_k x_k$ we have $\|y\|^2 = \left(\sum\limits_{k=1}^{n} \alpha_k x_k, \sum\limits_{j=1}^{n} \alpha_j x_j \right) = 0$. Hence $y = \sum\limits_{k=1}^{n} \alpha_k x_k = 0$ and vectors $x_1, \dots, x_n$ are linearly dependent.

2.19° and **2.20°** Write the square of the norm as a scalar product and use its properties.

2.24 (2) Show that $(M^\perp)^\perp = \overline{\operatorname{span}(M)}$.

2.26° $L^\perp = \{x \in L_2(\mathbb{R}) \mid x(t) = 0,\ t \leq 0 \pmod m\}$.

2.28° (1)–(7) $\{0\}$. Use problem 2.4.

2.29° (1) $\{0\}$; (2), (4) the set of odd functions; (3) the set of even functions.

2.30° (1) $\overline{\operatorname{span}(\{\cos kt \mid k \geq 0\})}$, i.e. the set of all even functions;

(2) $\overline{\text{span}(\{1, \sin kt \mid k \geq 1\})}$, i.e. the set of functions of the form $x(t) + c$, where $x(t)$ is an odd function, and $c \in \mathbb{K}$ is constant; (3) $\overline{\text{span}(\{e^{ikt} \mid k \in \mathbb{Z}, k \leq 4\})}$; (4) $\overline{\text{span}(\{e^{-ikt} \mid k \in \mathbb{Z}, k \leq 2\})}$.

2.31° (1) $\{x \in l_2 \mid x_1 + x_2 = 0\}$, (2) $\overline{\text{span}(\{e_k \mid k > n\})}$; (3) $\overline{\text{span}(\{e_{2k-1} \mid k \in \mathbb{N}\})}$.

2.33* span($\{\sinh t\}$). Each function $x \in W_2^1([a, b])$ belongs to $L_2([a, b])$ and has a "generalized derivative" $x' \in L_2([a, b])$ in the sense that there exists a sequence $x_n \in C^1([a, b])$ such that $x_n \to x$ and $x_n' \to x'$ in $L_2([a, b])$. Let $x \perp \varphi_k$, $k \in \mathbb{Z}$, in $W_2^1([-\pi, \pi])$, where $\left\{\varphi_k = \frac{e^{ikt}}{\sqrt{2\pi}} : k \in \mathbb{Z}\right\}$ is an orthonormal basis in $L_2([-\pi, \pi])$. Expand x' in this basis:

$$x'(t) = \sum_{k \in \mathbb{Z}} a_k \varphi_k(t).$$

Here and further the series converges in $L_2([-\pi, \pi])$. Then

$$x(t) = \sum_{k \in \mathbb{Z} \setminus \{0\}} \frac{a_k}{ik} \varphi_k(t) + a_0 \frac{t}{\sqrt{2\pi}} + b_0 =$$

$$= \sum_{k \in \mathbb{Z} \setminus \{0\}} \Big(\frac{a_k}{ik} + \frac{(-1)^k \cdot i a_0}{k}\Big) \varphi_k(t) + \sqrt{2\pi} b_0 \varphi_0(t)$$

(we used the expansion of t in the basis $\{\varphi_k\}$). Since $\varphi_k'(t) = ik\varphi_k(t)$, then for all $k \in \mathbb{Z}$ we have $\int_{-\pi}^{\pi} \big(x(t)\overline{\varphi_k(t)} - x'(t) ik \overline{\varphi_k(t)}\big) dt = 0$. Using the expansions of x and x' into a series and the orthonormality of the system $\{\varphi_k\}$, for $k \in \mathbb{Z} \setminus \{0\}$ we obtain that $\frac{a_k}{ik} + \frac{(-1)^k \cdot i a_0}{k} - ik a_k = 0$, and for $k = 0$ we get that $b_0 = 0$. Hence $a_k = \frac{(-1)^k}{1+k^2} a_0$, $k \in \mathbb{Z} \setminus \{0\}$ and

$$x(t) = \sum_{k \in \mathbb{Z} \setminus \{0\}} \frac{(-1)^k \cdot i k a_0}{1 + k^2} \varphi_k(t) = \frac{a_0}{2} \cdot \sum_{k \in \mathbb{Z}} (-1)^k \Big(\frac{1}{1 - ik} - \frac{1}{1 + ik}\Big) \varphi_k(t).$$

Taking into account the expansions

$$e^t = \sum_{k \in \mathbb{Z}} \frac{2 \sinh \pi \cdot (-1)^k}{\sqrt{2\pi}(1 - ik)} \varphi_k(t), \ e^{-t} = \sum_{k \in \mathbb{Z}} \frac{-2 \sinh \pi \cdot (-1)^k}{\sqrt{2\pi}(1 + ik)} \varphi_k(t),$$

we get that $x(t) = \frac{a_0 \sqrt{2\pi}}{2 \sinh \pi} \cdot \frac{e^t - e^{-t}}{2} = \frac{a_0 \sqrt{2\pi}}{2 \sinh \pi} \sinh t$, where $a_0 \in \mathbb{C}$ is arbitrary.

2.34 Consider $\left\{\chi_{\{c\}} \mid c \in \mathbb{R}\right\}$.

2.35 To prove necessity, assume that $b > a + 2\pi$ and consider the function $\chi_{[a,a+2\pi]}(t) - \chi_{[b-2\pi,b]}(t)$. To prove sufficiency, use the Stone-Weierstrass theorem and the fact that the set $C([a, b])$ is dense in $L_2([a, b])$.

2.37 It is easy to check that the system is orthonormal. Let M_{N+1} be the set of functions which are constant on half-intervals

$$\left[\frac{i-1}{2^{N+1}}, \frac{i}{2^{N+1}}\right), \quad 1 \leq i \leq 2^{N+1}, \ N \in \mathbb{N} \cup \{0\}.$$

Then M_{N+1} is a linear subspace of dimension 2^{N+1}. Note that

$$\{x_0, x_{kn} \mid 1 \leq k \leq 2^n, \ 0 \leq n \leq N\} \subset M_{N+1},$$

and these $1 + (1 + 2 + \ldots + 2^N) = 2^{N+1}$ functions are linearly independent since they are orthonormal. Therefore

$$M_{N+1} = \operatorname{span}(\{x_0, x_{kn} \mid 1 \leq k \leq 2^n, \ 0 \leq n \leq N\}).$$

Since the set $\bigcup\limits_{N \in \mathbb{N}} M_{N+1}$ is dense in $L_2([0, 1])$, it follows that the system of Haar functions is complete.

2.38 (2)–(5) Use integration by parts.

2.39 (1) Show that the function $\chi_{[0,\frac{1}{4})} - \chi_{[\frac{1}{4},\frac{3}{4})} + \chi_{[\frac{3}{4},1]}$ is orthogonal to all x_n.
(2) Use the fact that

$$\operatorname{span}(\{P_0(t), P_1(t), \ldots, P_n(t)\}) = \operatorname{span}(\{1, t, \ldots, t^n\}), \ n \geq 0,$$

and the set of polynomials is dense in $L_2([-1, 1])$.

2.40 Use the fact that the sequence $\{\chi_{[a,b]} - \frac{b-a}{n}\chi_{[b,b+n]} : n \geq 1\}$ converges in $L_2(\mathbb{R})$ to $\chi_{[a,b]}$.

2.41 (1), (2) $\{0\}$. Let $y = (a_0, a_1, \ldots, a_k, \ldots) \in l_2$ belong to the orthogonal complement. Consider the function $f(t) = \sum\limits_{k=0}^{\infty} a_k t^k$ (the power series converges at least for $|t| < 1$ due to the Cauchy–Schwarz inequality). In items (1), (2) we have $f(\alpha) = 0$, $\alpha \in (0, 1)$ and $f(\frac{1}{n}) = 0$, $n \geq 2$, respectively. It follows from each of these conditions that $f \equiv 0$.

2.42° No, if M is not a subspace.

2.43 Since the system $\{e_n : n \geq 1\}$ is an orthonormal basis in l_2, then

$$\{x \in l_2 \mid (x, e_n) = 0, \ n \geq 2\} = \operatorname{span}(\{e_1\})$$

and $H \cap \operatorname{span}(\{e_1\}) = \{0\}$. Check that $\rho\big(x_1, \overline{\operatorname{span}(M)}\big) = 1$.

2.44* The necessity follows from Theorem 2.6. To prove sufficiency, assume that H is incomplete. Then there exists $z \in \widetilde{H} \backslash H$, where $\widetilde{H}$ is the complement of H. Since the space H is separable, there exists a countable subset of H, which is dense in $\widetilde{H}$. By orthogonalization we get a countable orthonormal system $\{e_n : n \geq 1\} \subset H$, which is a basis in $\widetilde{H}$. Let $z = \sum\limits_{j=1}^{\infty} \alpha_j e_j$, where $\alpha_j = (z, e_j)$, $j \geq 1$. Without loss of generality $\alpha_1 \neq 0$. Define $x_k = \overline{\alpha_1} e_k - \overline{\alpha_k} e_1$, $k \geq 2$. Then

$$(z, x_k) = \alpha_1 (z, e_k) - \alpha_k (z, e_1) = \alpha_1 \alpha_k - \alpha_k \alpha_1 = 0, \ k \geq 2.$$

Let us verify that $\{x_k : k \geq 2\}$ is a complete system in H. Indeed, for each $x \in \widetilde{H}$ we have $x = \sum\limits_{j=1}^{\infty} \beta_j e_j$, where $\beta_j = (x, e_j)$, $j \geq 1$. Therefore if $(x, x_k) = 0$, $k \geq 2$, then $\alpha_1 (x, e_k) - \alpha_k (x, e_1) = \alpha_1 \beta_k - \alpha_k \beta_1 = 0$, i.e. $\beta_k = \frac{\beta_1}{\alpha_1} \alpha_k$, $k \geq 2$. Hence $x = \frac{\beta_1}{\alpha_1} z$ and $x \in H$ only for $x = 0$. By the orthogonalization of the system $\{x_k : k \geq 2\}$ we get an orthonormal system $M = \{y_k : k \neq 2\} \subset H$, which is complete, because $\{y_k : k \geq 2\}^{\perp} = \{x_k : k \geq 2\}^{\perp}$ in H. Finally, M is not a basis in H, because $\overline{\text{span}}(M) \perp z$, therefore $e_1 \in H \setminus \overline{\text{span}(M)}$.

2.45 The necessity follows from problem 1.76°. To prove sufficiency, consider the set $\text{span}(\{x_\alpha : \alpha \in A\})$, where $\{x_\alpha : \alpha \in A\}$ is an orthonormal basis.

2.46 Not necessarily. In the space $H = L_2([0, 1])$ consider

$$e_1(t) = 1, \ e_n(t) = -\chi_{[\frac{1}{n}, \frac{1}{n-1}]}(t), \ t \in [0, 1], \ n \geq 2, \ c_n = 1, \ n \geq 1.$$

2.47 Due to orthonormality

$$(x, e_k) = \left(\sum_{n=1}^{\infty} x_n e_n, e_k \right) = \sum_{n=1}^{\infty} x_n (e_n, e_k) = x_k, \ k \geq 1.$$

2.48 The first two statements are obvious. Assume that for some $x \in C([0, 1])$ there exists a sequence $\{c_n : n \geq 0\}$ such that $x(u) = \sum\limits_{n=0}^{\infty} c_n u^n$ in $L_2([0, 1])$. If S_n is a partial sum of this series, then $(S_n, \chi_{[0,t]}) \to (x, \chi_{[0,t]})$, $n \to \infty$, for each $t \in [0, 1]$, i.e.

$$\sum_{n=0}^{\infty} \frac{c_n}{n+1} t^{n+1} = \int_0^t x(u) du = y(t), \ t \in [0, 1]$$

(the series is pointwise convergent). It follows that $y \in C^\infty((0, 1))$ as the sum of the power series, therefore $x \in C^\infty((0, 1))$. There is no contradiction with the Theorem 2.6, because the system in the condition of the problem is not orthonormal.

2.49 For all $n, m \in \mathbb{N} \cup \{0\}$ we have

$$\int_D z^n \bar{z}^m dt ds = \begin{cases} \frac{\pi}{n+1}, & n = m, \\ 0, & n \neq m, \end{cases} \quad \text{where } z = t + is.$$

Therefore if $x(z) = \sum_{n=0}^{\infty} a_n z^n$ for $|z| < 1$, then

$$\int_D |x(z)|^2 dt ds = \pi \sum_{n=0}^{\infty} \frac{|a_n|^2}{n+1}.$$

Hence $x \in A_2(D)$ if and only if $\sum_{n=0}^{\infty} \frac{|a_n|^2}{n+1} < \infty$ (the last inequality guarantees that the radius of convergence of the power series is at least 1). It follows that the correspondence $x \leftrightarrow a = (a_0, a_1, \ldots)$ is an isometric isomorphism between $A_2(D)$ and $l_{2,\alpha}$, where $\alpha = (\sqrt{\pi}, \sqrt{\frac{\pi}{2}}, \sqrt{\frac{\pi}{3}}, \ldots)$ (see problem 2.6°). This isomorphism maps polynomials to eventually zero sequences in the space $l_{2,\alpha}$. Since the completion of the set of eventually zero sequences is $l_{2,\alpha}$, the completion of the set of polynomials is $A_2(D)$.

2.50 See hint to problem 2.49. The result of the orthogonalization process is the sequence $\varphi_n(z) = \frac{1}{\sqrt{\pi n!}} z^n$, $n \geq 0$. Completion of $P(\mathbb{C})$ consists of functions $x(z) = \sum_{n=0}^{\infty} a_n z^n$ for which the series converges for $z \in \mathbb{C}$ and $\int_{\mathbb{C}} |x(z)|^2 e^{-|z|^2} dt ds < +\infty$.

2.51 Use Fubini's theorem.

2.52 Use the parallelogram law (problem 2.8°).

2.53 The statement follows from problems 2.52 and 1.101. In $H = l_2$ the set $M = \{(1 + \frac{1}{n}) e_n : n \geq 1\}$ has no element with the smallest norm.

2.54 $y = x_0 + r \frac{x - x_0}{\|x - x_0\|}$.

2.55* *Necessity.* Since $\lambda y + (1 - \lambda) z \in M$, $\lambda \in [0, 1]$, then

$$\|x - y\|^2 \leq \|x - \lambda y - (1 - \lambda) z\|^2 = \|x - y + (1 - \lambda)(y - z)\|^2.$$

Hence $2(x - y, y - z) + (1 - \lambda)\|y - z\|^2 \geq 0$, $\lambda \in [0, 1)$. We obtain the desired inequality as $\lambda \to 1-$. Sufficiency follows from

$$\|x - z\|^2 = \|x - y\|^2 + 2(x - y, y - z) + \|y - z\|^2 \geq \|x - y\|^2.$$

2.56 If $y = \mathrm{pr}_{L^\perp} x$, then $x = ta + y$, where $t \in \mathbb{K}$. Since $y \in L^\perp$, we have $(x - ta, a) = 0$. Hence $t = \frac{(x,a)}{\|a\|^2}$ and

$$\rho(x, L^\perp) = \|x - y\| = |t| \cdot \|a\| = \frac{|(x,a)|}{\|a\|}.$$

2.57 $M_n^\perp = \left\{ (\underbrace{t, t, \ldots, t}_{n}, 0, 0, \ldots) \,\middle|\, t \in \mathbb{K} \right\}$. It follows from problem 2.56 that $\rho(x_0, M_n) = \rho\big(x_0, (M_n^\perp)^\perp\big) = \frac{1}{\sqrt{n}}, n \geq 1$.

2.58 No. Consider $M = \{x \in l_2 \mid x_1 + x_2 = 0\}$ and $N = \mathrm{span}(\{e_1\})$.

2.59 $M + N = \left\{ (x_1, x_2, x_3, \frac{x_4}{3}, x_5, \frac{x_6}{5}, x_7, \frac{x_8}{7}, \ldots) \mid x \in l_2 \right\}$. The set $M + N$ is dense in l_2, since $\mathrm{span}(\{e_n : n \geq 1\}) \subset (M + N)$. But $M + N \neq l_2$, because $(1, 1, \frac{1}{3}, \frac{1}{3}, \frac{1}{5}, \frac{1}{5}, \ldots) \in l_2 \setminus (M + N)$.

2.60 Prove that the set $M + N$ is dense in $\{x \in l_2 \mid x_1 = 0\}$ and

$$(0, 0, \sin 1, 0, \sin \tfrac{1}{2}, 0, \sin \tfrac{1}{3}, \ldots) \notin (M + N).$$

2.61 It is obvious that the set $M + N$ is linear. To prove that $M + N$ is closed consider a convergent sequence $\{x_n + y_n : n \geq 1\}$, where $x_n \in M$, $y_n \in N$, $n \geq 1$. Since

$$\|(x_n + y_n) - (x_m + y_m)\|^2 = \|(x_n - x_m) + (y_n - y_m)\|^2 =$$
$$= \|x_n - x_m\|^2 + \|y_n - y_m\|^2 + (x_n - x_m, y_n - y_m) + (y_n - y_m, x_n - x_m) \geq$$
$$\geq \|x_n - x_m\|^2 + \|y_n - y_m\|^2 - 2\|x_n - x_m\| \cdot \|y_n - y_m\|(1 - \varepsilon) \geq$$
$$\geq \varepsilon\|x_n - x_m\|^2 + \varepsilon\|y_n - y_m\|^2, \; n, m \geq 1,$$

then $\{x_n : n \geq 1\}$ and $\{y_n : n \geq 1\}$ are Cauchy sequences, therefore they converge. The sets M and N are closed, so there exist $x \in M$, $y \in N$ such that $x_n \to x$, $y_n \to y$, $n \to \infty$. Hence $x_n + y_n \to x + y \in M + N$, $n \to \infty$.

2.62 No.

2.63 Let $\{y_n : n \geq 1\}$ be an orthonormal system in $L_2([b, c])$. We construct the desired extension $\{z_n : n \geq 1\}$ as follows: $z_n(t) = x_n(t)$ for $t \in [a, b]$, $z_1(t) = y_1(t)$, $z_n(t) = \alpha_{n,1} y_1(t) + \ldots + \alpha_{n,n-1} y_{n-1}(t) + y_n(t)$, $n \geq 2$, for $t \in [b, c]$, where $\alpha_{n,j}$ are determined from the system $(z_j, z_n) = 0$, $1 \leq j \leq n - 1$.

Problems of Chap. 3

3.9° (1), (2) $\|f\| = 1$; (3), (4) nonlinear, continuous; (5) $\|f\| = 1$; (6) $\|f\| = \frac{1}{2}$; (7) $\|f\| = 1$; (8) nonlinear, discontinuous; (9) nonlinear, continuous; (10) $\|f\| = \frac{1}{2}$; (11), (12) $\|f\| = \frac{2}{\pi}$; (13) nonlinear, continuous; (14) $\|f\| = 2$; (15) $\|f\| = 1$; (16) $\|f\| = \sum\limits_{k=1}^{n} |\lambda_k|$.

3.10° (1), (2) $\|f\| = 1$; (3) $\|f\| = \sup\limits_{n\geq 1} |y_n|$; (4) nonlinear, continuous; (5) $\|f\| = 1$; (6) $\|f\| = \frac{\pi}{\sqrt{6}}$; (7) $\|f\| = \frac{1}{\sqrt[4]{2}}$; (8) $\|f\| = 2$; (9) $\|f\| = 1$ for $p = 1$, $\|f\| = \left(\sum\limits_{n=1}^{\infty} \frac{1}{n^q}\right)^{1/q}$ for $1 < p < +\infty$; (10) $\|f\| = \frac{1}{2}$ for $p = 1$, $\|f\| = \left(\frac{1}{2^q-1}\right)^{1/q}$ for $1 < p \leq +\infty$; (11) nonlinear, discontinuous; (12) nonlinear, continuous; (13) $\|f\| = 1$; (14) $\|f\| = 1$ for $p = 1$, $\|f\| = 2^{\frac{1}{q}}$ for $1 < p \leq +\infty$; (15) $\|f\| = 2$ for $p = 1$, $\|f\| = (1 + 2^q)^{\frac{1}{q}}$ for $1 < p \leq +\infty$.

3.11° (1) $\|f\| = \frac{1}{\sqrt{3}}$; (2) $\|f\| = 1$; (3) $\|f\| = \sqrt{2}$; (4) nonlinear, continuous; (5) $\|f\| = 1$ for $p = 1$, $\|f\| = \left(\frac{1}{q+1}\right)^{1/q}$ for $1 < p \leq +\infty$; (6) $\|f\| = 1$ for $p = 1$, $\|f\| = \left(\frac{1}{2}B\left(\frac{q+1}{2}, \frac{1}{2}\right)\right)^{1/q}$ for $1 < p \leq +\infty$, where B is the beta function; (7) $\|f\| = \frac{1}{2}$ for $p = 1$, $\|f\| = \left(\frac{1}{(q+1)2^{q+1}}\right)^{1/q}$ for $1 < p \leq +\infty$; (8) $\|f\| = 3$ for $p = 1$, $\|f\| = (1 + 3^q)^{\frac{1}{q}}$ for $1 < p \leq +\infty$; (9) $\|f\| = \frac{\pi}{2}$.

3.12° (1) $\|f\| = 1$; consider the sequence $x_n(t) = (1 - t)^n$; (2) linear, discontinuous; (3) $\|f\| = 2$; (4)–(6) $\|f\| = 1$; (7) $\|f\| = \frac{1}{\sqrt{\alpha(2-\alpha)}}$ for $0 < \alpha < 2$, linear, discontinuous for $\alpha \geq 2$; to prove this, substitute $u = t^\alpha$ and consider the sequence

$$x_n(u) = u^{\frac{1}{\alpha}-1} \cdot \chi_{[\frac{1}{n},1]}(u);$$

(8) $f(x) = x(0)$, $\|f\| = 1$; (9) $\|f\| = 1$; (10)–(12) linear, discontinuous; in item (12) consider the sequence $x^{(n)} = (\operatorname{sign} y_1, \ldots, \operatorname{sign} y_n, 0, 0, \ldots)$, $n \geq 1$.

3.15° (1), (2) $\|f\| = 1$; (3) $\|f\| = 8$; (4) $\|f\| = \frac{4}{3}$; (5) $\|f\| = 2$; (6) $\|f\| = \frac{1}{2}$; (7) $\|f\| = \frac{7}{2}$; (8), (9) $\|f\| = 2$; (10) $\|f\| = \pi$; (11) $\|f\| = 3$; (12) $\|f\| = 6$; (13) $\|f\| = \frac{\pi^2}{6}$; (14) $\|f\| = \frac{1}{20}$; (15) $\|f\| = 1$; (16) $\|f\| = \sum\limits_{k=1}^{n} |\lambda_k|$; (17) $\|f\| = 1$.

3.16° (1) $\|f\| = 1$; (2) $\|f\| = \frac{\pi}{2}$; (3) $\|f\| = 1$; (4) $\|f\| = \frac{3}{4}$; (5) $\|f\| = \sqrt{2}$; (6) $\|f\| = \left(\frac{\pi^2}{6}\right)^{2/7}$; (7) $\|f\| = 1$; (8) $\|f\| = 4$ for $p = 1$, $\|f\| = (2^q + 4^q)^{\frac{1}{q}}$ for $1 < p \leq +\infty$; (9) $\|f\| = \frac{1}{2}$ for $p = 1$, $\|f\| = \left(\frac{1}{2^q-1}\right)^{1/q}$ for $1 < p \leq +\infty$;

(10) $\|f\| = \sqrt{2}$; (11) $\|f\| = \pi$; (12) $\|f\| = \sqrt{\frac{\pi}{2}}$; (13) $\|f\| = 2$ for $p = 1$, $\|f\| = \left(\frac{2^{q+1}-1}{q+1} + 2^q\right)^{1/q}$ for $1 < p \leq +\infty$; (14) $\|f\| = 1$ for $p = 1$, $\|f\| = \left(\frac{\pi}{q}\right)^{\frac{1}{2q}}$ for $1 < p \leq +\infty$; (15) $\|f\| = \frac{\pi}{2}$.

3.17° (1) if $p = 1$, then $\alpha \geq 0$ and $\|f\| = 1$; if $1 < p < +\infty$, then $\alpha > \frac{1}{q}$ and $\|f\| = \left(\sum\limits_{n=1}^{\infty} \frac{1}{n^{\alpha q}}\right)^{1/q}$; (2) if $p = 1$, then $\alpha \leq 1$ and $\|f\| = |2^\alpha - 1|$; if $1 < p < +\infty$, then $\alpha < 1 - \frac{1}{q}$ and $\|f\| = \left(\sum\limits_{n=1}^{\infty} \left((n+1)^\alpha - n^\alpha\right)^q\right)^{1/q}$; (3) if $p = 1$, then $\alpha \leq 0$ and $\|f\| = 1$; if $1 < p < +\infty$, then $\alpha < \frac{1}{q}$, $\|f\| = \left(\frac{1}{1-\alpha q}\right)^{1/q}$; (4) if $p = 1$, then $\alpha \geq 0$ and $\|f\| = \left(\frac{\alpha}{e}\right)^\alpha$ for $\alpha > 0$, $\|f\| = 1$ for $\alpha = 0$; if $1 < p < +\infty$, then $\alpha > -\frac{1}{q}$ and $\|f\| = \frac{1}{q^\alpha}\left(\frac{1}{q}\Gamma(\alpha q + 1)\right)^{1/q}$, where Γ is the gamma function; (5) if $p = 1$, then $\alpha \geq 0$ and $\|f\| = 1$; if $1 < p < +\infty$, then $\alpha > -\frac{1}{q}$ and $\|f\| = \left(\frac{1}{2}B\left(\frac{\alpha q+1}{2}, \frac{1}{2}\right)\right)^{1/q}$, where B is the beta function.

3.18° (1) Let X be an NLS of dimension $n < \infty$, $\{e_1, \ldots, e_n\}$ be a basis in X, f be a linear functional on X. Then

$$f(x) = \sum_{k=1}^{n} a_k x_k, \quad x = \sum_{k=1}^{n} x_k e_k,$$

where $a_k = f(e_k) \in \mathbb{K}$. Continuity follows from the fact that convergence in a finite-dimensional NLS is equivalent to coordinate-wise convergence (see problem 1.72). (2) Lines in $\mathbb{R}^2$ (correspondingly, planes in $\mathbb{R}^3$), which pass through the origin.

3.19 Use the fact that $x = \sum\limits_{n=1}^{\infty} x_n e_n$, $x \in c_0$.

3.20 Use the fact that

$$x = e_0 \lim_{n\to\infty} x_n + \sum_{k=1}^{\infty} (x_k - \lim_{n\to\infty} x_n) e_k, \quad x \in c,$$

where $e_0 = (1, \ldots, 1, \ldots)$.

3.21 It is necessary and sufficient that the element $a \in l_1$ which defines the functional (see problem 3.19) has finitely many nonzero coordinates.

3.22* (1) If $f \in (C^1([a,b]))^*$, then there exist $\alpha \in \mathbb{K}$ and $g \in BV_0([a,b])$ such that $f(x) = \alpha x(a) + \int_a^b x'(u)dg(u)$, $x \in C^1([a,b])$. To prove this, apply the formula

$$x(t) = x(a) + \int_a^t x'(u)du, \ t \in [a,b], \ x \in C^1([a,b]),$$

and show that the functional F, defined by the formula $F(y) = f\left(\int_a^t y(u)du\right)$, $y \in C([a,b])$, belongs to $(C([a,b]))^*$.

(2) $f(x) = \sum_{k=0}^{n-1} \alpha_k x^{(k)}(a) + \int_a^b x^{(n)}(u)dg(u)$, $x \in C^n([a,b])$, where α_0, α_1, ..., $\alpha_{n-1} \in \mathbb{K}$ and $g \in BV_0([a,b])$. Use the Taylor's formula with the integral remainder.

3.23 $\|f\| = \int_a^b |p(t)|dt$. (1) Apply the formula $f(x) = \int_a^b x(t)dP(t)$, where P is the antiderivative of p. (2) Consider the sequence of functionals $f_n(x) = \int_a^b p_n(t)x(t)dt$, $x \in C([a,b])$, where $p_n \in C([a,b])$, $n \geq 1$. Show that if $p_n \to p$ in $L_1([a,b])$, then $\|f_n - f\| \to 0$, therefore $\|f_n\| \to \|f\|$, $n \to \infty$.

3.24 (1) $f(x) = \int_0^1 x(t)dt - x(0)$; (2) $f(x) = x(0)$; (3) $f(x) = \int_0^1 x(t)dt$.

3.26 (1) (a) No; (b) yes; (2) (a) no; (b) yes; (3) all of them.

3.28 $\|f\| = \sup\limits_{x \neq 0} \frac{|f(x)|}{\|x\|} \geq \sup\limits_{x \in L} \frac{1}{\|x\|} = \frac{1}{\inf\limits_{x \in L} \|x\|}$. Conversely, since $\frac{x}{f(x)} \in L$ for all $x \notin \operatorname{Ker} f$, then $\|f\| = \sup\limits_{x \notin \operatorname{Ker} f} \frac{|f(x)|}{\|x\|} = \sup\limits_{x \notin \operatorname{Ker} f} \frac{1}{\|\frac{x}{f(x)}\|} \leq \sup\limits_{y \in L} \frac{1}{\|y\|}$.

3.29 (2) *Sufficiency.* For any sequence $\{y_n : n \geq 1\} \subset X \setminus \{0\}$ such that $y_n \to 0$, $n \to \infty$, put $x_n = \frac{y_n}{\sqrt{\|y_n\|}}$, $n \geq 1$. Then $x_n \to 0$, $n \to \infty$, and the sequence $\{f(x_n) : n \geq 1\}$ is bounded. Therefore $f(y_n) = \sqrt{\|y_n\|}\, f(x_n) \to 0$, $n \to \infty$.

(3) *Necessity.* Let the sequence $\{x_n : n \geq 1\}$ be a Cauchy sequence. Since

$$|f(x_n) - f(x_m)| \leq \|f\| \cdot \|x_n - x_m\| \to 0, \ n, m \to \infty,$$

then the sequence $\{f(x_n) : n \geq 1\}$ is a Cauchy sequence in $\mathbb{K}$, therefore it is bounded. The sufficiency follows from the sufficiency of condition (2).

(4) *Sufficiency.* If a set $A \subset X$ is open, then the set $-A = \{-x \mid x \in A\}$ is also open. Therefore for any $c \in \mathbb{R}$ the set $\{x \mid f(x) > c\} = \{x \mid f(-x) < -c\}$ is open, hence for all $c_1, c_2 \in \mathbb{R}$, $c_1 < c_2$, the set $f^{-1}\big((c_1, c_2)\big) = \{x \mid f(x) >$

$c_1\} \cap \{x \mid f(x) < c_2\}$ is open. Since every open set in $\mathbb{R}$ is a union of intervals, then the preimage of every open sets is an open set.

3.30 It follows from the condition of the problem that for each $n \in \mathbb{N}$ there exists $x_n \in X$ such that $\|x_n\| = 1$ and $|f(x_n)| \geq n$. For any $\alpha \in \mathbb{K}$ and $z \in X$ put $y_n = \frac{\alpha - f(z)}{f(x_n)} x_n + z, n \geq 1$. Then $y_n \to z, n \to \infty$, and $f(y_n) = \alpha, n \geq 1$.

3.31 By condition of the problem $g(x) = 0$ for all x such that $f(x) = 0$. Therefore if the statement of the problem is incorrect, then there exist $y, z \in X \setminus \operatorname{Ker} f$, for which $\frac{g(y)}{f(y)} \neq \frac{g(z)}{f(z)}$. Put $t = \frac{y}{f(y)} - \frac{z}{f(z)}$. Then $f(t) = \frac{f(y)}{f(y)} - \frac{f(z)}{f(z)} = 0$ and $g(t) = \frac{g(y)}{f(y)} - \frac{g(z)}{f(z)} \neq 0$, so $t \in \operatorname{Ker} f \setminus \operatorname{Ker} g$, a contradiction.

3.32 *Necessity.* If f is a linear combination of functionals $f_1, \ldots, f_n$ and $f_k(x) = 0$, $1 \leq k \leq n$, then $f(x) = 0$. *Sufficiency.* Without loss of generality, for each $1 \leq i \leq n$ there exists $z_i \in \bigcap_{k \neq i} \operatorname{Ker} f_k \setminus \operatorname{Ker} f_i$ (if it is false for some i, then

$$\bigcap_{k \neq i} \operatorname{Ker} f_k = \bigcap_{k=1}^{n} \operatorname{Ker} f_k \subset \operatorname{Ker} f$$

and it is enough to show that f is a linear combination of functionals f_k, $k \neq i$). Put $c_i = \frac{f(z_i)}{f_i(z_i)}$, $1 \leq i \leq n$. Let us show that $f(x) = \sum_{i=1}^{n} c_i f_i(x)$ for each $x \in X$. Indeed, if $y = x - \sum_{i=1}^{n} \frac{f_i(x)}{f_i(z_i)} z_i$ then for each $1 \leq k \leq n$ we have

$$f_k(y) = f_k\left(x - \frac{f_k(x)}{f_k(z_k)} z_k\right) - \sum_{i \neq k} \frac{f_i(x)}{f_i(z_i)} f_k(z_i) = f_k(x) - \frac{f_k(x)}{f_k(z_k)} f_k(z_k) = 0,$$

hence $y \in \bigcap_{k=1}^{n} \operatorname{Ker} f_k \subset \operatorname{Ker} f$. Thus $f(y) = 0$ and

$$f(x) = f\left(y + \sum_{i=1}^{n} \frac{f_i(x)}{f_i(z_i)} z_i\right) = \sum_{i=1}^{n} \frac{f_i(x)}{f_i(z_i)} f(z_i) = \sum_{i=1}^{n} c_i f_i(x).$$

3.33° (1) In all normed linear spaces. For example, consider $f = 0$. (2) It is necessary and sufficient that $\dim X \leq 1$.

3.34 If $\operatorname{Ker} f = \{0\}$ or $\operatorname{Ker} g = \{0\}$, then it follows from item (2) of problem 3.33° that $\dim X \leq 1$. Let $y \in \operatorname{Ker} f \setminus \{0\}$, $z \in \operatorname{Ker} g \setminus \{0\}$. Then by the condition of the problem $g(y) \neq 0$ and $f(z) \neq 0$. Verify that for every $x \in X$ we have $x - \frac{g(x)}{g(y)} y - \frac{f(x)}{f(z)} z \in \operatorname{Ker} f \cap \operatorname{Ker} g$ whence $x = \frac{g(x)}{g(y)} y + \frac{f(x)}{f(z)} z$. Thus $\{y, z\}$ is a basis of X and $\dim X = 2$.

3.35 Let $A \subset X$ and for each $x \in A$ there exists $r_x > 0$ such that $B(x, r_x) \subset A$. Show that $B(f(x), r_x \cdot \|f\|) \subset f(A)$.

3.36* *Sufficiency.* Let Ker f be a closed set, but $f \notin X^*$. By problem 3.30 there exists a sequence $\{x_n : n \geq 1\} \subset X$ such that $x_n \to 0$, $n \to \infty$, and $|f(x_n)| \geq 1$, $n \geq 1$. Fix $z \in X \backslash \text{Ker}\, f$. By item (2) of problem 3.2 there exist $y_n \in \text{Ker}\, f$ and $\lambda_n \in \mathbb{K}$ such that $x_n = y_n + \lambda_n z$, $n \geq 1$. Since $|f(x_n)| = |\lambda_n| \cdot |f(z)| \geq 1$, $n \geq 1$, then $\frac{x_n}{\lambda_n} \to 0$, $n \to \infty$. Therefore $\frac{y_n}{\lambda_n} \to -z$, $n \to \infty$, so $z \in \text{Ker}\, f$, which contradicts to the choice of z.

3.37 Use problems 3.30 and 3.36*.

3.38* For $x \in \text{Ker}\, f$ the equality is obvious. From now on we will assume that $x \notin \text{Ker}\, f$. Note that

$$\|f\| \cdot \rho(x, \text{Ker}\, f) = \inf_{y \in \text{Ker}\, f} \|x - y\| \cdot \|f\| \geq \inf_{y \in \text{Ker}\, f} |f(x - y)| = |f(x)|.$$

On the other hand, any element $z \in X \setminus \text{Ker}\, f$ has the form $z = y + \lambda x$, where $y \in \text{Ker}\, f$, $\lambda \in \mathbb{K} \setminus \{0\}$. Therefore

$$\frac{|f(z)|}{\|z\|} = \frac{|\lambda| \cdot |f(x)|}{|\lambda| \cdot \|x - (-\frac{y}{\lambda})\|} \leq \frac{|f(x)|}{\rho(x, \text{Ker}\, f)},$$

whence $\|f\| \cdot \rho(x, \text{Ker}\, f) \leq |f(x)|$.

3.39 Note that $|f(x_0 - y_0)| = |f(x_0)| + |f(y_0)|$ and apply problem 3.38*.

3.40* (1) (a) $\Rightarrow$ (b) Let $z \in \overline{B}(0, 1)$ be such that $|f(z)| = \|f\| \neq 0$. Then $\|z\| = 1$. For any $x \in X$ we have $y = x - \frac{f(x)}{f(z)} z \in \text{Ker}\, f$ and

$$\|x - y\| = \left\| \frac{f(x)}{f(z)} z \right\| = \frac{|f(x)|}{|f(z)|} = \frac{\|f\| \cdot \rho(x, \text{Ker}\, f)}{\|f\|} = \rho(x, \text{Ker}\, f)$$

due to problem 3.38*. (c) $\Rightarrow$ (a) Let $x \in X \backslash \text{Ker}\, f$ and $y \in \text{Ker}\, f$ be such that $\rho(x, \text{Ker}\, f) = \|x - y\| \neq 0$. Then for $z = \frac{x-y}{\|x-y\|}$ we have $\|z\| = 1$ and

$$|f(z)| = \frac{|f(x) - 0|}{\|x - y\|} = \frac{\|f\| \cdot \rho(x, \text{Ker}\, f)}{\rho(x, \text{Ker}\, f)} = \|f\|$$

due to problem 3.38*.

(2) It is necessary and sufficient that there exists there exists no more than one element $z \in \overline{B}(0, 1)$ such that $f(z) = \|f\|$.

3.41 Due to problem 3.40*, it is sufficient to take $L = \text{Ker}\, f$, where in item (1) for $f \in X^*$ there is no element $z \in X$ such that $\|z\| = 1$ and $f(z) = \|f\|$, and in item (2) for $f \in X^*$ there exist infinitely many such elements.

(1) $f(x)=\int\limits_0^{\frac{1}{2}} x(t)dt - \int\limits_{\frac{1}{2}}^1 x(t)dt$, $x\in C([0,1])$; $f(x)=\sum\limits_{n=1}^{\infty}\frac{x_n}{n^2}$, $x\in c_0$;

(2) $f(x)=x(0)$, $x\in C([0,1])$; $f(x)=x_1$, $x\in c_0$.

3.42 (1) Since f is linear and nonnegative, it follows that $f(x)\le f(y)$ for all $x,y\in C([a,b])$ such that $x(t)\le y(t)$, $t\in[a,b]$. For any function $x\in\overline{B}(0,1)$ we have $-1=-e_0(t)\le x(t)\le e_0(t)=1$, $t\in[a,b]$, thus $-f(e_0)\le f(x)\le f(e_0)$. therefore $|f(x)|\le f(e_0)$, $x\in\overline{B}(0,1)$, and the equality is achieved at $x=e_0$.

(2) Every real function is the difference of two nonnegative functions, whence if $x(t)\in\mathbb{R}$, $t\in[a,b]$, then $f(x)\in\mathbb{R}$. Similarly, if $\frac{1}{i}x(t)\in\mathbb{R}$, $t\in[a,b]$, then $f(\frac{1}{i}x)=\frac{1}{i}f(x)\in\mathbb{R}$. It follows that $f(\operatorname{Re}x)=\operatorname{Re}f(x)$ and $f(\operatorname{Im}x)=\operatorname{Im}f(x)$ for all $x\in C([a,b])$.

Fix an arbitrary function $x\in C([a,b])$ and choose $\alpha\in\mathbb{C}$ such that $|\alpha|=1$ and $f(\alpha x)=|f(x)|$. Then

$$|f(x)|=f(\alpha x)=\operatorname{Re}f(\alpha x)=f(\operatorname{Re}(\alpha x))\le\|x\|\,f(e_0),$$

because $\operatorname{Re}(\alpha x)(t)\le|x(t)|\le\|x\|e_0(t)$, $t\in[a,b]$. The equality is achieved at $x=e_0$.

(3) By Theorem 3.4 for the functional f there exists a function $g\in BV_0([a,b])$ such that $f(x)=\int\limits_a^b x(t)dg(t)$, $x\in C([a,b])$, moreover $\|f\|=\|g\|_V$. Since

$$\|f\|=f(e_0)=\int\limits_a^b dg(t)=g(b),$$

then $g(b)=V(g,[a,b])$. Hence $g(b)\ge 0$. For any $a\le c_1<c_2\le b$ we have

$$g(b)=V(g,[a,b])\ge|g(c_1)|+|g(c_2)-g(c_1)|+|g(b)-g(c_2)|\ge|g(b)|.$$

It follows that $g(c_1)$, $g(c_2)-g(c_1)$ and $g(b)-g(c_2)$ are nonnegative real numbers. Therefore g is a nondecreasing real function on $[a,b]$. Thus f is a nonnegative functional.

3.43 g is a nondecreasing real function on $[a,b]$.

3.44 Use that by Jordan's theorem every function of bounded variation on $[a,b]$ is the difference of two nondecreasing functions on $[a,b]$.

3.45 (2) If $x(t)\ge 0$, $t\in[a,b]$, then $f(x)=f(\sqrt{x})^2\ge 0$, so f is a nonnegative functional. Due to item (1) of problem 3.42, f is continuous and $\|f\|=f(e_0)$, where $e_0(t)=1$, $t\in[a,b]$. Since $f(e_0^2)=f(e_0)=f(e_0)^2$, then $f(e_0)=0$ or $f(e_0)=1$. Therefore $\|f\|=f(e_0)=1$, because f is nonzero.

(3) Let $e_1(t) = t$, $t \in [a, b]$. Since f is a nonnegative functional and

$$ae_0(t) \leq e_1(t) \leq be_0(t), \ t \in [a, b],$$

then $a = af(e_0) \leq f(e_1) \leq bf(e_0) = b$. Hence there exists $\tau \in [a, b]$ such that $f(e_1) = \tau$. Since f is multiplicative, if $e_k(t) = t^k$, $t \in [a, b]$, $k \geq 1$, then $f(e_k) = f(e_1)^k = \tau^k$. Also we have $f(e_0) = 1$, so it follows from the linearity of f that $f(p) = p(\tau)$ for any polynomial p. Finally, for any function $x \in C([a, b])$ there exists a sequence of polynomials $\{p_n : n \geq 1\}$ such that $p_n \to x$, $n \to \infty$, in $C([a, b])$. Since f is continuous, then $f(x) = \lim_{n\to\infty} f(p_n) = \lim_{n\to\infty} p_n(\tau) = x(\tau)$.

3.46 Put $f_1(x) = h((x, 0))$, $x \in X_1$, $f_2(y) = h((0, y))$, $y \in X_2$.

3.47 Let $\{x_{\alpha_n} : n \geq 1\}$ be an arbitrary countable subset of the Hamel basis $\{x_\alpha : \alpha \in A\}$ in the NLS X. Put $f(x_{\alpha_n}) = \|x_{\alpha_n}\|n$, $n \geq 1$, define f arbitrarily on other elements of the Hamel basis, then define f on X by linearity.

3.48 We will show that for any $x \in \operatorname{Ker} f$ and for any $\varepsilon > 0$ there exists $y \in M \cap \operatorname{Ker} f$ such that $\|x - y\| < \varepsilon$. For $f = 0$ the statement is obvious, from now on we assume that $f \neq 0$. Firstly, consider the case $\mathbb{K} = \mathbb{R}$. Let an element $u \notin \operatorname{Ker} f$ be such that $\|u\| = \frac{\varepsilon}{2}$ and $f(u) > 0$. Then $f(x + u) = f(u) > 0$, $f(x - u) = -f(u) < 0$. Since the set M is dense in X, and the functional f is continuous, then in sufficiently small neighborhoods of the points $x + u$ and $x - u$ there exist elements $y_+, y_- \in M$ such that $\|x - y_+\| < \varepsilon$, $\|x - y_-\| < \varepsilon$, and $f(y_+) > 0$ and $f(y_-) < 0$. Then $\alpha f(y_+) + (1-\alpha) f(y_-) = 0$ for some $\alpha \in (0, 1)$. Put $y = \alpha y_+ + (1 - \alpha)y_-$. Then $y \in M$ due to convexity of the set M, $f(y) = 0$ and $\|x - y\| \leq \alpha\|x - y_+\| + (1 - \alpha)\|x - y_-\| < \varepsilon$.

In the case $\mathbb{K} = \mathbb{C}$ the reasonings are similar. Consider an element $u \notin \operatorname{Ker} f$ such that $\|u\| = \frac{\varepsilon}{2}$. Then for $\xi = \frac{1}{2} + i\frac{\sqrt{3}}{2}$ the convex hull of the values $f(x + \xi^k u) = \xi^k f(u)$, $k = 0, 1, 2$, contains 0. In sufficiently small neighborhoods of the points $x + \xi^k u$ there exist elements $y_k \in M$, $k = 0, 1, 2$, such that $\|x - y_k\| < \varepsilon$, $k = 0, 1, 2$, and the convex hull of the values $f(y_k)$, $k = 0, 1, 2$, contains 0. Then in the convex hull of y_0, y_1, y_2 there exists an element y, for which $f(y) = 0$, $y \in M$ and $\|x - y\| < \varepsilon$.

3.49* The set $M' = \{y - x \mid y \in M\}$ is convex and dense in X. Apply the statement of problem 3.48 and induction on k to obtain that the set $M' \cap \left(\bigcap_{i=1}^{k} \operatorname{Ker} f_i\right)$ is convex and dense in $\bigcap_{i=1}^{k} \operatorname{Ker} f_i$ for $k = 1, 2 \ldots, n$. Therefore there exists an element $z \in M' \cap \left(\bigcap_{i=1}^{n} \operatorname{Ker} f_i\right)$ such that $\|z\| < \varepsilon$. Correspondingly $y = z + x \in M$, $\|y - x\| < \varepsilon$ and $f_i(y) = f_i(z) + f_i(x) = f_i(x)$, $1 \leq i \leq n$.

3.50 Apply problem 3.49* for the case when $X = C([a, b])$, M is the set of polynomials and $f_i(x) = x(t_i)$, $1 \leq i \leq n$.

Problems of Chap. 4

4.4° $F(x) = \alpha x_1 + \beta x_2 + a_3 x_3$, $x \in \mathbb{R}^3$, where $a_3 \in \mathbb{R}$ is arbitrary. For $p = 1$ the extension preserves the norm if $|a_3| \le \max\{|\alpha|, |\beta|\}$, and for $1 < p \le +\infty$ if $a_3 = 0$. For $1 < p \le +\infty$ such an extension is always unique, and for $p = 1$ it is unique if and only if $\alpha = \beta = 0$.

4.5° $F((x_1, x_2)) = a_1 x_1 + a_2 x_2$, $(x_1, x_2) \in \mathbb{R}^2$, where $a_1 + a_2 = \alpha$. For $1 \le p < +\infty$ the extension preserves the norm if $a_1 = a_2 = \frac{\alpha}{2}$, and for $p = +\infty$ if $a_1 \in [0, \alpha]$. Such an extension is unique for $1 \le p < +\infty$.

4.6° $F((x_1, x_2)) = (1 - 2b)x_1 + bx_2$, $(x_1, x_2) \in \mathbb{R}^2$, where $b \in \mathbb{R}$ is arbitrary. The extension preserves the norm if and only if $b = \frac{2}{5}$.

4.7° (1) If $f(x) = (x, a)$, $x \in G$, where $a \in G$ is fixed, then $F(x) = (x, b)$, $x \in H$, where $b \in H$ is fixed and $b - a \in G^{\perp}$. The extension preserves the norm only if $b = a$. (2) $F(x) = (x, a + \alpha h)$, $x \in H$, where $\alpha \in \mathbb{K}$ is fixed; $\|F\| = \|f\|$ if and only if $a + \alpha h = \mathrm{pr}_G a$, i.e., for $\alpha = -\frac{(a,h)}{\|h\|^2}$. (3) $F(x) = \left(x, a + \sum\limits_{k=1}^{n} \alpha_k h_k\right)$, $x \in H$, where $\alpha_k \in \mathbb{K}$, $1 \le k \le n$ are fixed; $\|F\| = \|f\|$ if and only if $\alpha_k = -(a, h_k)$, $k = 1, \ldots, n$.

4.8° $F(x) = \int\limits_T h(t)x(t)d\mu(t)$, $x \in L_p(T)$, where $h \in L_q(T)$ is such that $h = a$ almost everywhere on A. For $1 < p < +\infty$, the extension preserves the norm if and only if $h = 0$ almost everywhere on $T \setminus A$, and for $p = 1$ if $\operatorname*{ess\,sup}\limits_{t \in T\setminus A} |h(t)| \le \operatorname*{ess\,sup}\limits_{t \in A} |a(t)|$.

4.9 If $f(x) = \sum\limits_{n=1}^{\infty} a_n x_n$, $x \in c_0$, where $a \in l_1$ is fixed, and F is an extension of f to c, then $F(x) = a_0 \lim\limits_{n \to \infty} x_n + \sum\limits_{n=1}^{\infty} a_n x_n$, $x \in c$, where $a_0 \in \mathbb{K}$ is arbitrary. $\|F\| = \|f\|$ if and only if $a_0 = 0$.

4.10° $f(x) = 2x(0)$.

4.11° (1) $f(x) = -x(0) + 3x(1)$; (2) $f(x) = 2x(0)$.

4.12° Such a functional is not unique. For example, all functionals of the form

$$f(x) = \left(\int\limits_A t^{\alpha+1} dt\right)^{-1} \cdot \int\limits_A t^{\alpha} x(t)dt, \ x \in L_2([0, 1]),$$

where $\alpha \ge 0$, satisfy the condition of the problem.

4.13 (2) No. For example, $c^* = c_0^* = l_1$ (see problems 3.19 and 3.20).

4.14 $\Phi^* = l_2$.

4.15° (1) $F(x) = x(0)$. (2) $F(x) = x(1)$. (3) The extension does not exist. *Method I*. The set of polynomials of the form $p_1(t) = (t+1)p(t)$, $p \in P$, is dense in $C([0,1])$ and $f(p_1) = 0$, but zero functional is not an extension of f. *Method II*. Let $x_n \in C([-1,1])$ be such that $x_n(-1) = n$, $x_n(t) = 0$ for $t \in [0,1]$ and x_n is linear on $[-1,0]$, and $p_n \in P$ be such that $\|x_n - p_n\|_{C([-1,1])} \le 1$. Then the sequence $\{p_n : n \ge 1\}$ is bounded in $C([0,1])$, but $f(p_n) \ge n - 1$; (4) The extension does not exist. The set of polynomials of the form $p_1(t) = p(t^{N+1})$, $p \in P$, is dense in $C([0,1])$ and $f(p_1) = c_0 p_1(0)$, but $F(x) = c_0 x(0)$ is not an extension of f.

4.16° (4) Use that the function $x_0(t) = \sin\frac{1}{t}$, $t \in [-1,1]\backslash\{0\}$, does not belong to the closed linear span of the set of functions $x \in L_\infty([-1,1])$ such that $x = 0$ almost everywhere on $[-1,0]$ or x is continuous at point 0.

4.17° (2) Show that the functional F from item (1) cannot be represented by the formula from problem 3.13°.

4.18 Apply the Hahn–Banach theorem to the functional

$$f(x) = \lim_{n\to\infty} x_{2n-1},\ x \in G = \left\{x \in l_\infty \mid \exists \lim_{n\to\infty} x_{2n-1} \in \mathbb{R}\right\}.$$

4.19 Show that the functional from item (1) of problem 4.3 cannot be represented by the formula from problem 3.14°.

4.20 (1) Consider a linear set $G = \{Tx - x \mid x \in l_\infty\}$. Let us check that $\rho(e, \overline{G}) = 1$. Indeed, for $x = 0$ we have $\|Tx - x - e\| = 1$. Assume that $\|Tz - z - e\| = a < 1$ for some $z = (z_1, z_2, \ldots) \in l_\infty$. Then $|z_{k+1} - z_k - 1| \le a$, $k \ge 1$, whence $z_{k+1} - z_k \ge 1 - a$, $k \ge 1$, and $z_{k+1} - z_1 \ge k(1-a)$ for all $k \ge 1$, which is impossible for $z \in l_\infty$. By Corollary 4.1 of the Hahn–Banach theorem for the subspace $\overline{G}$ and the element $e \notin \overline{G}$ there exists $f \in l_\infty^*$ such that $\|f\| = 1$, $f(e) = \rho(e, \overline{G}) = 1$ and $f(y) = 0$, $y \in \overline{G}$. Hence for every $x \in l_\infty$ we have $f(Tx - x) = 0$, i.e. $f(Tx) = f(x)$.

(2) (d) If $x_n \ge 0$, $n \ge 1$, then for $y = \|x\|\, e - x$ we have $\|y\| \le \|x\|$. Therefore $f(y) \le \|f\| \cdot \|y\| \le \|x\|$ and $f(x) = f(\|x\|\, e - y) = \|x\| \cdot 1 - f(y) \ge 0$.

(e) It follows from (c) and (d) that if for $x \in l_\infty$ and $a, b \in \mathbb{R}$ there exists $N \in \mathbb{N}$ such that $a \le x_n \le b$ for all $n \ge N$, then $f(x - ae) \ge 0$ and $f(be - x) \ge 0$, therefore $a \le f(x) \le b$. Since

$$\forall \varepsilon > 0\ \exists N \in \mathbb{N}\ \forall n \ge N : \liminf_{n\to\infty} x_n - \varepsilon \le x_n \le \limsup_{n\to\infty} x_n + \varepsilon,$$

then $\liminf\limits_{n\to\infty} x_n - \varepsilon \le f(x) \le \limsup\limits_{n\to\infty} x_n + \varepsilon$ for any $\varepsilon > 0$.

(f) Follows from (e).

4.21 (1), (2) For $x = (1, -1, 1, -1, \ldots)$ and $y = (-1, 1, -1, 1, \ldots)$ due to properties (f) and (c) we have $\operatorname*{Lim}_{n\to\infty} (x_n + y_n) = 0$, $\operatorname*{Lim}_{n\to\infty} x_n y_n = -1$ and $\operatorname*{Lim}_{n\to\infty} x_n = \operatorname*{Lim}_{n\to\infty} y_n$. Hence $\operatorname*{Lim}_{n\to\infty} x_n = \operatorname*{Lim}_{n\to\infty} y_n = 0$.

4.22 Consider the set $G = \{T_s x - x \mid x \in B([0, +\infty)), s > 0\}$ and use reasonings similar to the solution of problem 4.20.

4.23 The extension exists due to the Hahn–Banach theorem. If $F_1, F_2 \in X^*$ are two extensions of $f \in G^*$, then $\frac{1}{2}(F_1 + F_2)$ is also an extension of f and

$$\|f\| \le \left\|\frac{1}{2}(F_1 + F_2)\right\| \le \frac{1}{2}\big(\|F_1\| + \|F_2\|\big) = \|f\|.$$

Therefore $\|F_1\| = \|F_2\| = \|\frac{1}{2}(F_1 + F_2)\|$. Since X^* is strictly convex, it follows that $F_1 = F_2$.

4.24* If $f \in c_0^*$, then for some $a = (a_1, \ldots, a_n, \ldots) \in l_1$ we have $f(x) = \sum_{n=1}^{\infty} x_n a_n$, $x \in c_0$. Then $F(x) = \sum_{n=1}^{\infty} x_n a_n$, $x \in l_\infty$, is an extension of f, and $\|F\| = \|f\| = \|a\|_1$ (see problems 3.13° and 3.19). Let $G \in l_\infty^*$ be another extension of f. Then there exists $x = (x_1, \ldots, x_n, \ldots) \in l_\infty$, for which $G(x) \ne F(x)$. Without loss of generality $\|x\| = 1$, and also $G(x) > F(x)$, if $\mathbb{K} = \mathbb{R}$, and $\operatorname{Re} G(x) > \operatorname{Re} F(x)$, if $\mathbb{K} = \mathbb{C}$. Put $y_n = \frac{|a_n|}{a_n}$ for $a_n \ne 0$ and $y_n = 0$ for $a_n = 0$. Consider the sequence

$$y^{(n)} = (y_1, \ldots, y_n, x_{n+1}, x_{n+2}, \ldots),\ n \ge 1.$$

We have $\|y^{(n)}\| \le 1$, and since $y^{(n)} - x \in c_0$, then

$$G(y^{(n)}) - F(y^{(n)}) = G(x) + f(y^{(n)} - x) - F(x) - f(y^{(n)} - x) = G(x) - F(x),$$

$n \ge 1$. It is easy to check that $F(y^{(n)}) \to \|a\|_1$, therefore

$$G(y^{(n)}) \to \|a\|_1 + G(x) - F(x),\ n \to \infty.$$

Hence $\|G\| > \|a\|_1 = \|f\|$.

4.25 (1) Put $f_k(x) = a_k$, $x = \sum_{k=1}^{n} a_k x_k \in G = \operatorname{span}(\{x_1, \ldots, x_n\})$, $k = 1, \ldots, n$, and apply the Hahn–Banach theorem.

(2) No. Consider in the space $L_2([0, 1])$ the elements $x_n(t) = \chi_{[r_n,1]}(t)$, $t \in [0, 1]$, $n \ge 1$, where $\{r_n : n \ge 1\} = \mathbb{Q} \cap (0, 1)$.

(3) The answers are positive.

(4) It is necessary and sufficient that $x_{\alpha_0} \notin \overline{\operatorname{span}(\{x_\alpha \mid \alpha \in A\backslash\{\alpha_0\}\})}$ for every $\alpha_0 \in A$.

(5) Due to problem 3.32, for each $1 \le i \le n$ there exists $z_i \in \bigcap_{k \ne i} \operatorname{Ker} f_k \setminus \operatorname{Ker} f_i$. Put $x_i = \frac{z_i}{f(z_i)}$, $1 \le i \le n$.

4.27 *Necessity.*

$$\left| \sum_{k=1}^{m} \lambda_k c_k \right| = \left| f \left(\sum_{k=1}^{m} \lambda_k x_k \right) \right| \le M \left\| \sum_{k=1}^{m} \lambda_k x_k \right\|.$$

Sufficiency. Let $\{x_{n_k} : k \ge 1\} \subset \{x_n : n \ge 1\}$ be a sequence of linearly independent elements, for which $G = \operatorname{span}(\{x_{n_k} : k \ge 1\}) = \operatorname{span}(\{x_n : n \ge 1\})$. Put

$$f(x) = \sum_{k=1}^{m} \alpha_k c_{n_k} \text{ for } x = \sum_{k=1}^{m} \alpha_k x_{n_k} \in G.$$

We will show that $f(x_n) = c_n$, $n \ge 1$. Indeed, since $x_n \in \operatorname{span}(\{x_{n_k} : k \ge 1\})$, then $x_n = \sum_{k=1}^{m} \beta_k x_{n_k}$ for some $m \in \mathbb{N}$ and $\beta_1, \ldots, \beta_m \in \mathbb{K}$. Then

$$\left| c_n - \sum_{k=1}^{m} \beta_k c_{n_k} \right| \le M \left\| x - \sum_{k=1}^{m} \beta_k x_{n_k} \right\| = 0,$$

whence $f(x_n) = \sum_{k=1}^{m} \beta_k f(x_{n_k}) = \sum_{k=1}^{m} \beta_k c_{n_k} = c_n$. Apply the Hahn–Banach theorem to f and use that $|f(x)| \le M\|x\|$, $x \in G$.

4.28° (1), (2) Yes.

4.29 Use problem 4.25. (2) Yes. (3) $\dim X = \dim X^*$.

4.31 Use problem 4.30°.

4.33 Use problem 4.32°.

4.34 Use Corollary 4.1 of the Hahn–Banach theorem.

4.36 $\varphi(y) = (\lim_{n\to\infty} y_n, y_1, y_2, \ldots)$.

4.37 Apply problems 3.19 and 3.20.

4.38* Let G be a subspace of a reflexive space X. For every functional $f \in X^*$ the restriction $f \upharpoonright G$ belongs to G^*, hence for every $F \in G^{**}$ one can define a functional $F_0(f) = F(f \upharpoonright G)$, $f \in X^*$. It is easy to check that $F_0 \in X^{**}$. Since X is reflexive, there exists an element $x_0 \in X$ such that $F_0(f) = f(x_0)$, $f \in X^*$. Let us show that $x_0 \in G$. Assume that $x_0 \notin G$. Then by Corollary 4.1 of Hahn–Banach theorem there exists $h \in X^*$ such that $h(x) = 0$, $x \in G$, and

$h(x_0) \neq 0$. Then $h(x_0) = F_0(h) = F(h \restriction G) = F(0) = 0$, a contradiction. Thus for $F \in G^{**}$ there exists $x_0 \in G$ such that $F(f \restriction G) = f(x_0)$, $f \in X^*$. Since for every $g \in G^*$ by the Hahn–Banach theorem there exists $f \in X^*$ such that $f \restriction G = g$, then $F(g) = g(x_0)$, $g \in G^*$. Therefore $F = \varphi(x_0)$ and $G^{**} = \varphi(G)$, where $\varphi : G \to G^{**}$ is the canonical embedding.

4.39* *Necessity.* Let $\Lambda \in X^{***}$, i.e. Λ is a continuous linear functional on X^{**}. Put $h(x) = \Lambda(\varphi(x))$, $x \in X$, where $\varphi : X \to X^{**}$ is a canonical embedding. Since Λ and φ are linear and continuous, then $h \in X^*$. Due to reflexivity of X every element of X^{**} has the form $\varphi(x) = F_x$ for some $x \in X$. Therefore $\Lambda(F_x) = h(x) = F_x(h)$, $F_x \in X^{**}$. Hence the space X^* is reflexive.

Sufficiency. Since $\varphi : X \to X^{**}$ is a linear isometry, then $\varphi(X) = \{F_x,\ x \in X\}$ is a subspace of X^{**}. Suppose that $\varphi(X) \neq X^{**}$. Then by Corollary 4.1 of the Hahn–Banach theorem there exists a nonzero functional $\Lambda \in X^{***}$ such that $\Lambda(F_x) = 0$, $x \in X$. Since the space X^* is reflexive, there exists $h \in X^* \setminus \{0\}$ such that $\Lambda(F) = F(h)$, $F \in X^{**}$. Then $\Lambda(F_x) = F_x(h) = h(x) = 0$ for all $x \in X$, a contradiction with the choice of h. Therefore $\varphi(X) = X^{**}$, so the space X is reflexive.

4.40* On the unit sphere $S^*(0, 1)$ of the dual space X^* there exists a countable set $\{f_n : n \geq 1\}$ which is dense in $S^*(0, 1)$. Let $y_n \in X$ be such that $\|y_n\| = 1$ and $|f_n(y_n)| \geq \frac{1}{2}$, $n \geq 1$. Put $G = \overline{\text{span}(\{y_n : n \geq 1\})}$ and verify that $G = X$. If this is false, then by Corollary 4.1 of the Hahn–Banach theorem there exists a functional $f \in X^*$ such that $\|f\| = 1$ and $f \restriction G = 0$. Then $f \in S^*(0, 1)$, but $\|f_n - f\| \geq |f_n(y_n) - f(y_n)| \geq \frac{1}{2}$, $n \geq 1$, which contradicts the density of $\{f_n : n \geq 1\}$ in $S^*(0, 1)$. Now it follows from problem 1.93 that $X = G$ is a separable space.

The converse statement is not true (consider $X = l_1$).

4.41 (1) Yes, due to Theorem 3.2. (2) Yes, because the set $\varphi(X)$ is closed in X^{**}.

4.42 Assume that there exists a function $x_0 \in C([0, 1])$, for which

$$F(g) = \int_0^1 x_0(t)dg(t) = \int_0^{1/2} dg(t), \quad g \in BV_0([0, 1]),$$

and show that $x_0(t) = \chi_{(0, \frac{1}{2})}(t)$, $t \in (0, 1)\backslash\{\frac{1}{2}\}$, which is impossible.

4.45 (2) An orthogonal complement to M. (3) Use the fact that for $x \notin \overline{\text{span}(M)}$ by Corollary 4.1 of the Hahn–Banach theorem there exists $f \in M^{\perp}$ such that $f(x) \neq 0$.

(4) Since X is a reflexive space, then

$$M^{\perp\perp} = \{\varphi(x),\ x \in X \mid f(x) = 0 \text{ for all } f \in M^{\perp}\}.$$

(5) No. If $X = c$, $M = c_0$, then

$$M^{\perp\perp} = \{x = (x_0, x_1, \dots) \in l_\infty \mid x_0 = 0\},$$
$$\varphi(M) = \{x = (x_0, x_1, \dots) \in l_\infty \mid x_0 = 0, \lim_{n\to\infty} x_n = 0\}.$$

4.46 No. For example, $X = c$, $L = \{y = (y_0, y_1, \dots) \in l_1 \mid y_0 = 0\}$ or $X = Y^*$, $L = \varphi(Y)$, where Y is a nonreflexive Banach space, $\varphi : Y \to Y^{**}$ is the canonical embedding.

4.47 (2) If M is not total in X^*, then by Corollary 4.1 of the Hahn–Banach theorem there exists a nonzero functional $F \in X^{**}$ such that $F(f) = 0$, $f \in M$. Since the space X is reflexive, then $F(f) = f(x)$ for some $x \in X \setminus \{0\}$. Then $f(x) = F(f) = 0$, $f \in M$, but $x \neq 0$.

(3) Let $X = Y^*$. Consider the set $M = \varphi(Y)$, where $\varphi : Y \to Y^{**}$ is the canonical embedding. If $x \in Y^*$ is such that $f(x) = 0$ for all $f \in M$, then $x(y) = 0$ for all $y \in M$, so $x = 0$. Therefore if (b) implies (a), then $\varphi(Y)$ is a total set in Y^{**}. But $\varphi(Y)$ is a subspace of Y^{**}, hence $\varphi(Y) = Y^{**}$, i.e. the space Y is reflexive. Due to problem 4.39*, the space X is also reflexive.

4.48 (2) $\overline{\text{span}(M)} = \left\{ \sum_{n=1}^{\infty} a_n \delta_{t_n} \mid t_n \in [a, b],\ a_n \in \mathbb{K},\ n \geq 1,\ \sum_{n=1}^{\infty} |a_n| < +\infty \right\}.$
By Theorem 3.4 every functional $f \in (C([a, b]))^*$ has the form $f(x) = \int_a^b x(t)dg(t)$, $x \in C([a, b])$, where $g \in BV_0([a, b])$, and $\|f\| = \|g\|_V$. Put

$$g_1 = \big(g(a+) - g(a)\big)\chi_{(a,b]} + \sum_{s\in(a,b]} \big(g(s) - g(s-)\big)\chi_{[s,b]}$$

(the function g_1 is defined correctly, because the sum contains at most countable number of nonzero terms, and $\sum_{s\in(a,b]} \big|g(s) - g(s-)\big| < +\infty$). Check that $g_1 \in BV_0([a, b])$, $g_2 = g - g_1 \in BV_0([a, b]) \cap C([a, b])$ and $\|g\|_V = \|g_1\|_V + \|g_2\|_V$. Further, show that $\rho(f, \text{span}(M)) = \|g_2\|_V$, whence $f \in \overline{\text{span}(M)} \Leftrightarrow g_2 = 0$.

4.49 For every $r > 0$ and $y \in S(0, r)$ by Corollary 4.2 of the Hahn–Banach theorem there exists a functional $f_y \in X^*$ such that $\|f_y\| = 1$ and $f_y(y) = r$. Verify that $\overline{B}(0, r) = \bigcap_{y\in S(0,r)} \{x \in X \mid f_y(x) \leq r\}$.

4.50 (1) Let $\{y_n : n \geq 1\} \subset X \setminus \{0\}$ be a dense set in X and $f_n \in X^*$, $n \geq 1$, be functionals for which $\|f_n\| = 1$ and $f_n(y_n) = \|y_n\|$. Then

$$\overline{B}(x_0, r) = \bigcap_{n=1}^{\infty} \big\{x \in X \,\big|\, |f_n(x) - f_n(x_0)| \leq r\big\}$$

for all $x_0 \in X$ and $r > 0$. Indeed, if $\|x - x_0\| \le r$, then $|f_n(x) - f_n(x_0)| \le r, n \ge 1$. If $\|x - x_0\| = r + \delta$, where $\delta > 0$, we can choose k such that $\|x - x_0 - y_k\| < \frac{\delta}{2}$. Then $f_k(y_k) = \|y_k\| > r + \frac{\delta}{2}$ and $|f_k(x) - f_k(x_0) - f_k(y_k)| \le \frac{\delta}{2}$, so $|f_k(x) - f_k(x_0)| > r$.

(2) In a real NLS $\{x \in X \mid |f(x) - b| \le a\} =$

$$= \{x \in X \mid f(x) \le a + b\} \cap \{x \in X \mid (-f)(x) \le a - b\}, \ a, b \in \mathbb{R}.$$

4.51 (1) The necessity follows from Corollary 4.1 of the Hahn–Banach theorem, and the sufficiency follows from problem 3.2.

4.52 (1) Since p is homogeneous, we have $p(0) = 0$ and $p(-x) = p(x)$, $x \in X$. Therefore $0 = p(0) \le p(x) + p(-x) = 2p(x)$, $x \in X$.

The seminorm $p : X \to \mathbb{R}$ is a norm, if $p(x) = 0 \Rightarrow x = 0$.

4.53 (1), (2) Consider $X = \mathbb{R}$, $p(x) = 2x + |x|$, $x \in \mathbb{R}$.

4.54 (2) To prove the inequality $|f(x)| \le p(x)$ for $f(x) \ne 0$, consider $y = \frac{|f(x)|}{f(x)} x$ and note that $|f(x)| = f(y) = g(y) \le p(y) = p(x)$.

4.55 Apply problems 4.51 and 4.54.

(1) If $M_0 = \{x \in X \mid g(x) = a\}$, where $g \in X^*_{\mathbb{R}}$, $a \in \mathbb{R}$, then

$$iM_0 = \left\{x \in X \mid g\left(\frac{x}{i}\right) = a\right\} = \{x \in X \mid h(x) = a\},$$

where $h(x) = -g(ix)$, $x \in X$.

(2) *Necessity.* If $M = \operatorname{Ker} f$, where $f \in X^*$, then $M_0 = \operatorname{Ker} g$, where $g = \operatorname{Re} f$.

Sufficiency. If $M_0 = \operatorname{Ker} g$, where $g \in X^*_{\mathbb{R}}$, then $M = \operatorname{Ker} f$, where

$$f(x) = g(x) - ig(ix), \ x \in X.$$

4.56* *Necessity.* Let $x_1, x_2 \in \overline{B}(0, 1)$ and $f(x_1) = f(x_2) = \|f\|$. Then $\frac{x_1+x_2}{2} \in \overline{B}(0, 1)$ and $f(\frac{x_1+x_2}{2}) = \|f\|$. Hence $\|x_1\| = \|x_2\| = \|\frac{x_1+x_2}{2}\| = 1$, and since X is strictly normed, then $x_1 = x_2$.

Sufficiency. Let us show that if there exist different elements $x_1, x_2 \in X$, for which $\|x_1\| = \|x_2\| = \|\frac{x_1+x_2}{2}\| = 1$, then there exists a functional $f \in X^*$, for which $f(x_1) = f(x_2) = \|f\|$. Firstly, consider the case $\mathbb{K} = \mathbb{R}$. By Corollary 4.1 of the Hahn–Banach theorem for the element x_1 and the subspace $G = \{\alpha(x_1 - x_2) \mid \alpha \in \mathbb{R}\}$ there exists $f \in X^*$ such that $\|f\| = 1$, $f \restriction G = 0$ and $f(x_1) = \rho(x_1, G)$. Let us verify that $\rho(x_1, G) = \inf_{\alpha \in \mathbb{R}} \|y_\alpha\| = 1$, where $y_\alpha = x_1 - \alpha(x_1 - x_2) = (1-\alpha)x_1 + \alpha x_2$, $\alpha \in \mathbb{R}$. Indeed, for $\alpha \in [0, 1]$ we have $\|y_\alpha\| \le (1-\alpha)\|x_1\| + \alpha\|x_2\| = 1$ and $1 = \|y_{1/2}\| \le \frac{1}{2}(\|y_\alpha\| + \|y_{1-\alpha}\|) \le 1$. Hence $\|y_\alpha\| = 1$ for all $\alpha \in [0, 1]$. For $\alpha \in \mathbb{R} \setminus [0, 1]$ we have

$$\|y_\alpha\| \ge \Big| |1 - \alpha| \cdot \|x_1\| - |\alpha| \cdot \|x_2\| \Big| \ge 1.$$

Therefore $f(x_1) = \rho(x_1, G) = 1$. Since $x_2 - x_1 \in G$, then $f(x_2 - x_1) = 0$ and $f(x_2) = f(x_1) = 1 = \|f\|$. In the case $\mathbb{K} = \mathbb{C}$ it follows from the reasonings above that there exists $g \in X_R^*$ such that $g(x_1) = g(x_2) = \|g\| = 1$. Put $f(x) = g(x) - ig(ix)$, $x \in X$. Since $|g(x)| \leq \|x\|$ for all $x \in X$, then by problem 4.54 we have $|f(x)| \leq \|x\|$ for all $x \in X$. Therefore $f \in X^*$ and $\|f\| \leq 1$. Since

$$1 = g(x_1) = \text{Re} f(x_1) \leq |f(x_1)| \leq \|f\| \leq 1,$$

then $\|f\| = 1$ and $f(x_1) = g(x_1) = 1$. Similarly $f(x_2) = 1$, so $f(x_2) = f(x_1) = \|f\|$.

4.57 Consider the functional $f(x) = 0$, $x \in G = \{0\}$, and apply Theorem 4.4.

4.58 Obviously, $p(x) \in \mathbb{R}$ and $p(\lambda x) = \lambda p(x)$ for all $x \in M(\mathbb{R})$ and $\lambda \geq 0$. Let $x, y \in M(\mathbb{R})$ and $\varepsilon > 0$. Choose $\alpha_1, \ldots, \alpha_k, \beta_1, \ldots, \beta_l$ such that

$$\sup_{t\in\mathbb{R}} \frac{1}{k} \sum_{i=1}^{k} x(t + \alpha_i) < p(x) + \varepsilon \text{ and } \sup_{t\in\mathbb{R}} \frac{1}{l} \sum_{j=1}^{l} x(t + \beta_j) < p(y) + \varepsilon.$$

For all $t \in \mathbb{R}$ we have

$$\frac{1}{kl} \sum_{i=1}^{k} \sum_{j=1}^{l} x(t + \alpha_i + \beta_j) = \frac{1}{l} \sum_{j=1}^{l} \left(\frac{1}{k} \sum_{i=1}^{k} x(t + \alpha_i + \beta_j) \right) < p(x) + \varepsilon,$$

because for fixed t and β_j

$$\frac{1}{k} \sum_{i=1}^{k} x(t + \alpha_i + \beta_j) \leq \sup_{t\in\mathbb{R}} \frac{1}{k} \sum_{i=1}^{k} x(t + \alpha_i) < p(x) + \varepsilon.$$

Similarly

$$\frac{1}{kl} \sum_{i=1}^{k} \sum_{j=1}^{l} y(t + \alpha_i + \beta_j) < p(y) + \varepsilon, \ t \in \mathbb{R}.$$

Hence

$$p(x + y) \leq \sup_{t\in\mathbb{R}} \frac{1}{kl} \sum_{i=1}^{k} \sum_{j=1}^{l} (x + y)(t + \alpha_i + \beta_j) < p(x) + p(y) + 2\varepsilon.$$

Since $\varepsilon > 0$ is arbitrary, then $p(x + y) \leq p(x) + p(y)$.

4.59* (1) Since $-F(x) = F(-x) \le p(-x)$, then

$$-p(-x) \le F(x) \le p(x), \ x \in M(\mathbb{R}).$$

(a) If $x(t) \ge 0$, $t \in \mathbb{R}$, then $p(-x) \le 0$, hence $F(x) \ge 0$.

(b) Without loss of generality $s > 0$. Consider the function $y(t) = x(t+s) - x(t)$, $t \in \mathbb{R}$, and show that $F(x(\cdot + s)) - F(x) = F(y) = 0$. For $\alpha_j = js$, $1 \le j \le k$, we have

$$p(y) \le \pi(y; \alpha_1, \ldots, \alpha_k) = \sup_{t \in \mathbb{R}} \frac{1}{k} \sum_{j=1}^{k} \big(x(t + (j+1)s) - x(t + js)\big) =$$
$$= \sup_{t \in \mathbb{R}} \frac{1}{k} \big(x(t + (k+1)s) - x(t)\big) \to 0, \ k \to \infty,$$

therefore $p(y) \le 0$. Similarly $p(-y) \le 0$, hence $F(y) = 0$.

(c) Obviously, $p(e) = 1$ and $p(-e) = -1$, therefore $F(e) = 1$.

(d) Let us verify that the property (d) is a consequence of (a)–(c) and of linearity of the functional F. For $A \subset [0, 1]$ denote by x_A the function with period 1 such that $x_A(t) = \chi_A(t)$ for $t \in [0, 1]$. Due to (b) the number $F(x_{\{a\}}) = c$ does not depend on a. For any finite set $\{a_1, \ldots, a_k\} \subset (0, 1]$ we have $0 \le x_{\{a_1,\ldots,a_k\}}(t) = \sum_{i=1}^{k} x_{\{a_i\}} \le 1$, $t \in \mathbb{R}$. Thus due to (a), (c) and linearity of F we have

$$0 \le F(x_{\{a_1,\ldots,a_k\}}) = \sum_{i=1}^{k} F(x_{\{a_i\}}) = kc \le 1, \ k \in \mathbb{N}.$$

Hence $c = 0$, $F(x_{\{a\}}) = 0$ and $F(x_{[a,b]}) = F(x_{(a,b]})$ for all $0 \le a < b \le 1$. For $n \ge 1$ and $0 \le k \le n$ due to (b), (c) and linearity of F we have

$$F\big(x_{[0,\frac{k}{n}]}\big) = \frac{k}{n}\Big(F\big(x_{[0,\frac{1}{n}]}\big) + F\big(x_{(\frac{1}{n},\frac{2}{n}]}\big) + \ldots + F\big(x_{(\frac{n-1}{n},1]}\big)\Big) = \frac{k}{n} F(e) = \frac{k}{n}.$$

Hence $F(x_{[0,b]}) = b$ for $b \in [0, 1] \cap \mathbb{Q}$. Since for $0 \le b' < b'' \le 1$ it follows from (a) that $F(x_{[0,b']}) \le F(x_{[0,b'']})$, then $F(x_{[0,b]}) = b$ for $b \in [0, 1]$. Finally, for $0 \le a < b \le 1$ we have

$$F(x_{[a,b]}) = F(x_{(a,b]}) = F(x_{[0,b]}) - F(x_{[0,a]}) = b - a.$$

(2) In view of (a) and (d), for any partition of the interval $[0, 1]$ the number $F(x)$ lies between the lower and upper Darboux sums, therefore $F(x)$ lies between the lower and upper integrals of the function x over the interval $[0, 1]$.

(3) Let $M_1(\mathbb{R}) \subset M(\mathbb{R})$ be the set of bounded real functions on $\mathbb{R}$ with period 1, which are Lebesgue integrable on $[0, 1]$. Put $f(x) = \int\limits_0^1 x(t)dt$, $x \in M_1(R)$. For every $x \in M_1(\mathbb{R})$ and $\alpha_1, \ldots, \alpha_k > 0$ we have

$$\sup_{t\in\mathbb{R}} \frac{1}{k}\sum_{j=1}^{k} x(t+\alpha_j) \geq \frac{1}{k}\sum_{j=1}^{k}\int_0^1 x(t+\alpha_j)dt = \int_0^1 x(t)dt = f(x),$$

therefore $f(x) \leq p(x)$, $x \in M_1(R)$. By Theorem 4.4 (Hahn–Banach theorem) there exists a linear functional $F : M(\mathbb{R}) \to \mathbb{R}$ such that $F(x) = f(x)$ for $x \in M_1(R)$ and $F(x) \leq p(x)$ for all $x \in M(\mathbb{R})$.

4.60 Put $\mu(A) = F(x_A)$, where x_A is a function with period 1 such that $x_A(t) = \chi_A(t)$ for $t \in [0, 1]$, $A \subset [0, 1]$, F is a functional from problem 4.59*.

Problems of Chap. 5

5.5° (2) Use Corollary 4.3 of the Hahn–Banach theorem.

5.6° Use the fact that in a finite-dimensional space strong convergence is equivalent to the coordinate-wise convergence (see problem 1.72).

5.7 Note that the set $\{\chi_{[s,t]} \mid s, t \in \mathbb{R},\ s < t\}$ is total in $L_q(\mathbb{R})$ and apply Theorem 5.5.

5.8 (1) Use that $c_0^* = l_1$ and the set $\{e_n : n \geq 1\}$ is total in the space l_1. (2) $x^{(n)} \xrightarrow{w} x$ in c if and only if the following conditions are fulfilled:

(a) $\exists C > 0\ \forall n \geq 1 : \|x^{(n)}\| \leq C$;
(b) $\forall k \geq 1 : x_k^{(n)} \to x_k, n \to \infty$, and $\lim\limits_{k\to\infty} x_k^{(n)} \to \lim\limits_{k\to\infty} x_k, n \to \infty$.

5.9° Here and further we denote by x_0 the limit of convergent sequences in the spaces $C([0, 1])$, L_p and by $x^{(0)}$ the limit in l_p. (1) Weakly, $x^{(0)} = 0$; (2) strongly, $x^{(0)} = 0$; (3) does not converge; (4) strongly, $x^{(0)} = 0$; (5) strongly, $x^{(0)} = (1, \frac{1}{2}, \ldots, \frac{1}{k}, \ldots)$; (6) strongly, $x^{(0)} = (\frac{1}{2}, \frac{2}{2^2}, \ldots, \frac{k}{2^k}, \ldots)$; (7) strongly, $x^{(0)} = (1, 0, \frac{1}{3^2}, 0, \frac{1}{5^2}, \ldots)$; (8) does not converge; (9) strongly, $x_0 = 0$; (10) for $1 \leq p < 2$ strongly, $x_0 = 0$; for $p = 2$ weakly, $x_0 = 0$; for $2 < p \leq +\infty$ does not converge; (11)–(14) weakly, $x_0 = 0$; (15) for $1 \leq p < 2$ strongly, $x_0 = 0$; for $p = 2$ weakly, $x_0 = 0$; for $2 < p \leq +\infty$ does not converge; (16) does not converge; (17) strongly, $x_0 = 0$; (18) for $1 \leq p < 4$ strongly, $x_0 = 0$; for $p = 4$ weakly, $x_0 = 0$; for $4 < p \leq +\infty$ does not converge; (19) weakly, $x_0 = 0$; (20) for $1 \leq p < 2$ does not converge; for $p = 2$ weakly, $x_0 = 0$; for $2 < p \leq +\infty$ strongly, $x_0 = 0$.

5.10° (1) Weakly, $x_0 = 0$; (2) strongly, $x_0(t) = 1$; (3) does not converge; (4) strongly, $x_0 = 0$; (5), (6) weakly, $x_0 = 0$; (7) does not converge.

5.11° (1) Weakly, $x_0 = 0$; (2) weakly, $x_0 = 0$; (3) weakly, $x_0 = 0$; (4) does not converge; consider $f(x) = \int\limits_{\mathbb{R}} x(t)dt$, $x \in L_1(\mathbb{R})$; (5) does not converge.

5.12° There is no strong convergence. Denote by f_0 the weak-$*$ limit of the sequence $\{f_n : n \geq 1\}$. (1), (2) $f_0 = 0$; (3) $f_0 = 0$. Consider a total set $\{t^k,\ t \in [0,1] : k \geq 0\}$; (4) $f_0(x) = x(1)$; (5) $f_0(x) = x'(0)$. To prove the absence of strong convergence, consider $(f_n - f_0)(x_n)$, where $x_n(t) = \frac{1}{n}\sin nt$, $t \in [-1, 1]$, $n \geq 1$.

5.13 X is not a Banach space.

5.14 Let $X = \{x \in l_\infty \mid \exists n \in \mathbb{N} : x_k = 0,\ k \geq n\}$ (the set of sequences which has finitely many nonzero terms) endowed with the norm $\|\cdot\|_\infty$. Put $f_n(x) = nx_n$, $x = (x_1, x_2, \ldots) \in X$, $n \geq 1$. Verify that $f_n \in X^*$, $f_n \xrightarrow{w-*} 0$ and $\|f_n\| = n$, $n \geq 1$.

5.15° Prove that weak convergence in X^* implies weak-$*$ convergence.

5.16 (1) The image of the sequence $\{f_n : n \geq 1\}$ under the isometric isomorphism between c_0^* and l_1 (see problem 3.19) is the sequence $\{e_n : n \geq 1\} \subset l_1$. Put $F(x) = \sum\limits_{k=1}^{\infty} x_k$, $x \in l_1$. We have $F \in l_1^*$ and $F(e_n) \to 1 \neq 0 = F(0)$, $n \to \infty$, hence $e_n \overset{w}{\nrightarrow} 0$ in l_1, therefore $f_n \overset{w}{\nrightarrow} 0$ in c_0^*.

(2) The image of the sequence $\{f_n : n \geq 1\}$ under the isometric isomorphism between l_1^* and l_∞ (see Corollary 3.1 of Theorem 3.1) is the sequence

$$\{a^{(n)} = (\underbrace{0, \ldots, 0}_{n}, 1, 1, \ldots) : n \geq 1\} \subset l_\infty.$$

Consider a functional $F \in l_\infty^*$ such that $F(a) = \lim\limits_{k\to\infty} a_k$ for $a \in c$.

(3) The images of functionals f_n, $n \geq 1$, and f_0 under the isometric isomorphism between $(C([0,1]))^*$ and $BV_0([0,1])$ (see Theorem 3.4) are functions $g_n = \chi_{[\frac{1}{n},1]}$, $n \geq 1$, and $g_0 = \chi_{(0,1]}$. Consider the functional $F(g) = \sum\limits_{k=1}^{\infty}\big(g(\frac{1}{k}) - g(\frac{1}{k}-)\big)$, $g \in BV_0([0,1])$.

5.17° (4) Note that

$$\|x_n - x\|^2 = \|x_n\|^2 - (x_n, x) - (x, x_n) + \|x\|^2.$$

Since $(x_n, x) \to (x, x) = \|x\|^2$ and $(x, x_n) \to \|x\|^2$, $n \to \infty$, then

$$0 \leq \limsup_{n\to\infty} \|x_n - x\|^2 = \limsup_{n\to\infty} \|x_n\|^2 - \|x\|^2 \leq 0.$$

5.18° (1) $x_n = y_n = e_n, n \geq 1$; (2) $x_n = e_n$, $y_n = e_{n+1}, n \geq 1$, where $\{e_n : n \geq 1\}$ is an orthonormal system in H.

5.19° Put $y_n = \sum\limits_{k=1}^{n} x_k$, $n \geq 1$. (2) $\Rightarrow$ (3) The weakly convergent sequence $\{y_n : n \geq 1\}$ is bounded. Therefore there exists $C > 0$ such that $\|y_n\|^2 = \sum\limits_{k=1}^{n} \|x_k\|^2 \leq C$, $n \geq 1$. (3) $\Rightarrow$ (1) For all $m > n$ we have $\|y_m - y_n\|^2 = \sum\limits_{k=n+1}^{m} \|x_k\|^2 \to 0$, $n, m \to \infty$, therefore the sequence $\{y_n : n \geq 1\}$ is a Cauchy sequence.

5.20 (2) For $x \neq 0$ by Corollary 4.2 of the Hahn–Banach theorem there exists $f \in X^*$ such that $\|f\| = 1$ and $f(x) = \|x\|$. Therefore

$$\|x\| = f(x) = \lim_{n\to\infty} f(x_n) \leq \liminf_{n\to\infty} \|x_n\|.$$

5.22° Apply Theorem 5.3 to the sequence $\{\varphi(x_n) : n \geq 1\} \subset X^{**}$, where $\varphi : X \to X^{**}$ is the canonical embedding.

5.23 Let $G = \overline{\text{span}(\{x_n \mid n \geq 1\})}$. Then G is a separable Banach space, moreover G is reflexive as a subspace of a reflexive space (see problem 4.38*). Therefore the space G^{**} is separable and due to problem 4.40* the space G^* is separable. Apply Theorem 5.4 to the sequence $\{\varphi(x_n) : n \geq 1\} \subset G^{**}$, where $\varphi : G \to G^{**}$ is the canonical embedding.

5.24 (1) Consider the sequence $\{x_n(t) = t^n,\ t \in [0, 1] : n \geq 1\}$. Show that $\int\limits_0^1 x_n(t)dg(t) \to g(1) - g(1-)$ for all functions $g \in BV_0([0, 1])$. Since the pointwise limit of the sequence $\{x_n : n \geq 1\}$ is a discontinuous function, this sequence does not converge weakly according to the criterion of weak convergence in $C([0, 1])$ (item (1) of problem 5.1). (2), (3) Show that the sequence $\left\{x^{(n)} = \sum\limits_{k=1}^{n} e_{2k} : n \geq 1\right\}$ is a weakly Cauchy sequence, but it does not converge weakly.

5.26 If $\|x_n - x\| \not\to 0$, $n \to \infty$, then there exist $\varepsilon > 0$ and a subsequence $\{y_j = x_{n_j} : j \geq 1\}$ such that $\|y_j - x\| \geq \varepsilon$, $j \geq 1$. Consider a sequence $\{f_j : j \geq 1\} \subset X^*$ such that $\|f_j\| = 1$ and $|f_j(y_j - x)| = \|y_j - x\| \geq \varepsilon$, $j \geq 1$. According to theorem 5.4 there exist a subsequence $\{f_{j_k} : k \geq 1\}$ and a functional $f \in X^*$ such that $f_{j_k} \xrightarrow{w-*} f$, $k \to \infty$. Then $f_{j_k}(y_{j_k} - x) = f_{j_k}(y_{j_k}) - f_{j_k}(x) \to f(x) - f(x) = 0$, a contradiction with the choice of f_j.

5.27 (2) Verify that the set $M = \{e_n \mid n \geq 1\} \subset l_2$ is strongly closed. This set is not weakly closed, because $e_n \xrightarrow{w} 0 \notin M$, $n \to \infty$ (see the solution of problem 5.2). (3) Use Corollary 4.1 of the Hahn–Banach theorem. (4) Let $x_n \in \overline{B}(x_0, r)$, $n \geq 1$ be such that $x_n \xrightarrow{w} y$. Then $x - x_0 \xrightarrow{w} y - x_0$. Use item (2) of problem 5.20.

5.28 For $x = 0$ the statement is obvious. If $x \neq 0$ and $\|x_n\| \to \|x\|$, $n \to \infty$, then without loss of generality $x_n \neq 0$, $n \geq 1$. Put $y = \frac{x}{\|x\|}$ and $y_n = \frac{x_n}{\|x_n\|}$, $n \geq 1$. Then $\|y\| = \|y_n\| = 1$ and $y_n \overset{w}{\to} y$, $n \to \infty$, because $f(y_n) = \frac{f(x_n)}{\|x_n\|} \to \frac{f(x)}{\|x\|} = f(y)$ for all $f \in X^*$. Let us show that $\|\frac{y_n+y}{2}\| \to 1$, $n \to \infty$. Then the definition of the uniformly convex Banach space will imply that $\|y_n - y\| \to 0$, $n \to \infty$, therefore

$$\|x_n - x\| \leq \|y_n - y\| \cdot \|x_n\| + \|y\| \cdot (\|x_n\| - \|x\|) \to 0, \ n \to \infty.$$

By Corollary 4.2 of the Hahn–Banach theorem, there exists a functional $f \in X^*$ such that $\|f\| = 1$ and $f(y) = \|y\| = 1$. We have $|f(\frac{y_n+y}{2})| \leq \|\frac{y_n+y}{2}\| \leq 1$, $n \geq 1$. But $f(\frac{y_n+y}{2}) = \frac{f(y_n)+f(y)}{2} \to f(y) = 1$, because $y_n \overset{w}{\to} y$, $n \to \infty$. Therefore $\|\frac{y_n+y}{2}\| \to 1$, $n \to \infty$, which completes the proof.

5.29* Assume that $x^{(n)} \overset{w}{\to} x^{(0)}$ and $\|x^{(n)} - x^{(0)}\| \not\to 0$. Then there exist $\varepsilon > 0$ and a sequence $\{y^{(j)} = x^{(n_j)} - x^{(0)} : j \geq 1\}$ such that $y^{(j)} \overset{w}{\to} 0$ and $\|y^{(j)}\| \geq \varepsilon$, $j \geq 1$. For $j_1 = 1$ choose $p_1 \in \mathbb{N}$ such that $\sum\limits_{k=p_1+1}^{\infty} |y_k^{(j_1)}| < \frac{\varepsilon}{3}$. Since the weak convergence in l_1 implies coordinate-wise convergence, then $\sum\limits_{k=1}^{p_1} |y_k^{(j)}| \to 0$, $j \to \infty$. Choose $j_2 > j_1$ such that $\sum\limits_{k=1}^{p_1} |y_k^{(j_2)}| < \frac{\varepsilon}{6}$, and choose $p_2 > p_1$ such that $\sum\limits_{k=p_2+1}^{\infty} |y_k^{(j_2)}| < \frac{\varepsilon}{6}$. Similarly, by induction we choose positive integers $1 = j_1 < j_2 < j_3 < \ldots$ and $p_1 < p_2 < p_3 < \ldots$ such that $\sum\limits_{k=1}^{p_{m-1}} |y_k^{(j_m)}| < \frac{\varepsilon}{6}$ and $\sum\limits_{k=p_m+1}^{\infty} |y_k^{(j_m)}| < \frac{\varepsilon}{6}$, $m \geq 2$. Put $N_1 = \{1, \ldots, p_1\}$ and $N_m = \{p_{m-1}+1, \ldots, p_m\}$, $m \geq 2$.

By construction, $\sum\limits_{k \notin N_m} |y_k^{(j_m)}| < \frac{\varepsilon}{3}$, $m \geq 1$, and since $\sum\limits_{k=1}^{\infty} |y_k^{(j_m)}| = \|y^{(j_m)}\| \geq \varepsilon$, then $\sum\limits_{k \in N_m} |y_k^{(j_m)}| \geq \frac{2\varepsilon}{3}$, $m \geq 1$.

For $k \in N_m$, $m \geq 1$, put $a_k = \frac{|y_k^{(j_m)}|}{y_k^{(j_m)}}$, if $y_k^{(j_m)} \neq 0$, and $a_k = 0$, if $y_k^{(j_m)} = 0$. Consider the functional $f(x) = \sum\limits_{k=1}^{\infty} x_k a_k$, $x \in l_1$. Since $|a_k| \leq 1$, $k \geq 1$, then $(a_1, a_2, \ldots) \in l_\infty$ and $f \in l_1^*$. Let us show that $f(y^{(j_m)}) \not\to 0$. Indeed, $\sum\limits_{k \in N_m} y_k^{(j_m)} a_k = \sum\limits_{k \in N_m} |y_k^{(j_m)}| \geq \frac{2\varepsilon}{3}$ and $\left|\sum\limits_{k \notin N_m} y_k^{(j_m)} a_k\right| \leq \sum\limits_{k \notin N_m} |y_k^{(j_m)}| < \frac{\varepsilon}{3}$, hence

$$|f(y^{(j_m)})| \geq \left|\sum_{k \in N_m} y_k^{(j_m)} a_k\right| - \left|\sum_{k \notin N_m} y_k^{(j_m)} a_k\right| \geq \frac{\varepsilon}{3}, \quad m \geq 1.$$

Thus $y^{(j_m)} \overset{w}{\not\to} 0$, a contradiction.

5.31 (1) Apply the Banach–Steinhaus theorem to the sequence of functionals

$$f_n(x) = \sum_{k=1}^{n} a_k x_k, \ x \in l_p, \ n \geq 1.$$

(2) Apply the Banach–Steinhaus theorem to the sequence of functionals

$$f_n(x) = \int_a^b h_n(t)x(t)dt, \ x \in L_p([a,b]), \ n \geq 1,$$

where $h_n(t) = h(t)$, if $|h(t)| \leq n$, and $h_n(t) = 0$, if $|h(t)| > n$, $n \geq 1$.

5.32 If the sequence $a = \{a_n : n \geq 1\}$ is such that the series $\sum_{n=1}^{\infty} a_n x_n$ converges for all $x \in c$ or all $x \in c_0$, then $a \in l_1$.

5.33 *Necessity.* Apply the Banach–Steinhaus theorem to the sequence $\{\varphi(x_n) : n \geq 1\}$, where $\varphi : X \to X^{**}$ is the canonical embedding.

5.34 $f(x) = \int_0^1 x(t)dt$, $x \in C([0,1])$. There is no convergence in norm. Check that $\|f_n - f\| = 2$, $n \geq 1$.

5.36 *Necessity.* The condition $\sup_{n\geq 1} V(F_n, [a,b]) < +\infty$ follows from the criterion of weak-$*$ convergence. For all $t \in [a,b]$ we have

$$|F_n(t)| = |F_n(t) - F_n(a)| \leq \sup_{n\geq 1} V(F_n, [a,b]),$$

therefore the sequence $\{F_n : n \geq 1\}$ is uniformly bounded. Let t_0 be any point of continuity of F_0 or $t_0 = b$. We will show that the bounded sequence $\{F_n(t_0) : n \geq 1\}$ has a unique limit point $F_0(t_0)$. It will follow that $F_n(t_0) \to F_0(t_0)$, $n \to \infty$. Consider any subsequence $\{F_{n_k} : k \geq 1\}$, for which the sequence $\{F_{n_k}(t_0) : k \geq 1\}$ converges. By Helly's selection principle (see, for example, [21], Ch. 10, Sec. 36, Theorem 5) there exists a subsequence $\{m_j = n_{k_j} : j \geq 1\}$ such that the functions F_{m_j} converge pointwise to some function Φ, in particular $\Phi(a) = \lim_{j\to\infty} F_{m_j}(a) = 0$. By Helly's convergence theorem (see, for example, [21], Ch. 10, Sec. 36, Theorem 4) we have $\Phi \in BV([a,b])$ and $f_{m_j} \overset{w-*}{\longrightarrow} \varphi$, where $\varphi(x) = \int_a^b x(t)d\Phi(t)$, $x \in C([a,b])$. Then $\varphi = f_0$ by the uniqueness of the weak-$*$ limit of the sequence. Let $F \in BV_0([a,b])$ be such that $F(t) = \Phi(t)$ at all points of continuity of the function F, and also at $t = a$ and $t = b$. Then

$$f_0(x) = \varphi(x) = \int_a^b x(t)dF(t) = \int_a^b x(t)dF_0(t), \ x \in C([a, b]).$$

It follows from Theorem 3.3 that $F = F_0$. Thus $F_{m_j}(t_0) \to \Phi(t_0) = F_0(t_0)$, $j \to \infty$.

Sufficiency. Fix $x_0 \in C([a, b])$. The sequence of numbers $\{f_n(x_0) : n \geq 1\}$ is bounded due to the uniform boundedness of variations. Let us show that this sequence has a unique limit point $f_0(x_0)$. It will follow that $f_n(x_0) \to f_0(x_0)$, $n \to \infty$. Let the subsequence $\{f_{n_k}(x_0) : k \geq 1\}$ be convergent. By Helly's selection principle, there exists a subsequence $\{m_j = n_{k_j} : j \geq 1\}$ such that the functions F_{m_j} converge pointwise to some function Φ, in particular $\Phi(a) = \lim_{j\to\infty} F_{m_j}(a) = 0$. By Helly's convergence theorem, $\Phi \in BV([a, b])$ and $f_{m_j} \xrightarrow{w-*} \varphi$, where

$$\varphi(x) = \int_a^b x(t)d\Phi(t) = \int_a^b x(t)dF_0(t) = f_0(x), \ x \in C([a, b]),$$

because $\Phi(t) = F_0(t)$ at all points of continuity of the function F_0, and also at $t = a$ and $t = b$. Therefore $f_{m_j}(x_0) \to f_0(x_0)$, $j \to \infty$.

5.37 (1) Each function $x \in C([0, 1])$ can be represented as $x = x_+ - x_-$, where x_+, x_- are nonnegative continuous functions on $[0, 1]$. Prove that $\lim_{n\to\infty} f_n(x_\pm)$ exists, and apply Theorem 5.3. (2) No. Consider the sequence of functions

$$\{p_n(t) = p_0(2^n t), \ t \in [0, 1] : n \geq 1\},$$

where $p_0(t) = (-1)^k$, $t \in [k, k + 1)$, $k \in \mathbb{N} \cup \{0\}$. (3) Yes.

5.38 (1), (2) If $f \neq 0$, then it is necessary and sufficient that the sequence $\{\alpha_n : n \geq 1\}$ converges.

5.39 (1) Consider the cases $|t - c| < \delta$ and $|t - c| \geq \delta$. (2) Since the functionals f_n are nonnegative, it follows from (1) that

$$|f_n(x) - x(c)f_n(e_0)| \leq \frac{2\|x\|}{\delta^2} f_n(y) + \varepsilon f_n(e_0)$$

for all $x \in C([a, b])$ and $n \geq 1$. Hence $\limsup_{n\to\infty} |f_n(x) - x(c)| \leq \varepsilon$, where $\varepsilon > 0$ is arbitrary.

5.40 According to problem 3.42, we have $\|f_n - f\| = (f_n - f)(e_0) \to 0, n \to \infty$.

5.41 Consider the functionals

$$f_n(x) = \sum_{k=1}^{\infty} a_{nk}x_k,\ n \geq 1,\ f(x) = \lim_{n\to\infty} x_n,\ x = (x_1, x_2, \ldots) \in c,$$

and verify that $f_n \xrightarrow{w-*} f$ if and only if conditions (1)–(3) are fulfilled. To do this, use the criterion of weak-$*$ convergence of functionals (Theorem 5.2) and note that the sequence $\{e_n : n \geq 0\}$, where $e_0 = (1, 1, \ldots)$, is total in c.

5.42* Let the sequence $\{a_n : n \geq 0\} \subset \mathbb{C}$ be fixed, $X_n = \sum_{k=0}^{n} x_k$ and $Y_n = \sum_{k=0}^{n} y_k$, $n \geq 0$. It is easy to verify that $Y_n = \sum_{k=0}^{n} a_{n-k}X_k$. Put

$$f_n(X) = \sum_{k=0}^{n} a_{n-k}X_k,\ X = (X_0, X_1, \ldots) \in c.$$

Then $f_n \in c^*$ and $Y_n = f_n(X)$. The series $\sum_{n=0}^{\infty} y_n$ converges for any convergent series $\sum_{n=0}^{\infty} x_n$ if and only if the sequence $\{f_n(X) : n \geq 1\}$ converges for all $X \in c$, i.e., if the sequence $\{f_n : n \geq 1\}$ converges weakly-$*$ in the space c^*. Verify that the latter condition is equivalent to the convergence of the series $\sum_{n=0}^{\infty} |a_n|$.

5.43 If $x_n = \sum_{k=1}^{n} a_k$, $y_n = \sum_{k=1}^{n} a_k b_k$, $n \geq 1$, then

$$y_n = f_n(x) := \sum_{k=1}^{n-1}(b_k - b_{k+1})x_k + b_n x_n,\ x = (x_1, x_2, \ldots)$$

(Abel's identity).

(1) The series $\sum_{n=1}^{\infty} a_n b_n$ converges for any convergent series $\sum_{n=1}^{\infty} a_n$ if and only if the sequence $\{f_n : n \geq 1\}$ converges weakly-$*$ in the space c^*.

(2) The series $\sum_{n=1}^{\infty} a_n b_n$ converges for any sequence $\{a_n : n \geq 1\} \subset \mathbb{C}$ with uniformly bounded partial sums $x_n = \sum_{k=1}^{n} a_n$, $n \geq 1$, if and only if the sequence $\{f_n : n \geq 1\}$ converges weakly-$*$ in the space l_∞^*.

Problems of Chap. 6

6.8° Use the fact that in a finite-dimensional space convergence in norm is equivalent to the coordinate-wise convergence (see problem 1.72).

6.9° If $\{e_1, \dots, e_m\} \subset \mathbb{R}^m$ and $\{g_1, \dots, g_n\} \subset \mathbb{R}^n$ are standard bases in $\mathbb{R}^m$ and $\mathbb{R}^n$, respectively, then $Ax = \sum_{j=1}^{n}\left(\sum_{k=1}^{m} a_{jk}x_k\right) g_j$, $x = \sum_{k=1}^{m} x_k e_k$, where $(a_{jk})_{j=1,k=1}^{n\ \ m}$ is a matrix such that $(a_{1k}, a_{2k}, \dots, a_{nk})$ are the coordinates of the vector Ae_k in the basis $\{g_1, \dots, g_n\}$.

(1) $\|A\| = \max\limits_{1\le k\le m} \sum_{j=1}^{n} |a_{jk}|$. We have $\|Ax\|_1 = \sum_{j=1}^{n}\left|\sum_{k=1}^{m} a_{jk}x_k\right| \le$

$$\le \sum_{j=1}^{n}\sum_{k=1}^{m} |a_{jk}|\cdot|x_k| = \sum_{k=1}^{m}\left(\sum_{j=1}^{n}|a_{jk}|\right)\cdot|x_k| \le \max_{1\le k\le m}\sum_{j=1}^{n}|a_{jk}|\cdot\|x\|_1,$$

equality is achieved at $x = e_{k_0}$, where $\max\limits_{1\le k\le m}\sum_{j=1}^{n}|a_{jk}| = \sum_{j=1}^{n}|a_{jk_0}|$.

(2) $\|A\| = \max\limits_{1\le j\le n}\sum_{k=1}^{m}|a_{jk}|$. (3) $\|A\| = \max\limits_{1\le j\le n,\ 1\le k\le m}|a_{jk}|$.

(4) $\|A\| = \max\limits_{x_k=\pm1,\ 1\le k\le m}\sum_{j=1}^{n}\left|\sum_{k=1}^{m} a_{jk}x_k\right|$. (5) $\|A\| = \max\limits_{1\le j\le n}\sqrt{\sum_{k=1}^{m} a_{jk}^2}$.

In items (2), (3), (5) consider the functionals $f_j(x) = \sum_{k=1}^{m} a_{jk}x_k$, $x \in \mathbb{R}^m$. Since $\|Ax\|_\infty = \max\limits_{1\le j\le n}|f_j(x)|$, $x \in \mathbb{R}^m$, then $\|A\| = \max\limits_{1\le j\le n}\|f_j\|$.

6.10° $\|A\| = \sup\limits_{n\ge1}|\alpha_n|$.

6.11° (1), (2) $\|A\| = \max\limits_{t\in[a,b]}|\alpha(t)|$.

6.12 $\|A\| = \|\alpha\|_\infty$.

6.13° $\|A\| = \left(\sum_{n=1}^{\infty}\frac{1}{n^2}\right)^{1/2} = \frac{\pi}{\sqrt6}$.

6.14° $\|A\| \le \left(\sum_{j,k=1}^{\infty}|\alpha_{jk}|^2\right)^{1/2}$.

6.15 For $t \in [a, b]$ consider the functionals

$$f_t(x) = (Ax)(t) = \int_a^b K(t,s)x(s)ds,\ \ x \in C([a,b]).$$

According to item (1) of problem 3.23, these functionals are linear and continuous, and $\|f_t\| = \int_a^b |K(t,s)|ds$. Since $\|Ax\| = \max_{t\in[a,b]} |f_t(x)|$, $x \in C([a,b])$, then

$$\|A\| = \max_{t\in[a,b]} \|f_t\| = \max_{t\in[a,b]} \int_a^b |K(t,s)|ds.$$

6.16 Put $M = \max_{s\in[a,b]} \int_a^b |K(t,s)|dt$. We have

$$\|Ax\|_1 = \int_a^b \left| \int_a^b K(t,s)x(s)ds \right| dt \le \int_a^b \int_a^b |K(t,s)| \cdot |x(s)|dsdt =$$

$$= \int_a^b \left(\int_a^b |K(t,s)|dt \right) \cdot |x(s)|ds \le \max_{s\in[a,b]} \int_a^b |K(t,s)|dt \cdot \|x\|_1 = M \cdot \|x\|_1.$$

Therefore $A \in \mathcal{L}(L_1([a,b]))$ and $\|A\| \le M$. To prove the inequality $\|A\| \ge M$ consider $s_0 \in [a,b]$, for which $M = \int_a^b |K(t,s_0)|dt$, and put $x_n = \chi_{[s_0-\frac{1}{n}, s_0+\frac{1}{n}]\cap[a,b]}$, $n \ge 1$. Using the uniform continuity of the kernel K, verify that $\frac{\|Ax_n\|_1}{\|x_n\|_1} \to M$, $n \to \infty$.

6.17° Use the Fubini's theorem and the Cauchy–Schwarz inequality.

6.18 $\|A\| = \max_{t\in[a,b]} \int_a^t |K(t,s)|ds$. See hint for problem 6.15.

6.19 $\|A\| = \int_{-\pi}^{\pi} |u(t)|dt$. Use problem 6.16.

6.20 (1) $\|A\| = 1$. (2) To prove that operator A is closed, use the theorem on term-by-term differentiation of a functional sequence. To prove the discontinuity of A, consider the sequence $\{t^n,\ t \in [0,1] : n \ge 1\}$.

6.21° Use problem 6.3.

6.22 (1) $\|I\| = 1$. (2) $\|I\| = (b-a)^{\frac{1}{r}-\frac{1}{p}}$ for $p < +\infty$, $\|I\| = (b-a)^{\frac{1}{r}}$ for $p = +\infty$. Apply Hölder's inequality.

6.24 A^n are integral operators with kernels K_n, which are defined by induction:

(1) $K_1(t,s) = K(t,s)$ and $K_n(t,s) = \int_a^b K(t,\tau)K_{n-1}(\tau,s)d\tau$, $(t,s) \in [a,b]^2$, $n \ge 2$;

(2) $K_1(t,s) = K(t,s)$ and $K_n(t,s) = \int\limits_s^t K(t,\tau)K_{n-1}(\tau,s)d\tau$, $a \le s \le t \le b$, $n \ge 2$.

6.25 $\|A\| = \frac{1}{\beta-\alpha}$ for $\gamma = \alpha$, $\|A\| = \left(\frac{(\alpha-\gamma)^{\alpha-\gamma}}{(\beta-\gamma)^{\beta-\gamma}}\right)^{\frac{1}{\beta-\alpha}}$ for $\gamma < \alpha$.

6.26° $\|A\| = \|y\| \cdot \|z\|$. Use problem 6.3.

6.27° $\|A\| = \|u\|_X \cdot \|v\|_{X^*}$. Use problem 6.3.

6.28° *Necessity.* Since $R(A)$ is a finite-dimensional subspace, it has a basis $\{y_1, \dots, y_n\} \subset Y$. Let $f_1(x), \dots, f_n(x) \in \mathbb{K}$ be the coordinates of Ax in this basis. For any $x', x'' \in X$ and $\alpha, \beta \in \mathbb{K}$ we have

$$A(\alpha x' + \beta x'') = \alpha Ax' + \beta Ax'' = \sum_{k=1}^{n} (\alpha f_k(x') + \beta f_k(x''))y_k.$$

Therefore $f_k(\alpha x' + \beta x'') = \alpha f_k(x') + \beta f_k(x'')$ due to the uniqueness of the expansion of a vector with respect to a given basis, hence the functionals $f_1, \dots, f_n$ are linear. The continuity of these functionals follows from the fact that in a finite-dimensional space convergence is equivalent to the coordinate-wise convergence.

6.30 $\|A\| = \max\limits_{1\le k\le n} |c_k|$. By the Pythagorean theorem and Bessel's inequality

$$\|Ax\|^2 = \sum_{k=1}^{n} \left|c_k(x, y_k)\right|^2 \le \max_{1\le k\le n} |c_k|^2 \cdot \sum_{k=1}^{n} \left|(x, y_k)\right|^2 \le \max_{1\le k\le n} |c_k|^2 \cdot \|x\|^2,$$

the equality is achieved at $x = y_{k_0}$, where $\max\limits_{1\le k\le n} |c_k| = |c_{k_0}|$.

6.31° (1) $\|A\| = \frac{1}{\beta+1}$; (2) $\|A\| = \frac{1}{2}e^3(1 - e^{-2})$; (3) $\|A\| = \frac{2}{\pi}$; (4) $\|A\| = 2$; (5) $\|A\| = \pi$; (6) $\|A\| = \frac{1}{\sqrt{(2\alpha+1)(2\beta+1)}}$; (7), (8) $\|A\| = \pi$; (9) $\|A\| = 1$. In items (1), (2), (4), (6) use problem 6.3, in item (3) use problem 6.15, in item (5), (7), (8) use problem 6.30.

6.32° (1)–(6) $\|A\| = 1$.

6.34° (2) No. Consider $X = C^1([a,b])$ endowed with the uniform norm, $Y = C([a,b])$, $(Ax)(t) = x'(t)$, $t \in [a,b]$, $x \in X$. (3) (a) $A = I$; (b) Let $X = Y = l_2$, $Ax = (x_1, \frac{x_2}{2}, \frac{x_3}{3}, \dots)$. Then the set $R(A)$ is dense in l_2, because it contains all the sequences with finitely many nonzero terms, but $(1, \frac{1}{2}, \frac{1}{3}, \dots) \notin R(A)$.

6.35° (1) It is enough to prove that the set $R(A)$ is closed. Let $y_n = Ax_n$ and $y_n \to y$, $n \to \infty$. Then the sequence $\{y_n : n \ge 1\}$ is a Cauchy sequence. Since $\|y_n - y_m\| \ge c\|x_n - x_m\|$, $n, m \ge 1$, then the sequence $\{x_n : n \ge 1\}$ is also a

Cauchy sequence, so it converges. Let $x_n \to x$, $n \to \infty$. Then $y_n = Ax_n \to Ax$, i.e. $y = Ax \in R(A)$.

6.36* For each $x \in X$ define $[x] = \{x + y \mid y \in \operatorname{Ker} A\}$. Since $\operatorname{Ker} A$ is a subspace, the quotient space $X/\operatorname{Ker} A = \{[x] \mid x \in X\}$ is an NLS endowed with the norm $\|[x]\| = \inf_{y \in \operatorname{Ker} A} \|x + y\|$ (see, for example, [21], Ch. 4, Sec. 13). Let $Px = [x]$, $P : X \to X/\operatorname{Ker} A$. Then $P \in \mathcal{L}(X, X/\operatorname{Ker} A)$ and $\|P\| \le 1$. Put $B[x] = Ax$, $x \in X$. The operator $B : X/\operatorname{Ker} A \to Y$ is well-defined and is an injection, because for $[x'] = [x'']$ we have $x' - x'' \in \operatorname{Ker} A$ and $Ax' = Ax''$, and for $[x'] \neq [x'']$ we have $x' - x'' \notin \operatorname{Ker} A$ and $Ax' \neq Ax''$. Then $B : X/\operatorname{Ker} A \to R(A)$ is a bijection and $\dim X/\operatorname{Ker} A = \dim R(A) < +\infty$. Hence $B \in \mathcal{L}(X/\operatorname{Ker} A, Y)$ by problem 6.8° and $A = BP \in \mathcal{L}(X, Y)$.

6.37° $\|P\| = 1$ for $G \neq \{0\}$, $\|P\| = 0$ for $G = \{0\}$.

6.38 *Sufficiency.* Let there exist $x_0 \in X$ and $r > 0$ such that $\|Ax\| < 1$ for all $x \in B(x_0, r)$. Then for all $u \in \overline{B}(0, r)$ we have

$$\|Au\| = \|A(x_0 + \tfrac{u}{2}) - A(x_0 - \tfrac{u}{2})\| \le \|A(x_0 + \tfrac{u}{2})\| + \|A(x_0 - \tfrac{u}{2})\| < 2,$$

whence the operator A is bounded and $\|A\| \le \frac{2}{r}$.

6.39 (3) If the operator A is unbounded, then there exist $x_n \in X$ such that $\|x_n\| = 1$ and $\|Ax_n\| \ge n$, $n \ge 1$. Then for $y_n = \frac{x_n}{\sqrt{n}}$ we have $y_n \to 0$, but the sequence $\{Ay_n : n \ge 1\}$ does not converge weakly, because it is unbounded.

6.40 *Sufficiency.* If A is not linear, then there exist $x, y \in X$ and $\alpha, \beta \in \mathbb{K}$ such that $A(\alpha x + \beta y) \neq \alpha Ax + \beta Ay$. By Corollary 4.3 of the Hahn–Banach theorem, there exists a functional $f \in X^*$ such that $f(A(\alpha x + \beta y)) \neq f(\alpha Ax + \beta Ay) = \alpha f(Ax) + \beta f(Ay)$, hence the functional f_0 is not linear, a contradiction. Let the operator A be linear. By the condition of the problem if $x_n \to x$, then $Ax_n \xrightarrow{w} Ax$, $n \to \infty$. It follows from item (3) of problem 6.39 that the operator A is bounded.

6.42 (1) $\alpha \ge 0$, $\|A\| = 1$. (2) $0 < \alpha \le 1$, $\|A\| = \frac{1}{\sqrt{\alpha}}$. Since

$$\|Ax\|^2 = \int_0^1 |x(t^\alpha)|^2 dt = \left| s = t^\alpha \right| = \int_0^1 \left| \frac{s^{\frac{1-\alpha}{2\alpha}}}{\sqrt{\alpha}} x(s) \right|^2 ds,$$

then the operator A and the operator of multiplication by the function $\frac{t^{\frac{1-\alpha}{2\alpha}}}{\sqrt{\alpha}}$ will be bounded or unbounded simultaneously and in the former case will have equal norms. (3) $0 < \alpha \le 2\beta + 1$, $\|A\| = \frac{1}{\sqrt{\alpha}}$.

6.44 (1) For each subset $S \subset \mathbb{N}$, consider the diagonal operator

$$A_S x = (\alpha_{1,S} x_1, \alpha_{2,S} x_2, \ldots), \ x \in l_p,$$

where $\alpha_{k,S} = 1$ for $k \in S$ and $\alpha_{k,S} = 0$ for $k \notin S$. Note that the set $\{A_S \mid S \subset \mathbb{N}\}$ is uncountable and according to problem 6.10° for any $S, T \subset \mathbb{N}$, $S \neq T$, we have $\|A_S - A_T\| = 1$. Thus the space $\mathcal{L}(l_p)$ is nonseparable.

(2) For each number $c \in (a, b]$ consider the operator of multiplication

$$A_c x = \chi_{[0,c]} \cdot x, \ \ x \in L_p([a, b]).$$

Since the set $\{A_c \mid c \in (a, b]\}$ is uncountable and according to problem 6.12 for any $a < c < d \leq b$ we have $\|A_d - A_c\| = \|\chi_{(c,d]}\|_\infty = 1$, the space $\mathcal{L}(L_p([a, b]))$ is nonseparable.

6.45 Fix $f \in X^*$ for which $\|f\| = 1$. The set of operators

$$\{A_y(x) = f(x)y, x \in X \mid y \in X\}$$

is a subspace of $\mathcal{L}(X)$, which is isometrically isomorphic to X. Since this subspace is nonseparable, the space $\mathcal{L}(X)$ is also nonseparable.

6.46° (1) No, if $\operatorname{Ker} A \neq \{0\}$. (2) Yes; yes.

6.48° Denote $c_A = \sup\limits_{x,y \in S(0,1)} |(Ax, y)|$. If $A \in \mathcal{L}(H)$, then for all $x, y \in H$ we have $|(Ax, y)| \leq \|A\| \cdot \|x\| \cdot \|y\|$, hence $c_A \leq \|A\| < +\infty$. On the other hand, if $c_A < +\infty$, then for any $x \in X$ and $y = Ax$ we have $\|Ax\|^2 = (Ax, y) \leq c_A\|x\| \cdot \|Ax\|$, hence $A \in \mathcal{L}(X)$ and $\|A\| \leq c_A$.

6.49° (1) Apply the Fubini's theorem and the Cauchy–Schwarz inequality to prove that $|(Ax, y)| \leq (c_1c_2)^{\frac{1}{2}} \|x\| \cdot \|y\|$, $x, y \in L_2(T, \mu)$, and use problem 6.48°.

6.50* (2) Use problem 1.85.

(3) Let G be a subspace of X, $u \notin G$ and $G' = \operatorname{span}(G \cup \{u\})$. Let us show that for $A \in \mathcal{L}(G, Y)$ there exists an extension $A' \in \mathcal{L}(G', Y)$ that preserves the norm. For every element $x \in G' \setminus G$ there exists a unique $\lambda \in \mathbb{R} \setminus \{0\}$ such that $x - \lambda u \in G$. Put $g = u - \frac{1}{\lambda}x \in G$, then $x = \lambda(u - g)$. If $A'u = y$, then $A'x = A'(\lambda(u - g)) = \lambda(y - Ag)$. The condition $\|A'\| = \|A\|$ is satisfied if and only if $\|\lambda(y - Ag)\| \leq \|A\| \cdot \|\lambda(u - g)\|$ for all $g \in G$ and $\lambda \in \mathbb{R} \setminus \{0\}$, i.e. $\|y - Ag\| \leq \|A\| \cdot \|u - g\|$ for all $g \in G$. Therefore A' preserves the norm if y belongs to the intersection of all closed balls $B_g = \overline{B}(Ag, \|A\| \cdot \|u - g\|)$, $g \in G$. Since Y is a space of type $\mathcal{M}$, this intersection is nonempty if $B_{g_1} \cap B_{g_2} \neq \varnothing$ for any $g_1, g_2 \in G$. For $r_1 = \|A\| \cdot \|u - g_1\|$, $r_2 = \|A\| \cdot \|u - g_2\|$ we have

$$r_1 + r_2 = \|A\| \cdot (\|u - g_1\| + \|u - g_2\|) \geq \|A\| \cdot \|g_1 - g_2\| \geq \|Ag_1 - Ag_2\|.$$

If $Ag_1 = Ag_2$, then the balls have a common center. Otherwise $r_1 + r_2 > 0$ and for $z = \frac{r_1}{r_1+r_2} \cdot Ag_2 + \frac{r_2}{r_1+r_2} \cdot Ag_1$ we have $\|z - Ag_1\| = \frac{r_1}{r_1+r_2} \cdot \|Ag_2 - Ag_1\| \leq r_1$ and $\|z - Ag_2\| = \frac{r_2}{r_1+r_2} \cdot \|Ag_1 - Ag_2\| \leq r_2$, whence $z \in B_{g_1} \cap B_{g_2}$. To show that there exists an extension of A to the entire space X that preserves the norm, consider the set of extensions A_P of A to the subspaces $G \subset P$ that preserve the norm, define a partial order by $A_P \preceq A_Q$, if A_Q is an extension of A_P, and apply Zorn's lemma.

6.51° *Method I.* Consider the functionals $f_y(x) = (Ax, y) = (x, By)$, $x, y \in H$. For all $x \in H$ and $y \in \overline{B}(0, 1)$ we have $|f_y(x)| \le \|Ax\| \cdot \|y\| \le \|Ax\|$, hence by the Banach–Steinhaus theorem there exists $C > 0$ such that $\|f_y\| = \|By\| \le C$ for all $y \in \overline{B}(0, 1)$. Therefore $B \in \mathcal{L}(H)$. Similarly $A \in \mathcal{L}(H)$.

Method II. Let us show that if $x_n \to x$ and $Ax_n \to z$, $n \to \infty$, then $z = Ax$. Indeed, for all $y \in H$ we have $(z, y) = \lim\limits_{n\to\infty} (Ax_n, y) = \lim\limits_{n\to\infty} (x_n, By) = (x, By) = (Ax, y)$, therefore $z = Ax$. Thus the graph of the operator A is closed, so $A \in \mathcal{L}(H)$ by Theorem 6.1 (Closed Graph Theorem). Similarly $B \in \mathcal{L}(H)$.

Problems of Chap. 7

7.5° Uniform convergence in $\mathcal{L}(X, \mathbb{K})$ is equivalent to convergence in norm in X^*, while strong and weak convergences are equivalent to weak-$*$ convergence in X^*.

7.6° Repeat the reasoning from the proof of Theorem 5.2.

7.7° In the formulation of problem 7.6°, replace $A_n x \to Ax$, $n \to \infty$, in X_2 with $A_n x \xrightarrow{w} Ax$, $n \to \infty$, in X_2.

7.8° Denote by A the limit of the sequence $\{A_n : n \ge 1\}$, if it exists. (1) For $p = 1$ does not converge; for $1 < p < +\infty$ weakly, $A = 0$. (2) For $1 \le p < +\infty$ strongly, $A = 0$; for $p = +\infty$ does not converge. (3), (4) Strongly, $A = 0$. (5) For $p = 1$ and $p = +\infty$ does not converge; for $1 < p < +\infty$ weakly, $A = 0$. (6) Strongly, $A = 0$. (7) Does not converge. (8) For $p = 1$ strongly, $Ax = \sum\limits_{n=1}^{\infty} x_n e_1$; for $1 < p \le +\infty$ does not converge. (9) Uniformly, $Ax = (x_1, \frac{x_2}{2}, \frac{x_3}{3}, \ldots)$. (10) Uniformly, $A = I$. (11) Strongly, $A = I$. (12) Uniformly, $A = 0$. (13) Uniformly, $Ax = (x_1, \frac{x_2}{2}, \frac{x_3}{3}, \ldots)$. (14) Uniformly, $Ax = (x_1, \frac{x_2}{2}, \ldots, \frac{x_k}{k}, 0, \ldots)$.

7.9* (3), (6) Weakly, $A = 0$. (11) Does not converge.

7.10 (1) Uniformly, strongly, weakly $\Leftrightarrow$ $\{\alpha_n : n \ge 1\}$ converges. (2) For $1 < p \le +\infty$ uniformly, strongly $\Leftrightarrow$ $\alpha_n \to 0$, weakly $\Leftrightarrow$ $\{\alpha_n : n \ge 1\}$ is bounded; for $p = 1$ uniformly, strongly, weakly $\Leftrightarrow$ $\alpha_n \to 0$. (3) For $1 \le p < +\infty$ strongly, weakly $\Leftrightarrow$ $\{\alpha_n : n \ge 1\}$ is bounded, uniformly $\Leftrightarrow$ $\alpha_n \to 0$; for $p = +\infty$ uniformly, strongly, weakly $\Leftrightarrow$ $\alpha_n \to 0$. (4) For $1 \le p < +\infty$ strongly, weakly $\Leftrightarrow$ $\{\alpha_n : n \ge 1\}$ converges, uniformly $\Leftrightarrow$ $\alpha_n \to 0$; for $p = +\infty$ uniformly, strongly, weakly $\Leftrightarrow$ $\alpha_n \to 0$. (5) For $1 < p \le +\infty$ uniformly, strongly $\Leftrightarrow$ $\alpha_n = 0$, $n \ge 1$, weakly $\Leftrightarrow$ $\{\alpha_n : n \ge 1\}$ is bounded; for $p = 1$ uniformly, strongly, weakly $\Leftrightarrow$ $\alpha_n = 0$, $n \ge 1$. (6) For $p = 1$ uniformly $\Leftrightarrow$ $\alpha_n = 0$, $n \ge 1$, strongly, weakly $\Leftrightarrow$ $\alpha_n \to 0$; for $1 < p < +\infty$ uniformly $\Leftrightarrow$ $\alpha_n = 0$, $n \ge 1$, strongly $\Leftrightarrow$ $\alpha_n \to 0$, weakly $\Leftrightarrow$ $\{\alpha_n : n \ge 1\}$ is bounded; for $p = +\infty$ uniformly, strongly $\Leftrightarrow$ $\alpha_n = 0$, $n \ge 1$, weakly $\Leftrightarrow \{\alpha_n : n \geq 1\}$ is bounded. (7) For $1 \leq p < +\infty$ uniformly $\Leftrightarrow \alpha_n = 0$, $n \geq 1$, strongly, weakly $\Leftrightarrow \{\alpha_n : n \geq 1\}$ is bounded; for $p = +\infty$ uniformly, strongly $\Leftrightarrow \alpha_n = 0$, $n \geq 1$, weakly $\Leftrightarrow \alpha_n \to 0$. (8) Uniformly, strongly, weakly $\Leftrightarrow \{\alpha_n : n \geq 1\}$ converges. (9) For $1 \leq p < +\infty$ uniformly $\Leftrightarrow \alpha_n \to 0$, strongly, weakly $\Leftrightarrow \{\alpha_n : n \geq 1\}$ converges; for $p = +\infty$ uniformly, strongly, weakly $\Leftrightarrow \alpha_n \to 0$. (10) For $1 \leq p < +\infty$ uniformly $\Leftrightarrow \alpha_n = 0$, $n \geq 1$, strongly, weakly $\Leftrightarrow \{\alpha_n : n \geq 1\}$ converges; for $p = +\infty$ uniformly, strongly $\Leftrightarrow \alpha_n = 0$, $n \geq 1$, weakly $\Leftrightarrow \alpha_n \to 0$.

7.11° Denote by A the limit of the sequence $\{A_n : n \geq 1\}$, if it exists. (1), (6), (8), (13), (16)–(18), (20) Uniformly, $A = 0$. (2)–(5), (7), (19) Does not converge. (9) Uniformly, $(Ax)(t) = \int\limits_0^t x(s) \sin s ds$. (10) Strongly, $A = I$. (11) Weakly, $A = 0$. (12) Strongly, $(Ax)(t) = x(0)t$. (14) Uniformly, $(Ax)(t) = \int\limits_0^{\frac{1}{2}} tx(s)ds$. (15) Weakly, $A = 0$. (21) Strongly, $A = I$.

7.12° (1) For $1 < p < +\infty$ weakly, $A = 0$; for $p = 1$ does not converge. (2), (3) Strongly, $A = 0$. (4) Uniformly, $A = 0$. (5)–(8) Weakly, $A = 0$. (9) Strongly, $A = 0$. (10) Strongly, $A = I$. (11) For $p = 1$ strongly, for $1 < p < +\infty$ uniformly, $A = 0$.

7.13° The limit is the operator of multiplication by some function $p \in C([a, b])$. (1), (2) $p_n \to p, n \to \infty$, in $C([a, b])$; (3) $p_n \xrightarrow{w} p, n \to \infty$, in $C([a, b])$ (i.e. the sequence $\{p_n : n \geq 1\}$ is bounded in $C([a, b])$ and converges pointwise to p).

7.14 (2) It is necessary and sufficient that the space X_2 is complete with respect to weak convergence.

7.17* (1) Use that $S_n x$ is a partial sum of the Fourier series of the function x.

(2) It follows from the estimates

$$\int\limits_{-\pi}^{\pi} |D_n(t)|dt = 2 \int\limits_0^{\pi} \frac{\left|\sin \frac{2n+1}{2} t\right|}{2 \sin \frac{t}{2}} dt \geq 2 \int\limits_0^{\pi} \frac{\left|\sin \frac{2n+1}{2} t\right|}{t} dt =$$

$$= 2 \int\limits_0^{\frac{(2n+1)\pi}{2}} \frac{|\sin u|}{u} du > 2 \sum_{k=1}^{n} \int\limits_{(k-1)\pi}^{k\pi} \frac{|\sin u|}{k\pi} du = \frac{4}{\pi} \sum_{k=1}^{n} \frac{1}{k} \sim \frac{4}{\pi} \ln n,$$

$n \to \infty$.

(3) Prove that $\|S_n\| = \frac{1}{\pi} \int\limits_{-\pi}^{\pi} |D_n(t)|dt$ in $\widetilde{C}$ and in $L_1([-\pi, \pi])$ (in the case of $L_1([-\pi, \pi])$ use problem 6.19), and apply the Banach–Steinhaus theorem 7.1.

(4) Fix $t_0 \in [-\pi, \pi]$. Consider the sequence of functionals $f_n(x) = (S_n x)(t_0)$ and use that $\|f_n\| = \|S_n\|$.

(5) For the sequence of functionals

$$f_n(y) = \frac{1}{\pi}\int_{-\pi}^{\pi}(S_n x)(t)y(t)dt = \frac{a_0\alpha_0}{2} + \sum_{k=1}^{n}(a_k\alpha_k + b_k\beta_k),$$

$y \in L_1([-\pi, \pi])$, $n \geq 1$, establish that $\|f_n\| = \frac{1}{\pi}\|S_n x\|_\infty$, $n \geq 1$. Then apply item (3) and the Banach–Steinhaus theorem 7.1.

7.18* Due to linearity, it is sufficient to prove the boundedness of the set of values $B(x, y)$ for $x, y \in \overline{B}(0, 1)$. Consider the operators $A_y \in \mathcal{L}(X_1, X_2)$, $y \in \overline{B}(0, 1)$, which are defined by the formula $A_y x = B(x, y)$, $x \in X_1$. For each $x \in X_1$ we have $B(x, \cdot) \in \mathcal{L}(X_1, X_2)$, therefore

$$\forall x \in X_1\ \exists C_x > 0\ \forall y \in \overline{B}(0, 1)\ :\ \|A_y x\| = \|B(x, y)\| \leq C_x.$$

By the Banach–Steinhaus theorem 7.1, there exists $C > 0$ such that $\|A_y\| \leq C$ for all $y \in \overline{B}(0, 1)$, therefore $\|B(x, y)\| \leq C$ for all $x, y \in \overline{B}(0, 1)$.

7.19 To prove the discontinuity of B on $X \times X$, consider a sequence $\{p_n : n \geq 1\} \subset X$, which converges in X to the function $x(t) = \frac{1}{\sqrt{t}}$, $t \in (0, 1]$. Apply Fatou's lemma to get that $\liminf\limits_{n\to\infty} B(p_n, p_n) = +\infty$.

7.20° (3) Consider $X = l_2$ and the operators A_n, B_n, defined by the formulae

$$A_n x = (x_n, x_{n+1}, \ldots),\ \ B_n x = (\underbrace{0, \ldots, 0}_{n-1}, x_1, x_2, \ldots),\ \ x \in l_2,\ n \geq 1.$$

Verify that $A_n \xrightarrow{w} 0$, $B_n \xrightarrow{w} 0$, $n \to \infty$, and $A_n B_n = I$, $n \geq 1$.

7.21° (3) For $X = l_2$, $A_n x = (x_n, x_{n+1}, \ldots)$, $x \in l_2$, and $x^{(n)} = e_n$ verify that $A_n \xrightarrow{s} 0$, $x^{(n)} \xrightarrow{w} 0$, $n \to \infty$, and $A_n x^{(n)} = e_1$, $n \geq 1$.

7.22° Establish that $\lim\limits_{n\to\infty}(A_n x, y)$ exists for all $x, y \in H$. Using the weak completeness of H, prove that for each $x \in H$ there exists an element $Ax \in H$ such that $A_n x \xrightarrow{w} Ax$. Then prove that $A \in \mathcal{L}(H)$.

7.23* Consider a bounded sequence of operators $\{A_n : n \geq 1\}$ and a set $L = \{x_n : n \geq 1\}$ which is dense in H. Using the diagonal method, construct a subsequence $\{A_{n_k} : k \geq 1\}$ such that for each x_n the sequence $\{A_{n_k} x_n : k \geq 1\}$ converges weakly, and use problem 7.22°.

7.24 Prove that the sequence $\{\varphi_n(A) : n \geq 0\}$ is a Cauchy sequence.

7.27 *Necessity.* If $A_n \xrightarrow{s} A$, then $R(A) \subset \overline{\bigcup_{n=1}^{\infty} R(A_n)}$, and the finite-dimensional spaces $R(A_n)$ are separable. *Sufficiency.* Let $\{e_n : n \geq 1\}$ be an orthonormal basis in H. For $A \in \mathcal{L}(H)$ consider $A_n x = \sum_{k=1}^{n} x_k A e_k$, $n \geq 1$, where $x = \sum_{k=1}^{\infty} x_k e_k \in H$.

7.28* (2) Firstly, prove the equality for functions from T_n, then for $\sin mt$, $\cos mt$, where $m > n$. Then take into account that for a fixed s the right-hand side and left-hand side of the equality are continuous linear functionals on $\widetilde{C}$, which coincide on $\bigcup_{n=1}^{\infty} T_n$.

(3) Use item (2) and estimates from the solutions of problem 7.17*.

7.29 (1) Linearity and nonnegativity of A imply that $(Ax)(t) \leq (Ay)(t)$, $t \in [a, b]$, for all $x, y \in C([a, b])$ such that $x(t) \leq y(t)$, $t \in [a, b]$. Since for any function $x \in \overline{B}(0, 1)$ we have

$$-1 = -e_0(t) \leq x(t) \leq e_0(t) = 1, \ t \in [a, b],$$

then $-(Ae_0)(t) \leq (Ax)(t) \leq (Ae_0)(t)$, $t \in [a, b]$. Therefore $\|Ax\| \leq \|Ae_0\|$, $x \in \overline{B}(0, 1)$. Equality is achieved at $x = e_0$.

(2) Put $y_s(t) = (t - s)^2$, $s, t \in [a, b]$. Similarly to item (1) of problem 5.39, for every $x \in C([a, b])$, $s \in [a, b]$ and $\varepsilon > 0$ we have

$$-\frac{2\|x\|}{\delta^2} y_s(t) - \varepsilon \leq x(t) - x(s) \leq \frac{2\|x\|}{\delta^2} y_s(t) + \varepsilon, \ t \in [a, b],$$

where the number $\delta = \delta(x, \varepsilon) > 0$ is such that $|x(t) - x(s)| < \varepsilon$ for all $t, s \in [a, b]$ for which $|t - s| < \delta$. Since the operators A_n are nonnegative, it follows that

$$\big|(A_n x)(s) - x(s) \cdot (A_n e_0)(s)\big| \leq \frac{2\|x\|}{\delta^2} (A_n y_s)(s) + \varepsilon (A_n e_0)(s), \ s \in [a, b].$$

Taking into account that

$$(A_n e_0)(s) \to 1, \ (A_n y_s)(s) = (A_n(e_2 - 2se_1 + s^2 e_0))(s) \to 0, \ n \to \infty,$$

uniformly for $s \in [a, b]$, we get that $\limsup_{n \to \infty} \|A_n x - x\| \leq \varepsilon$, where $\varepsilon > 0$ is arbitrary.

The answer to the question is positive.

7.30 (1) Use problem 7.29 or the fact that $\|B_n x - K_n x\| \to 0$, $n \to \infty$, $x \in C([0, 1])$, where $B_n x$ is the Bernstein polynomial.

(2) Use item (1) and problem 7.6°. In the space $L_\infty([0, 1])$ we have $K_n \overset{s}{\nrightarrow} I$, otherwise this space would be separable.

Problems of Chap. 8

8.4° It is known from linear algebra that statements (1), (2), (4) and (5) are equivalent. The equivalence of statements (2) and (3) follows from problem 6.8°.

8.5° Note that $R(A)$ is dense in l_2, because it contains all elements of the form $x = (x_1, \dots, x_n, 0, \dots)$, but $(1, \frac{1}{2}, \frac{1}{3}, \dots) \notin R(A)$.

8.6° $(I + A)^{-1} = I + \sum\limits_{n=1}^{\infty} (-1)^n A^n$.

8.7° For $C = B^{-1}A^{-1}$ verify that $(AB)C = C(AB) = I$ and apply Theorem 8.3.

8.8° (1) $(A + B)^{-1} = A^{-1} + \sum\limits_{n=1}^{\infty} (A^{-1}B)^n A^{-1}$. Use the equality $A + B = A(I + A^{-1}B)$, Theorem 8.4 and problem 8.7°.

8.9 The necessity follows from problem 8.7°.

Sufficiency. Apply Theorem 8.2. If $R(A^2) = X$, then $R(A) = X$. Let $m > 0$ be such that $\|A^2x\| \geq m\|x\|$, $x \in X$. Then $\|Ax\| \geq \frac{\|A^2x\|}{\|A\|} \geq \frac{m}{\|A\|}\|x\|$, $x \in X$.

8.10° $(A_i^{-1}y)(t) = y'(t)$, $t \in [0, 1]$, $y \in X_i$, $i = 0, 1$.

8.11 Apply the method of solving linear differential equations of the first order.

(1) $(A^{-1}y)(t) = \frac{1}{2}e^{\frac{3}{2}t} \int\limits_0^t e^{-\frac{3}{2}s} y(s)ds$; (2) $(A^{-1}y)(t) = e^{\frac{t^2}{2}} \int\limits_0^t e^{-\frac{s^2}{2}} y(s)ds$;

(3) $(A^{-1}y)(t) = e^{-e^t} \int\limits_0^t e^{e^s} y(s)ds$; (4) $(A^{-1}y)(t) = e^{-\frac{2}{3}t^3} \int\limits_0^t e^{\frac{2}{3}s^3} y(s)ds$;

(5) $(A^{-1}y)(t) = e^{-2te^{-t}-2e^{-t}} \cdot \int\limits_0^t e^{2se^{-s}+2e^{-s}-s} y(s)ds$;

(6) $(A^{-1}y)(t) = e^{-t}(t + 1) \int\limits_0^t \frac{e^s}{(s+1)^2} y(s)ds$;

(7) $(A^{-1}y)(t) = \int\limits_0^t \exp\left(-\int\limits_s^t \alpha(u)du\right) y(s)ds$, $t \in [0, 1]$, $y \in C([0, 1])$.

8.12° Let us find the algebraic inverse of the operator A. We will verify that for every $y \in C([0, 1])$ the equation

$$(Ax)(t) = x(t) - \int_0^1 f(t)g(s)x(s)ds = y(t), \; t \in [0, 1],$$

has a unique solution. If $\int\limits_0^1 g(s)x(s)ds = c$, then $x(t) = cf(t) + y(t)$, $t \in [0, 1]$. Hence $\int\limits_0^1 g(s)(cf(s) + y(s))ds = c$, therefore $c\int\limits_0^1 g(s)f(s)ds + \int\limits_0^1 g(s)y(s)ds = c$ and $c = \left(1 - \int\limits_0^1 g(s)f(s)ds\right)^{-1} \int\limits_0^1 g(s)y(s)ds$. Thus

$$(A^{-1}y)(t) = x(t) = \left(1 - \int_0^1 g(s)f(s)ds\right)^{-1} \int_0^1 f(t)g(s)y(s)ds + y(t).$$

The operator A^{-1} is linear and continuous on $C([0, 1])$ due to problem 6.15.

8.13 Let us verify that the equation $(Ax)(t) = x(t) - \int\limits_0^t f(t)g(s)x(s)ds = y(t)$, $t \in [0, 1]$, has a unique solution for any $y \in C([0, 1])$. Put $z(t) = \int\limits_0^t g(s)x(s)ds$, then $x(t) = f(t)z(t) + y(t)$. Note that $z \in C^1([0, 1])$,

$$z'(t) = g(t)x(t) = g(t)f(t)z(t) + g(t)y(t) \text{ and } z(0) = 0.$$

Solving the differential equation, we get

$$z(t) = \int_0^t \exp\left(\int_s^t g(u)f(u)du\right) g(s)y(s)ds$$

and $(A^{-1}y)(t) = x(t) = \int\limits_0^t \exp\left(\int\limits_s^t g(u)f(u)du\right) f(t)g(s)y(s)ds + y(t)$. The operator A^{-1} is linear and continuous on $C([0, 1])$ due to problem 6.18.

8.14° $\left((AB)^{-1}y\right)(t) = \frac{y(\sqrt{t})}{\sqrt{t}+1}$, $\left((BA)^{-1}y\right)(t) = \frac{y(\sqrt{t})}{t+1}$, $t \in [0, 1]$, $y \in C([0, 1])$.

8.15° Use that $B = (BA)A^{-1}$.

8.16 (1) Put $C = I + B(I - AB)^{-1}A$ and show that $(I - BA)C = C(I - BA) = I$.

(2) In the space $X = l_2$ consider $Ax = (x_2, x_3, \ldots)$, $Bx = (0, x_1, x_2, \ldots)$, $x \in l_2$. Verify that $AB = I$ and $\operatorname{Ker} BA \neq \{0\}$.

8.17 (1) Yes. Show that $\operatorname{Ker} A = \operatorname{Ker} B = \{0\}$.

(2) Yes. Since $\operatorname{Ker} BA \subset \operatorname{Ker} A$ and $R(AB) \subset R(A)$, then $\operatorname{Ker} A = \{0\}$ and $R(A) = X$. Therefore there exists an algebraic inverse operator $A^{-1} : X \to X$. Show that $A^{-1} = (BA)^{-1} \cdot B \in \mathcal{L}(X)$.

8.18° Use Theorem 8.2.

8.20 (1) Yes for $i = 0, 1$. (2) No for $i = 0$, and yes for $i = 1$. (3) No for $i = 0, 1$.

8.21° $(B^{-1}y)(t) = y(\sqrt[3]{t}), t \in [-1, 1], y \in C([-1, 1])$.

8.22

$$(A^{-1}y)(t) = -\frac{1}{4}\int_0^t e^{2s-2t}y(s)ds + \frac{1}{4}\int_0^t e^{2t-2s}y(s)ds +$$

$$+\frac{1}{4(e^2 - e^{-2})}\left(\int_0^1 e^{2s-2}y(s)ds - \int_0^1 e^{2-2s}y(s)ds\right)(e^{2t} - e^{-2t}),\ t \in [0, 1].$$

8.23 $(A^{-1}y)(t) = \int_0^t \left(\int_0^{t_2}\left(\int_0^{t_1} e^{t_1-s}y(s)ds\right)dt_1\right)dt_2, t \in [0, 1]$.

8.24° (1) Yes, $A^{-1} = A$. (2) Yes, $A^{-1}y = (y_1 - y_2 - y_3, y_2 + y_3, y_3, y_4, \ldots)$, $y \in l_2$. (3)–(5) No. (6) Yes, $A^{-1}y = (y_1 + y_2, y_2, y_3, \ldots)$, $y \in l_2$.(7) Yes, $A^{-1}y = \left(\frac{1}{2}(y_2 + y_3), \frac{1}{2}(y_1 - y_2), \frac{1}{2}(y_1 - y_3), y_4, y_5, \ldots\right)$, $y \in l_2$.

8.25° Consider any linearly independent functions $x_1, x_2, x_3 \in C([0, 1])$. Since $\dim R(A) = 2$, the functions Ax_1, Ax_2 and Ax_3 are linearly dependent. Therefore $\operatorname{Ker} A \neq \{0\}$.

8.26° No, because $\operatorname{Ker} A \neq \{0\}$. In particular, $\operatorname{Ker} A$ contains all odd functions from $C([0, 1])$.

8.27 Denote by X_i the space X endowed with the norm $\|\cdot\|_i$, $i = 1, 2$. Consider the identity operator $A : X_1 \to X_2$. Since A is bijective and $A \in \mathcal{L}(X_1, X_2)$ by the condition of the problem, then $A^{-1} \in \mathcal{L}(X_2, X_1)$ according to Theorem 8.1. Therefore $\|x\|_1 \le \|A^{-1}\|\cdot\|x\|_2$, $x \in X$. The completeness is essential. For example, consider $X = C([0, 1])$, $\|x\|_1 = \max\limits_{0\le t\le 1}|x(t)|$ and $\|x\|_2 = \int_0^1 |x(t)|dt$. Then the norm $\|\cdot\|_2$ can be estimated by the norm $\|\cdot\|_1$, but these norms are not equivalent.

8.28 It is necessary and sufficient that (1) $\alpha(t) \ge 0$, $t \in [a, b]$, and $\alpha > 0$ on a dense subset of $[a, b]$; (2) $\alpha(t) > 0$, $t \in [a, b]$. To prove the necessity, assume that $\alpha(t_0) = 0$ for some $t_0 \in [a, b]$ and consider the sequence $x_n(t) = e^{-n|t-t_0|}$, $t \in [a, b]$, $n \ge 1$. (3) $\alpha(t) > 0$, $t \in [a, b]$. Obviously, the norm $\|\cdot\|_\alpha$ can be estimated by the uniform norm $\|\cdot\|$. Therefore, as a consequence of problems 1.33° and 8.27, the space $(C([a, b]), \|\cdot\|_\alpha)$ is complete if and only if the norms $\|\cdot\|_\alpha$ and $\|\cdot\|$ are equivalent.

8.29 (1) Since $\mu(T) < +\infty$, then $L_\infty(T, \mu) \subset L_2(T, \mu)$ and $\|x\|_2 \le \sqrt{\mu(T)}\|x\|_\infty$, $x \in L_\infty(T, \mu)$. Therefore the set M which is closed in $L_2(T, \mu)$

is also closed in $L_\infty(T, \mu)$. Denote by M_2 and M_∞ the space M endowed with norms $\|\cdot\|_2$ and $\|\cdot\|_\infty$, respectively. The identity operator $A : M_\infty \to M_2$ is linear and continuous. According to Theorem 8.1 there exists $A^{-1} \in \mathcal{L}(M_2, X_\infty)$. Hence for $C = \|A^{-1}\|$ we have $\|x\|_\infty \le C \cdot \|x\|_2$, $x \in M$.

(2) Assume that $\dim M = +\infty$. Let $\{\varphi_n : n \ge 1\} \subset M$ be an orthonormal system. Define

$$f_N(y, t) = \sum_{n=1}^{N} y_n \varphi_n(t), \;\; y = (y_1, y_2, \dots) \in l_2, \;\; N \ge 1.$$

Due to item (1), we have $\|f_N(y, \cdot)\|_\infty \le C\|f_N(y, \cdot)\|_2 \le C\|y\|_2$, therefore $|f_N(y, t)| \le C\|y\|_2 \pmod{\mu}$. It follows that $\sum\limits_{n=1}^{N} \varphi_n^2(t) \le C \pmod{\mu}$. But then $N = \sum\limits_{n=1}^{N} \int\limits_T \varphi_n^2(t) d\mu(t) \le C\mu(T)$ for all $N \ge 1$, a contradiction.

8.30* (1) Since the expansion of an element into a series with respect to the Schauder basis is unique, it follows that the functionals α_i and operators P_k are linear. Since $\|P_k x - x\| \to 0$, $k \to \infty$, then $\|P_k x\| \to \|x\|$, whence $\|x\| \le |||x||| < +\infty$, $x \in X$. It is easy to check that $|||\cdot|||$ is a norm on X. Let us show that the space $(X, |||\cdot|||)$ is complete.

Let a sequence $\left\{x_n = \sum\limits_{i=1}^{\infty} \alpha_i(x_n) e_i : n \ge 1\right\}$ be a Cauchy sequence with respect to the norm $|||\cdot|||$. Then this sequence is a Cauchy sequence with respect to the norm $\|\cdot\|$, so it converges to some $x \in X$. For each fixed $k \ge 1$ the sequence $\left\{P_k x_n = \sum\limits_{i=1}^{k} \alpha_i(x_n) e_i : n \ge 1\right\}$ is also a Cauchy sequence with respect to the norm $\|\cdot\|$, so it converges to some y_k. Since $P_k x_n \in \operatorname{span}(\{e_1, \dots, e_k\})$, then the limits $\lim\limits_{n\to\infty} \alpha_i(x_n) = \beta_i$ exist for $1 \le i \le k$, and $y_k = \sum\limits_{i=1}^{k} \beta_i e_i$. Note that the numbers β_i do not depend on the choice of $k \ge i$. Now we will verify that $x = \sum\limits_{i=1}^{\infty} \beta_i e_i$, where the series converges in the norm $\|\cdot\|$. For any $\varepsilon > 0$ choose N such that $|||x_n - x_N||| < \frac{\varepsilon}{3}$ for $n \ge N$, and K such that $\|P_k x_N - x_N\| < \frac{\varepsilon}{3}$, $k \ge K$. Then for $k \ge K$ we have

$$\|y_k - x\| = \lim_{n\to\infty} \|P_k x_n - x_n\| \le$$
$$\le \lim_{n\to\infty} (\|P_k x_n - P_k x_N\| + \|P_k x_N - x_N\| + \|x_N - x_n\|) < \varepsilon.$$

Hence $x = \lim\limits_{k\to\infty} y_k = \sum\limits_{i=1}^{\infty} \beta_i e_i$. Due to the uniqueness of the expansion into a series with respect to the Schauder basis we have $\beta_i = \alpha_i(x)$, $i \ge 1$, and $y_k = P_k x$, $k \ge 1$.

Finally,

$$|||x_n - x||| = \sup_{k\geq 1} \|P_k x_n - P_k x\| \leq$$

$$\leq \limsup_{m\to\infty} \sup_{k\geq 1} \|P_k x_n - P_k x_m\| = \limsup_{m\to\infty} |||x_n - x_m||| \to 0, \ n \to \infty,$$

therefore the space $(X, |||\cdot|||)$ is complete.

(2) Use item (1) and problem 8.27.

(3) Due to item (2) there exists $C > 0$ such that $|||x||| = \sup_{k\geq 1} \|P_k x\| \leq C\|x\|$, $x \in X$. Hence $\|P_k\| \leq C, k \geq 1$.

(4) Use item (3) and note that $\sum_{i=1}^{k} \lambda_i e_i = P_k \left(\sum_{i=1}^{m} \lambda_i e_i \right)$ for all $m > k$.

8.31 (1) The sufficiency follows from the inequality

$$\|x_n - x\| \leq \sum_{i=1}^{k} |\alpha_i(x_n) - \alpha_i(x)| \cdot \|e_i\| +$$

$$+ \sup_{n\geq 1} \left\| \sum_{i=k+1}^{\infty} \alpha_i(x_n) e_i \right\| + < \left\| \sum_{i=k+1}^{\infty} \alpha_i(x) e_i \right\|, \ n \geq 1, \ k \geq 1.$$

Necessity. Condition (a) follows from item (3) of problem 8.30*. Condition (b) can be rewritten in notations from problem 8.30* as follows: $\sup_{n\geq 1} \|x_n - P_k x_n\| \to 0$, $k \to \infty$. We have

$$\|x_n - P_k x_n\| \leq \|x_n - x\| + \|x - P_k x\| + \|P_k x - P_k x_n\| \leq$$

$$\leq (1 + C)\|x_n - x\| + \|x - P_k x\|,$$

where $C > 0$ is such that $\|P_k\| \leq C, k \geq 1$.

For $\varepsilon > 0$ choose N such that $\|x_n - x\| < \frac{\varepsilon}{2(C+1)}$ for $n \geq N$ and choose K such that $\|x - P_k x\| < \frac{\varepsilon}{2}$ and $\|x_n - P_k x_n\| < \varepsilon$, $1 \leq n < N$, for $k \geq K$. Then $\sup_{n\geq 1} \|x_n - P_k x_n\| < \varepsilon$ for all $k \geq K$.

8.32 For $x = \sum_{i=1}^{\infty} \alpha_i(x) e_i$ put $Tx = \sum_{i=1}^{\infty} \alpha_i(x) v_i$. Verify that $|\alpha_i(x)| \leq 2C\|x\|$, $x \in X$, $i \geq 1$, the series $\sum_{i=1}^{\infty} \alpha_i(x) v_i$ converges in norm and $\|I - T\| < 1$. Therefore the inverse operator $T^{-1} \in \mathcal{L}(X)$ exists and for every $y \in X$ we have $y = \sum_{i=1}^{\infty} \alpha_i(T^{-1}y) v_i$.

Problems of Chap. 9

9.6° (5) It follows from the equalities

$$(A^{-1})^* A^* = (AA^{-1})^* = I, \ A^*(A^{-1})^* = (A^{-1}A)^* = I.$$

9.8 (1) $\Rightarrow$ (2) by Theorem 9.1; (2) $\Rightarrow$ (3) Put $C = A^*$.

9.10° (3) $\Rightarrow$ (2) It follows from the polarization identity (see problem 9.7°) that $4b(y, x) = b[x + y] - b[x - y] + ib[x - iy] - ib[x + iy], \ x, y \in H$. Thus if $b[x + y], b[x - y], b[x - iy], b[x + iy] \in \mathbb{R}$, then $b(y, x) = \overline{b(x, y)}$.

9.11° (1) Use item (1) of problem 9.7°. (2) $H = \mathbb{R}^2$, $Ax = \left(\begin{smallmatrix} 0 & 1 \\ -1 & 0 \end{smallmatrix}\right) x$.

9.12° (2) $H = \mathbb{C}$, $Az = iz$, $z \in \mathbb{C}$. (9) $m = -\|A\|$, $M = \|A\|$. (10) $H = \mathbb{R}^2$, $Ax = \left(\begin{smallmatrix} 1 & 0 \\ 0 & 0 \end{smallmatrix}\right) x$, $Bx = \left(\begin{smallmatrix} 0 & 0 \\ 0 & 1 \end{smallmatrix}\right) x$. (14) If $y = Ax \in R(A)$, then $(I - A)y = Ax - A^2x = 0$, so $y \in \operatorname{Ker}(I - A)$; if $y \in \operatorname{Ker}(I - A)$, then $y = Ay \in R(A)$.

9.13° (1) Apply the polarization identity for scalar product (see problem 2.13°).
(2) Let $\dim H = n$, $\{e_1, \ldots, e_n\}$ be an orthonormal basis in H and A be an isometric operator. Show that $\{Ae_1, \ldots, Ae_n\}$ is an orthonormal system, therefore it is a basis in H. (5) Consider $H = l_2$, $Ax = (0, x_1, x_2, \ldots)$, $x \in l_2$.

9.14° (4) $m = 2$, $\mathcal{A} = \left(\begin{smallmatrix} 0 & 1 \\ -1 & 0 \end{smallmatrix}\right)$.

9.15° (1) $A^*y = (\overline{\alpha_1}y_1, \overline{\alpha_2}y_2, \ldots)$; (2) $A^*y = (0, y_1, y_2, \ldots)$; (3), (6), (9), (10) $A^* = A$; (4) $A^*y = (y_j, 0, 0, \ldots)$; (5) $A^*y = (y_1 + \ldots + y_j, 0, 0, \ldots)$; (7) $A^*y = (\underbrace{0, \ldots, 0}_{j-1}, \overline{\alpha_j}y_1, \overline{\alpha_{j+1}}y_2, \overline{\alpha_{j+2}}y_3, \ldots)$; (8) $A^*y = (\overline{\alpha_1}y_3, \overline{\alpha_2}y_4, \overline{\alpha_3}y_5, \ldots)$; (11) $A^*y = (3y_1, -2y_1 + y_2, y_3, y_4, \ldots)$.

9.16° (1), (2), (4), (5) $A^* = A$; (3) $(A^*y)(t) = \overline{a(t - \tau)}y(t - \tau)$; (6) $(A^*y)(t) = \int_{\sqrt{t}}^{1} s^3 y(s)ds$; (7) $(A^*y)(t) = \int_0^1 t^3 s^2 y(s)ds$; (8) $(A^*y)(t) = \int_t^{\sqrt[5]{t}} tsy(s)ds$; (9) $(A^*y)(t) = \chi_{[0,1]}(t) \cdot \int_{-\sqrt{t}}^{\sqrt{t}} (s - 3t^2)y(s)ds$; (10) $(A^*y)(t) = \frac{1}{\alpha}t^{\frac{1-\alpha}{\alpha}} y(t^{\frac{1}{\alpha}})$; (11) $A^*u = \overline{(z, u)}y$.

9.17° (1) $(A^*y)(t) = \frac{y((t+1)^{1/7})}{\sqrt{7}(t+1)^{3/7}}$; (2) $(A^*y)(t) = \frac{y(t^{1/4})}{2t^{3/8}}$.

9.18° (1) $\alpha_n \in \mathbb{R}$, $n \geq 1$; (2) $\alpha_n \in \mathbb{C}$ are arbitrary, $n \geq 1$; (3), (4) $|\alpha_n| = 1$, $n \geq 1$; (5) $\alpha_n \in \{0, 1\}$, $n \geq 1$; (6) $\alpha_n \neq 0$, $n \geq 1$.

9.20° (1) $(A_s^{-1}x)(t) = (A_s^*x)(t) = x(t - s)$, $t \in \mathbb{R}$, $x \in L_2(\mathbb{R})$.

9.21° If and only if the subspace coincides with the entire space.

9.22° (1) $Px = \sum_{n=1}^{k} (x, e_n)e_n$, $x \in H$.

9.23 (1) $H = \mathbb{C}$, $Az = iz$, $z \in \mathbb{C}$; (2) $H = \mathbb{R}^2$, $Ax = \begin{pmatrix} 1 & 1 \\ 0 & 0 \end{pmatrix} x$; (3)–(5) $H = \mathbb{R}^2$, $Ax = \begin{pmatrix} 1 & 0 \\ 0 & -1 \end{pmatrix} x$; (6), (7) $H = l_2$, $Ax = (x_1, \frac{x_2}{2}, \frac{x_3}{3}, \ldots)$, $x \in l_2$. In item (6) verify that the set $R(A)$ contains all the sequences with finitely many nonzero terms, but $(1, \frac{1}{2}, \frac{1}{3}, \ldots) \notin R(A)$. In item (7) show that the set $\{(Ax, x) \mid x \in H, \ \|x\| = 1\}$ contains numbers $\frac{1}{n}$, $n \geq 1$, but does not contain 0. (8) $H = \mathbb{R}^2$, $Ax = \begin{pmatrix} 1 & 0 \\ 0 & 0 \end{pmatrix} x$, $Bx = \begin{pmatrix} 1 & 1 \\ 1 & 1 \end{pmatrix} x$; (9) $H = \mathbb{R}^2$, $Ax = \begin{pmatrix} 1 & 1 \\ 0 & 0 \end{pmatrix} x$; (10) $H = \mathbb{R}^2$, $Ax = \begin{pmatrix} 0 & 1 \\ 0 & 0 \end{pmatrix} x$; (11) $H = \mathbb{C}^2$, $Ax = \begin{pmatrix} 1 & 0 \\ 0 & -1 \end{pmatrix} x$, $Bx = \begin{pmatrix} 0 & i \\ -i & 0 \end{pmatrix} x$;

9.24 (1) Verify that the function $(x, y)_\varepsilon = \big((A + \varepsilon I)x, y\big)$, $x, y \in H$, is a scalar product on H for any $\varepsilon > 0$. Write the Cauchy–Schwarz inequality for this scalar product and pass to the limit in the inequality as $\varepsilon \to 0+$. (2) Use inequality from item (1) with $y = Ax$.

9.25 (1) No. Let $H = \mathbb{R}^2$, $Ax = \begin{pmatrix} 0 & 1 \\ 1 & 0 \end{pmatrix} x$, $e_1 = (1, 0)$, $e_2 = (0, 1)$. Then $e_1, e_2 \in G$, but $e_1 + e_2 \notin G$. (2) Yes. In view of item (2) of problem 9.24, if $x \in G$, then $Ax = 0$. It follows that $G = \operatorname{Ker} A$.

9.26 For every $n \geq 1$ we have

$$\|A^n x\|^2 = (A^n x, A^n x) = (A^{n-1}x, A^{n+1}x) \leq \|A^{n-1}x\| \cdot \|A^{n+1}x\|.$$

9.27 (1) It follows from Theorem 9.1.

(2) Set $c_A = \sup_{x \in S(0,1)} |(Ax, x)|$. Obviously, $c_A \leq \|A\|$. It is easy to check that for a self-adjoint operator A we have

$$\big(A(u + v), u + v\big) - \big(A(u - v), u - v\big) = 4\operatorname{Re}(Au, v), \ \ u, v \in H.$$

Now the parallelogram law implies that for $u, v \in S(0, 1)$ we have

$$4\operatorname{Re}(Au, v) \leq c_A(\|u + v\|^2 + \|u - v\|^2) = 2c_A(\|u\|^2 + \|v\|^2) = 4c_A.$$

For arbitrary $x, y \in S(0, 1)$ choose $\alpha \in \mathbb{C}$ such that $|\alpha| = 1$ and $\operatorname{Re}(Ax, \alpha y) = |(Ax, y)|$. Then $|(Ax, y)| = \operatorname{Re}(Ax, \alpha y) \leq c_A$, hence $\|A\| \leq c_A$.

9.29 Use problem 6.51°.

9.31° Verify that the operator $B = A^*A - AA^*$ is self-adjoint, and use item (3) of problem 9.27.

9.32 (1) Use problems 9.30° and 9.31°.

(2) Since

$$\|A^2x\|^2 = (A^2x, A^2x) = (A^*A^*AAx, x) = (A^*AA^*Ax, x) = \\ = (A^*Ax, A^*Ax) = \|A^*Ax\|^2, \ x \in H,$$

then $\|A^2\| = \|A^*A\|$, further use item (2) of problem 9.1.

(4) Use item (5) of problem 9.6°.

9.33* (1) $\Leftrightarrow$ (2) Follows from item (2) of problem 9.5.

(3) $\Rightarrow$ (4) Follows from item (1) of problem 9.32.

(4) $\Rightarrow$ (5) Since $Px - P^2x = 0$, $x \in H$, then $(I - P)x \in \operatorname{Ker} P$, whence $(I - P)x \perp Px$. Therefore

$$(Px, x) = (Px, Px + (I - P)x) = (Px, Px) = \|Px\|^2.$$

(6) $\Rightarrow$ (1) Let $G = \operatorname{Ker}(I - P)$. Let us show that $Px = \operatorname{pr}_G x$, $x \in H$. Since $(I - P)Px = 0$, then $Px \in G$. It remains to verify that $z = (x - Px) \perp G$. Fix $y \in G$. Since $Py = y$ and $Pz = Px - P^2x = 0$, then $\big(P(y + \alpha z), y + \alpha z\big) = (y, y) + \overline{\alpha}(y, z) \geq 0$ for all $\alpha \in \mathbb{K}$, which is possible only if $(y, z) = 0$.

9.34* (1) If $P_1 + P_2$ is an orthogonal projector, then

$$\|x\|^2 \geq \big((P_1 + P_2)x, x\big) = (P_1x, x) + (P_2x, x) = \|P_1x\|^2 + \|P_2x\|^2, \ x \in H.$$

For $x = P_2y$ we get $\|P_2y\|^2 \geq \|P_1P_2y\|^2 + \|P_2y\|^2$, so $P_1P_2y = 0$, $y \in H$. On the other hand, if $P_1P_2 = 0$, then $(P_1P_2)^* = P_2^*P_1^* = P_2P_1 = 0$. Thus

$$(P_1 + P_2)^2 = P_1^2 + P_2^2 = P_1 + P_2$$

and $(P_1 + P_2)^* = P_1 + P_2$, hence $P_1 + P_2$ is an orthogonal projector according to item (2) of problem 9.5.

(2) If P_1P_2 is an orthogonal projector, then $P_1P_2 = (P_1P_2)^* = P_2^*P_1^* = P_2P_1$. On the other hand, if $P_1P_2 = P_2P_1$, then $(P_1P_2)^2 = P_1P_2P_1P_2 = P_1^2P_2^2 = P_1P_2$ and $(P_1P_2)^* = P_2P_1 = P_1P_2$, hence P_1P_2 is an orthogonal projector according to item (2) of problem 9.5.

(5) $P_1AP_1 = AP_1$; (6) $AP_1 = P_1A$.

9.35° (3) Let $H = l_2$, $A_nx = (x_n, x_{n+1}, \ldots)$, $x \in l_2$. Verify that $A_n \xrightarrow{s} 0$ and $A_n^* \overset{s}{\nrightarrow} 0$, as $n \to \infty$.

9.36 (1) For all $x, y \in H$ we have

$$(Ax, y) = \lim_{n\to\infty} (A_nx, y) = \lim_{n\to\infty} (x, A_ny) = (x, Ay).$$

(3) Use item (4) of problem 5.17°.

(4) Since $A_n \xrightarrow{s} A$, then $A_n \xrightarrow{w} A$, whence $A_n^* \xrightarrow{w} A^*$ due to item (2) of problem 9.35°. Also we have $\|A_n x\| \to \|Ax\|$, thus $\|A_n^* x\| \to \|A^* x\|$, $x \in H$, due to problem 9.31°. Therefore item (3) implies that $A_n^* \xrightarrow{s} A^*$.

(5) Consider $H = l_2$, $A_n x = (\underbrace{0, \ldots, 0}_{n}, x_1, x_2, \ldots)$, $x \in l_2$, $n \geq 1$.

(6) Consider $H = l_2(\mathbb{Z}) =$

$$= \left\{ x = (\ldots, x_{-1}, x_0, x_1, \ldots) \,\middle|\, x_k \in \mathbb{R},\ k \in \mathbb{Z},\ \sum_{k \in \mathbb{Z}} x_k^2 < +\infty \right\},$$

and operators, defined by the formulae $(A_n x)_k = x_{k-n}$, $k \in \mathbb{Z}$, $n \geq 1$.

9.37 Use item (2) of problem 9.24.

9.38* Let $\|A_n\| \leq C$, $n \geq 1$. Since $A_n - A_m \geq 0$ for $n > m$, then due to item (2) of problem 9.24 we have

$$\|(A_n - A_m)x\|^2 \leq \|A_n - A_m\| \big((A_n - A_m)x, x\big) \leq$$
$$\leq 2C\big((A_n - A_m)x, x\big),\ x \in H,\ n > m \geq 1.$$

Verify that for each $x \in H$ the sequence $\{(A_n x, x) : n \geq 1\}$ is monotonic and bounded, so it converges. Then the sequence $\{A_n x : n \geq 1\}$ is a Cauchy sequence, so it also converges. Put $Ax = \lim_{n \to \infty} A_n x$, $x \in H$, and verify that $A \in \mathcal{L}(H)$.

9.39 (1) Use problem 9.38* or the criterion of strong convergence of operators (problem 7.6°). Verify that $P_n x \to Px$, $n \to \infty$, for all $x \in M = \left(\bigcup_{n=1}^{\infty} H_n\right) \cup \left(\bigcup_{n=1}^{\infty} H_n\right)^{\perp}$.

(2) Apply item (1) to the operators $I - P_n$, which are the orthogonal projectors onto $H_n^{\perp}$, $n \geq 1$, and prove that $\overline{\operatorname{span}\left(\bigcup_{n=1}^{\infty} H_n^{\perp}\right)} = \left(\bigcap_{n=1}^{\infty} H_n\right)^{\perp}$.

(3) Apply item (1) to the orthogonal projectors $P'_n = \sum_{k=1}^{n} P_k$.

(5) The monotonicity is essential. Consider the sequence $P_n x = (x, y_n) y_n$, $x \in H$, $n \geq 1$, where $\{y, y_n : n \geq 1\} \subset H$, $\|y_n\| = 1$, $y_n \neq y$, $n \geq 1$, $y_n \to y$, $n \to \infty$.

(6) Consider $H = l_2$, $P_n x = (x, y_n) y_n$, where $y_n = \left(\frac{1}{2}, 0, \ldots, 0, \underbrace{\frac{\sqrt{3}}{2}}_{n}, 0, \ldots\right)$, $n \geq 1$.

(7) Use item (3) of problem 9.36.

9.40* Consider the operator $Bx = (x, g) f$, $x \in H$, where $f, g \in H$ are arbitrary fixed elements.

9.41 In view of Theorem 8.2, it is sufficient to verify that the condition

$$\|(A - \lambda I)x\| \geq c\|x\|, \ x \in H,$$

where $c > 0$, implies that $R((A - \lambda I)) = H$. According to problem 6.35° we have $\operatorname{Ker}(A - \lambda I) = \{0\}$ and $R(A - \lambda I)$ is a subspace of H. If $R(A - \lambda I) \neq H$, then there exists $y \neq 0$, for which $y \perp R(A - \lambda I)$. Thus

$$((A - \lambda I)x, y) = (Ax, y) - \lambda(x, y) = (x, Ay - \overline{\lambda}y) = 0 \text{ for all } x \in X,$$

whence $Ay = \overline{\lambda}y$. Then $\overline{\lambda}(y, y) = (Ay, y) = (y, Ay) = \lambda(y, y)$, so $\lambda \in \mathbb{R}$ and $(A - \lambda I)y = Ay - \lambda y = 0$. But $\operatorname{Ker}(A - \lambda I) = \{0\}$, a contradiction.

9.42 (1) *Necessity.* Assume that there exists $A^{-1} \in \mathcal{L}(H)$, but

$$\forall n \geq 1 \; \exists u_n \in H \; : \; 0 \leq (Au_n, u_n) < \frac{1}{n}\|u_n\|^2.$$

Then for $x_n = \frac{u_n}{\|u_n\|}$ by item (2) of problem 9.24 we have $\|Ax_n\|^2 \leq \frac{\|A\|}{n}$, $n \geq 1$. Hence $y_n = Ax_n \to 0$, but $A^{-1}y_n = x_n \not\to 0$, $n \to \infty$, a contradiction. To prove sufficiency, use problem 9.41.

(2) Use item (1).

(3) Use item (1) and the inequality $mI \leq A \leq MI$.

(4) Prove that $\|(A - \lambda I)x\|^2 = \|(A - \alpha I)x\|^2 + |\beta|^2\|x\|^2$, $x \in H$.

(5) Use item (4) and problem 9.41.

9.43 *Necessity.* According to item (5) of problem 9.6° there exists $(A^*)^{-1} \in \mathcal{L}(H)$. By Theorem 8.2, there also exists $c > 0$ such that $\|Ax\|^2 = (A^*Ax, x) \geq c\|x\|^2$ and $\|A^*x\|^2 = (AA^*x, x) \geq c\|x\|^2$, $x \in H$. Since the operators A^*A and AA^* are self-adjoint, then $A^*A \geq cI$, $AA^* \geq cI$.

Sufficiency. According to item (1) of problem 9.42, the operators A^*A and AA^* are continuously invertible, hence due to item (2) of problem 8.17 the operator A is continuously invertible.

9.44 (1) If $(A^n)^* = A^n$, then $(A^{n+1})^* = (A^n \cdot A)^* = A^* \cdot (A^n)^* = A \cdot A^n = A^{n+1}$.

(2) Using item (1) of the current problem and item (3) of problem 9.1, we obtain by induction that $\left\|A^{2^k}\right\| = \|A\|^{2^k}$, $k \geq 1$. For any $n \geq 1$ choose k such that $2^k > n$. Then $\left\|A^{2^k}\right\| \leq \|A^n\| \cdot \|A\|^{2^k - n} \leq \|A\|^{2^k} = \left\|A^{2^k}\right\|$, hence both inequalities turn into equalities and $\|A^n\| = \|A\|^n$.

(3) If $n = 2m$, then $(A^n x, x) = (A^m x, A^m x) = \|A^m x\|^2 \geq 0$, $x \in H$. If $n = 2m + 1$, then $(A^n x, x) = \big(A(A^m x), A^m x\big) \geq 0$, $x \in H$, because $A \geq 0$.

9.45 Use problem 7.26 and item (2) of problem 9.44.

9.46° If $\mathbb{K} = \mathbb{R}$, then $A = A^*$ under condition $\gamma_k = \beta_k$, $k \geq 1$. If $\mathbb{K} = \mathbb{C}$, then $A = A^*$ under condition $\alpha_k \in \mathbb{R}$ and $\gamma_k = \overline{\beta_k}$, $k \geq 1$.

9.52* (1) (a) By induction we obtain that all operators T_n are polynomials of S, so they commute with S and with each other. Therefore

$$T_{n+1} - T_n = \tfrac{1}{2}(S + T_n^2) - \tfrac{1}{2}(S + T_{n-1}^2) = \tfrac{1}{2}(T_n + T_{n-1})(T_n - T_{n-1}),$$

whence by induction we obtain that T_n and $T_n - T_{n-1}$ are polynomials of S with nonnegative coefficients.

(b) The inequality $T_n \geq T_{n-1}$ follows from (a) and item (3) of problem 9.44.

(c) Use problem 9.38*. To prove the convergence $T_n^2 \xrightarrow{s} T^2$, $n \to \infty$, show that T commutes with S, so it also commutes with T_n, $n \geq 1$. Therefore we have $T_n^2 - T^2 = (T_n + T)(T_n - T)$ and

$$\|T_n^2 x - T^2 x\| \leq \|T_n + T\| \cdot \|T_n x - Tx\| \leq 2\|T_n x - Tx\|, \ x \in H.$$

(2) For $A \neq 0$ check that the operator $S = I - \frac{A}{\|A\|}$ is nonnegative and $\|S\| \leq 1$, apply item (1) and put $B = \sqrt{\|A\|}(I - T)$.

(3) Let B be a square root of A, constructed in solution of item (2), $C \in \mathcal{L}(H)$, $C \geq 0$ and $C^2 = A$. Then B commutes with all operators that commute with A, and since $AC = C^3 = CA$, then B commutes with C. Therefore

$$(B + C)(B - C) = B^2 - C^2 = A - A = 0.$$

Fix $x \in H$ and put $y = (B - C)x$. Since $(By, y) \geq 0$, $(Cy, y) \geq 0$ and

$$(By, y) + (Cy, y) = \big((B + C)y, y\big) = \big((B + C)(B - C)x, y\big) = 0,$$

then $(By, y) = (Cy, y) = 0$. Due to item (3) of problem 9.27, we obtain $By = Cy = 0$. Thus $\|y\|^2 = \big((B - C)x, y\big) = \big(x, (B - C)y\big) = 0$, so $Bx = Cx$ for every $x \in H$.

9.53 (1) $\sqrt{A}x = (\sqrt{\alpha_1}x_1, \sqrt{\alpha_2}x_2, \ldots)$; (2), (4), (6) $\sqrt{A} = A$;
(3) $\sqrt{A}x = (x_1 + x_2, x_1 + 2x_2, x_3, 0, \ldots)$; (5) $(\sqrt{A}x)(t) = \sqrt{t}x(t)$;
(7) $(\sqrt{A}x)(t) = \sqrt{\frac{2}{e^2-1}} \int\limits_0^1 e^{t+s}x(s)ds$.

9.54* (1) Suppose that $A = B^2$. Then $\operatorname{Ker} B \subset \operatorname{Ker} A = \operatorname{span}(\{e_1\})$ and $\operatorname{Ker} B \neq \{0\}$, hence $\operatorname{Ker} B = \operatorname{span}(\{e_1\})$. Thus $ABe_2 = B^3 e_2 = BAe_2 = Be_1 = 0$. Therefore $Be_2 \in \operatorname{span}(\{e_1\})$ and $B^2 e_2 = 0 \neq Ae_2 = e_1$, a contradiction. (2) If $B^2 = A$, then $(B^*)^2 = A^*$, where A^* is the operator from item (1).

9.55 (1) A and $\sqrt{B}$ commute, therefore

$$(ABx, x) = \big(\sqrt{B}A\sqrt{B}x, x\big) = \big(A\sqrt{B}x, \sqrt{B}x\big) \geq 0, \ x \in H.$$

(4) Prove that $AC \leq AD \leq BD$.

9.56 Use item (3) of problem 9.1.

9.57° (1) $\Leftrightarrow$ (2) Use problem 9.30°.
(2) $\Leftrightarrow$ (3) Use the equality $(Ax, x) = \big(\sqrt{A}x, \sqrt{A}x\big) = \big\|\sqrt{A}x\big\|^2$, $x \in H$.

9.58 (1), (2) No. $H = \mathbb{C}^2$, $A = \left(\begin{smallmatrix} 1 & 0 \\ 0 & 0 \end{smallmatrix}\right)$, $B = \left(\begin{smallmatrix} 1 & 0 \\ 0 & -1 \end{smallmatrix}\right)$. (3) Yes.

9.59 If $B \geq 0$, then

$$\|Ax\| = \|Bx\| \Leftrightarrow (A^*Ax, x) = (Ax, Ax) = (Bx, Bx) = (B^2x, x).$$

Since the operators A^*A and B^2 are self-adjoint, due to item (3) of problem 9.27 the last equality holds for all $x \in H$ if and only if $A^*A = B^2$, i.e. $B = \sqrt{A^*A}$ (the square root of the nonnegative operator A^*A exists and is unique according to problem 9.52*).

9.60 (1) If $0 \leq A \leq I$, then $\|A\| \leq 1$ due to item (2) of problem 9.27, $\big\|\sqrt{A}\big\| \leq 1$ due to problem 9.56 and $0 \leq \sqrt{A} \leq I$ again due to item (2) of problem 9.27. Then $A = \sqrt{A} \cdot \sqrt{A} \leq I \cdot \sqrt{A} = \sqrt{A}$ according to item (3) of problem 9.55.
(2) If $A \geq I$, then

$$\big\|\sqrt{A}x\big\|^2 = \big(\sqrt{A}x, \sqrt{A}x\big) = (Ax, x) \geq (x, x) = \|x\|^2, \ x \in H.$$

Hence by problem 9.41 there exists the inverse operator $\big(\sqrt{A}\big)^{-1} = A^{-\frac{1}{2}} \in \mathcal{L}(H)$, therefore there exists $A^{-1} = \big(A^{-\frac{1}{2}}\big)^2 \in \mathcal{L}(H)$. Moreover $A^{-1} \leq I$, because $(A^{-1}x, x) = \big\|A^{-\frac{1}{2}}x\big\|^2 \leq \|x\|^2 = (x, x)$, $x \in H$. According to item (1) we have $A^{-1} \leq A^{-\frac{1}{2}} \leq I$, thus $I = A^{-1} \cdot A \leq \sqrt{A} = A^{-\frac{1}{2}} \cdot A \leq A = I \cdot A$ according to item (3) of problem 9.55.

9.61* Similarly to the solution of item (2) of problem 9.60, we obtain that there exist operators $A^{-\frac{1}{2}}, A^{-1}, B^{-1} \in \mathcal{L}(H)$. Since $B \geq A$, then for all $x \in H$ we have

$$\big((B - A)A^{-\frac{1}{2}}x, A^{-\frac{1}{2}}x\big) = \big(A^{-\frac{1}{2}}(B - A)A^{-\frac{1}{2}}x, x\big) =$$
$$= \big(A^{-\frac{1}{2}}BA^{-\frac{1}{2}}x, x\big) - (x, x) \geq 0.$$

Therefore $C = A^{-\frac{1}{2}}BA^{-\frac{1}{2}} \geq I$. Similarly to the solution of item (2) of problem 9.60, we get that $C^{-1} = A^{\frac{1}{2}}B^{-1}A^{\frac{1}{2}} \leq I$. Hence for all $x \in H$ and for $y = A^{-\frac{1}{2}}x$ we have

$$\big((A^{-1} - B^{-1})x, x\big) = \big((A^{-1} - B^{-1})A^{\frac{1}{2}}y, A^{\frac{1}{2}}y\big) =$$
$$= \big(A^{\frac{1}{2}}(A^{-1} - B^{-1})A^{\frac{1}{2}}y, y\big) = (y, y) - \big(A^{\frac{1}{2}}B^{-1}A^{\frac{1}{2}}y, y\big) \geq 0.$$

9.62 Use the fact that $\| \, |A|x \, \| = \|Ax\|$, $x \in H$.

9.65 (1) *Necessity.* Denote by P_1 the orthogonal projector onto H_1. Let the operator U be partially isometric. For $x \in H_1$ we have

$$(U^*Ux, x) = (Ux, Ux) = \|Ux\|^2 = \|x\|^2 = (P_1x, x),$$

and for $x \in H_1^\perp = \operatorname{Ker} U$ we have $(U^*Ux, x) = 0 = (P_1x, x)$.

Sufficiency. Let $U^*U = P_1$. Then $\|Ux\|^2 = (U^*Ux, x) = (P_1x, x) = \|P_1x\|^2$, $x \in H$, i.e. $\|Ux\| = \|x\|$, $x \in H_1$, and $\|Ux\| = 0$, $x \in H_1^\perp$.

9.66 *Existence.* Define the operator U on $R(|A|)$ by the formula $U|A|x = Ax$, $x \in H$ (this definition is correct, because $\operatorname{Ker} |A| = \operatorname{Ker} A$ as a consequence of item (4) of problem 9.62, so if $|A|x = |A|y$, then $Ax = Ay$). Then $A = U|A|$, and since $\| \, |A|x \, \| = \|Ax\|$, $x \in H$, then $U : R(|A|) \to H$ is an isometric operator. Extend U to $\overline{R(|A|)}$ by continuity, and then to H in such a way that $Ux = 0$ for $x \in \left(R(|A|)\right)^\perp$. We obtain a partially isometric operator $U : H \to H$, for which $\operatorname{Ker} U = \left(R(|A|)\right)^\perp = \operatorname{Ker} |A|$ (the last equality follows from problem 9.30°, because the operator $|A|$ is self-adjoint).

Uniqueness. Let $A = UB$ be a polar decomposition. Since

$$\overline{R(B)} = (\operatorname{Ker} B)^\perp = (\operatorname{Ker} U)^\perp,$$

then $U : \overline{R(B)} \to H$ is an isometric operator, thus $\|Ax\| = \|UBx\| = \|Bx\|$, $x \in H$. Now it follows from problem 9.59 that $B = |A|$. The equality $U|A|x = Ax$, $x \in H$, uniquely defines U on the set $(\operatorname{Ker} U)^\perp = \overline{R(|A|)}$, therefore it uniquely determines U.

9.67 Since $A = U|A|$, then $A^* = |A|U^*$. Due to item (1) of problem 9.65, the operator U^*U is an orthogonal projector onto $(\operatorname{Ker} U)^\perp = \overline{R(|A|)}$, hence $U^*U|A| = |A|$ and $A^* = U^*U|A|U^*$. The operator $U|A|U^*$ is nonnegative, moreover

$$U|A|U^* \cdot U|A|U^* = U|A|^2U^* = U|A| \cdot |A|U^* = AA^*,$$

so $U|A|U^* = \sqrt{AA^*} = |A^*|$. Thus $A^* = U^*U|A|U^* = U^*|A^*|$, whence $A = |A^*|U$.

9.68 (1) $Bx = |A|x = (|\alpha_1|x_1, |\alpha_2|x_2, \ldots)$, $Ux = (\beta_1x_1, \beta_2x_2, \ldots)$, $x \in l_2$, where $\beta_i = \frac{\alpha_i}{|\alpha_i|}$ for $\alpha_i \neq 0$ and $\beta_i = 0$ for $\alpha_i = 0$. (2) $|A|$ is an orthogonal projector onto $\overline{\operatorname{span}(\{e_n : n \geq 2\})}$, $U = A$. (3) $|A| = U = A$.

9.69 (1) (a) $\operatorname{Ker} A = \{0\}$; (b) $\operatorname{Ker} A = \operatorname{Ker} A^* = \{0\}$.

(2) Since $AA^* = A^*A$, then $|A^*| = |A|$. According to problem 9.67, we have $A = |A|U = U|A|$.

(3) If operators A and C commute, then A^* commutes with $C^* = C^{-1}$, therefore it commutes with C. Hence C commutes with A^*A, and in view of item (2) of problem 9.52* it commutes with $|A| = \sqrt{A^*A}$. For $y = |A|x \in R(|A|)$ we have

$$CUy = CU|A|x = CAx = ACx = U|A|Cx = UC|A|x = UCy.$$

For $y \in \big(R(|A|)\big)^{\perp} = \operatorname{Ker}|A| = \operatorname{Ker}U$ we have $CUy = 0 = UCy$, because $|A|y = 0$ implies that $C|A|y = |A|Cy = 0$, i.e. $Cy \in \operatorname{Ker}|A| = \operatorname{Ker}U$. It follows that $CUy = UCy$ for all $y \in H$.

(4) No. Consider the operator A from item (2) of problem 9.68 and verify that $C = A$ does not commute with $|A|$.

9.70 (1) Due to item (5) of problem 9.42, there exist $(A + \lambda I)^{-1}, (A + \overline{\lambda} I)^{-1} \in \mathcal{L}(H)$. Since $U^{-1} = (A + \overline{\lambda} I)(A + \lambda I)^{-1}$ and $U^* = (A + \lambda I)^{-1}(A + \overline{\lambda} I)$, then $U^{-1} = U^*$, because the operator $A + \overline{\lambda} I$ commutes with $A + \lambda I$, so it commutes with $(A + \lambda I)^{-1}$. According to item (2) of problem 9.4, the operator U is unitary.

(2) Due to item (2) of problem 9.4, we have

$$U^{-1} = (A - iI)(A + iI)^{-1} = U^* = (A^* + iI)^{-1}(A^* - iI).$$

Hence $(A^* + iI)(A - iI) = (A^* - iI)(A + iI)$, thus $A = A^*$. (3) We have

$$\begin{aligned} A^* &= -i(U^* - I)^{-1}(U^* + I) = -i(U^{-1} - I)^{-1}(U^{-1} + I) = \\ &= -i(I - U)^{-1}U \cdot U^{-1}(I + U) = i(U + I)(U - I)^{-1} = A, \end{aligned}$$

because $(U^{-1} - I)^{-1} = (I - U)^{-1}U$ and the operators $U + I$ and $(U - I)^{-1}$ commute.

9.72° The formulae remain the same, but, firstly, the operator A' acts from l_q to l_q, where q is the adjoint index to p, and, secondly, in items (1), (7), (8) one has to replace $\overline{\alpha_k}$ with α_k.

9.73 $A'y = (0, y_1, y_2, \ldots)$, where (1) $A' : l_1 \to l_1$; (2) $A' : l_2 \to l_\infty$; (3) $A' : l_1 \to l_\infty$.

9.74 (1) $A' : l_2 \to l_\infty$, $Ax = x$, $x \in l_2$.

(2) $A' : BV_0([0,1]) \to BV_0([0,2])$, $(A'g)(t) = \begin{cases} g(t), & t \in [0,1], \\ g(1), & t \in [1,2]; \end{cases}$

(3) $A' : L_q([0,1]) \to L_q([0,2])$, $(A'y)(t) = \begin{cases} y(t), & t \in [0,1], \\ 0, & t \in [1,2]; \end{cases}$

(4) $A' : BV_0([0,1]) \to BV_0([0,1])$, $(A'g)(t) = \int\limits_0^t a(u)dg(u)$, $t \in [0,1]$;

(5) $A' : BV_0([0,1]) \to BV_0([0,1])$, $(A'g)(t) = \int_0^t (g(1) - g(u))du$, $t \in [0,1]$;

(6) $A' : Y^* \to X^*$, $A'g = g(y)f$, $g \in Y^*$.

9.75° (5) Assume that there exists $A^{-1} \in \mathcal{L}(X)$. Then $AA^{-1} = A^{-1}A = I$, hence $(A^{-1})'A' = A'(A^{-1})' = I'$, where I' is the identity operator in X^*. Therefore Theorem 8.3 implies that $(A')^{-1} = (A^{-1})' \in \mathcal{L}(X^*)$.

9.76 (1) If $x \in \operatorname{Ker} A$, then $(A'g)(x) = g(Ax) = 0$ for all $g \in X_2^*$. Conversely, if $x \notin \operatorname{Ker} A$, then $Ax \neq 0$ and by Corollary 4.2 of the Hahn–Banach theorem there exists $g \in X_2^*$ such that $(A'g)(x) = g(Ax) \neq 0$.

(2) If $y = \lim_{n\to\infty} Ax_n \in \overline{R(A)}$, then for all $f \in \operatorname{Ker} A'$ we have

$$f(y) = \lim_{n\to\infty} f(Ax_n) = \lim_{n\to\infty} (A'f)(x_n) = 0.$$

Conversely, if $y \notin \overline{R(A)}$ then by Corollary 4.1 of the Hahn–Banach theorem there exists $f \in X_2^*$ such that $f(Ax) = 0$ for all $x \in X_1$ and $f(y) \neq 0$. Since $(A'f)(x) = f(Ax) = 0$ for all $x \in X_1$, then $A'f = 0$, so $f \in \operatorname{Ker} A'$.

(3) The necessity follows from Theorem 8.2 and item (5) of problem 9.75°.

Sufficiency. Due to problem 6.35°, the set $R(A)$ is closed, $\operatorname{Ker} A = \{0\}$ and $\operatorname{Ker} A' = \{0\}$. Now item (2) implies that $R(A) = X_2$, therefore the operator A is continuously invertible by Theorem 8.2.

(4) *Sufficiency.* Due to item (3) and Theorem 8.2, it is sufficient to show that

$$\exists c > 0 \, \forall x \in X_1 \; : \; \|Ax\|_2 \geq c\|x\|_1.$$

Fix $x \in X_1$. According to the Corollary 4.2 of the Hahn–Banach theorem, there exists $g \in X_1^*$ such that $\|g\| = 1$ and $\|x\|_1 = g(x)$. For $f = (A')^{-1}g \in X_2^*$ we have

$$\|x\|_1 = (A'f)(x) = f(Ax) \leq \|f\| \cdot \|Ax\|_2 = \\ = \|(A')^{-1}g\| \cdot \|Ax\|_2 \leq \|(A')^{-1}\| \cdot \|Ax\|_2.$$

Problems of Chap. 10

10.11 Let $\{x_1, \ldots, x_n\}$ be a finite $\frac{\varepsilon}{2}$-net for M. Put $M_k = M \cap \overline{B}(x_k, \frac{\varepsilon}{2})$, $1 \leq k \leq n$.

10.12 (1)–(3) Yes. (4) No. In the space $X = \mathbb{R}$ consider sets

$$M = \mathbb{N}, \; N = \left\{ -n + \frac{1}{n} \mid n \geq 2 \right\}.$$

10.13° (1) Use proof by contradiction and the Bolzano-Weierstrass criterion;

(2) If M is not a compact set, then there exists a sequence $\{x_n : n \geq 1\} \subset M$, which consists of different elements and has no limit points. Put

$$r_n = \frac{1}{3} \inf\{\rho(x_n, x_m) \mid m \in \mathbb{N},\ m \neq n\},\ n \geq 1.$$

Let $f(x) = (-1 + \frac{1}{n})(1 - \frac{\rho(x, x_n)}{r_n})$ for $x \in \overline{B}(x_n, r_n)$, $n \geq 1$, and $f(x) = 0$ for $x \in M \setminus \left(\bigcup_{n=1}^{\infty} \overline{B}(x_n, r_n) \right)$. Verify that the function f is lower semicontinuous on M, but f does not attain its minimum value on this set.

10.14 No. For example, consider the set $M = \mathbb{N}$ in the Banach space $X = \mathbb{R}$.

10.15 *Necessity.* For any $\varepsilon > 0$ consider a finite $\frac{\varepsilon}{2}$-net $\{x_1, \ldots, x_k\}$ for the set M. Since $L_n \subset L_{n+1}$, $n \geq 1$, and the set $\bigcup_{n=1}^{\infty} L_n$ is dense in X, then for each $x \in X$ we have $\rho(x, L_n) \to 0$, $n \to \infty$. Therefore there exists N such that $\rho(x_i, L_n) < \frac{\varepsilon}{2}$ for all $1 \leq i \leq k$ and $n \geq N$. For each $x \in M$ there exists $1 \leq i \leq k$ such that $\|x - x_i\| < \frac{\varepsilon}{2}$, whence $\rho(x, L_n) \leq \|x - x_i\| + \rho(x_i, L_n) < \varepsilon$. Thus $\sup_{x \in M} \rho(x, L_n) < \varepsilon$, $n \geq N$.

Sufficiency. Fix $\varepsilon > 0$. Let us show that there exists a finite ε-net for the set M. Choose N such that $\sup_{x \in M} \rho(x, L_N) < \frac{\varepsilon}{2}$. Then for every $x \in M$ there exists $y(x) \in L_N$ such that $\|x - y(x)\| < \frac{\varepsilon}{2}$. Since the set M is bounded, the set $M_N = \{y(x) \mid x \in M\}$ in the finite-dimensional space L_N is bounded, hence M_N is precompact. Therefore in the space L_N there exists a finite $\frac{\varepsilon}{2}$-net for the set M_N, which by construction is an ε-net for the set M.

10.16° (1), (2) Apply the statement of problem 10.15 for the subspaces $L_n = \text{span}(\{e_1, \ldots, e_n\})$. (3) Apply the statement of problem 10.15 for the subspaces $L_n = \text{span}(\{e_0, e_1, \ldots, e_n\})$, where $e_0 = (1, \ldots, 1, \ldots)$.

10.17° The set M is precompact in H if and only if M is bounded and

$$\forall\, \varepsilon > 0\ \exists\, N = N(\varepsilon) \in \mathbb{N}\ \forall\, x = \sum_{n=1}^{\infty} x_n e_n \in M\ :\ \sum_{n=N+1}^{\infty} |x_n|^2 < \varepsilon.$$

10.18 In items (1)–(9) use the Ascoli–Arzelà theorem. In item (5) to prove the equicontinuity, note that for all $t_1, t_2 \in [a, b]$, $t_1 < t_2$, we have

$$|x(t_2) - x(t_1)|^2 = \left| \int_{t_1}^{t_2} x'(s) ds \right|^2 \leq \int_{t_1}^{t_2} ds \cdot \int_{t_1}^{t_2} |x'(s)|^2 ds \leq k_2(t_2 - t_1)$$

due to the Newton–Leibniz formula and the Cauchy–Schwarz inequality. Compact sets are the set in item (2), and the sets from items (9) and (10) under the condition that they are closed.

10.19 Use the Ascoli–Arzelà theorem. Precompact sets are the sets in items (3), (5), (8)–(10), while only the set in item (5) is compact. In item (3) prove that the set $\{n + 2\pi k \mid n \in \mathbb{N},\ k \in \mathbb{Z}\}$ is dense in $\mathbb{R}$, and show that the closure of the set $\{\sin(t + n) \mid n \geq 1\}$ contains all functions of the form $\sin(t + a)$, $a \in \mathbb{R}$.

10.20 For precompactness it is necessary and sufficient that A is (1), (2) bounded; (3) bounded from below; (4) arbitrary. For compactness in items (1)–(4) it is necessary and sufficient that A is compact.

10.21° (1), (4) Yes. (2), (3) No.

10.22° Use problem 10.16°.

10.23° (1), (2) It is necessary and sufficient that $a_n \to +\infty$, $n \to \infty$.

10.24° The converse statement is false.

10.26 The set M is precompact in $C^k([a, b])$ if and only if M is uniformly bounded and the set $\{x^{(k)} \mid x \in M\}$ is equicontinuous.

10.27 Repeat the reasonings from the proof of the Ascoli-Arzelà theorem for the space $C([a, b])$.

10.28 It is necessary and sufficient that the convergence in X_1 implies the convergence in X_2 (i.e., the embedding operator $A : X_1 \to X_2$, $Ax = x$, is continuous).

10.29° (1)–(4) Yes.

10.30 *Necessity.* The set M is bounded, therefore there exists $C > 0$ such that $|x(t)| \leq C$, $x \in M$, $t \in T$. For any $\varepsilon > 0$ there exists a finite $\frac{\varepsilon}{3}$-net $\{x_1, \ldots, x_m\}$ for the set M. Consider the mapping $F = (x_1, \ldots, x_m) : T \to \mathbb{R}^m$. Since $F(T) \subset [-C, C]^m$, then $F(T)$ can be partitioned into disjoint sets $S_1, \ldots, S_n$ with diameters less than $\frac{\varepsilon}{3}$. Then for $T_k = F^{-1}(S_k)$ we have $T = \bigcup_{k=1}^{n} T_k$ and on each set T_k each of the functions $x_1, \ldots, x_m$ varies at most by $\frac{\varepsilon}{3}$. For each $x \in M$ there exists $1 \leq i \leq m$ such that $\|x - x_i\| < \frac{\varepsilon}{3}$. Then for any $1 \leq k \leq n$ and $t_1, t_2 \in T_k$ we have $|x(t_1) - x(t_2)| \leq |x(t_1) - x_i(t_1)| + |x_i(t_1) - x_i(t_2)| + |x_i(t_2) - x(t_2)| < \varepsilon$.

Sufficiency. Let $x(t) \in [-C, C]$, $x \in M$, $t \in M$. For $\varepsilon > 0$ consider a finite partition $T = \bigcup_{k=1}^{n} T_k$ such that on each set T_k each function from M varies at most by $\frac{\varepsilon}{2}$, and also consider an $\frac{\varepsilon}{2}$-net $\{a_1, \ldots, a_m\}$ for the segment $[-C, C]$. Check that the finite set of functions, which attain one of the values $a_1, \ldots, a_m$ on each of the sets $T_1, \ldots, T_n$, is an ε-net for M.

10.31 The set M is precompact in the space $B(T, Y)$ if and only if for every $\varepsilon > 0$ there exists a finite partition $T = \bigcup_{k=1}^{n} T_k$ such that

$$\forall\, 1 \leq k \leq n\ \forall x \in M\ \forall t_1, t_2 \in T_k\ :\ \|x(t_1) - x(t_2)\| \leq \varepsilon.$$

10.34° (4) If $A \in S_0(X_1, X_2)$, then

$$\exists f_1, \ldots, f_n \in X_1^* \, \exists e_1, \ldots, e_n \in X_2 \; : \; Ax = \sum_{k=1}^{n} f_k(x)e_k, \; x \in X_1,$$

and without loss of generality $f_1, \ldots, f_n$ and $e_1, \ldots, e_n$ are linearly independent. Therefore for all $l \in X_2^*$ we have $(A'l)(x) = l(Ax) = \sum\limits_{k=1}^{n} f_k(x)l(e_k), x \in X_1$, from which $R(A') = \text{span}(\{f_1, \ldots, f_n\})$ and $\dim R(A') = \dim R(A) = n$.

10.35° (1) Yes. (2) No, because it is possible that A is unbounded.

10.36° (1) Not compact. Verify that the sequence $\{Ae_n : n \geq 1\}$ does not contain a convergent subsequence. (2)–(4) Compact.

10.37° (1)–(4) $\alpha_n \to 0, n \to \infty$. The solution is quite similar to the solution of problem 10.3.

10.38° $|\alpha_n|^2 + |\beta_n|^2 + |\gamma_n|^2 \to 0, n \to \infty$.

10.39° (1) No. Consider $X_1 = X_2 = l_2$, $A_n x = \sum\limits_{k=1}^{n} x_k e_k$ and $Ax = x$. (2) No. Consider $X_1 = X_2 = \{x \in l_2 \,|\, \exists n \in \mathbb{N} \; : \; x_k = 0, \; k \geq n\}$ endowed with the norm $\|\cdot\|_2$, $A_n x = \sum\limits_{k=1}^{n} \frac{x_k}{k} e_k$, $n \geq 1$, and $Ax = \sum\limits_{k=1}^{\infty} \frac{x_k}{k} e_k$. The operator A is not compact, because the sequence $x^{(n)} = (1, \frac{1}{2}, \ldots, \frac{1}{n}, 0, \ldots)$, $n \geq 1$, is bounded, but the sequence $\{Ax^{(n)} : n \geq 1\}$, which converges to $(1, \frac{1}{4}, \ldots, \frac{1}{n^2}, \ldots)$ coordinate-wise, does not contain a convergent subsequence in the space X_2.

10.40° Use item (4) of problem 10.18.

10.41° (1), (8) Compact because the operator is finite-dimensional. (2) Compact according to problem 10.40°. (3) Compact according to problem 10.4. (4)–(7) Not compact. In items (4), (5), (7) show that for $x_n(t) = t^n$, $t \in [0, 1]$, $n \geq 1$, the sequence $\{Ax_n : n \geq 1\}$ does not contain a convergent subsequence. In item (6) use that $A = 3I + B$, where $(Bx)(t) = \int\limits_0^1 e^{ts} x(s)ds$, and the operator B is compact according to problem 10.4.

10.42° (1) Apply the Ascoli–Arzelà theorem. (2) Approximate K by continuous functions in $L_2([a, b]^2)$ and use item (1) of the current problem and item (3) of Theorem 10.2.

Remark A more general statement is true. If $L_2(T, \mu)$ is a separable Hilbert space and

$$(Ax)(t) = \int_T K(t, s)x(s)ds, \; t \in T, \; x \in L_2(T, \mu),$$

where $K \in L_2(T \times T, \mu \times \mu)$, then $A \in S_\infty(L_2(T, \mu))$, because A is a Hilbert–Schmidt operator (see Chap. 12).

10.43° (1), (2) Compact because the operator is finite-dimensional. (3) Not compact as the sum of identity operator and finite-dimensional operator. (4), (5) Compact according to item (2) of problem 10.42°.

10.44° (1), (2) It is necessary and sufficient that $a = 0$. (1) If $a(t_0) \neq 0$, then for $x_n(t) = e^{-n|t-t_0|}$, $t \in [0, 1]$, $n \geq 1$, the sequence $\{Ax_n : n \geq 1\}$ does not contain a convergent subsequence. (2) For the sequence $x_n(t) = e^{int}$, $t \in [0, 1]$, $n \geq 1$, verify that $x_n \overset{w}{\to} 0$, $n \to \infty$, in $L_2([0, 1])$. Then $\|Ax_n\|_2 = \|a\|_2 \to 0$, $n \to \infty$, by item (5) of Theorem 10.2. Hence $a(t) = 0 \,(\mathrm{mod}\, m)$.

10.45 (3) Use (1) of the current problem, item (4) of Theorem 10.2 and the continuity of the operator of embedding $C([a, b])$ into $L_p([a, b])$.

10.46 (1)–(4) No. In items (1), (2) show that the sequence $\{Ae_n : n \geq 1\}$ does not contain a convergent subsequence. In items (3), (4) show that for $x_n(t) = e^{int}$, $t \in [0, 1]$, $n \geq 1$, the sequence $\{Ax_n : n \geq 1\}$ does not contain a convergent subsequence.

10.47° It is necessary and sufficient that $\alpha_n \to +\infty$, $n \to \infty$. Firstly, verify that $l_{2,\alpha} \subset c_0$ for $\alpha = \{\alpha_n : n \geq 1\} \subset [1, +\infty)$. The image of the unit ball $B(0, 1)$ of the space $l_{2,\alpha}$ is the set $\left\{(x_1, x_2, \ldots) \mid \sum\limits_{n \geq 1} \alpha_n |x_n|^2 < 1\right\}$. Apply the criterion of compactness in c_0 (item (2) of problem 10.16°) to verify whether this set is compact.

10.48 (1), (2) No. (3) Yes.

10.50 (1) $Ax = (0, x_1, 0, x_3, 0, x_5, \ldots)$.

10.51° (1) No. (2), (3) Yes.

10.52 (1) Yes. Assume that $A \notin \mathcal{L}(H)$. Then there exists an orthonormal sequence $\{e_n : n \geq 1\}$ such that $\|Ae_n\| \geq n$, $n \geq 1$. Indeed, there exists a vector e_1 such that $\|e_1\| = 1$ and $\|Ae_1\| \geq 1$, and for $k \geq 1$ the operator A is unbounded on the subspace $H_k = \{e_1, \ldots, e_k\}^\perp$, hence there exists a vector $e_{k+1} \in H_k$ such that $\|e_{k+1}\| = 1$ and $\|Ae_{k+1}\| \geq k + 1$. Let $Bx = \sum\limits_{k=1}^{\infty} \frac{\alpha_k}{\sqrt{k}} e_k$ for $x = \sum\limits_{k=1}^{\infty} \alpha_k e_k$ and $Bx = 0$ for $x \in \{e_n : n \geq 1\}^\perp$. Then $B \in S_\infty(H)$ and $\|ABe_n\| \geq \sqrt{n} \to +\infty$, $n \to \infty$, hence $AB \notin \mathcal{L}(H)$.

(2) Yes. Fix $z \in H \setminus \{0\}$. If $B_y A \in \mathcal{L}(H)$ for all one-dimensional operators of the form $B_y x = (x, y)z$, $x \in H$, $y \in H$, then A maps any strongly convergent sequence into a weakly convergent one, so $A \in \mathcal{L}(H)$ according to item (3) of problem 6.39.

10.53 The Hausdorff criterion implies that a precompact set contains a countable dense subset.

10.54° If $c_0 \neq 0$, then it is not possible. If $c_0 = 0$, one can take $A = 0$.

10.55 (1) No. Use Banach inverse operator theorem (Theorem 8.1) and item (2) of problem 10.33°. (2) Yes.

10.56° It is necessary and sufficient that $\dim X < +\infty$.

Necessity. In the case $\dim X = +\infty$ use the theorem on the almost orthogonal vector (see, for example, [8], Ch. 7, Sec. 1, Theorem 1.1) to establish that there exists a sequence $\{x_n : n \geq 1\} \subset X$ such that $\|x_n\| = 1$, $n \geq 1$, and $\|x_n - x_m\| \geq \frac{1}{2}$ for $n \neq m$. Then $\|Ax_n - Ax_m\| \geq \frac{c}{2}$, $n \neq m$, and the sequence $\{Ax_n : n \geq 1\}$ does not contain a convergent subsequence.

10.57° Assume that the statement of the problem is false. Then there exists $c > 0$ such that $\|Ax\| \geq c\|x\|$ for all $x \in X_1$. Similarly to the solution of problem 10.56° this leads to a contradiction with the condition $A \in S_\infty(X_1, X_2)$.

10.58 (1) If the precompact set M is not nowhere dense, then its closure $\overline{M}$ is a compact set and contains some ball $B(x, r)$, so it contains the closed ball $\overline{B}(x, \frac{r}{2})$. Then $\overline{B}(x, \frac{r}{2})$ is a compact set in X, whence $\dim X < +\infty$.

(2) Let $B_n = B(0, n)$, $n \geq 1$, be balls in the space X_1. Then $R(A) = \bigcup\limits_{n=1}^{\infty} A(B_n)$ and the sets $A(B_n)$ are nowhere dense in X_2 due to (1).

10.59 Use item (2) of problem 10.58 and problem 10.45.

10.60* Since $R(A)$ is a closed set, then $R(A)$ is a Banach space and $A \in S_\infty(X_1, R(A))$. If $\dim R(A) = \infty$, then according to item (2) of problem 10.58 $R(A)$ is a set of the first category in $R(A)$, which is impossible by the Baire category theorem (see, for example, [26], Theorem III.8 or [21], Ch. 2, Sec. 7, Theorem 2).

10.61 Let H be a Hilbert space, $A \in S_\infty(H)$, $M \subset R(A)$ be a subspace of H and P be an orthogonal projector onto M. Then $PA \in S_\infty(H)$, $M = R(PA)$ and $\dim M < +\infty$ due to problem 10.60*.

10.62 Let $X_1 = C^1([a, b])$, $X_2 = C([a, b])$, A be an operator of embedding X_1 into X_2, which is compact according to item (1) of problem 10.45. Verify that $A \in S_\infty(M_1, M_2)$, where we denote by M_1 and M_2 the set M endowed with the norms of spaces X_1 and X_2, respectively, and use problem 10.60*.

10.63 *Necessity.* Let $\{x_1^{(n)}, \ldots, x_{m(n)}^{(n)}\}$ be a finite $\frac{1}{n}$-net for the set $A(B(0, 1))$, $L_n = \operatorname{span}\big(\{x_1^{(n)}, \ldots, x_{m(n)}^{(n)}\}\big)$, P_n be an orthogonal projector onto L_n. Consider the sequence $\{P_n A : n \geq 1\}$.

10.64° If $A \in S_\infty(H)$ or $A^* \in S_\infty(H)$, then $A^*A \in S_\infty(H)$ and $AA^* \in S_\infty(H)$ according to item (4) of Theorem 10.2. Let $A^*A \in S_\infty(H)$. Then for any bounded sequence $\{x_n : n \geq 1\}$ there exists a convergent subsequence $\{A^*Ax_{n_k} : k \geq 1\}$. This subsequence is a Cauchy sequence. By the Cauchy–Schwarz inequality

$$\|Ay\|^2 = (Ay, Ay) = (A^*Ay, y) \leq \|A^*Ay\| \cdot \|y\|, \;\; y \in H,$$

thus the sequence $\{Ax_{n_k} : k \geq 1\}$ is also a Cauchy sequence, so it converges. Therefore $A \in S_\infty(H)$. It follows from the previous reasonings that if $AA^* = (A^*)^*A^* \in S_\infty(H)$, then $A^* \in S_\infty(H)$.

10.65 (1) In view of problem 10.64° it is enough to show that the operator A^*A or $(A^*A)^*A^*A = A^*AA^*A$ is compact. Since the operators A and A^* commute, then $A^*AA^*A = A^*A^*A^2 \in S_\infty(H)$ by item (4) of Theorem 10.2. (2) If $A^n \in S_\infty(H)$ and $2^k > n$, then $A^{2^k} = A^n \cdot A^{2^k-n} \in S_\infty(H)$. Since all powers of a normal operator are normal operators, it follows from item (1) that the operators $A^{2^{k-1}}, A^{2^{k-2}}, \ldots, A^2, A$ are compact.

10.67° Let P_n be the orthogonal projector onto span$(\{e_1, \ldots, e_n\})$, $n \geq 1$. Show that $AP_n \rightrightarrows A$, $n \to \infty$.

10.68 It is sufficient to show that the set $A(\overline{B}(0,1))$ is closed. Let $\{x_n : n \geq 1\} \subset \overline{B}(0,1)$ and $y_n = Ax_n \to y$, $n \to \infty$. It follows from problem 5.23 that there exist a subsequence $\{x_{n_k} : k \geq 1\}$ and $x \in X_1$ such that $x_{n_k} \xrightarrow{w} x$, moreover $x \in \overline{B}(0,1)$ due to item (2) of problem 5.20. By item (5) of Theorem 10.2 we have $Ax_{n_k} \to Ax$, $k \to \infty$, hence $y = Ax \in A(\overline{B}(0,1))$.

10.69 (1) The function $x \mapsto \|x\|$ is continuous on the set $A(\overline{B}(0,1))$, which is compact due to problem 10.68. Therefore there exists $h \in \overline{B}(0,1)$, for which $\|Ah\| = \max_{\|x\| \leq 1} \|Ax\| = \|A\|$.

(2) Consider $X = C([0,1])$ and $(Ax)(t) = \int_0^1 x(s)ds - x(1)$, $t \in [0,1]$, $x \in C([0,1])$.

10.70 (1) The necessity follows from item (5) of Theorem 10.2. *Sufficiency.* For any bounded sequence $\{x_n : n \geq 1\} \subset X_1$ according to problem 5.23 there exist a subsequence $\{x_{n_k} : k \geq 1\}$ and an element $x \in X_1$ such that $x_{n_k} \xrightarrow{w} x$, $k \to \infty$. Then $Ax_{n_k} \to Ax$, $k \to \infty$. Hence $A \in S_\infty(X_1, X_2)$. (2) Let $X_1 = C([0,1])$, $X_2 = L_2([0,1])$, A be the embedding operator from X_1 to X_2. According to item (3) of problem 10.46, the operator A is not compact. Apply the criterion of weak convergence in $C([0,1])$ (item (1) of problem 5.1) and the Lebesgue's dominated convergence theorem to show that if $x_n \xrightarrow{w} x$, then $Ax_n \to Ax$, $n \to \infty$.

10.71 Use item (1) of problem 10.70. If $x_n \xrightarrow{w} x$ in $L_p([a,b])$, then $Ax_n \xrightarrow{w} Ax$ in $C([a,b])$, $n \to \infty$. Apply the criterion of weak convergence in $C([a,b])$ (item (1) of problem 5.1) and the Lebesgue's dominated convergence theorem to show that $Ax_n \to Ax$ in $L_p([a,b])$.

10.72 Use item (2) of problems 9.24 and item (1) of 10.70.

10.73 (1) Use problem 10.72. (2) Use item (1).

10.74* *Necessity.* Let $\{f_n : n \geq 1\} \subset X_2^*$ and $\|f_n\| \leq 1$, $n \geq 1$. Let us show that the sequence $\{A'f_n : n \geq 1\}$ contains a convergent subsequence. Let $B(0,1)$ be the unit ball in X_1. Then $K = \overline{A(B(0, 1))} \subset X_2$ is a compact set, because $A \in S_\infty(X_1, X_2)$. By the Ascoli-Arzelà theorem (problem 10.27), the set of functions $\{f_n : n \geq 1\}$ is precompact in the space $C(K)$. Indeed, the functions $\{f_n : n \geq 1\}$ are uniformly bounded on K, because $|f_n(y)| \leq \|y\|$, $y \in K$, and the set K is bounded. Also the set $\{f_n : n \geq 1\}$ is equicontinuous on K, because $|f_n(y_1) - f_n(y_2)| \leq \|y_1 - y_2\|$, $y_1, y_2 \in K$. Therefore there exists a subsequence $\{f_{n_k} : k \geq 1\}$ which converges in $C(K)$. Since $(A'f)(x) = f(Ax)$, $x \in X_1$, $f \in X_2^*$, then

$$\|A'f\| = \sup\big\{|(A'f)(x)| \,\big|\, x \in B(0, 1)\big\} = \sup\big\{|f(y)| \,\big|\, y \in A(B(0, 1))\big\}.$$

Therefore the sequence $\{A'f_{n_k} : k \geq 1\}$ is a Cauchy sequence in X_1^*, so it converges, because the space X_1^* is complete.

Sufficiency. Let $\{x_n : n \geq 1\} \subset X_1$, $\|x_n\| \leq 1$, $n \geq 1$, $B(0, 1)$ be the unit ball in X_2^*. Then $K = \overline{A'(B(0, 1))} \subset X_1^*$ is a compact set. Apply the Ascoli–Arzelà theorem to verify that the set of functions $\{F_{x_n} : n \geq 1\}$, where $F_x(f) = f(x)$, $x \in X_1$, $f \in X_1^*$, is precompact in the space $C(K)$. Further, verify that

$$\|Ax\| = \sup\big\{|F_x(g)| \,\big|\, g \in A'(B(0, 1))\big\}, \ x \in X_1.$$

It follows that if a subsequence $\{F_{x_{n_k}} : k \geq 1\}$ converges in $C(K)$, then the sequence $\{Ax_{n_k} : k \geq 1\}$ is a Cauchy sequence in X_2. Thus it converges, because the space X_2 is complete.

10.75 (1) If $A \in \mathcal{L}(X, l_1)$ and $x_n \overset{w}{\to} x$, $n \to \infty$, then $Ax_n \overset{w}{\to} Ax$, $n \to \infty$ in l_1. According to problem 5.29*, we have $Ax_n \to Ax$, $n \to \infty$, therefore $A \in S_\infty(X, l_1)$ as a result of item (1) of problem 10.70.

(2) According to problem 10.74* we have $A \in S_\infty(c_0, X)$ if and only if $A' \in S_\infty(X^*, c_0^*)$. It remains to note that $c_0^* = l_1$ (problem 3.19) and the space X^* is reflexive (problem 4.39*), and to use item (1).

10.76 (1) By the Banach–Steinhaus theorem, there exists $C > 0$ such that $\|A\| \leq C$ and $\|A_n\| \leq C$, $n \geq 1$. By the Hausdorff criterion, for any $\varepsilon > 0$ there exists a finite $\frac{\varepsilon}{4C}$-net $\{x_1, \dots, x_m\} \subset X$ for M. Choose N such that $\|(A_n - A)x_i\| \leq \frac{\varepsilon}{2}$, $1 \leq i \leq m$, $n \geq N$. For each $x \in M$ there exists x_i such that $\|x - x_i\| < \frac{\varepsilon}{4C}$. Then for $n \geq N$ we have

$$\|(A_n - A)x\| \leq \|(A_n - A)x_i\| + \|(A_n - A)(x - x_i)\| \leq \frac{\varepsilon}{2} + 2C \cdot \frac{\varepsilon}{4C} = \varepsilon.$$

(2) Apply (1) for the case $Y = \mathbb{K}$.

10.77* (1) The equivalence of weak-$*$ convergence and the convergence in the metric ρ follows from Theorem 5.2, and the compactness of the space (B, ρ) follows from Theorem 5.4.

(2) The necessity follows from item (2) of problem 10.76.

Sufficiency. Consider the set of functions $\widetilde{M} = \{F_x \mid x \in M\} \subset C(B)$, where $F_x(f) = f(x)$, $f \in B$. It is easy to check that $X \ni x \mapsto F_x \in C(B)$ is an isometry. Let us verify that the set $\widetilde{M}$ satisfies the conditions of the Ascoli–Arzelà theorem (problem 10.27). If the set $\widetilde{M}$ is not uniformly bounded, then there exist $x_n \in M$ such that $\|F_{x_n}\| = \|x_n\| \geq n$, $n \geq 1$. By Corollary 4.2 of the Hanh–Banach theorem, there exist $f_n \in X^*$ such that $\|f_n\| = 1$ and $f_n(x_n) = \|x_n\|$. Then for $g_n = \frac{1}{n} f_n$ we have $g_n \to 0$, but $|g_n(x_n)| \geq 1$, $n \geq 1$, a contradiction with the uniform convergence of $g_n(x)$ to 0 for $x \in M$. Now assume that the set $\widetilde{M}$ is not equicontinuous. Then there exist $\varepsilon > 0$, $x_n \in M$ and $f_n, g_n \in B$ such that $\rho(f_n, g_n) \to 0$, $n \to \infty$, and $|F_{x_n}(f_n) - F_{x_n}(g_n)| > \varepsilon$, $n \geq 1$. Since the space (B, ρ) is compact, there exists a convergent subsequence of the sequence $\{f_n : n \geq 1\}$. Without loss of generality, the sequence $\{f_n : n \geq 1\}$ converges. Then $f_n - g_n \overset{w-*}{\longrightarrow} 0$, $n \to \infty$, but $|F_{x_n}(f_n) - F_{x_n}(g_n)| = |(f_n - g_n)(x_n)| > \varepsilon$, a contradiction with the uniform convergence of $(f_n - g_n)(x)$ to 0 for $x \in M$. By the Ascoli-Arzelà theorem, the set $\widetilde{M}$ is precompact. Therefore for any bounded sequence $\{x_n : n \geq 1\} \subset M$ there exists a convergent subsequence $\{F_{x_{n_k}} : k \geq 1\}$ in $C(B)$. This subsequence is a Cauchy sequence in $C(B)$. The corresponding subsequence $\{x_{n_k} : k \geq 1\}$ is a Cauchy sequence in X, so it converges.

10.78 *Necessity.* If $f_n \overset{w-*}{\longrightarrow} f$, then $f_n(y)$ converges to $f(y)$ uniformly for $y \in A(\overline{B}(0, 1))$ as a consequence of item (2) of problem 10.76. Thus $(A' f_n)(x) = f_n(Ax)$ converges to $(A' f)(x) = f(Ax)$ uniformly for $x \in \overline{B}(0, 1)$, hence $\|A' f_n - A' f\| \to 0$, $n \to \infty$.

Sufficiency. By Theorem 5.4 for any sequence of functionals $\{f_n : n \geq 1\} \subset \overline{B}(0, 1)$ in X_2^* there exist a subsequence $\{f_{n_k} : k \geq 1\}$ and a functional $f \in X_2^*$ such that $f_{n_k} \overset{w-*}{\longrightarrow} f$, $k \to \infty$. Then $A' f_{n_k} \to A' f$ in X_1^*. Therefore $A' \in S_\infty(X_2^*, X_1^*)$ and $A \in S_\infty(X_1, X_2)$ as a result of problem 10.74*.

10.79 The necessity of condition (b) follows from item (1) of problem 10.76.

Sufficiency. For any $\varepsilon > 0$ choose n such that $\|A_n x - x\| < \frac{\varepsilon}{2}$, $x \in M$. Then a finite $\frac{\varepsilon}{2}$-net for the precompact set $A_n(M)$ will be an ε-net for the set M. According to the Hausdorff criterion, the set M is precompact.

10.80 (1) Use the Hölder inequality.

(2) We have

$$\|x_h\|_1 \leq \int_a^b \frac{1}{2h} \int_{t-h}^{t+h} |x(s)| ds dt = \frac{1}{2h} \int_a^b \int_{-h}^{h} |x(u + t)| du dt =$$

$$= \frac{1}{2h} \int_a^b \int_{-h}^{h} |x(u + t)| dt du = \frac{1}{2h} \int_{-h}^{h} \int_{a-u}^{b-u} |x(s)| ds du \leq \|x\|_1.$$

Indeed, $\int\limits_{a-u}^{b-u} |x(s)| ds \leq \|x\|_1$, $u \in [-h, h]$, because $x(s) = 0$ when $s \notin [a, b]$.

(3) Let $p > 1$ and q be the conjugate index to p. By Hölder's inequality

$$|x_h(t)| = \frac{1}{2h}\left|\int\limits_{t-h}^{t+h} x(s)ds\right| \le \frac{1}{2h}\left(\int\limits_{t-h}^{t+h} |x(s)|^p ds\right)^{\frac{1}{2}} (2h)^{\frac{1}{q}},$$

i.e. $|x_h(t)|^p \le \frac{1}{2h}\int\limits_{t-h}^{t+h} |x(s)|^p ds = (A_h|x|^p)(t)$. Due to item (2) we have

$$\int\limits_a^b |x_h(t)|^p dt \le \left\|A_h|x|^p\right\|_1 \le \left\||x|^p\right\|_1 = \int\limits_a^b |x(t)|^p dt.$$

(4) Firstly, consider the case $x \in C([a, b])$, then use the density of this set in $L_p([a, b])$ and item (3).

(5) Verify that the set $A_h(B(0, 1))$, where $B(0, 1)$ is the ball in $L_p([a, b])$, satisfies the conditions of the Ascoli-Arzelà theorem. To prove the equicontinuity, use the estimate

$$|x_h(t_2) - x_h(t_1)| = \frac{1}{2h}\left|\int\limits_{t_1+h}^{t_2+h} x(s)ds - \int\limits_{t_1-h}^{t_2-h} x(s)ds\right| \le \frac{\|x\|_p}{h}|t_2 - t_1|^{\frac{1}{q}},$$

$t_1, t_2 \in [a, b]$.

(6) Apply the Ascoli-Arzelà theorem.

10.81 Use problem 10.79 and items (4) and (7) of problem 10.80.

10.82 *Necessity.* Consider operators $T_h \in \mathcal{L}\big(L_p([a, b])\big)$, $h > 0$, which are defined by the formulae $(T_h x)(t) = x(t + h)$, $t \in [a, b]$. For $x \in C([a, b])$ we have

$$\|T_h x - x\|_p \to 0,\ h \to 0+,$$

the set $C([a, b])$ is dense in $L_p([a, b])$ and $\|T_h - I\| \le 2$, hence $T_h \xrightarrow{s} I$, $h \to 0+$. As a result of item (1) of problem 10.76, the convergence $\|T_h x - x\|_p \to 0$, $h \to 0+$, is uniform for $x \in M$.

Sufficiency. Condition (b) of the current problem implies condition (b) of problem 10.81, hence the set M is compact according to this problem.

10.83° No. The operators B_n are compact as finite-dimensional operators. If the convergence is uniform, then item (3) of Theorem 10.2 implies that the identity operator in $C([0, 1])$ is compact, a contradiction.

10.84* For each $x \in X$ put $[x] = \{x + u \mid u \in \operatorname{Ker} A\}$. Since $\operatorname{Ker} A$ is a subspace, then the factor-space $X/\operatorname{Ker} A = \{[x] \mid x \in X\}$ is a Banach space with the norm

$\|[x]\| = \inf_{u \in \operatorname{Ker} A} \|x + u\|$ (see, for example, [21], Ch. 4, Sec. 13). Let us construct operators $C \in \mathcal{L}(Y, X/\operatorname{Ker} A)$ and $\widetilde{A} \in S_\infty(X/\operatorname{Ker} A, Z)$ such that $B = \widetilde{A}C$. It will follow that $B \in S_\infty(Y, Z)$.

Put $\widetilde{A}[x] = Ax$, $x \in X$, and $Cy = [x]$, if $Ax = By$ (such an element x exists because $R(B) \subset R(A)$). The operators $\widetilde{A}$ and C are defined correctly because if $[x'] = [x'']$, then $x' - x'' \in \operatorname{Ker} A$ and $Ax' = Ax''$, and if $[x'] \neq [x'']$, then $x' - x'' \notin \operatorname{Ker} A$ and $Ax' \neq Ax''$. Obviously, the operators $\widetilde{A}$ and C are linear and $B = \widetilde{A}C$. If $\{[x_n] : n \geq 1\}$ is a bounded sequence in $X/\operatorname{Ker} A$, then we can choose $x_n \in [x_n]$ such that $\{x_n : n \geq 1\}$ is a bounded sequence in X, hence there exists a convergent subsequence $\{Ax_{n_k} = \widetilde{A}[x_{n_k}] : k \geq 1\}$. Therefore $\widetilde{A} \in S_\infty(X/\operatorname{Ker} A, Z)$. Finally, if $y_n \to y$ and $Cy_n = [x_n] \to [x]$, $n \to \infty$, then $\widetilde{A}[x_n] \to \widetilde{A}[x]$ and $\widetilde{A}[x_n] = By_n \to By$, i.e. $Cy = [x]$. Thus the graph of the operator C is closed, therefore $C \in \mathcal{L}(Y, X/\operatorname{Ker} A)$ by the Closed Graph Theorem 6.1.

10.85 For any bounded sequence $\{x_n : n \geq 1\} \subset X_1$ choose a sequence $\{y_n : n \geq 1\} \subset M$ such that $\|x_n - y_n\| \to 0$, $n \to \infty$. Check that if the subsequence $\{Ay_{n_k} : k \geq 1\}$ converges, then the subsequence $\{Ax_{n_k} : k \geq 1\}$ also converges.

10.86 (1) Consider $\theta \in [a, b]$, for which $x(\theta) = \frac{1}{b-a} \int\limits_a^b x(s)ds$.

(2) Due to item (1) and the Hölder inequality

$$|x(t)| \leq \int\limits_a^b |x'(s)|ds + \frac{1}{b-a} \int\limits_a^b |x(s)|ds \leq$$

$$\leq c_1 \|x'\|_p + c_2 \|x\|_p \leq (c_1 + c_2) \|x\|_{W_p^1}, \quad t \in [a, b].$$

(3) It follows from item (2) that every sequence $\{x_n : n \geq 1\} \subset C^1([a, b])$, which is a Cauchy sequence in $W_p^1([a, b])$, is also a Cauchy sequence in $C([a, b])$, so it converges in $C([a, b])$. By the limit transition, we obtain that $\|x\|_C \leq c\|x\|_{W_p^1}$ for all $x \in W_p^1([a, b])$.

(4) Due to item (6) of problem 10.18, the restriction of the operator A to $C^1([a, b])$ belongs to $S_\infty\big(C^1([a, b]), C([a, b])\big)$. Therefore, due to item (3) of the current problem and problem 10.85, we have $A \in S_\infty\big(W_p^1([a, b]), C([a, b])\big)$.

10.87 (1) Use problem 6.47.

(2) Consider the sequence $x_n(t) = \frac{1}{n} e^{int}$, $t \in [a, b]$, $n \geq 1$, which is bounded in $W_p^1([a, b])$. Verify that $Dx_n \xrightarrow{w} 0$ in $L_p([a, b])$ and $\|Dx_n\|_p = (b - a)^{1/p} \not\to 0$, $n \to \infty$, so the sequence $\{Dx_n : n \geq 1\}$ does not have a convergent subsequence.

10.88 According to problem 10.70, it is sufficient to show that if $x_n \xrightarrow{w} 0$, then $Ax_n \to 0$, $n \to \infty$. For μ-almost all $t \in T$ we have $K(t, \cdot) \in L_q(T, \mu)$, hence $(Ax_n)(t) = \int\limits_T K(t, s) x_n(s) d\mu(s) \to 0 \,(\operatorname{mod} \mu)$, $n \to \infty$. The weakly convergent sequence is bounded. Let $\int_T |K(t,s)|^q d\mu(s) \le a \pmod{\mu}$ and $\|x_n\|_p \le b$, $n \ge 1$. By Hölder's inequality, for μ-almost all $t \in T$ we have

$$|(Ax_n)(t)| \le \left(\int_T |K(t,s)|^q d\mu(s) \right)^{1/q} \cdot \|x_n(s)\|_p \le a^{1/q} b \in L_p(T,\mu).$$

Therefore $\|Ax_n\|_p \to 0$, $n \to \infty$, by Lebesgue's dominated convergence theorem.

10.89 (2) By Hölder's inequality for all $x \in L_p(T,\mu)$, $y \in L_q(T,\mu)$

$$\left| \int_T (Ax)(t)y(t)d\mu(t) \right| \le \int_T \int_T |K(t,s)x(s)y(t)| d\mu(s) d\mu(t) \le$$

$$\le \left(\int_T \int_T |K(t,s)| \cdot |x(s)|^p d\mu(t) d\mu(s) \right)^{\frac{1}{p}} \times$$

$$\times \left(\int_T \int_T |K(t,s)| \cdot |y(t)|^q d\mu(s) d\mu(t) \right)^{\frac{1}{q}} \le c_1^{\frac{1}{p}} c_2^{\frac{1}{q}} \|x\|_p \|y\|_q,$$

whence $\|Ax\|_p = \sup\limits_{\|y\|_q = 1} \int_T (Ax)(t)y(t)d\mu(t) \le c_1^{\frac{1}{p}} c_2^{\frac{1}{q}} \|x\|_p$.

(3) In the space $L_p([0,1])$, $1 \le p < +\infty$, consider the kernel

$$K(t,s) = \begin{cases} 2^n, \ (t,s) \in [\frac{1}{2^n}, \frac{1}{2^{n-1}})^2, \ n \ge 1, \\ 0 \quad \text{otherwise}, \end{cases}$$

and the sequence $x_n(t) = 2^{\frac{n}{p}} \cdot \chi_{[\frac{1}{2^n}, \frac{1}{2^{n-1}})}(t)$, $t \in [0,1]$, $n \ge 1$, which is bounded in $L_p([0,1])$.

(4) The operators $(A_n x)(t) = \int_T K_n(t,s)x(s)d\mu(s)$, $t \in T$, $n \ge 1$, where

$$K_n(t,s) = K(t,s) \cdot \chi_{\{|K(t,s)| \le n\}},$$

belong to $S_\infty(L_p(T,\mu))$ according to problem 10.88. It follows from (2) that

$$\|A_n - A\| \le c_{1,n}^{\frac{1}{p}} c_2^{\frac{1}{q}} \to 0, \ n \to \infty,$$

hence $A \in S_\infty(L_p(T,\mu))$ by item (3) of Theorem 10.2.

10.90 Verify that the operator A satisfies the conditions of item (4) of problem 10.89.

10.91 (1) The function $[a, b] \ni t \mapsto K(t, \cdot) \in L_1([a, b])$ is continuous on $[a, b]$, so it is uniformly continuous. Verify that the set $A(B(0, 1))$, where $B(0, 1)$ is the ball in $C([a, b])$, satisfies the conditions of the Ascoli-Arzelà theorem.

(2) Apply the Ascoli-Arzelà theorem.

(3) Verify that the operator A satisfies the conditions of item (4) of problem 10.89.

10.92 The inclusion $A \in \mathcal{L}(L_2(\mathbb{R}))$ follows from item (2) of problem 10.89. The operator A is compact if and only if $u = 0$. Consider the sequence $x_n(t) = \chi_{[n,n+1)}(t)$, $t \in \mathbb{R}$, $n \geq 1$, and use item (5) of Theorem 10.2.

10.93 The operator A is compact only for $u = 0$.

10.94 (1) Use item (2) of problem 10.89.

(2) Let $K(t, s) = u(t + s)$, $t, s \in [0, +\infty)$. Verify that $K \in L_2([0, +\infty)^2)$, and use remark to the solution of item (2) of problem 10.42°.

10.95 (1) For all $x, y \in L_2([0, +\infty))$ by the Cauchy–Schwarz inequality we have

$$|(Ax, y)| \leq \left(\int_0^{+\infty} \int_0^{+\infty} \frac{|x(s)|^2}{t+s} \left(\frac{s}{t} \right)^{\frac{1}{2}} dtds \right)^{\frac{1}{2}} \cdot \left(\int_0^{+\infty} \int_0^{+\infty} \frac{|y(t)|^2}{t+s} \left(\frac{t}{s} \right)^{\frac{1}{2}} dsdt \right)^{\frac{1}{2}} =$$

$$= \left(\pi \int_0^{+\infty} |x(s)|^2 ds \right)^{\frac{1}{2}} \cdot \left(\pi \int_0^{+\infty} |y(t)|^2 dt \right)^{\frac{1}{2}} = \pi \|x\|_2 \cdot \|y\|_2.$$

(2) Consider the sequence $x_n(t) = \sqrt{n} \chi_{[\frac{1}{n}, \frac{2}{n}]}(t)$, $t \geq 0$, $n \geq 1$. Verify that $Ax_n \to 0 \, (\mathrm{mod}\, m)$ and $\|Ax_n\|_2 = \|Ax_1\|_2 \not\to 0$, $n \to \infty$, hence the sequence $\{Ax_n : n \geq 1\}$ does not contain a convergent subsequence.

10.96 (1) Let $K(t, s) = \chi_{\{0 \leq s \leq t \leq 1\}}$. Construct a sequence of functions $\{K_n \in C([0, 1]^2) : n \geq 1\}$ such that $\sup\limits_{s \in [0,1]} \int\limits_0^1 |K_n(t, s) - K(t, s)| dt \to 0$, $n \to \infty$. The operators $(A_n x)(t) = \int\limits_0^1 K_n(t, s) x(s) ds$, $t \in [0, 1]$, $n \geq 1$, are compact according to problem 10.4 and $A_n \rightrightarrows A$, $n \to \infty$, according to item (1) of problem 10.89. Therefore $A \in S_\infty(L_1([0, 1]))$ by item (3) of Theorem 10.2.

(2) Let $B_n = \left\{ t \in [0, 1) \;\middle|\; [2^n t] \text{ is an even number} \right\}$, $n \geq 1$. Put

$$K(t, s) = \sum_{n=1}^{\infty} \chi_{B_n}(t) \cdot \chi_{[\frac{1}{n+1}, \frac{1}{n})}(s)$$

(for any $(t, s) \in [0, 1]^2$ at most one term is nonzero). The integral operator A with kernel K is not compact. Indeed, the sequence $x_n(t) = n(n+1)\chi_{[\frac{1}{n+1}, \frac{1}{n})}(t)$, $n \geq 1$, is bounded in $L_1([0, 1])$ and $(Ax_n)(t) = \chi_{B_n}(t)$. It is easy to check that for $i \neq j$ we have $\|Ax_i - Ax_j\|_1 = m(B_i \triangle B_j) = \frac{1}{2}$, hence the sequence $\{Ax_n : n \geq 1\}$ does not contain a convergent subsequence.

10.97 (2) Consider $K(t, s) = f(t - s)$, where $f(t) = (1 - t^2)\chi_{[-1,1]}(t)$. For $x_n(t) = f(t - n)$ verify that $(Ax_n)(t) = (Af)(t - n)$, $t \in \mathbb{R}$, $n \geq 1$, and show that the sequence $\{Ax_n : n \geq 1\}$ does not contain a convergent subsequence.

10.98 (2) Show that for

$$x_n(t) = \begin{cases} nt, & 0 \leq t \leq \frac{1}{n}, \\ 1, & \frac{1}{n} < t \leq 1, \end{cases} \quad n \geq 1,$$

the sequence $\{Ax_n : n \geq 1\}$ does not contain a convergent subsequence.

(3) Note that $\|Ax\|_p = \sup\limits_{\|y\|_q=1} \int\limits_0^\infty (Ax)(t)y(t)dt$, where q is the conjugate index to p. For $\|y\|_q = 1$ we have

$$\left|\int\limits_0^\infty (Ax)(t)y(t)dt\right| \leq \int\limits_0^\infty |(Ax)(t)| \cdot |y(t)|dt \leq \int\limits_0^\infty \int\limits_0^t \frac{1}{t}|x(s)| \cdot |y(t)|dsdt =$$

$$= \int\limits_0^\infty \int\limits_0^1 |x(ts)| \cdot |y(t)|dsdt = \int\limits_0^1 \int\limits_0^\infty |x(ts)| \cdot |y(t)|dtds \leq$$

$$\leq \int\limits_0^1 \left(\int\limits_0^\infty |x(ts)|^p dt\right)^{\frac{1}{p}} \|y\|_q ds = \int\limits_0^1 s^{-\frac{1}{p}} \|x\|_p ds = \frac{p}{p-1}\|x\|_p.$$

Therefore $A \in \mathcal{L}(L_p([0, +\infty)))$ and $\|A\| \leq \frac{p}{p-1}$.

(4) Consider the sequence $x_n(t) = \frac{1}{n^{1/p}} \cdot \chi_{[0,n]}(t)$, $t \geq 0$, $n \geq 1$. Verify that $Ax_n \to 0 \,(\text{mod}\, m)$, $n \to \infty$, and $\|Ax_n\|_p \geq 1$, $n \geq 1$, so the sequence $\{Ax_n : n \geq 1\}$ does not contain a convergent subsequence.

Problems of Chap. 11

11.6° (2) Let $A(x_1, \ldots, x_m) = (x_2, 0, \ldots, 0)$. Then $A^2 = 0$, whence $r(A) = 0$.

11.7° $\|R_z(A)\| = \left\|-\sum\limits_{n=0}^\infty z^{-(n+1)}A^n\right\| \leq \sum\limits_{n=0}^\infty |z|^{-(n+1)}\|A\|^n = \frac{1}{|z|-\|A\|}$.

11.8° Since A commutes with $A-\lambda I$, then A commutes with $R_\lambda(A) = (A-\lambda I)^{-1}$ for $\lambda \in \rho(A)$.

11.9° $\sigma_p(A) = \{a_n : n \geq 1\}$, $\sigma(A) = \overline{\{a_n : n \geq 1\}}$, $\sigma_c(A) = \varnothing$, $\sigma_r(A) = \sigma(A)\backslash\sigma_p(A)$, $R_\lambda(A)y = \left(\frac{y_1}{\alpha_1-\lambda}, \frac{y_2}{\alpha_2-\lambda}, \ldots\right)$, $y \in l_\infty$, $r(A) = \sup_{n\geq 1}|a_n|$. For $\lambda \in \sigma(A)\backslash\sigma_p(A)$ show that the distance from the element $(1, 1, \ldots)$ to the set $R(A-\lambda I)$ equals 1, therefore $\overline{R(A-\lambda I)} \neq l_\infty$ and $\lambda \in \sigma_r(A)$.

11.10° (1) $\sigma_p(A) = \sigma(A) = \{1\}$, $r(A) = 1$, $R_\lambda(A)y = \left(\frac{y_1}{1-\lambda} - \frac{y_2}{(1-\lambda)^2}, \frac{y_2}{1-\lambda}, \frac{y_3}{1-\lambda}, \ldots\right)$, $y \in l_2$; (2) $\sigma_p(A) = \sigma(A) = \left\{1, \frac{5\pm\sqrt{33}}{2}\right\}$, $r(A) = \frac{5+\sqrt{33}}{2}$,

$$R_\lambda(A)y = \left(\frac{(4-\lambda)y_1-2y_2}{\lambda^2-5\lambda-2}, \frac{(1-\lambda)y_2-3y_1}{\lambda^2-5\lambda-2}, \frac{y_3}{1-\lambda}, \frac{y_4}{1-\lambda}, \ldots\right), \quad y \in l_2;$$

(3) $\sigma_p(A) = \sigma(A) = \{-2, 1, 1 \pm \sqrt{3}i\}$, $r(A) = 2$,

$$R_\lambda(A)y = \left(\frac{-\lambda^2 y_1+8y_2+2\lambda y_3}{\lambda^3+8}, \frac{-\lambda y_1-\lambda^2 y_2+2y_3}{\lambda^3+8}, \frac{-4y_1-4\lambda y_2-\lambda^2 y_3}{\lambda^3+8}, \frac{y_4}{1-\lambda}, \frac{y_5}{1-\lambda}, \ldots\right),$$
$$y \in l_2;$$

(4) $\sigma_p(A) = \sigma(A) = \{0, 1\}$, $r(A) = 1$, $R_\lambda(A)y = \left(\frac{y_1}{1-\lambda}, \ldots, \frac{y_k}{1-\lambda}, -\frac{y_{k+1}}{\lambda}, -\frac{y_{k+2}}{\lambda}, \ldots\right)$, $y \in l_2$; (5) $\sigma_p(A) = \sigma(A) = \{0\}$, $r(A) = 0$,

$$R_\lambda(A)y = \left(-\frac{y_1}{\lambda}, -\frac{y_1}{\lambda^2} - \frac{y_2}{\lambda}, -\frac{y_1}{\lambda^3} - \frac{y_2}{\lambda^2} - \frac{y_3}{\lambda}, -\frac{y_4}{\lambda}, -\frac{y_5}{\lambda}, \ldots\right), \quad y \in l_2;$$

(6) $\sigma_p(A) = \sigma(A) = \{0, -1, 5\}$, $r(A) = 5$,

$$R_\lambda(A)y = \left(\frac{(4-\lambda)y_1-2y_2}{\lambda^2-5\lambda}, \frac{(1-\lambda)y_2-2y_1}{\lambda^2-5\lambda}, -\frac{y_3}{1+\lambda}, -\frac{y_4}{1+\lambda}, \ldots\right), \quad y \in l_2;$$

(7) $\sigma_p(A) = \sigma(A) = \{0, -1, 3\}$, $r(A) = 3$,

$$R_\lambda(A)y = \left(\frac{(1-\lambda)y_1-2y_2}{\lambda^2-2\lambda-3}, \frac{(1-\lambda)y_2-2y_1}{\lambda^2-2\lambda-3}, \frac{y_3}{1-\lambda}, \frac{y_4}{1-\lambda}, \ldots\right), \quad y \in l_2;$$

(8) $\sigma_p(A) = \sigma(A) = \{0, 2 \pm 3i\}$, $r(A) = \sqrt{13}$,

$$R_\lambda(A)y = \left(\frac{(1-\lambda)y_1-5y_2}{\lambda^2-4\lambda+13}, \frac{(3-\lambda)y_2+2y_1}{\lambda^2-4\lambda+13}, -\frac{y_3}{\lambda}, -\frac{y_4}{\lambda}, \ldots\right), \quad y \in l_2;$$

(9) $\sigma_p(A) = \sigma(A) = \{1, \lambda_1, \lambda_2\}$, where $\lambda_{1,2} = \frac{1}{2}\left(\alpha + \delta \pm \sqrt{(\alpha-\delta)^2 + 4\beta\gamma}\right)$, $r(A) = \max\{1, |\lambda_1|, |\lambda_2|\}$,

$$R_\lambda(A)y = \left(\frac{(\delta-\lambda)y_1-\beta y_2}{(\alpha-\lambda)(\delta-\lambda)-\beta\gamma}, \frac{(\alpha-\lambda)y_2-\gamma y_1}{(\alpha-\lambda)(\delta-\lambda)-\beta\gamma}, \frac{y_3}{1-\lambda}, \frac{y_4}{1-\lambda}, \ldots\right), \quad y \in l_2;$$

(10) $\sigma_p(A) = \sigma(A) = \{1, 2, 4\}$, $r(A) = 4$,

$$R_\lambda(A)y = \left(\frac{(5-\lambda)y_1 - y_3}{(\lambda-4)^2}, \frac{y_2}{2-\lambda}, \frac{(3-\lambda)y_3 + y_1}{(\lambda-4)^2}, \frac{y_4}{1-\lambda}, \frac{y_5}{1-\lambda}, \dots\right), \ y \in l_2.$$

11.11° (1) $\|A\| = 1$, $r(A) = 0$, $\sigma_p(A) = \varnothing$, $\sigma(A) = \{0\}$. To find $r(A)$, calculate A^n and show that $\|A^n\| = \frac{1}{n!}$, $n \geq 1$. (2) $\|A\| = 1$, $r(A) = 0$, $\sigma_p(A) = \sigma(A) = \{0\}$. Verify that $A^{100} = 0$. (3) $\|A\| = 2$, $r(A) = 1$, $\sigma_p(A) = \{|\lambda| < 1\}$, $\sigma(A) = \{|\lambda| \leq 1\}$. To find $r(A)$, calculate A^n and show that $\|A^n\| = n + 1$, $n \geq 1$. If $Ax = \lambda x$, $x \neq 0$, then $x = x_1(1, \frac{\lambda}{2}, \frac{\lambda^2}{3}, \dots, \frac{\lambda^n}{n+1}, \dots)$, where $x_1 \neq 0$. The condition $x \in l_2$ is fulfilled if and only if $|\lambda| < 1$. (4) $\|A\| = 4$, $r(A) = 1$, $\sigma_p(A) = \sigma(A) = \{|\lambda| \leq 1\}$.

11.12 (1) $\sigma_p(A) = \varnothing$, $\sigma(A) = \sigma_r(A) = [a + 1, b + 1]$, $\sigma_c(A) = \varnothing$,

$$(R_\lambda(A)y)(t) = \frac{y(t)}{t+1-\lambda}, \ t \in [a, b], \ y \in C([a, b]);$$

(2) $\sigma_p(A) = \sigma_r(A) = \varnothing$, $\sigma(A) = \sigma_c(A) = [e^a, e^b]$,

$$(R_\lambda(A)y)(t) = \frac{y(t)}{e^t-\lambda}, \ t \in [a, b], \ y \in L_2([a, b]);$$

(3) $\sigma_p(A) = \varnothing$, $\sigma(A) = \sigma_r(A) = [-1, 1]$, $\sigma_c(A) = \varnothing$,

$$(R_\lambda(A)y)(t) = \frac{y(t)}{\sin t-\lambda}, \ t \in [0, 2\pi], \ y \in C([0, 2\pi]);$$

(4) $\sigma_p(A) = \sigma(A) = \{0, 1\}$, $\sigma_c(A) = \sigma_r(A) = \varnothing$,

$$(R_\lambda(A)y)(t) = \frac{y(t)}{1-\lambda} \chi_{[0, \frac{1}{2}]}(t) - \frac{y(t)}{\lambda} \chi_{(\frac{1}{2}, 1]}(t), \ t \in [0, 1], \ y \in L_2([0, 1]);$$

(5) $\sigma_p(A) = \{\lambda \in a([0, 1]) \mid$ the set $a^{-1}(\{\lambda\})$ has an interior point$\}$, $\sigma_c(A) = \varnothing$, $\sigma(A) = a([0, 1])$, $\sigma_r(A) = \sigma(A) \setminus \sigma_p(A)$,

$$(R_\lambda(A)y)(t) = \frac{y(t)}{a(t)-\lambda}, \ t \in [0, 1], \ y \in C([0, 1]).$$

(6) $\sigma_p(A) = \{\lambda \in a([0, 1]) \mid$ the set $a^{-1}(\{\lambda\})$ has positive Lebesgue measure$\}$, $\sigma(A) = a([0, 1])$, $\sigma_r(A) = \varnothing$, $\sigma_c(A) = \sigma(A) \setminus \sigma_p(A)$,

$$(R_\lambda(A)y)(t) = \frac{y(t)}{a(t)-\lambda}, \ t \in [0, 1], \ y \in L_2([0, 1]).$$

11.13 $\sigma_p(A) = \big\{\lambda \in \mathbb{C} \bigm| \mu(\{t \in T \mid a(t) = \lambda\}) > 0\big\}$.

11.14° (1) If $\lambda \in \mathbb{C} \setminus \{0\}$, then

$$A^{-1} - \frac{1}{\lambda} I = -\frac{1}{\lambda} A^{-1}(A - \lambda I).$$

Deduce that $\frac{1}{\lambda} \in \rho(A^{-1}) \setminus \{0\}$ if and only if $\lambda \in \rho(A) \setminus \{0\}$.

11.15* (3) If $P(z) = \text{const}$, the statement is obvious. Let P be a polynomial of degree $n \geq 1$, $\mu \in \mathbb{C}$ and $P(z) - \mu = c(z - \lambda_1) \dots (z - \lambda_n)$, where $c \neq 0$. Then $P(A) - \mu I = c(A - \lambda_1 I) \dots (A - \lambda_n I)$. If $\lambda_1, \dots, \lambda_n \in \rho(A)$, then all operators $A - \lambda_i I$ are continuously invertible, therefore the operator $P(A) - \mu I$ is also continuously invertible. Now assume that $\lambda_i \in \sigma(A)$ for some $1 \leq i \leq n$ and $P(z) - \mu = (z - \lambda_i)Q_i(z)$. Since

$$P(A) - \mu I = Q_i(A)(A - \lambda_i I) = (A - \lambda_i I)Q_i(A),$$

then $\text{Ker}(A - \lambda_i I) \subset \text{Ker}(P(A) - \mu I)$ and $R(P(A) - \mu I) \subset R(A - \lambda_i I)$. The operator $A - \lambda_i I$ is not continuously invertible, therefore by Theorem 8.1 this operator is not bijective, i.e. $\text{Ker}(A - \lambda_i I) \neq \{0\}$ or $R(A - \lambda_i I) \neq X$. It follows that $\text{Ker}(P(A) - \mu I) \neq \{0\}$ or $R(P(A) - \mu I) \neq X$, so the operator $P(A) - \mu I$ is not continuously invertible. Therefore $\mu \in \sigma(P(A))$ if and only if $\mu = P(\lambda)$ for some $\lambda \in \sigma(A)$.

(4) We will consider $P(z) \neq \text{const}$ and use notations from the solution of item (3). If $\lambda_i \in \sigma_p(A)$, then $\text{Ker}(A - \lambda_i I) \neq \{0\}$, hence $\text{Ker}(P(A) - \mu I) \neq \{0\}$, so $\mu \in \sigma_p(P(A))$. Now assume that $\mu \in \sigma_p(P(A))$, i.e.

$$(P(A) - \mu I)x = c(A - \lambda_1 I) \dots (A - \lambda_n I)x = 0$$

for some $x \neq 0$. Choose the largest $i \leq n$, for which $(A - \lambda_i I) \dots (A - \lambda_n I)x = 0$. Then $\lambda_i \in \sigma_p(A)$. Thus $\mu \in \sigma_p(P(A))$ if and only if $\mu = P(\lambda)$ for some $\lambda \in \sigma_p(A)$.

11.16 (1) Since $P(A) = 0$, then $\{P(\lambda) \mid \lambda \in \sigma(A)\} = \{0\}$ according to item (3) of problem 11.15*, i.e. $\sigma(A) \subset \{\lambda \in \mathbb{C} \mid P(\lambda) = 0\}$.

(2) Let λ be a root of the polynomial P and $P(z) = (z - \lambda)Q(z)$. Then $P(A) = (A - \lambda I)Q(A)$. Since $Q(A) \neq 0$, there exists $x \in X$ such that $y = Q(A)x \neq 0$ and $P(A)x = (A - \lambda I)y = 0$. Then y is an eigenvector of the operator A, which corresponds to the eigenvalue λ.

11.17° Use item (1) of problem 11.16.

(1) $A = 0$; (2) $A = I$; (3) $X = \mathbb{C}^2$, $A(x_1, x_2) = (x_1, 0)$.

11.18 The operator $A^2 - A$ is self-adjoint and $\sigma(A^2 - A) = \{0\}$ due to item (3) of problem 11.15*. Therefore $\|A^2 - A\| = r(A^2 - A) = 0$, i.e. $A^2 = A$ and according to item (2) of problem 9.5 the operator A is an orthogonal projector.

11.19° $\sigma(A) = \sigma(B) = \{0, 1\}$. Verify that $A^2 = A$ and $B^2 = B$ and use item (2) of problem 11.16.

11.20° $\sigma(A) = \sigma_p(A) = \{-1, 1\}$. The eigenvectors that correspond to the eigenvalues $\lambda = 1$ and $\lambda = -1$ are even and odd functions, respectively.

11.21 (1) If the operators A and A^* commute, then for any $\lambda \in C$ the operators $A - \lambda I$ and $(A - \lambda I)^* = A^* - \overline{\lambda} I$ also commute, i.e. the operator $A - \lambda I$ is normal. Due to item (1) of problem 9.32 we have $\operatorname{Ker}(A - \lambda I) = (R(A - \lambda I))^{\perp}$, therefore if $\overline{R(A - \lambda I)} \neq H$, then $\operatorname{Ker}(A - \lambda I) \neq \{0\}$ and $\lambda \in \sigma_p(A)$.

(2) According to item (2) of problem 9.32, we have $\|A^2\| = \|A\|^2$. Show by induction that the operators A^{2^k} are normal and $\|A^{2^k}\| = \|A\|^{2^k}$, $k \geq 1$.

11.22° $\sigma_p(A) = \sigma(A) = \{a + b, a - b\}$, $(R_\lambda(A)y)(t) = \frac{(a-\lambda)y(t) - by(-t)}{(a-\lambda)^2 - b^2}$.

11.23 Let $M \subset \mathbb{C}$ be a nonempty compact set. Show that there exists a sequence $\{\alpha_n : n \geq 1\} \subset M$, for which $\overline{\{\alpha_n : n \geq 1\}} = M$, and consider the operator from problem 11.1.

11.24° It follows from the condition of the problem that $\|A\| = \|A^{-1}\| = 1$. If $\lambda \in \sigma(A)$, then $\frac{1}{\lambda} \in \sigma(A^{-1})$ by item (1) of problem 11.14°. Hence $|\lambda| \leq r(A) \leq 1$ and $|\frac{1}{\lambda}| \leq r(A^{-1}) \leq 1$, i.e. $|\lambda| = 1$.

11.25° Use problem 11.24°.

11.26° Let $\lambda \notin \{0, 1\}$, $(P - \lambda I)x = y$. Since $x = x_1 + x_2$ and $y = y_1 + y_2$, where $x_1, y_1 \in R(P)$, $x_2, y_2 \in (R(P))^{\perp}$, then $x_1 - \lambda x_1 - \lambda x_2 = y_1 + y_2$, whence $x_1 = \frac{y_1}{1-\lambda}$, $x_2 = -\frac{1}{\lambda} y_2$.

11.27 (1) Due to item (4) of problem 9.76, the operator $A - \lambda I \in \mathcal{L}(X)$ is continuously invertible if and only if the operator $(A - \lambda I)' = A' - \lambda I'$ is continuously invertible, where I' is the identity operator in X^*. Therefore $\rho(A') = \rho(A)$, hence $\sigma(A') = \sigma(A)$.

(2) Let $\lambda \in \sigma_r(A)$ and $B = A - \lambda I$. Then $\overline{R(B)} \neq X$. By Corollary 4.1 of the Hahn–Banach theorem, there exists $f \in X^*$, $f \neq 0$, such that $f(Bx) = (B'f)(x) = 0$ for all $x \in X$. Thus $B'f = 0$, i.e. $\lambda \in \sigma_p(A')$.

(3) Let $\lambda \in \sigma_p(A)$ and $B = A - \lambda I$. There exists $x \in X$, $x \neq 0$, such that $Bx = 0$. Then $(B'f)(x) = f(Bx) = 0$ for all $f \in X^*$. By Corollary 4.2 of the Hahn–Banach theorem, there exists $g \in X^*$ such that $g(x) \neq 0$. Then $g \notin \overline{R(B')}$. Therefore $\overline{R(B')} \neq X^*$, i.e. $\lambda \in \sigma_p(A') \cup \sigma_r(A')$.

11.28 (1) $\sigma_p(A) = \{|\lambda| < 1\}$, $\sigma(A) = \{|\lambda| \leq 1\}$, $\sigma_r(A) = \varnothing$, $\sigma_c(A) = \{|\lambda| = 1\}$, $\sigma_p(B) = \varnothing$, $\sigma(B) = \{|\lambda| \leq 1\}$, $\sigma_r(B) = \{|\lambda| < 1\}$, $\sigma_c(B) = \{|\lambda| = 1\}$;

(2) $\sigma_p(A) = \{|\lambda| < 1\}$, $\sigma(A) = \{|\lambda| \leq 1\}$, $\sigma_r(A) = \varnothing$, $\sigma_c(A) = \{|\lambda| = 1\}$, $\sigma_p(B) = \varnothing$, $\sigma(B) = \sigma_r(B) = \{|\lambda| \leq 1\}$, $\sigma_c(B) = \varnothing$;

(3) $\sigma_p(A) = \sigma(A) = \{|\lambda| \leq 1\}$, $\sigma_c(A) = \sigma_r(A) = \varnothing$, $\sigma_p(B) = \varnothing$, $\sigma(B) = \sigma_r(B) = \{|\lambda| \leq 1\}$, $\sigma_c(B) = \varnothing$.

Repeat the reasonings, similar to the solution of problem 11.4, and use problem 11.28. In item (2) we have to prove separately that $\{|\lambda| = 1\} \subset \sigma_r(B)$. To do this, verify that $a = (\lambda^{-1}, \lambda^{-2}, \lambda^{-3}, \ldots) \notin \overline{R(B - \lambda I)}$ for $|\lambda| = 1$. If $x = (x_1, x_2, \ldots)$ and $(B - \lambda I)x = y = (y_1, y_2, \ldots)$, then $y_1 = -\lambda x_1$ and $y_n = x_{n-1} - \lambda x_n$, $n \geq 2$, whence $x_n = -\left(\sum\limits_{j=1}^{n} y_j \lambda^j\right) \lambda^{-n-1}$, $n \geq 1$. Assume that $y \in R(B - \lambda I)$ and $\|y - a\|_\infty \leq \frac{1}{2}$. Then $\operatorname{Re}(y_j \lambda^j) \geq \operatorname{Re}(\lambda^{-j} \cdot \lambda^j) - \|y - a\|_\infty \geq \frac{1}{2}$, $j \geq 1$, and $|x_n| = \left|\sum\limits_{j=1}^{n} y_j \lambda^j\right| \geq \frac{n}{2}$, $n \geq 1$, which is impossible for $x \in l_\infty$.

11.29 (1) $\sigma_p(A) = \{|\lambda| < 1\}$, $\sigma(A) = \{|\lambda| \leq 1\}$, $\sigma_r(A) = \varnothing$, $\sigma_c(A) = \{|\lambda| = 1\}$, $r(A) = 1$, $R_\lambda(A)y = x$, where $x_n = -\sum\limits_{k=0}^{\infty} \frac{y_{n+2k}}{\lambda^{k+1}}$, $n \geq 1$, $y \in l_p$.

(2) $\sigma_p(A) = \varnothing$, $\sigma(A) = \{|\lambda| \leq 1\}$, $\sigma_r(A) = \{|\lambda| < 1\}$, $\sigma_c(A) = \{|\lambda| = 1\}$, $r(A) = 1$, $R_\lambda(A)y = x$, where $x_n = -\sum\limits_{k=0}^{[n/2]} \frac{y_{n-2k}}{\lambda^{k+1}}$, $n \geq 1$, $y \in l_p$.

11.30° If $\dim H > 1$, then $\sigma_p(A) = \sigma(A) = \{0, (z, y)\}$, $R_\lambda(A)v = \frac{1}{\lambda}\left(\frac{(v,y)z}{(z,y)-\lambda} - v\right)$; if $\dim H = 1$, then $\sigma_p(A) = \sigma(A) = \{(z, y)\}$, $R_\lambda(A)v = \frac{v}{(z,y)-\lambda}$; $r(A) = |(z, y)|$, $A^n x = (x, y)(z, y)^{n-1} z$, $n \geq 1$.

11.32° (1) $(A^n x)(t) = \int\limits_0^t \frac{(t-s)^{n-1}}{(n-1)!} x(s)ds$, $t \in [0, 1]$, $n \geq 1$, $r(A) = 0$, $\sigma(A) = \{0\}$.

(2) $(R_\lambda(A)y)(t) = -\frac{1}{\lambda} y(t) - \frac{1}{\lambda^2} \int\limits_0^t e^{-\frac{t-s}{\lambda}} y(s)ds$, $t \in [0, 1]$, $y \in L_2([0, 1])$.

Apply the formula $R_\lambda(A) = -\sum\limits_{n=0}^{\infty} \lambda^{-(n+1)} A^n$, which is valid for $|\lambda| > r(A) = 0$.

(3) $\big((A + A^*)y\big)(t) = \int\limits_0^1 y(s)ds$, $t \in [0, 1]$, $y \in L_2([0, 1])$.

11.33 $\sigma(A) = \{0\}$. Since the function K is continuous on the compact set

$$\{(t, s) \mid 0 \leq s \leq t \leq 1\},$$

K is bounded. Let $|K(t, s)| \leq C$, $0 \leq s \leq t \leq 1$. Establish by induction that $(A^n x)(t) = \int\limits_0^t K_n(t, s)x(s)ds$, $t \in [0, 1]$, where

$$K_1(t, s) = K(t, s), \quad K_n(t, s) = \int_s^t K_1(t, u)K_{n-1}(u, s)du, \quad n \geq 2, \; 0 \leq s \leq t \leq 1.$$

Then verify by induction that

$$|K_n(t,s)| \le \frac{C^n(t-s)^{n-1}}{(n-1)!} \le \frac{C^n}{(n-1)!}, \quad 0 \le s \le t \le 1.$$

Therefore $\|A^n\| \le \frac{C^n}{(n-1)!}$, $n \ge 1$, whence $r(A) = 0$.

11.34 $\sigma(A) = \{0\}$. Show that $A^2 = 0$.

11.35 (1) The operator $A_2 - \lambda I_2 = B^{-1}(A_1 - \lambda I_1)B$, where I_1 and I_2 are the identity operators in X_1 and X_2, is continuously invertible if and only if the operator $A_1 - \lambda I_1$ is continuously invertible.

(2) If $A_1 x = \lambda x$ and $x \ne 0$, then $A_2(B^{-1}x) = B^{-1}A_1BB^{-1}x = \lambda(B^{-1}x)$. Hence $\sigma_p(A_1) \subset \sigma_p(A_2)$, similarly $\sigma_p(A_2) \subset \sigma_p(A_1)$.

(3) In view of items (1) and (2), we have $\sigma_c(A_1) \cup \sigma_r(A_1) = \sigma_c(A_2) \cup \sigma_r(A_2)$. Since $A_2 - \lambda I_2 = B^{-1}(A_1 - \lambda I_1)B$, then $y \in R(A_2 - \lambda I_2)$ if and only if $By \in R(A_1 - \lambda I_1)$. Therefore $\overline{R(A_2 - \lambda I_2)} \ne X_2$ if and only if $\overline{R(A_1 - \lambda I_1)} \ne X_1$.

11.36 Denote by X_i the space X endowed with the norm $\|\cdot\|_i$, and by A_i the operator A on the space X_i, $i = 1, 2$. Consider the identity operator $B : X_2 \to X_1$. If the norms $\|\cdot\|_1$ and $\|\cdot\|_2$ are equivalent, then the operator B is continuously invertible and $A_2 = B^{-1}A_1B$. It remains to apply problem 11.35.

11.37 *Method I.* Repeat the reasonings, similar to the solution of problem 11.5.

Method II. Consider the operator B_s in $L_2(\mathbb{R})$, which is defined by the formula $(B_s x)(t) = x(\frac{t}{s})$, $t \in \mathbb{R}$. Then $A_1 = B_s^{-1}A_sB_s$ for any $s \in \mathbb{R}\backslash\{0\}$. Use problems 11.35 and 11.5.

11.38 (1) Use the inequality $\|(AB)^n\| \le \|A^n\| \cdot \|B^n\|$.

(2) Since

$$\forall \varepsilon > 0 \, \exists N \in \mathbb{N} \, \forall n \ge N \ : \ \|A^n\|^{\frac{1}{n}} < r(A) + \varepsilon, \ \|B^n\|^{\frac{1}{n}} < r(B) + \varepsilon,$$

then for all $n > 2N$ we have

$$\|(A+B)^n\| \le \sum_{k=0}^{n} \binom{n}{k} \|A^k\| \cdot \|B^{n-k}\| \le$$

$$\le (r(A) + r(B) + 2\varepsilon)^n + \sum_{k=0}^{N} \binom{n}{k} \|A^k\| \cdot \|B^{n-k}\| + \sum_{k=n-N}^{n} \binom{n}{k} \|A^k\| \cdot \|B^{n-k}\| \le$$

$$\le (r(A)+r(B)+2\varepsilon)^n + \sum_{k=0}^{N} \binom{n}{k} \|A^k\| \cdot (r(B)+\varepsilon)^{n-k} + \sum_{k=n-N}^{n} \binom{n}{k} (r(A)+\varepsilon)^k \cdot \|B^{n-k}\|.$$

It follows that $\frac{\|(A+B)^n\|}{(r(A)+r(B)+2\varepsilon)^n} \le 1 + cq^n$ for some $c > 0$ and $0 < q < 1$. By letting $n \to \infty$, we get $r(A+B) \le r(A) + r(B) + 2\varepsilon$.

The condition $AB = BA$ is essential. For example, for $X = \mathbb{C}^2$, $Ax = \left(\begin{smallmatrix} 0 & 1 \\ 0 & 0 \end{smallmatrix}\right) x$, $Bx = \left(\begin{smallmatrix} 0 & 0 \\ 1 & 0 \end{smallmatrix}\right) x$ we have $r(A) = r(B) = 0$ and $r(AB) = r(A+B) = 1$.

11.39° Prove by induction on n that if $\|x\| \le 1$, then $|(A^n x)(t)| \le t^{2^n - 1}$, $t \in [0, \frac{1}{2}]$, $n \ge 1$. Then $\|A^n\| \le \left(\frac{1}{2}\right)^{2^n - 1}$, $n \ge 1$, and $\sqrt[n]{\|A^n\|} \to 0$, $n \to \infty$.

11.41 Use problem 9.41.

11.42 It is enough to consider $\lambda \in \mathbb{R}$, because $\sigma(A) \subset \mathbb{R}$ by Theorem 11.3. We will show that $\operatorname{Ker}(A - \lambda I) = \{0\}$. Indeed, if $Ax = \lambda x$ for some $x \ne 0$, then $(x, (A - \lambda I)y) = ((A - \lambda I)x, y) = 0$ for all $y \in H$, whence $R(A - \lambda I) \subset \{x\}^\perp$ and $R(A - \lambda I) \ne H$, a contradiction. Since $\operatorname{Ker}(A - \lambda I) = \{0\}$ and $R(A - \lambda I) = H$, then $\lambda \in \rho(A)$ by Theorem 8.1.

11.43 Check that $\|A\| = 1$, so $r(A) \le 1$. Since for $e_0(t) = 1$, $t \in [0, 1]$, we have $Ae_0 = e_0$, then $1 \in \sigma_p(A) \subset \sigma(A)$, so $r(A) \ge 1$.

11.46 Since $(AB)^n = A(BA)^{n-1}B$, then

$$\sqrt[n]{\|(AB)^n\|} \le \sqrt[n]{\|A\|} \sqrt[n]{\|(BA)^{n-1}\|} \sqrt[n]{\|B\|}, \quad n \ge 1.$$

Now Theorem 11.2 implies that $r(AB) \le r(BA)$, similarly $r(BA) \le r(AB)$. Also the equality $r(AB) = r(BA)$ follows immediately from problem 11.47*.

11.47* For $\lambda \in \rho(BA)\backslash\{0\}$ consider the operator $C = \frac{1}{\lambda}(A(BA - \lambda I)^{-1}B - I)$. Verify that $(AB - \lambda I)C = C(AB - \lambda I) = I$, whence $(BA - \lambda I)^{-1} = C \in \mathcal{L}(X)$, i.e. $\lambda \in \rho(AB)\backslash\{0\}$. Similarly, if $\lambda \in \rho(AB)\backslash\{0\}$, then $\lambda \in \rho(BA)\backslash\{0\}$. Therefore $\rho(AB)\backslash\{0\} = \rho(BA)\backslash\{0\}$, so $\sigma(AB)\backslash\{0\} = \sigma(BA)\backslash\{0\}$.

11.48 It follows from the condition of the problem that $\sigma(AB) = \{\lambda + c \,|\, \lambda \in \sigma(BA)\}$, while according to problem 11.47* we have $\sigma(AB)\backslash\{0\} = \sigma(BA)\backslash\{0\}$. Assume that $c \ne 0$. If $\lambda \in \sigma(AB)$ and $\lambda \ne 0$, then $\lambda \in \sigma(BA)$ and $\lambda + c \in \sigma(AB)$, and if $\lambda \in \sigma(AB)$ and $\lambda \ne c$, then $\lambda - c \in \sigma(BA)$ and $\lambda - c \in \sigma(AB)$. Hence for $\lambda_0 \in \sigma(AB)$ we obtain by induction that $\lambda_0 + nc \in \sigma(AB)$ for every $n \ge 1$ or $\lambda_0 - nc \in \sigma(AB)$ for every $n \ge 1$. However, the set $\sigma(AB)$ is bounded, a contradiction.

Problems of Chap. 12

12.3° Sets in items (2), (3), (6), (7).

12.4° (1), (2) A is compact, because it is finite-dimensional, $\sigma(A) = \{0, \frac{1}{5}\}$.

12.5° (1), (2) A is compact, $\sigma(A) = \left\{0, \int\limits_a^b u(s)v(s)ds\right\}$.

12.6° (1) $\sigma(A) = \{0, 1\}$; (2) $\sigma(A) = \{0, \frac{1}{6}\}$; (3) $\sigma(A) = \{0, \sin 1 - \cos 1\}$;
(4) $\sigma(A) = \{0, \frac{1}{2}\}$; (5) $\sigma(A) = \{0, \frac{1}{\alpha+1}\}$; (6) $\sigma(A) = \{0, 1\}$; (7) $\sigma(A) = \{0, 1, 2\}$;
(8) $\sigma(A) = \{0, \frac{\ln 2}{2}\}$.

12.7° (1) $\sigma(A) = \{0, \frac{1}{\alpha+1}\}$; (2) $\sigma(A) = \{0, \frac{1+e^2}{4}\}$; (3) $\sigma(A) = \{0, \frac{1}{7}\}$;
(4) $\sigma(A) = \{0, \frac{1-e^{-2}}{2}\}$; (5) $\sigma(A) = \{0, \frac{1}{5}\}$; (6) $\sigma(A) = \{0, 1 - \ln 2\}$;
(7) $\sigma(A) = \{0, \frac{1}{\alpha+\beta+1}\}$; (8) $\sigma(A) = \{0, e^{2/3} - e^{1/3}\}$.

12.8° (1) $\sigma(A) = \{0, -\frac{4}{5}\}$; (2) $\sigma(A) = \{0, \frac{1}{10}, \frac{9}{10}\}$; (3) $\sigma(A) = \left\{0, \frac{3\pm\sqrt{21}}{2}\right\}$;
(4) $\sigma(A) = \{0, \pm\frac{1}{4}\}$.

12.9° A is compact in items (1)–(3), (5), (6) and (8).
(1) $\sigma(A) = \{0, \pi, 2\pi\}$, $\|A\| = 2\pi$; (2) $\sigma(A) = \{0, -\pi i, \pi i\}$, $\|A\| = \pi$;
(3) $\sigma(A) = \{0, -\frac{\pi}{2}, \frac{\pi}{2}\}$, $\|A\| = \frac{\pi}{2}$; (4) $\sigma(A) = \{2, \frac{8}{3}, \frac{14}{9}\}$, $\|A\| = \frac{8}{3}$;
(5) $\sigma(A) = \left\{0, \frac{9}{10} - \frac{\sqrt{19}}{5}, \frac{9}{10} + \frac{\sqrt{19}}{5}\right\}$, $\|A\| = \frac{9}{10} + \frac{\sqrt{19}}{5}$;
(6) $\sigma(A) = \left\{0, -\frac{2i}{\sqrt{15}}, \frac{2i}{\sqrt{15}}\right\}$, $\|A\| = \frac{2}{\sqrt{15}}$;
(7) $\sigma(A) = \{3i, (3 - \pi\sqrt{2})i, (3 + \pi\sqrt{2})i\}$, $\|A\| = 3 + \pi\sqrt{2}$;
(8) $\sigma(A) = \{0, \frac{8}{15}, -\frac{2}{15}\}$, $\|A\| = \frac{8}{15}$; (9) $\sigma(A) = \{1, \frac{5}{3}, \frac{11}{5}\}$, $\|A\| = \frac{11}{5}$;
(10) $\sigma(A) = \{-1, -1 + \pi\}$, $\|A\| = \pi - 1$.
To find the norm $\|A\|$, in items (1), (3)–(5), (8)–(10) use that the operator A is self-adjoint, therefore $r(A) = \|A\|$ by Theorem 11.3. In items (2), (6), (7) use that the operator iA is self-adjoint.

12.10° A is compact in items (2) and (3).
(1) $\sigma(A) = \{-2, 2, 4, 5\}$, $\|A\| = 5$; (2) $\sigma(A) = \{0, \pm 1, \pm 2\}$, $\|A\| = 2$;
(3) $\sigma(A) = \{-1, 0, 2, 3\}$, $\|A\| = 3$; (4) $\sigma(A) = \{-1, 2\}$, $\|A\| = 2$.

12.11 $\sigma(A) = \{0, 2, 3\}$, $r(A) = 3$, $\|A\| = 2\sqrt{3}$. Show that

$$A^*Ax = (4x_1 + 2\sqrt{2}x_2, 2\sqrt{2}x_1 + 11x_2, 0, \ldots),$$

find $\sigma(A^*A)$ and $r(A^*A)$. Since the operator A^*A is self-adjoint, then we have $\|A\| = \sqrt{\|A^*A\|} = \sqrt{r(A^*A)}$ due to item (2) of problem 9.1 and Theorem 11.3.

12.12 Let $(A - \lambda I)x_n = y_n \to y$. Put $L = \mathrm{Ker}(A - \lambda I)$. For each $n \geq 1$ choose $z_n \in X$ such that $z_n - x_n \in L$ and $\|z_n\| = \min\{\|z\| \mid z - x_n \in L\}$ (the minimum is achieved because $\dim L < +\infty$). Then $(A - \lambda I)z_n = (A - \lambda I)x_n = y_n \to y$, $n \to \infty$. Let us show that the sequence $\{z_n : n \geq 1\}$ is bounded. It will follow that some subsequence $\{Az_{n_k} : k \geq 1\}$ converges. Since $z_{n_k} = \frac{1}{\lambda}(Az_{n_k} - y_{n_k})$, the sequence $\{z_{n_k} : k \geq 1\}$ also converges to some $z \in X$ and $(A - \lambda I)z = y$.

If the sequence $\{z_n : n \geq 1\}$ is unbounded, then without loss of generality $z_n \neq 0$, $n \geq 1$, and $\|z_n\| \to +\infty$, $n \to \infty$. Put $u_n = \frac{z_n}{\|z_n\|}$, $n \geq 1$. Then $\|u_n\| = 1$, $n \geq 1$, and $(A - \lambda I)u_n = \frac{y_n}{\|z_n\|} \to 0$, $n \to \infty$. There exists a convergent subsequence $\{Au_{n_k} : k \geq 1\}$, and $u_{n_k} = \frac{1}{\lambda}\big(Au_{n_k} - \frac{y_{n_k}}{\|z_{n_k}\|}\big)$ converges to some $u \in X$ as $k \to \infty$. Then $\|u\| = 1$ and $(A - \lambda I)u = 0$, hence $u \in L$. On the other hand, by the choice of z_n we have $\big\|z_n - \|z_n\| \cdot u\big\| \geq \big\|z_n\big\|$, therefore $\|u_n - u\| \geq 1$, which contradicts the convergence of u_{n_k} to u.

12.13° The diagonal operator $Ax = (\alpha_1 x_1, \alpha_2 x_2, \ldots)$, $x \in l_2$, is compact under the condition $\alpha_n \to 0$, $n \to \infty$, due to problem 10.3 and $\sigma_p(A) = \{\alpha_n : n \geq 1\}$ according to problem 11.1.

12.16° $K(t, s) = \sum\limits_{n=1}^{\infty} \frac{2}{\pi n^2} \sin nt \sin ns$, $(t, s) \in [0, \pi]^2$.

12.17° (2) Since $A\varphi_n = \lambda_n \varphi_n$, $n \geq 1$, then $\{\lambda_n : n \geq 1\} \subset \sigma_p(A)\backslash\{0\}$. Let $Ay = \mu y$, where $\mu \neq 0$, $y \neq 0$. Then $y = \frac{1}{\mu} Ay \in \overline{\text{span}(\{\varphi_n : n \geq 1\})}$. Therefore $y = \sum\limits_{n=1}^{\infty} (y, \varphi_n)\varphi_n$ and $Ay = \sum\limits_{n=1}^{\infty} \lambda_n (y, \varphi_n)\varphi_n$. Hence $\lambda_n (y, \varphi_n) = \mu(y, \varphi_n)$, $n \geq 1$, and if $(y, \varphi_{n_0}) \neq 0$, then $\mu = \lambda_{n_0}$.

Show that $A^* x = \sum\limits_{n=1}^{\infty} \overline{\lambda_n}(x, \varphi_n)\varphi_n$, $x \in H$. The operator A is self-adjoint if $\lambda_n \in \mathbb{R}$, $n \geq 1$; is nonnegative if $\lambda_n \geq 0$, $n \geq 1$; is always normal; is compact if $\lambda_n \to 0$, $n \to \infty$; is a Hilbert–Schmidt operator if $\sum\limits_{n=1}^{\infty} |\lambda_n|^2 < +\infty$.

12.18 (2) If $x = \sum\limits_{n=1}^{\infty} (x, \varphi_n)\varphi_n + Px$, then

$$(A - \lambda I)x = \sum_{n=1}^{\infty} (\lambda_n - \lambda)(x, \varphi_n)\varphi_n - \lambda P x.$$

Therefore $(A - \lambda I)x = y$ if and only if $(\lambda_n - \lambda)(x, \varphi_n) = (y, \varphi_n)$ for all $n \geq 1$, and $-\lambda Px = Py$, whence $x = \sum\limits_{n=1}^{\infty} \frac{(y, \varphi_n)}{\lambda_n - \lambda} \varphi_n - \frac{1}{\lambda} P y$.

12.19 Repeat the reasonings, similar to the solution of item (2) of problem 12.18.

12.20 *Uniqueness.* Let B be a nonnegative compact operator such that $B^2 = A$. For $\lambda > 0$ we have $(A - \lambda I)x = (B + \sqrt{\lambda} I)(B - \sqrt{\lambda} I)x$, $x \in H$, and the operator $B + \sqrt{\lambda} I$ is continuously invertible, because $\sigma(B) \subset [0, +\infty)$. Therefore $\sqrt{\lambda} \in \sigma_p(B)$ if and only if $\lambda \in \sigma_p(A)$, moreover $\text{Ker}(B - \sqrt{\lambda} I) = \text{Ker}(A - \lambda I)$. It follows that $Bx = \sum\limits_{\lambda \in \sigma_p(A)} \sqrt{\lambda} P_\lambda x$, $x \in H$, where P_λ are the orthogonal projectors onto $\text{Ker}(A - \lambda I)$, $\lambda \geq 0$.

12.21° (1) Verify that $K \in L_2([-1,1]^2)$ and $K(t,s) = \overline{K(s,t)}$, $t, s \in [-1,1]$.

(2) Verify that $(Ax)(t) = \sum\limits_{n=1}^{\infty} \frac{1}{n^2}(x,\varphi_n)\varphi_n(t)$, where $\varphi_n(t) = \sin n\pi t$, $n \geq 1$, is an orthonormal system in $L_2([-1,1])$. As a result of problem 12.17°, it follows that $\sigma(A) = \{0, \lambda_n = \frac{1}{n^2} : n \geq 1\}$, $r(A) = \|A\| = 1$, $\varphi_n(t)$ are eigenfunctions that correspond to eigenvalues λ_n, $n \geq 1$; $\psi_0(t) = \frac{1}{\sqrt{2}}$, $\psi_n(t) = \cos \pi n t$, $n \geq 1$, are eigenfunctions that correspond to the number $\lambda = 0$;

(3) $A = \sum\limits_{n=1}^{\infty} \frac{1}{n^2}(\cdot,\varphi_n)\varphi_n$.

12.22 (1) Use problem 12.17°.

(2) $\sigma_p(A) = \{2\pi c_n : n \in \mathbb{Z}\}$, $\sigma(A) = \sigma_p(A) \cup \{0\}$, $r(A) = \|A\| = 2\pi \max\limits_{n\in\mathbb{Z}} |c_n|$.

(3) $A = \sum\limits_{n\in\mathbb{Z}} 2\pi c_n(\cdot,\varphi_n)\varphi_n$.

12.23 (2) $\sigma(A) = \{0, \pi a_0, \pm\pi a_n : n \geq 1\}$, $r(A) = \|A\| = \pi \cdot \max\limits_{n\geq 0} |a_n|$; (3)

$$A = \lambda_0(\cdot,\varphi_0)\varphi_0 + \sum_{n=1}^{\infty} \big(\lambda_n^{(1)}(\cdot,\varphi_n^{(1)})\varphi_n^{(1)} + \lambda_n^{(2)}(\cdot,\varphi_n^{(2)})\varphi_n^{(2)}\big).$$

12.24° (1) $\sigma(A) = \{0, \pi\lambda_n, 1 \leq n \leq N\}$, $\varphi_n^{(1)}(t) = \frac{1}{\sqrt{\pi}}\cos nt$, $\varphi_n^{(2)}(t) = \frac{1}{\sqrt{\pi}}\sin nt$, $1 \leq n \leq N$;

(2) $\sigma(A) = \{0, \frac{\pi}{8}, \frac{\pi}{2}, \frac{3\pi}{4}\}$, $\varphi_1(t) = \frac{1}{\sqrt{\pi}}\cos 4t$, $\varphi_2(t) = \frac{1}{\sqrt{\pi}}\cos 2t$, $\varphi_3(t) = \frac{1}{\sqrt{2\pi}}$;

(3) $\sigma(A) = \{0, \frac{\pi}{n^2} : n \geq 1\}$, $\varphi_n(t) = \frac{1}{\sqrt{\pi}}\cos nt$, $n \geq 1$;

(4) $\sigma(A) = \{0, \frac{\pi}{n^2} : n \geq 1\}$, $\varphi_n(t) = \frac{1}{\sqrt{\pi}}\sin(2n+1)t$, $n \geq 1$;

(5) $\sigma(A) = \{0, \frac{\pi}{(2n+1)^2} : n \geq 0\}$, $\varphi_n(t) = \frac{1}{\sqrt{\pi}}\cos(2n+1)t$, $n \geq 0$;

(6) $\sigma(A) = \{0, \frac{\pi}{(2n+1)^2} : n \geq 0\}$,

$$\varphi_n^{(1)}(t) = \frac{1}{\sqrt{\pi}}\cos(2n+1)t,\ n \geq 0,\ \varphi_n^{(2)}(t) = \frac{1}{\sqrt{\pi}}\sin(2n+1)t,\ n \geq 0.$$

12.25 (1) $\sigma(A) = \{0, \frac{1}{n^2\pi^2} : n \geq 1\}$, $\varphi_n(t) = \sqrt{2}\sin n\pi t$, $n \geq 1$;

(2) $\sigma(A) = \{0, \frac{1}{(n-\frac{1}{2})^2-1} : n \geq 1\}$, $\varphi_n(t) = \sqrt{\frac{2}{\pi}}\sin(n-\frac{1}{2})t$, $n \geq 1$;

(3) $\sigma(A) = \{0, \frac{1}{1-(n-\frac{1}{2})^2} : n \geq 1\}$, $\varphi_n(t) = \sqrt{\frac{2}{\pi}}\cos(n-\frac{1}{2})t$, $n \geq 1$;

(4) $\sigma(A) = \{0, 1, -\frac{1}{n^2\pi^2} : n \geq 1\}$,

$$\varphi_0(t) = \sqrt{\frac{2}{e^2-1}}e^t,\ \varphi_n(t) = \sqrt{\frac{2}{1+n^2\pi^2}}(\sin n\pi t + n\pi\cos n\pi t),\ n \geq 1;$$

(5) $\sigma(A) = \{0, \frac{\sin 1}{n^2\pi^2-1} : n \geq 1\}$, $\varphi_n(t) = \sqrt{2}\sin n\pi t$, $n \geq 1$;

(6) $\sigma(A) = \{0, \frac{2}{n^2\pi^2}, n \geq 1;\ \frac{2}{(n-\frac{1}{2})^2\pi^2}, n \geq 1\}$,

$$\varphi_n^{(1)}(t) = \sin n\pi t,\ n \geq 1;\ \varphi_n^{(2)}(t) = \cos(n - \tfrac{1}{2})\pi t,\ n \geq 1.$$

12.26* $\|A\| = \frac{2}{\pi}$. Show that $(A^*Ax)(t) = \int_0^1 (1 - \max(t, s))x(s)ds$, find $\sigma(A^*A)$ and $r(A^*A)$. Since the operator A^*A is self-adjoint, we have $\|A\| = \sqrt{\|A^*A\|} = \sqrt{r(A^*A)}$ due to item (2) of problem 9.1 and Theorem 11.3.

12.27 Let $\sigma_p(A)\setminus\{0\} = \{\lambda_n : 1 \leq n \leq N\}$, where $N \leq +\infty$ and $|\lambda_1| \geq |\lambda_2| \geq \ldots$, $L_n = \mathrm{Ker}(A-\lambda_n I)$, P_n be the orthoprojector onto L_n and $A_n = \sum_{i=1}^{n} \lambda_n P_n \in S_0(H)$, $1 \leq n \leq N$. Show that $A - A_n = (A - A_n)^*$ and $\sigma(A - A_n) = \{0, \lambda_i : i \geq n+1\}$. Therefore $\|A - A_n\| = |\lambda_{n+1}| \to 0$, i.e. $A_n \rightrightarrows A$, $n \to \infty$, for $N = +\infty$ and $A = A_N \in S_0(H)$ for $N \in \mathbb{N}$.

12.28* According to problem 11.15* we have

$$\sigma(A^n) = \{\lambda^n \mid \lambda \in \sigma(A)\}, \quad \sigma_p(A^n) = \{\lambda^n \mid \lambda \in \sigma_p(A)\}.$$

Assume that $\mu \in \sigma(A)\setminus\{0\}$ and $\mu \notin \sigma_p(A)$. Since $A^k \in S_\infty(X)$, then $A^n \in S_\infty(X)$, $n \geq k$. Hence

$$\mu^n \in \sigma(A^n) \setminus \{0\} = \sigma_p(A^n) \setminus \{0\} = \{\lambda^n \mid \lambda \in \sigma_p(A)\} \setminus \{0\},$$

therefore for each $n \geq k$ there exists $\lambda_n \in \sigma_p(A)$ such that $\lambda_n \neq \mu$ and $\lambda_n^n = \mu^n$. It follows that there are infinitely many eigenvalues of the operator A on the circle $\{\lambda \in \mathbb{C} \mid |\lambda| = |\mu|\}$, so there are infinitely many eigenvalues of the compact operator A^k on the circle $\{\lambda \in \mathbb{C} \mid |\lambda| = |\mu|^k\}$, which is impossible.

It follows from the equality $\sigma(A^k) = \{\lambda^k \mid \lambda \in \sigma(A)\}$ that $0 \in \sigma(A)$ and for every $\varepsilon > 0$ the set $\{\lambda \in \sigma(A) \mid |\lambda| > \varepsilon\}$ is finite, therefore $\sigma(A)$ is a countable set without nonzero limit points. Finally, if $\lambda \neq 0$ and $\dim \mathrm{Ker}(A - \lambda I) = +\infty$, then $\dim \mathrm{Ker}(A^k - \lambda^k I) = +\infty$, which is impossible.

12.29 (1) According to problem 12.28* $1 \notin \sigma_p(A)$ implies that $1 \in \rho(A)$. Therefore $\|x\| = \|(A - I)^{-1}(A - I)x\| \leq C\|(A - I)^{-1}\|$ for all $x \in M$. (2) Show that $1 \notin \sigma_p(A_h)$, where A_h is the operator from problem 10.80.

12.30° (1) Note that

$$\|Ax - A_n x\|^2 = \sum_{k=n+1}^{\infty} |s_k(x, \varphi_k)|^2 \leq \sup_{k \geq n+1} |s_k|^2 \cdot \|x\|^2,\ n \geq 1,\ x \in H.$$

12.31° $A^*y = \sum\limits_{n=1}^{\infty} \overline{s_n}(y, \psi_n)\varphi_n,\ y \in H;$

$AA^*y = \sum\limits_{n=1}^{\infty} |s_n|^2(y, \psi_n)\psi_n,\ y \in H;\ A^*Ax = \sum\limits_{n=1}^{\infty} |s_n|^2(x, \varphi_n)\varphi_n,\ x \in H.$

12.32 Since the operator A^*A is compact and nonnegative, then

$$A^*A = \sum_{n=1}^{N} s_n^2(\cdot, \varphi_n)\varphi_n,$$

where s_n are positive numbers and $\{\varphi_n : 1 \le n \le N\}$ is an orthonormal sequence. Then $|A| = \sum\limits_{n=1}^{\infty} s_n(\cdot, \varphi_n)\varphi_n$. Put $\psi_n = \frac{1}{s_n}A\varphi_n,\ 1 \le n \le N$. Let us verify that $\{\psi_n : 1 \le n \le N\}$ is an orthonormal sequence. Indeed, for all $n, k \ge 1$

$$(\psi_n, \psi_k) = \frac{1}{s_n s_k}(A\varphi_n, A\varphi_k) = \frac{1}{s_n s_k}(A^*A\varphi_n, \varphi_k) = \frac{s_n}{s_k}(\varphi_n, \varphi_k) = \delta_{nk}.$$

Note that $x = \sum\limits_{n=1}^{N}(x, \varphi_n)\varphi_n + Px,\ x \in H$, where P is the orthogonal projector onto $\operatorname{Ker}|A|$, and $\operatorname{Ker} A = \operatorname{Ker}|A|$, because $(Ax, Ax) = (A^*Ax, x) = (|A|x, |A|x),\ x \in H$. Therefore $Ax = \sum\limits_{n=1}^{N}(x, \varphi_n)A\varphi_n = \sum\limits_{n=1}^{N} s_n(x, \varphi_n)\psi_n$.

12.33* By problem 12.32 we have

$$A = \sum_{n=1}^{N} s_n(\cdot, \varphi_n)\psi_n,\ A^* = \sum_{n=1}^{N} s_n(\cdot, \psi_n)\varphi_n,$$

where $\{\varphi_n : 1 \le n \le N\}$, $\{\psi_n : 1 \le n \le N\}$ are orthonormal sequences and $\{s_n : 1 \le n \le N\}$ are positive numbers, moreover one can assume that $s_1 \ge s_2 \ge s_3 \ge \dots$ Thus

$$\|A^*x\|^2 = \sum_{n=1}^{N} s_n^2|(x, \psi_n)|^2 \le \|Ax\|^2 = \sum_{n=1}^{N} s_n^2|(x, \varphi_n)|^2,\ x \in H.$$

For definiteness assume that

$$s_1 = \ldots = s_{n_1} > s_{n_1+1} = \ldots = s_{n_2} > s_{n_2+1} = \ldots$$

Denote

$$L_k = \operatorname{span}(\{\varphi_{n_{k-1}+1}, \ldots, \varphi_{n_k}\}),\ M_k = \operatorname{span}(\{\psi_{n_{k-1}+1}, \ldots, \psi_{n_k}\}),$$

$k \ge 1$ (here $n_0 = 0$). Then

$$\|A^*x\|^2 = \sum_{k\ge1} s_{n_k}^2 \|\mathrm{pr}_{M_k} x\|^2 \le \|Ax\|^2 = \sum_{k\ge1} s_{n_k}^2 \|\mathrm{pr}_{L_k} x\|^2, \ x \in H.$$

For all $x \in M_1$ we have $s_{n_1}^2 \|x\|^2 \le \sum\limits_{k\ge1} s_{n_k}^2 \|\mathrm{pr}_{L_k} x\|^2 \le s_{n_1}^2 \|x\|^2$, whence $\mathrm{pr}_{L_1} x = x$. Therefore $M_1 \subset L_1$, and since $\dim L_1 = \dim M_1$, then $L_1 = M_1$. By induction we obtain that $L_k = M_k$, $k \ge 1$. Thus $\|A^*x\| = \|Ax\|$, $x \in H$, so $AA^* = A^*A$ due to problem 9.31°.

12.34 It follows from Theorem 2.2 and Bessel's inequality that

$$(Ax, x) = \sum_{n=1}^{N^+} \lambda_n^+ |(x, \varphi_n^+)|^2 + \sum_{n=1}^{N^-} \lambda_n^- |(x, \varphi_n^-)|^2 \le \lambda_1^+ \|x\|^2, \ x \in H.$$

Similarly $(Ax, x) \le \lambda_n^+ \|x\|^2$ if the condition $(x, \varphi_k^+) = 0$ is fulfilled for $1 \le k \le n-1$.

12.35 (1) Let $H_n = \mathrm{span}(\{\varphi_1^+, \dots, \varphi_n^+\})$, $1 \le n \le N^+$. For $x \in H_n$ we have $(Ax, x) = \sum\limits_{k=1}^{n} \lambda_k^+ |(x, \varphi_k^+)|^2 \ge \lambda_n^+ \cdot \sum\limits_{k=1}^{n} |(x, \varphi_k^+)|^2 = \lambda_n^+ \|x\|^2$. If $\dim L = n-1$, then there exists $x \in H_n$ such that $x \perp L$ and $\|x\| = 1$. For this x we have $(Ax, x) \ge \lambda_n^+$, hence $\max\limits_{x \in L^\perp,\, \|x\|=1} (Ax, x) \ge \lambda_n^+$. For $L = H_{n-1}$ the equality is achieved according to problem 12.34.

(2) If $\dim L = n$, then there exists $x \in L$ such that $x \perp H_{n-1}$ and $\|x\| = 1$. According to problem 12.34, for this x we have $(Ax, x) \le \lambda_n^+$, thus $\min\limits_{x \in L,\, \|x\|=1} (Ax, x) \le \lambda_n^+$. Equality is achieved for $L = H_n$.

12.36* Assume that the operator A has at least n positive eigenvalues (taking into account their multiplicities) and φ_k^+ are the eigenvectors that correspond to eigenvalues λ_k^+, $1 \le k \le n$. Put $H_n = \mathrm{span}(\{\varphi_1^+, \dots, \varphi_n^+\})$. Note that $\dim(H_n \cap G) \ge n-1$ and for $x \in H_n \cap G$ we have

$$(PAPx, x) = (APx, Px) = (Ax, x) \ge \lambda_n^+ \|x\|^2$$

(the inequality was established in the solution of problem 12.35). It follows that the operator PAP has at least $n-1$ positive eigenvalues and for $(n-1)$-dimensional subspace $L_{n-1} \subset (H_n \cap G)$ we have

$$\mu_{n-1}^+ = \max_{L \in \mathcal{L}_{n-1}} \min_{x \in L,\, \|x\|=1} (PAPx, x) \ge \min_{x \in L_{n-1},\, \|x\|=1} (PAPx, x) \ge \lambda_n^+$$

due to item (2) of problem 12.35.

Now assume that the operator PAP have at least n positive eigenvalues and ψ_k^+ are the eigenvectors that correspond to the eigenvalues μ_k^+, $1 \le k \le n$. Put $G_n = \mathrm{span}(\{\psi_1^+, \dots, \psi_n^+\})$. Since $PAP\psi_k^+ = \mu_k^+\psi_k^+ \in G$, $1 \le k \le n$, then $G_n \subset G$. For $x \in G_n$ we have

$$(Ax, x) = (APx, Px) = (PAPx, x) \ge \mu_n^+ \|x\|^2.$$

Since $\dim G_n = n$, then the operator PAP has at least n positive eigenvalues and

$$\lambda_n^+ = \max_{L \in \mathcal{L}_n} \min_{x \in L,\, \|x\|=1} (Ax, x) \ge \min_{x \in G_n,\, \|x\|=1} (Ax, x) \ge \mu_n^+$$

due to item (2) of problem 12.35.

12.37 (1) Compare the canonical forms (see problem 12.32) of operators A and A^*.

(2) Verify that $\{s_n^2(A),\ 1 \le n \le N_A\}$ is the set of positive eigenvalues of the operator A^*A, written in the order of decreasing taking into account their multiplicities, and use item (1) of problem 12.35.

(3) To prove the second inequality, note that due to item (1) we have

$$s_n(AB) = s_n((AB)^*) = s_n(B^*A^*) \le \|B^*\|\, s_n(A^*) = \|B\|\, s_n(A).$$

(4) Use item (2).

(5) Consider $T \in S_0(H)$ such that $\dim R(T) \le n-1$. Put $L = (\mathrm{Ker}\, T)^\perp$. Then $\dim L = \dim R(T) \le n-1$ and according to item (2)

$$s_n(A) \le \max_{x \in L^\perp,\, \|x\|=1} \|Ax\| = \max_{x \in \mathrm{Ker}\, T,\, \|x\|=1} \|(A-T)x\| \le \|A - T\|.$$

If $A = \sum_{k=1}^{N_A} s_k(A)(\cdot, \varphi_k)\psi_k$ is the canonical form of the operator, then the equality is achieved for $T_{n-1} = \sum_{k=1}^{n-1} s_k(A)(\cdot, \varphi_k)\psi_k$. Indeed, $s_n(A) = s_1(A - T_{n-1})$, whence $s_n(A) = \|A - T_{n-1}\|$ due to item (2).

(6) According to item (5) there exist operators $T, S \in S_0(H)$ such that

$$s_m(A) = \|A - T\|,\ s_n(B) = \|B - S\|,\ \mathrm{rank}\, T \le m-1,\ \mathrm{rank}\, S \le n-1.$$

Then $\mathrm{rank}(T + S) \le m + n - 2$ and item (5) implies that

$$s_{m+n-1}(A+B) \le \|(A+B) - (T+S)\| \le$$
$$\le \|A - T\| + \|B - S\| = s_m(A) + s_n(B).$$

(7) According to item (5) there exists an operator $T \in S_0(H)$, $\operatorname{rank} T \le n-1$, for which $s_n(B) = \|B - T\|$. Then

$$s_n(A) \le \|A - T\| \le \|A - B\| + \|B - T\| = \|A - B\| + s_n(B).$$

Similarly $s_n(B) \le \|A - B\| + s_n(A)$.

12.38 By problem 12.32 we have $A = \sum_{n=1}^{\infty} s_n(A)(\cdot, \varphi_n)\psi_n$, where $\{\varphi_n : n \ge 1\}$ and $\{\psi_n : n \ge 1\}$ are orthonormal bases (it is possible that $s_n(A) = 0$ for some n). Then $\|A\|_2^2 = \sum_{n=1}^{\infty} \|A\varphi_n\|^2 = \sum_{n=1}^{\infty} s_n^2(A)$.

12.39 (1)–(3) Use items (1)–(3) of problem 12.37, respectively.

(4) Let $A = \sum_{n=1}^{N} s_n(A)(\cdot, \varphi_n)\psi_n$, $N \le +\infty$, be the canonical form (see problem 12.32). By Hölder's inequality for $p > 1$ we have

$$|(Ae_k, g_k)| = \sum_{n=1}^{N} s_n(A)|(e_k, \varphi_n)(\psi_n, g_k)| \le$$
$$\le \left(\sum_{n=1}^{N} s_n^p(A)|(e_k, \varphi_n)(\psi_n, g_k)|\right)^{1/p} \left(\sum_{n=1}^{N} |(e_k, \varphi_n)(\psi_n, g_k)|\right)^{1/q},$$

where q is the conjugate index to p. Since

$$\sum_{n=1}^{N} |(e_k, \varphi_n)(\psi_n, g_k)| \le \sum_{n=1}^{N} \frac{|(e_k, \varphi_n)|^2 + |(\psi_n, g_k)|^2}{2} \le \frac{\|e_k\|^2 + \|g_k\|^2}{2} = 1,$$

then $|(Ae_k, g_k)|^p \le \sum_{n=1}^{N} s_n^p(A)|(e_k, \varphi_n)(\psi_n, g_k)|$ for all $1 \le p < +\infty$. Hence

$$\sum_{k=1}^{\infty} |(Ae_k, g_k)|^p \le \sum_{k=1}^{\infty}\sum_{n=1}^{N} s_n^p(A)|(e_k, \varphi_n)(\psi_n, g_k)| =$$
$$= \sum_{n=1}^{N} s_n^p(A) \left(\sum_{k=1}^{\infty} |(e_k, \varphi_n)(\psi_n, g_k)|\right) \le$$
$$\le \sum_{n=1}^{N} s_n^p(A) \frac{\|\varphi_n\|^2 + \|\psi_n\|^2}{2} = \sum_{n=1}^{N} s_n^p(A).$$

Equality is achieved for $e_n = \varphi_n$, $g_n = \psi_n$, $1 \le n \le N$.

(5) For any orthonormal bases $\{e_n : n \geq 1\}$ and $\{g_n : n \geq 1\}$ by Minkowski inequality in view of item (4) we have $\left(\sum\limits_{n=1}^{\infty} \left|((A+B)e_n, g_n)\right|^p\right)^{1/p} \leq$

$$\leq \left(\sum_{n=1}^{\infty} |(Ae_n, g_n)|^p\right)^{1/p} + \left(\sum_{n=1}^{\infty} |(Be_n, g_n)|^p\right)^{1/p} \leq \|A\|_p + \|B\|_p.$$

Therefore $A + B \in S_p(H)$ and $\|A + B\|_p \leq \|A\|_p + \|B\|_p$.

(6) To verify the axioms of norm, use items (2) and (5). To prove completeness of the space, consider a Cauchy sequence $\{A_n : n \geq 1\}$ in $S_p(H)$. Due to item (2) this sequence is a Cauchy sequence with respect to the operator norm, therefore it converges to some operator $A \in S_\infty(H)$. According to item (7) of problem 12.37, for each $k \geq 1$ we have $s_k(A_n) \to s_k(A)$, $n \to \infty$. Show that $A \in S_p(H)$ and $\|A_n - A\|_p \to 0$, $n \to \infty$.

(7) Let $A = \sum\limits_{n=1}^{\infty} s_n(A)(\cdot, \varphi_n)\psi_n$ be the canonical form. Put

$$A_k = \sum_{n=1}^{k} s_n(A)(\cdot, \varphi_n)\psi_n, \ k \geq 1.$$

Verify that $\|A_k - A\|_p \to 0$, $k \to \infty$.

12.40 (1) The absolute convergence of the series follows from item (4) of problem 12.39. If $A = \sum\limits_{n=1}^{N} s_n(A)(\cdot, \varphi_n)\psi_n$ is the canonical form, then

$$\text{tr A} = \sum_{k=1}^{\infty} (Ae_k, e_k) = \sum_{k=1}^{\infty} s_n(A)(e_k, \varphi_n)(\psi_n, e_k) =$$

$$= \sum_{n=1}^{N} \sum_{k=1}^{\infty} s_n(A)(e_k, \varphi_n)(\psi_n, e_k) = \sum_{n=1}^{N} s_n(A)(\psi_n, \varphi_n)$$

does not depend on the choice of the basis $\{e_n : n \geq 1\}$. It is possible to change of order of summation in the double series because the series converges absolutely. Indeed,

$$\sum_{n=1}^{N} s_n(A) \left(\sum_{k=1}^{\infty} |(e_k, \varphi_n)(\psi_n, e_k)|\right) \leq \sum_{n=1}^{N} s_n(A) \frac{\|\varphi_n\|^2 + \|\psi_n\|^2}{2} \leq \|A\|_1.$$

(2) Since $A \geq 0$, then $\{s_n(A) : 1 \leq n \leq N\}$ are the positive eigenvalues of the operator A and in the canonical form $\psi_n = \varphi_n$, $1 \leq n \leq N$. It follows from the solution of item (1) that $\operatorname{tr} A = \sum_{n=1}^{N} s_n(A)(\psi_n, \varphi_n) = \sum_{n=1}^{N} s_n(A) = \sum_{n=1}^{\infty} \lambda_n$.

(3) For any orthonormal bases $\{e_n : n \geq 1\}$, $\{g_n : n \geq 1\}$ by the Cauchy–Schwarz inequality we have

$$\sum_{n=1}^{\infty} |(ABe_n, g_n)| = \sum_{n=1}^{\infty} |(Be_n, A^* g_n)| \leq \sum_{n=1}^{\infty} \|(Be_n\| \cdot \|A^* g_n)\| \leq$$

$$\leq \left(\sum_{n=1}^{\infty} \|(Be_n\|^2\right)^{1/2} \left(\sum_{n=1}^{\infty} \|(A^* e_n\|^2\right)^{1/2} = \|A\|_2 \cdot \|B\|_2.$$

Therefore $AB \in S_1$ and $\|AB\|_1 \leq \|A\|_2 \cdot \|B\|_2$ due to item (4) of problem 12.39.

(4) Due to problem 12.32 we have $A = \sum_{n=1}^{\infty} s_n(A)(\cdot, \varphi_n)\psi_n$, where $\{\varphi_n : n \geq 1\}$ and $\{\psi_n : n \geq 1\}$ are orthonormal bases (it is possible that $s_n(A) = 0$ for some n). Then

$$\operatorname{tr} \mathrm{AB} = \sum_{n=1}^{\infty} (AB\psi_n, \psi_n) = \sum_{n=1}^{\infty} s_n(A)(B\psi_n, \varphi_n),$$

$$\operatorname{tr} \mathrm{BA} = \sum_{n=1}^{\infty} (BA\varphi_n, \varphi_n) = \sum_{n=1}^{\infty} s_n(A)(B\psi_n, \varphi_n).$$

(5) Verify that for $A = \sum_{n=1}^{N} s_n(A)(\cdot, \varphi_n)\psi_n$ one can take $B = \sum_{n=1}^{N} \sqrt{s_n(A)}$ $(\cdot, \varphi_n)\psi_n$ and $C = \sum_{n=1}^{N} \sqrt{s_n(A)}(\cdot, \varphi_n)\varphi_n$.

12.41 *Sufficiency.* If $\{\varphi_n(t) : n \geq 1\}$ is an orthonormal basis in $L_2(T, \mu)$, then $\left\{e_{n,m}(t, s) = \varphi_n(t)\overline{\varphi_m(s)} : n, m \geq 1\right\}$ is an orthonormal basis in $L_2(T \times T, \mu \times \mu)$ due to problem 2.51. Verify that

$$(A\varphi_n, \varphi_m)_{L_2(T,\mu)} = (K, e_{n,m})_{L_2(T\times T,\mu\times\mu)}, \ n.m \geq 1,$$

and use the Parseval's identity for the function K and the basis $\{e_{n,m}\}$.

Necessity. Let $K(t, s) = \sum_{n,m=1}^{\infty} (A\varphi_n, \varphi_m)\varphi_n(t)\overline{\varphi_m(s)}$. Verify that the series converges in $L_2(T \times T, \mu \times \mu)$ and $(Ax)(t) = \int_T K(t, s)x(s)d\mu(s)$, $x \in L_2(T, \mu)$.

12.42 (2) Use problem 2.33*.

(3) To calculate the Hilbert–Schmidt norm, use the orthonormal bases

$$\left\{\frac{\sinh t}{\sqrt{\sinh 2\pi}}, \frac{e^{ikt}}{\sqrt{2\pi(1+k^2)}}, \ k\in\mathbb{Z}\right\} \text{ and } \left\{\frac{e^{ikt}}{\sqrt{2\pi}}, \ k\in\mathbb{Z}\right\}$$

in the spaces $W_2^1([-\pi,\pi])$ and $L_2([-\pi,\pi])$, respectively.

Problems of Chap. 13

13.5° (1) $x(t) = e^{\arctan t} - 1$; (2) $x(t) = 1+t$; (3) $x(t) = \cosh t$; (4) $x(t) = 1 + \frac{2e(e-1)}{e^2+1}\cdot e^{-t}$; (5) $x(t) = \arctan t + \frac{\pi^2}{32-8\pi}$; (6) $x(t) = 1$; (7) $x(t) = 2$; (8) $x(t) = 1 + t\ln 2$.

13.6° (1) $x_1(t) = 1$; $x_2(t) = 1 + \frac{1}{(t+1)(t+2)}$; (2) $x_1(t) = t$; $x_2(t) = \frac{10t}{9} + \frac{1}{12}$; (3) $x_1(t) = 1$; $x_2(t) = 1 + \frac{1}{10}\left(\frac{\pi}{4}\arctan\frac{\pi t}{4} - \frac{1}{2t}\ln\left(1+\frac{\pi^2t^2}{16}\right)\right)$.

13.7° (1) $R_\lambda(t,s) = e^{t^2-s^2+\lambda(t-s)}$, $x(t) = e^{2t^2+2t}(1+2t)$;
(2) $R_\lambda(t,s) = e^{(\lambda+1)(t-s)}$, $x(t) = \frac{1}{5}e^{3t} - \frac{1}{5}\cos t + \frac{2}{5}\sin t$;
(3) $R_\lambda(t,s) = \frac{1+t^2}{1+s^2}e^{\lambda(t-s)}$, $x(t) = e^t(1+t^2)$;
(4) $R_\lambda(t,s) = 3^{t-s}e^{\lambda(t-s)}$, $x(t) = 3^t(1-e^{-t})$;
(5) $R_\lambda(t,s) = \frac{2+\cos t}{2+\cos s}e^{\lambda(t-s)}$, $x(t) = e^t\sin t + (2+\cos t)e^t\ln\frac{3}{2+\cos t}$;
(6) $R_\lambda(t,s) = \frac{\cosh t}{\cosh s}e^{\lambda(t-s)}$, $x(t) = \cosh t(1-e^{-t})$;
(7) $R_\lambda(t,s) = e^{(\lambda-1)(t-s)}$, $x(t) = e^{\frac{t^2}{2}}(t+1) - 1$.

13.8 (1) $R_\lambda(t,s) = \frac{2e^{t+s}}{2-(e^2-1)\lambda}$, $|\lambda| < \frac{2}{e^2-1}$, $x(t) = \frac{2\pi e^t}{(e-1)(\pi^2+1)} + \sin\pi t$;
(2) $R_\lambda(t,s) = \frac{2\sin t\cos s}{2-\lambda}$, $|\lambda| < 2$, $x(t) = -\frac{\pi}{6}\sin t + \cos t$;
(3) $R_\lambda(t,s) = \frac{te^{s+1}}{e-2\lambda}$, $|\lambda| < \frac{e}{2}$, $x(t) = 4et + (3t^2+1)e^{-t}$;
(4) $R_\lambda(t,s) = \frac{5t^2s^2}{5-2\lambda}$, $|\lambda| < \frac{5}{2}$, $x(t) = 10(e-\frac{5}{e})t^2 + e^t$;
(5) $R_\lambda(t,s) = \frac{3(1+t)(1-s)}{3-2\lambda}$, $|\lambda| < \frac{3}{2}$, $x(t) = -\frac{6}{\pi}(1+t) + \pi\cos\pi t$;
(6) $R_\lambda(t,s) = \frac{3ts}{3-2\lambda} + \frac{5t^2s^2}{5-2\lambda}$, $|\lambda| < \frac{3}{2}$, $x(t) = \frac{10}{3}t^2 + 9t + 3$;
(7) $R_\lambda(t,s) = \frac{1}{1-\lambda} + \frac{3(2t-1)(2s-1)}{3-\lambda}$, $|\lambda| < 1$, $x(t) = 3t^2 + \frac{3}{5}t + \frac{7}{10}$;
(8) $R_\lambda(t,s) = \sin t\cos s + \cos 2t\sin 2s$, $\lambda\in\mathbb{C}$, $x(t) = \cos t$;
(9) $R_\lambda(t,s) = \frac{1}{1-\pi\lambda}(\sin t\sin s + \cos 2t\cos 2s)$, $|\lambda| < \frac{1}{\pi}$, $x(t) = 2\cos 2t$;
(10) $R_\lambda(t,s) = \frac{1}{1-\pi\lambda}\cos t\cos s + \frac{3}{1-3\pi\lambda}\sin 2t\sin 2s$, $|\lambda| < \frac{1}{3\pi}$, $x(t) = 6\cos t$;
(11) $R_\lambda(t,s) = \frac{2}{1-4\pi\lambda} + \frac{1}{1-\pi\lambda}\cos t\cos s$, $|\lambda| < \frac{1}{4\pi}$, $x(t) = 2t - \frac{2\pi}{3}$;
(12) $R_\lambda(t,s) = \frac{2}{2+(e^\pi+1)\lambda}e^t\cos s$, $|\lambda| < \frac{2}{e^\pi+1}$, $x(t) = \frac{t}{4} - \frac{e^t}{3(e^\pi+1)}$.

13.9 (3) Verify that $K_{i+j}(t,s) = \int_a^b K_i(t,u)K_j(u,s)du$ for all $i, j \geq 1$ and $\max_{a\leq t,s\leq b} |K_n(t,s)| \leq M^n(b-a)^{n-1}$, $n \geq 1$. Therefore for $|\lambda| < (M(b-a))^{-1}$ the series $\sum_{n=1}^{\infty} |\lambda^{n-1}K_n(t,s)|$ converges uniformly for $t, s \in [a,b]$, whence

$$\int_a^b R_\lambda(t,u)R_\lambda(u,s)du = \int_a^b \sum_{i=1}^{\infty} \lambda^{i-1}K_i(t,u) \sum_{j=1}^{\infty} \lambda^{j-1}K_j(u,s)du =$$

$$= \sum_{n\geq 2} \lambda^{n-2} \sum_{i+j=n} \int_a^b K_i(t,u)K_j(u,s)du = \sum_{n\geq 2}(n-1)\lambda^{n-2}K_n(t,s).$$

13.10° Here and further C, C_k are arbitrary complex constants.

(1) $x(t) = \frac{1}{1-2\lambda}$, $\lambda \in \mathbb{C}\setminus\{\frac{1}{2}\}$; $x \in \varnothing$, $\lambda = \frac{1}{2}$;

(2) $x(t) = \frac{59-14\lambda}{7(5-2\lambda)}t^2 + t^4$, $\lambda \in \mathbb{C}\setminus\{\frac{3}{2}, \frac{5}{2}\}$; $x(t) = Ct + \frac{19}{7}t^2 + t^4$, $\lambda = \frac{3}{2}$; $x \in \varnothing$, $\lambda = \frac{5}{2}$.

(3) $x(t) = \frac{3\pi\lambda}{2(2\lambda+3)}\sin t + \cos 2t$, $\lambda \in \mathbb{C}\setminus\{-\frac{3}{2}, -\frac{3}{4}\}$; $x(t) = -\frac{3\pi}{4}\sin t + C\cos t + \cos 2t$, $\lambda = -\frac{3}{4}$; $x \in \varnothing$, $\lambda = -\frac{3}{2}$.

(4) $x(t) = \frac{2\lambda}{12\lambda^2-5}(5\sqrt[3]{t} + 6\lambda) + 1 - 6t^2$, $\lambda \in \mathbb{C}\setminus\left\{\pm\sqrt{\frac{5}{12}}\right\}$; $x \in \varnothing$, $\lambda = \pm\sqrt{\frac{5}{12}}$.

(5) $x(t) = \frac{5(2\lambda-3)}{3(5-2\lambda)}t^4 + t^2$, $\lambda \in \mathbb{C}\setminus\{\frac{1}{2}, \frac{5}{2}\}$; $x(t) = -\frac{5}{6}t^4 + Ct^3 + t^2$, $\lambda = \frac{1}{2}$; $x \in \varnothing$, $\lambda = \frac{5}{2}$.

(6) $x(t) = \frac{20\lambda}{1-2\lambda}t^2 + 7t^4 + 3$, $\lambda \in \mathbb{C}\setminus\{\frac{5}{4}, \frac{1}{2}\}$; $x(t) = Ct - \frac{50}{3}t^2 + 7t^4 + 3$, $\lambda = \frac{5}{4}$; $x \in \varnothing$, $\lambda = \frac{1}{2}$.

(7) $x(t) = \frac{2t^2}{3-2\lambda} + \frac{2t}{3+2\lambda}$, $\lambda \in \mathbb{C}\setminus\{\pm\frac{3}{2}\}$; $x \in \varnothing$, $\lambda = \pm\frac{3}{2}$.

(8) $x(t) = \frac{12\lambda}{3-4\lambda}\sin 2t + \pi - 2t$, $\lambda \in \mathbb{C}\setminus\{-\frac{3}{2}, \frac{3}{4}\}$; $x(t) = -2\sin 2t + C\cos 2t + \pi - 2t$, $\lambda = -\frac{3}{2}$; $x \in \varnothing$, $\lambda = \frac{3}{4}$.

(9) $x(t) = \frac{3\pi\lambda}{8\lambda^2-9}(2\lambda\cos 2t + \frac{3}{2}\sin 2t) + \sin t$, $\lambda \in \mathbb{C}\setminus\{\pm\frac{3}{2\sqrt{2}}\}$; $x \in \varnothing$, $\lambda = \pm\frac{3}{2\sqrt{2}}$.

(10) $x(t) = \frac{\pi\lambda}{2}\sin 3t + \cos t$, $\lambda \in \mathbb{C}$.

(11) $x(t) = -\frac{\pi^2\lambda\cos t}{6(1+2\lambda)} + 1 - \frac{2t}{\pi}$, $\lambda \in \mathbb{C}\setminus\{\pm\frac{1}{2}\}$; $x(t) = C + (\frac{\pi^2}{8}C - \frac{\pi^2}{24})\cos t + 1 - \frac{2t}{\pi}$, $\lambda = \frac{1}{2}$; $x \in \varnothing$, $\lambda = -\frac{1}{2}$.

(12) $x(t) = \frac{2}{2-\pi\lambda} + \cos 4t$, $\lambda \in \mathbb{C}\setminus\{\frac{2}{\pi}, \frac{4}{\pi}\}$; $x(t) = C_1\cos 2t + C_2\sin 2t - 1 + \cos 4t$, $\lambda = \frac{4}{\pi}$; $x \in \varnothing$, $\lambda = \frac{2}{\pi}$.

(13) $x(t) = \cos 3t$, $\lambda \in \mathbb{C}\setminus\{\frac{1}{\pi}\}$; $x(t) = C_1\cos t + C_2\cos 2t + \cos 3t$, $\lambda = \frac{1}{\pi}$.

(14) $x(t) = \frac{\cos t}{1-\pi\lambda}$, $\lambda \in \mathbb{C}\setminus\{\frac{1}{2\pi}, \frac{1}{\pi}\}$; $x(t) = 2\cos t + C\sin 2t$, $\lambda = \frac{1}{2\pi}$; $x \in \varnothing$, $\lambda = \frac{1}{\pi}$.

(15) $x(t) = \frac{\sin t}{1-\pi\lambda}$, $\lambda \in \mathbb{C} \setminus \{\frac{1}{3\pi}, \frac{1}{\pi}\}$; $x(t) = \frac{3}{2}\sin t + C\cos 2t$, $\lambda = \frac{1}{3\pi}$; $x \in \varnothing$, $\lambda = \frac{1}{\pi}$.

13.11 (1) $x(t) = \sum\limits_{k=1}^{100} \frac{2(-1)^k k \sin(\sqrt{2}\pi)\sin kt}{(2-k^2)(2\sqrt{k}-\pi)} + \sin\sqrt{2}t$. (2) $x(t) = \frac{30}{11} + \frac{45}{11}t^2$.

(3) $x(t) = \frac{8\cos t + (16-4\pi)\sin t}{\pi^2 - 8\pi + 12}$. (4) $x(t) = -\frac{16}{5}t^2 + \frac{18}{5}t + 1$. (5) $x(t) = -\frac{2}{5}t^2 + \frac{6}{5}t + \frac{4}{15}$.

13.12° Apply the Fredholm alternative. (1) $\lambda \in \mathbb{C} \setminus \{\frac{1}{b-a}\}$. (2) If $a \neq -b$, then $\lambda \in \mathbb{C} \setminus \left\{\frac{4}{e^{-a^4} - e^{-b^4}}\right\}$, if $a = -b$, then $\lambda \in \mathbb{C}$.

13.13° (1) $\int\limits_{-1}^{1} t^3 y(t)dt = 0$, $\lambda \in \mathbb{C} \setminus \{\frac{5}{2}\}$; (2) $\int\limits_{-1}^{1} ty(t)dt = 0$, $\lambda \in \mathbb{C} \setminus \{\frac{1}{2-\ln 9}\}$;

(3) $\int\limits_{-1}^{1} \frac{y(t)}{\sqrt[3]{t}}dt = 0$, $\lambda \in \mathbb{C} \setminus \{\frac{5}{6}\}$.

13.14° (1) $q = 0$, $3p + 5r = 0$; (2) $p + 3r = 0$, $q = 0$; (3) $q = 0$; (4) $p = q = 0$; (5) $5p + 7q = 0$; (6) $5p + 3q = 0$; (7) $p + 3r = 0$, $q = 0$; (8) $p = 0$; (9) $3p + 5q = 0$;

(10) $p = q = 0$; (11) $\pi p + 4q = 0$; (12) $2q + r = 0$; (13) $p = 0$.

13.15° For all functions $y \in C([0, \pi])$.

13.16° $p \in (\frac{1}{3}, 3)$. The homogeneous adjoint equation $f(t) = \overline{\lambda}\int\limits_0^1 (ps - t)f(s)ds$, $t \in [0, 1]$, should have only a zero solution for all $\lambda \in \mathbb{R}$.

13.17° (1) $\mathbb{C} \setminus \{\pm\frac{2}{\pi}, \frac{1}{2\pi}\}$; (2) $\mathbb{C} \setminus \{\pm\frac{i}{\pi}\}$; (3) $\mathbb{C} \setminus \{-1, \frac{1}{8}\}$.

13.18 (1) $\mathbb{C} \setminus \{-\frac{3}{2}, \frac{3}{8}\}$. (2) $\mathbb{C} \setminus \{\frac{1}{2}, \frac{3}{2}\}$. (3) $\mathbb{C} \setminus \{-\frac{7}{3}, \frac{7}{11}\}$; (4) $\mathbb{C} \setminus \{\frac{35}{16}\}$.

13.19 Use that by the Taylor's formula with the integral remainder

$$x^{(n-k)}(t) = \sum_{j=0}^{k-1} c_{n-k+j}\frac{t^j}{j!} + \int_0^t \frac{(t-s)^{k-1}}{(k-1)!}x^{(n)}(s)ds, \ 1 \le k \le n.$$

13.20° (1) $\lambda = \frac{3}{7}$, $\varphi(t) = \sqrt{\frac{3}{7}}t$. (2) $\mu_n = \frac{2}{\pi}(n^2 + n)$, $\varphi_n(t) = \sqrt{\frac{2}{\pi}}\sin nt$, $n \in \mathbb{N}$. (3) $\mu_1 = 2$, $\mu_2 = 6$, $\varphi_1(t) = 3t - 1$, $\varphi_2(t) = \sqrt{3}(t-1)$. (4) $\mu_{1,2} = \pm\frac{1}{\sqrt{3}}$, $\varphi_{1,2}(t) = \sqrt{6}(t - \frac{1}{2}) \mp \frac{i}{\sqrt{2}}$. (5) $\mu_{1,2} = \pm\frac{1}{\pi}$, $\varphi_{1,2}(t) = \frac{\cos t}{\sqrt{2\pi}} \pm \frac{\sin t}{\sqrt{2\pi}}$. (6) $\mu_n = \frac{1}{2\pi\alpha^n}$, $n \ge 0$, $\varphi_n(t) = \frac{1}{\sqrt{2\pi}}e^{int}$, $n \ge 0$.

13.21 (1) $\mu_1 = \frac{1}{\pi}$, $\varphi_1(t) = \frac{1}{\sqrt{\pi}}\sin t$; $x(t) = \cos t + \frac{\sin t}{1-\pi\lambda}$ for $\lambda \in \mathbb{C} \setminus \{\mu_1\}$; $x \in \varnothing$ for $\lambda = \mu_1$.

(2) $\mu_1 = \frac{1}{\pi}$, $\mu_2 = \frac{2}{\pi}$, $\varphi_1(t) = \frac{1}{\sqrt{2\pi}}$, $\varphi_2^{(1)}(t) = \frac{\cos 2t}{\sqrt{\pi}}$, $\varphi_2^{(2)}(t) = \frac{\sin 2t}{\sqrt{\pi}}$; $x(t) = \frac{2\sin 2t}{2-\pi\lambda}$ for $\lambda \in \mathbb{C} \setminus \{\mu_1, \mu_2\}$; $x(t) = C + 2\sin 2t$ for $\lambda = \mu_1$; $x \in \varnothing$ for $\lambda = \mu_2$.

(3) $\mu_n = n^2\pi^2$, $\varphi_n(t) = \sqrt{2}\sin n\pi t$, $n \geq 1$; $x(t) = \frac{\pi^2}{4\pi^2-\lambda}\sin 2\pi t + \frac{\pi^2}{9\pi^2-\lambda}\sin 3\pi t$ for $\lambda \in \mathbb{C} \setminus \{\mu_n : n \geq 1\}$; $x(t) = \frac{\sin 2\pi t}{4-n^2} + \frac{\sin 3\pi t}{9-n^2} + C\sin n\pi t$ for $\lambda = \mu_n, n \geq 1, n \neq 2, n \neq 3$; $x \in \varnothing$ for $\lambda \in \{\mu_2, \mu_3\}$.

(4) $\mu_0 = 1$, $\mu_n = -n^2\pi^2, n \geq 1$,

$$\varphi_0(t) = \sqrt{\frac{2}{e^2-1}}e^t,\ \varphi_n(t) = \sqrt{\frac{2}{1+n^2\pi^2}}(\sin n\pi t + n\pi\cos n\pi t),\ n \geq 1;$$

$x(t) = \frac{\pi^2}{\pi^2+\lambda}(\sin\pi t + \pi\cos\pi t)$ for $\lambda \in \mathbb{C} \setminus \{\mu_n : n \geq 0\}$;
$x(t) = Ce^t + \frac{\pi^2}{\pi^2+1}(\sin\pi t + \pi\sin\pi t)$ for $\lambda = \mu_0$;
$x(t) = \frac{1}{1-n^2}(\sin\pi t + \pi\cos\pi t) + C(\sin n\pi t + n\pi\cos n\pi t)$ for $\lambda = \mu_n, n \geq 2$;
$x \in \varnothing$ for $\lambda = \mu_1$.

(5) $\mu_n^{(1)} = \frac{n^2\pi^2}{2}$, $\mu_n^{(2)} = \frac{(2n-1)^2\pi^2}{8}$, $n \geq 1$,

$$\varphi_n^{(1)}(t) = \sin n\pi t,\ \varphi_n^{(2)}(t) = \cos(n - \tfrac{1}{2})\pi t,\ n \geq 1;$$

$x(t) = \frac{32\lambda}{\pi}\sum_{k=1}^{\infty}\frac{(-1)^{k+1}\cos(k-\frac{1}{2})\pi t}{\pi^2(2k-1)^3-8\lambda(2k-1)} + 1$ for $\lambda \in \mathbb{C} \setminus \{\mu_n^{(1)}, \mu_n^{(2)} : n \geq 1\}$;
$x(t) = \frac{16n^2}{\pi}\sum_{k=1}^{\infty}\frac{(-1)^{k+1}\cos(k-\frac{1}{2})\pi t}{(2k-1)^3-4n^2(2k-1)} + C\sin\pi nt + 1$ for $\lambda = \mu_n^{(1)}, n \geq 1$;
$x \in \varnothing$ for $\lambda \in \{\mu_n^{(2)},\ n \geq 1\}$.

(6) $\mu_n = \frac{4+(2n-1)^2\pi^2}{8(e^2+1)}$, $\varphi_n(t) = \sqrt{2}\sin(n - \frac{1}{2})\pi t, n \geq 1$;
$x(t) = \frac{2\lambda}{\pi^2}\sum_{k=1}^{\infty}\frac{(-1)^{k-1}\sin(k-\frac{1}{2})\pi t}{(k-\frac{1}{2})^2(\mu_k-\lambda)} + t$ for $\lambda \in \mathbb{C} \setminus \{\mu_n : n \geq 1\}$;
$x \in \varnothing$ for $\lambda \in \{\mu_n : n \geq 1\}$.

(7) $\mu_n = 4n^2 - 1$, $\varphi_n(t) = \frac{2}{\sqrt{\pi}}\sin 2nt, n \geq 1$; $x(t) = \frac{8\lambda}{\pi}\sum_{k=1}^{\infty}\frac{k\sin 2kt}{(4k^2-1)(4k^2-1-\lambda)} + \cos t$ for $\lambda \in \mathbb{C} \setminus \{\mu_n : n \geq 1\}$; $x \in \varnothing$ for $\lambda \in \{\mu_n : n \geq 1\}$.

(8) $\mu_n = (n - \frac{1}{2})^2 - 1$, $\varphi_n(t) = \sqrt{\frac{2}{\pi}}\sin(n - \frac{1}{2})t, n \geq 1$; $x(t) = \frac{5}{5-4\lambda}\sin\frac{3}{2}t$ for $\lambda \in \mathbb{C} \setminus \{\mu_n : n \geq 1\}$; $x(t) = \frac{5}{9-4(n-\frac{1}{2})^2}\sin\frac{3}{2}t + C\sin(n - \frac{1}{2})t$ for $\lambda = \mu_n, n \geq 1$, $n \neq 2$; $x \in \varnothing$ for $\lambda = \mu_2$.

(9) $\mu_n = 1 - (n - \frac{1}{2})^2$, $\varphi_n(t) = \sqrt{\frac{2}{\pi}}\cos(n - \frac{1}{2})t, n \geq 1$;
$x(t) = \frac{2\lambda}{\pi}\sum_{k=1}^{\infty}\frac{(-1)^k(k-\frac{1}{2})-1}{1-(k-\frac{1}{2})^2-\lambda}\cdot\frac{\cos(k-\frac{1}{2})t}{(k-\frac{1}{2})^2} + t$ for $\lambda \in \mathbb{C} \setminus \{\mu_n : n \geq 1\}$; $x \in \varnothing$ for $\lambda \in \{\mu_n : n \geq 1\}$.

(10) $\mu_n = \frac{n^2\pi^2-1}{\sin 1}$, $\varphi_n(t) = \sqrt{2}\sin n\pi t, n \geq 1$;
$x(t) = \frac{8\lambda}{\pi}\sum_{k=1}^{\infty}\frac{k\sin 1\sin 2k\pi t}{(4k^2-1)(4k^2\pi^2-1-\lambda\sin 1)} + \cos\pi t$ for $\lambda \in \mathbb{C} \setminus \{\mu_n : n \geq 1\}$;

$x(t) = \frac{8((2n-1)^2\pi^2-1)}{\pi^3} \sum_{k=1}^{\infty} \frac{k \sin 2k\pi t}{(4k^2-1)(4k^2-(2n-1)^2)} + C \sin(2n-1)\pi t + \cos \pi t$ for $\lambda = \mu_{2n-1}, n \geq 1$; $x \in \varnothing$ for $\lambda = \mu_{2n}, n \geq 1$.

(11) $\mu_n = (n - \frac{1}{2})^2\pi^2$, $\varphi_n(t) = \sqrt{2}\sin(n - \frac{1}{2})\pi t$, $t \in [0, 1]$, $n \geq 1$;

$x(t) = \frac{2\lambda}{\pi} \sum_{k=1}^{\infty} \frac{(-1)^k \sin(k-\frac{1}{2})\pi t}{(k^2-k-\frac{3}{4})((k-\frac{1}{2})^2\pi^2-\lambda)} + \sin \pi t$ for $\lambda \in \mathbb{C} \setminus \{\mu_n : n \geq 1\}$; $x \in \varnothing$ for $\lambda \in \{\mu_n : n \geq 1\}$.

13.22 (1) $x''(t) = -\frac{\pi^2}{4}x(t)$, $x(0) = 0$, $x(2) = 1$; $x(t) = \sin \frac{\pi t}{4}$.

(2) $x''(t) = 2e^t$, $x(0) = x(1) = 0$; $x(t) = 2e^t - 2 + (2 - 2e)t$.

(3) $x(t) = 3\pi^2 \sin^3 \pi t - 6\pi^2 \sin \pi t \cos^2 \pi t$.

(4) $x'(t) = 0$, $x(t) = -1$.

13.23 (1) $x(t) = 16 \int_0^1 \frac{x(s)ds}{(|t-s|+4)^3} - 8y''(t)$; (2) $x(t) = \frac{1}{2} \int_1^2 \sin|t-s|x(s)ds + \frac{y''(t)}{2}$;

(3) $x(t) = \frac{1+i}{4} \int_0^1 \frac{e^{i|t-s|}}{1+|t-s|} \left(\frac{2}{(1+|t-s|)^2} - \frac{2i}{1+|t-s|} - 1 \right) x(s)ds - \frac{1+i}{4} y''(t)$;

(4) $x(t) = -\frac{1}{4} \int_0^1 x(s)ds + \frac{y'(t)}{4}$.

13.24 (1) $x(t) = -\int_0^t x(s)ds + 1$. (2) $x(t) = -\int_0^t e^{ts-t^2} sx(s)ds + e^{t-t^2}$.

(3) $x(t) = \int_0^t \sin(t-s)x(s)ds - \cos t$.

13.25 (1) $x(t) = \int_0^t x(s)ds + 2t$; (2) $x(t) = -\int_1^t \frac{x(s)ds}{3t+3t\sqrt[3]{t^2}} + \frac{4+3\sqrt[3]{t^2}}{3+3\sqrt[3]{t^2}}$;

(3) $x(t) = -\int_0^t \frac{2t}{t^2+t+1} x(s)ds + \frac{2t^3+t^2}{t^2+t+1}$.

13.26 $\mu_0 = \frac{1}{\pi^2}$, $\varphi_0(t) = \frac{1}{\sqrt{2\pi}}$, $\mu_n^{\pm} = \pm\frac{(2n-1)^2}{4}$, $\varphi_n^+(t) = \frac{1}{\sqrt{\pi}} \sin(2n-1)t$, $\varphi_n^-(t) = \frac{1}{\sqrt{\pi}} \cos(2n-1)t$, $n \geq 1$.

(1) $x(t) = \frac{1}{1-\pi^2\lambda} + \frac{2\sin t}{1-4\lambda} - \frac{\cos t}{1+4\lambda}$ for $\lambda \in \mathbb{C} \setminus \{\mu_0, \mu_n^{\pm} : n \geq 1\}$;

$x(t) = \frac{4}{4-(2n-1)^2\pi^2} + \frac{2\sin t}{1-(2n-1)^2} - \frac{\cos t}{1+(2n-1)^2} + C\sin(2n-1)t$ for $\lambda = \mu_n^+, n \geq 2$;

$x(t) = \frac{4}{4+(2n-1)^2\pi^2} + \frac{2\sin t}{1+(2n-1)^2} - \frac{\cos t}{1-(2n-1)^2} + C\cos(2n-1)t$ for $\lambda = \mu_n^-$, $n \geq 2$;

$x \in \varnothing$ for $\lambda \in \{\mu_0, \mu_1^{\pm}\}$.

(2) $x(t) = \frac{9}{9-4\lambda} \sin 3t + \cos 2t$ for $\lambda \in \mathbb{C} \setminus \{\mu_0, \mu_n^{\pm} : n \geq 1\}$;

$x(t) = \frac{9}{9-(2n-1)^2} \sin 3t + C\sin(2n-1)t + \cos 2t$ for $\lambda = \mu_n^+, n \geq 1, n \neq 2$;

$x(t) = \frac{9}{9+(2n-1)^2} \sin 3t + C\cos(2n-1)t + \cos 2t$ for $\lambda = \mu_n^-, n \geq 1$;

$x(t) = C + \frac{9\pi^2}{9\pi^2-4} \sin 3t + \cos 2t$ for $\lambda = \mu_0$; $x \in \varnothing$ for $\lambda = \mu_2^+$.

(3) $x(t) = \frac{16\lambda}{\pi}\sum_{k=1}^{\infty}\frac{\sin(2k-1)t}{(2k-1)^3-4(2k-1)\lambda} + \operatorname{sign} t$ for $\lambda \in \mathbb{C} \setminus \{\mu_0, \mu_n^{\pm} : n \geq 1\}$;

$x(t) = -\frac{4}{\pi}(2n-1)^2\sum_{k=1}^{\infty}\frac{\sin(2k-1)t}{(2k-1)^3+(2k-1)(2n-1)^2} + C\cos(2n-1)t + \operatorname{sign} t$ for $\lambda = \mu_n^-, n \geq 1$;

$x(t) = C + \frac{16}{\pi}\sum_{k=1}^{\infty}\frac{\sin(2k-1)t}{(2k-1)^3\pi^2-4(2k-1)} + \operatorname{sign} t$ for $\lambda = \mu_0$; $x \in \varnothing$ for $\lambda \in \{\mu_n^+ : n \geq 1\}$.

(4) $x(t) = \frac{\pi^4\lambda}{3(1-\pi^2\lambda)} + 16\lambda\sum_{k=1}^{\infty}\frac{\cos(2k-1)t}{(2k-1)^4+4(2k-1)^2\lambda} + t^2$ for $\lambda \in \mathbb{C}\setminus\{\mu_0, \mu_n^{\pm} : n \geq 1\}$;

$x(t) = \frac{(2n-1)^2\pi^4}{3(4-(2n-1)^2\pi^2)} + 4(2n-1)^2\sum_{k=1}^{\infty}\frac{\cos(2k-1)t}{(2k-1)^4+(2n-1)^2(2k-1)^2} + C\sin(2n-1)t + t^2$ for $\lambda = \mu_n^+,\ n \geq 1$;

$x \in \varnothing$ for $\lambda \in \{\mu_0, \mu_n^- \parallel n \geq 1\}$.

13.27 $\mu_0 = \frac{1}{\pi^2}$, $\varphi_0(t) = \frac{1}{\sqrt{2\pi}}$, $\mu_n = -\frac{(2n-1)^2}{4}$, $\varphi_n^{(1)}(t) = \frac{1}{\sqrt{\pi}}\cos(2n-1)t$, $\varphi_n^{(2)}(t) = \frac{1}{\sqrt{\pi}}\sin(2n-1)t$, $n \geq 1$.

(1) $x(t) = \frac{1}{1-\pi^2\lambda} + \frac{\sin t}{1+4\lambda} - 2\cos 2t$ for $\lambda \in \mathbb{C} \setminus \{\mu_n : n \geq 0\}$;

$x(t) = \frac{4}{4+(2n-1)^2\pi^2} + \frac{\sin t}{1+(2n-1)^2} + C_1\sin(2n-1)t + C_2\cos(2n-1)t - 2\cos 2t$ for $\lambda = \mu_n, n \geq 2$; $x \in \varnothing$ for $\lambda \in \{\mu_0, \mu_1\}$.

(2) $x(t) = \frac{\sin t}{1+4\lambda} + \frac{9\sin 3t}{9+4\lambda} + \sin 2t$ for $\lambda \in \mathbb{C} \setminus \{\mu_n : n \geq 0\}$;

$x(t) = \frac{\sin t}{1-(2n-1)^2} + \frac{9\sin 3t}{9-(2n-1)^2} + C_1\sin(2n-1)t + C_2\cos(2n-1)t + \sin 2t$ for $\lambda = \mu_n, n \geq 3$;

$x(t) = C + \frac{\pi^2\sin t}{\pi^2+4} + \frac{9\pi^2\sin 3t}{9\pi^2+4} + \sin 2t$ for $\lambda = \mu_0$; $x \in \varnothing$ for $\lambda \in \{\mu_1, \mu_2\}$.

(3) $x(t) = -\frac{16\lambda}{\pi}\sum_{k=1}^{\infty}\frac{\sin(2k-1)t}{(2k-1)^3+4(2k-1)\lambda} + \operatorname{sign} t$ for $\lambda \in \mathbb{C} \setminus \{\mu_n : n \geq 0\}$;

$x(t) = C - \frac{16}{\pi}\sum_{k=1}^{\infty}\frac{\sin(2k-1)t}{(2k-1)^3\pi^2+4(2k-1)} + \operatorname{sign} t$ for $\lambda = \mu_0$; $x \in \varnothing$ for $\lambda \in \{\mu_n : n \geq 1\}$.

(4) $x(t) = \frac{\pi^4\lambda}{3(1-\pi^2\lambda)} + 16\lambda\sum_{k=1}^{\infty}\frac{\cos(2k-1)t}{(2k-1)^4+4(2k-1)^2\lambda} + t^2$ for $\lambda \in \mathbb{C} \setminus \{\mu_n : n \geq 0\}$;

$x \in \varnothing$ for $\lambda \in \{\mu_n : n \geq 0\}$.

13.29 Apply Theorem 13.3.

13.30 (1) $\lambda \geq \frac{1}{2}$. Show that x should satisfy the differential equation $x''(t) = (1-2\lambda)x(t)$. Bounded on $\mathbb{R}$ solutions of this equation have the form $x(t) = C$ for $\lambda = \frac{1}{2}$ and $x(t) = C_1\cos\sqrt{2\lambda-1} + C_2\sin\sqrt{2\lambda-1}$ for $\lambda > \frac{1}{2}$. Check that these solutions satisfy the initial integral equation.

(2) Uniqueness follows from the fact that the corresponding homogeneous equation has only a zero solution in the space $L_\infty(\mathbb{R})$.

Problems of Chap. 14

14.17° (1) Yes. (2), (3) No.

14.19 (1) Let $p > 1$ and q be the conjugate index to p. By Hölder's inequality

$$|f_\varepsilon(x)|^p = \left|\int_{\mathbb{R}} (\omega_\varepsilon(x-y))^{\frac{1}{p}+\frac{1}{q}} f(y)dy\right|^p \le \int_{\mathbb{R}} \omega_\varepsilon(x-y)|f(y)|^p dy\cdot$$

$$\cdot\left(\int_{\mathbb{R}} \omega_\varepsilon(x-y)dy\right)^{\frac{p}{q}} = \int_{\mathbb{R}} \omega_\varepsilon(x-y)|f(y)|^p dy,\ x \in \mathbb{R}.$$

Therefore for all $1 \le p < +\infty$ we have

$$\|f_\varepsilon\|_p^p \le \int_{\mathbb{R}}\int_{\mathbb{R}} \omega_\varepsilon(x-y)|f(y)|^p dydx =$$

$$= \int_{\mathbb{R}} |f(y)|^p \left(\int_{\mathbb{R}} \omega_\varepsilon(x-y)dx\right) dy = \|f\|_p^p.$$

14.20 Consider the functions f_ε, $\varepsilon > 0$, defined in problem 14.18°. Since $f \in L_2(\mathbb{R})$, then $f_\varepsilon \to f$ in $L_2(\mathbb{R})$ as $\varepsilon \to 0+$ according to item (2) of problem 14.19. Therefore there exists a subsequence $g_n = f_{\varepsilon_n}$, for which $g_n \to f \pmod{m}$ as $n \to \infty$.

14.23 Repeat the reasonings, similar to the solution of problem 14.2.

14.25° Use problem 14.24°.

14.26° (1) Singular, $\operatorname{supp} f = \{0, 1\}$; (2) singular, $\operatorname{supp} f = \{0\}$; (3) regular, $\operatorname{supp} f = \mathbb{R}$; (4) singular, $\operatorname{supp} f = [-1, 1]$; (5) singular, $\operatorname{supp} f = \{-2, 0, 2\}$; (6)–(8) regular, $\operatorname{supp} f = \mathbb{R}$; (9) singular, $\operatorname{supp} f = [0, 2\pi]$; (10) singular, $\operatorname{supp} f = \{0, 1, 3\}$.

14.28 Let us show that for each $n \ge 1$ there exists $C_n \ge 0$ such that

$$|\langle f, \varphi\rangle| \le C_n\|\varphi\|_n, \text{ where } \|\varphi\|_n := \max_{x\in[-n,n]} |\varphi(x)|,$$

for all $\varphi \in \mathcal{D}(\mathbb{R})$ such that $\operatorname{supp}\varphi \subset [-n, n]$. Due to item (1) of problem 14.18° there exists a nonnegative function $\eta \in \mathcal{D}(\mathbb{R})$ such that $\eta(x) = 1$, $x \in [-n, n]$. If $\operatorname{supp}\varphi \subset [-n, n]$, then $\|\varphi\|_n \cdot \eta \pm \varphi \in \mathcal{D}(\mathbb{R})$ are nonnegative functions. Therefore $\|\varphi\|_n \cdot \langle f, \eta\rangle \pm \langle f, \varphi\rangle \ge 0$, so for $C_n = \langle f, \eta\rangle \ge 0$ we get $|\langle f, \varphi\rangle| \le C_n\|\varphi\|_n$.

According to item (2) of problem 14.19 the set $\{\varphi \in \mathcal{D}(\mathbb{R}) \mid \operatorname{supp}\varphi \subset [-n,n]\}$ is dense in the set $C_0([-n,n]) := \{g \in C(\mathbb{R}) \mid \operatorname{supp} g \subset [-n,n]\}$. Therefore f can be extended by continuity to a continuous linear functional F_n on $C_0([-n,n])$. By Theorem 3.5 there exists a measure μ_n on $\mathcal{B}([-n,n])$ such that $F_n(g) = \int\limits_{[-n,n]} g d\mu_n$, $g \in C_0([-n,n])$. Verify that $\mu_m(B) = \mu_n(B)$ for all $m > n$ and $B \in \mathcal{B}([-n,n])$. Deduce that there exists a measure μ on $\mathcal{B}(\mathbb{R})$ such that $\mu(B) = \mu_n(B)$ for $B \in \mathcal{B}([-n,n])$, $n \geq 1$. Thus for $\varphi \in \mathcal{D}(\mathbb{R})$ if $\operatorname{supp}\varphi \subset [-n,n]$, then

$$\langle f, \varphi\rangle = F_n(\varphi) = \int\limits_{[-n,n]} \varphi d\mu_n = \int\limits_{\mathbb{R}} \varphi d\mu.$$

14.31 (1), (2), (4) Use item (2) of problem 14.30°. (5) Use that

$$\langle t^n e^{ixt}, \varphi(x)\rangle = \frac{i^{n+1}}{t}\langle e^{ixt}, \varphi^{(n+1)}(x)\rangle,\ \varphi \in \mathcal{D}(\mathbb{R}),\ t > 0.$$

14.32 For all $\varphi \in \mathcal{D}(\mathbb{R})$ we have

$$\left\langle \frac{1}{x \pm i\varepsilon}, \varphi(x)\right\rangle = \int\limits_{\mathbb{R}} \frac{(x \mp i\varepsilon)\varphi(x)}{x^2+\varepsilon^2}dx = \int\limits_{\mathbb{R}} \frac{x^2}{x^2+\varepsilon^2}\cdot\frac{\varphi(x)}{x}dx \mp i\int\limits_{\mathbb{R}} \frac{\varepsilon\varphi(x)}{x^2+\varepsilon^2}dx.$$

Show that the first term tends to v. p. $\int\limits_{\mathbb{R}} \frac{\varphi(x)}{x}dx$ as $\varepsilon \to 0+$. According to item (1) of problem 14.31, the second term tends to $\mp i\pi\varphi(0)$.

14.34 $c\delta_0$, $c \in \mathbb{C}$. Use problem 14.23 and repeat the reasonings, similar to the solution of problem 14.7.

14.37° (1) $\operatorname{sign} x$; (2) $2\sum\limits_{k\in\mathbb{Z}}(-1)^k\delta_{k\pi}$.

14.39° (1) $\delta_a^{(n-1)}$; (2) $\delta_0^{(n-1)} + a\delta_0^{(n-2)} + \ldots + a^{n-1}\delta_0 + a^n\theta(x)e^{ax}$; (3) $2\delta_0^{(n-1)}$; (4) $\operatorname{sign} x$, $n=1$; $2\delta_0^{(n-2)}$, $n \geq 2$; (5) $2\sum\limits_{k\in\mathbb{Z}}(-1)^{k+1}\delta^{(n-1)}_{(k+\frac{1}{2})\pi}$; (6) $\sum\limits_{k\in\mathbb{Z}}\delta_k^{(n-1)}$.

14.40° (1) $f' = \theta(x)\cos x$, $f'' = \delta_0 - \theta(x)\sin x$, $f''' = \delta_0' - \theta(x)\cos x$;

(2) $f' = \delta_0 - \theta(x)\sin x$, $f'' = \delta_0' - \theta(x)\cos x$, $f''' = \delta_0'' - \delta_0 + \theta(x)\sin x$;

(3) $f' = 2x\chi_{[-2,2]}(x) + 4\delta_{-2} - 4\delta_2$, $f'' = 2\chi_{[-2,2]}(x) - 4\delta_{-2} - 4\delta_2 + 4\delta'_{-2} - 4\delta'_2$, $f''' = 2\delta_{-2} - 2\delta_2 - 4\delta'_{-2} - 4\delta'_2 + 4\delta''_{-2} - 4\delta''_2$;

(4) $f' = \chi_{[0,1)}(x) + 2x\chi_{[1,+\infty)}(x)$, $f'' = 2\chi_{[1,+\infty)}(x) + \delta_0 + \delta_1$, $f''' = 2\delta_1 + \delta_0' + \delta_1'$;

(5) $f' = 2(x+1)\chi_{[-1,0)}(x) + 2x\chi_{[0,+\infty)}(x)$, $f'' = 2\chi_{[-1,+\infty)}(x) - 2\delta_0$, $f''' = 2\delta_{-1} - 2\delta_0'$;

(6) $f' = \cos x\,\chi_{[-\pi,\pi]}(x)$, $f'' = -\sin x\,\chi_{[-\pi,\pi]}(x) - \delta_{-\pi} + \delta_\pi$,
$f''' = -\cos x\,\chi_{[-\pi,\pi]}(x) - \delta'_{-\pi} + \delta'_\pi$;

(7) $f' = 2x\chi_{(-\infty,1)}(x) + 2(x-2)\chi_{[1,2)}(x)$, $f'' = 2\chi_{(-\infty,2)}(x) - 4\delta_1$, $f''' = -2\delta_2 - 4\delta_1'$;

(8) $f' = -\sin x \cdot \operatorname{sign}(\cos x)\chi_{[0,2\pi]}(x) + \delta_0 - \delta_{2\pi}$,
$f'' = -|\cos x|\chi_{[0,2\pi]}(x) + 2\delta_{\frac{\pi}{2}} + 2\delta_{\frac{3\pi}{2}} + \delta_0' - \delta_{2\pi}'$,
$f''' = \sin x \cdot \operatorname{sign}(\cos x)\chi_{[0,2\pi]}(x) - \delta_0 + \delta_{2\pi} + 2\delta'_{\frac{\pi}{2}} + 2'\delta_{\frac{3\pi}{2}} + \delta_0'' - \delta_{2\pi}''$.

14.41° (1) Use item (2) of problem 14.30°.
(4) Verify that for all $\varphi \in \mathcal{D}(\mathbb{R})$ we have

$$\left\langle \frac{1}{x^2}\chi_{[\varepsilon,2\varepsilon)}(x) - \frac{1}{\varepsilon}\delta_\varepsilon + \frac{1}{2\varepsilon}\delta_{2\varepsilon}, \varphi(x)\right\rangle = \int_\varepsilon^{2\varepsilon} \frac{\varphi'(x)}{x}dx =$$

$$= \int_1^2 \frac{\varphi'(\varepsilon u)}{u}du \to \varphi'(0)\ln 2 = -\ln 2\langle \delta_0'', \varphi(x)\rangle, \ \varepsilon \to 0+.$$

14.43 $\alpha'\theta + \alpha(0)\delta_0$.

14.44 (5) For all $\varphi \in \mathcal{D}(\mathbb{R})$ we have

$$\langle \alpha\delta_0^{(n)}, \varphi\rangle = \langle \delta_0^{(n)}, \alpha\varphi\rangle = (-1)^n\langle \delta_0, (\alpha\varphi)^{(n)}\rangle =$$

$$=(-1)^n\left\langle \delta_0, \sum_{k=0}^n \binom{n}{k}\alpha^{(k)}\varphi^{(n-k)}\right\rangle = (-1)^n\sum_{k=0}^n \binom{n}{k}\alpha^{(k)}(0)\varphi^{(n-k)}(0) =$$

$$=(-1)^n\sum_{k=0}^n \binom{n}{k}\alpha^{(k)}(0)\langle \delta_0, \varphi^{(n-k)}\rangle = \sum_{k=0}^n (-1)^k\binom{n}{k}\alpha^{(k)}(0)\langle \delta_0^{(n-k)}, \varphi\rangle.$$

14.45 Let $\sum\limits_{k=0}^n c_k\delta_0^{(k)} = 0$ in $\mathcal{D}'(\mathbb{R})$. In view of items (3),4) of problem 14.44 we have $x^n \cdot \sum\limits_{k=0}^n c_k\delta_0^{(k)} = (-1)^n c_n\delta_0 = 0$ in $\mathcal{D}'(\mathbb{R})$, thus $c_n = 0$. Similarly we get successively that $c_{n-1} = 0, \dots, c_0 = 0$.

14.47 Let f be a 2π-periodic function such that $f(x) = \frac{x}{2} - \frac{x^2}{4\pi}$, $x \in [0, 2\pi]$. We have $f(x) = \frac{\pi}{6} - \frac{1}{2\pi}\sum\limits_{k\in\mathbb{Z}\setminus\{0\}} \frac{1}{k^2}e^{ikx}$. The Fourier series converges uniformly on $\mathbb{R}$, so it converges in $\mathcal{D}'(\mathbb{R})$. Since the operation of differentiation is continuous in $\mathcal{D}'(\mathbb{R})$, the series in $\mathcal{D}'(\mathbb{R})$ can be differentiated term by term. Therefore $f'' = \frac{1}{2\pi}\sum\limits_{k\in\mathbb{Z}\setminus\{0\}} e^{ikx}$ in $\mathcal{D}'(\mathbb{R})$. On the other hand, due to relation between the generalized and classical derivatives we get $f'' = -\frac{1}{2\pi} + \sum\limits_{k\in\mathbb{Z}} \delta_{2\pi k}$ in $\mathcal{D}'(\mathbb{R})$.

14.48 Let g be a 2π-periodic function such that $g(x) = |x|$, $x \in [-\pi, \pi]$. Verify that $g(x) = \frac{\pi}{2} - \frac{4}{\pi} \sum\limits_{k=0}^{\infty} \frac{\cos(2k+1)x}{(2k+1)^2}$, $x \in \mathbb{R}$, and apply to g the reasonings, similar to the solution of problem 14.47.

14.49° $c_1 x + c_2$.

14.51 (1) $c\delta_1$; (2) $c_1\delta_{-1} + c_2\delta_1$; (3) $c\delta_0 + \mathcal{P}\frac{1}{x^2}$; (4) $c_1 + c_2\theta(x) + c_3\delta_0$; (5) $c_0 + c_1 x + c_2\theta(x+1) + c_3(x+1)\theta(x+1)$; (6) $c_1\delta_0 + c_2\delta_1$; (7) $c\delta_0 + \mathcal{P}\frac{1}{x}$; (8) $c_1\delta_0 + c_2\delta_0'$; (9) $c_1 + c_2\theta(x) - \mathcal{P}\frac{1}{x}$; (10) $c_1 + c_2\theta(x) + c_3\delta_0 - \mathcal{P}\frac{1}{x}$; (11) $c_0 + c_1 x + c_2 x^2 + c_3(x+1)^2\theta(x+1)$.

14.52*. (1) $y_{\text{gen}} = \sum\limits_{k\in\mathbb{Z}} c_k\delta_{k\pi}$. Let $\xi \in \mathcal{D}(\mathbb{R})$ be such that $\xi = 1$ in the neighborhood of 0 and $\operatorname{supp}\xi \subset [-1, 1]$. For $\varphi \in \mathcal{D}(\mathbb{R})$ consider the function

$$\psi(x) = \frac{1}{\sin x}\left(\varphi(x) - \sum_{k\in\mathbb{Z}} \varphi(k\pi)\xi(x - k\pi)\right), \quad x \notin \{k\pi \mid k \in \mathbb{Z}\}.$$

This function is infinitely differentiable for $x \neq k\pi$, and in some neighborhood of the point $k\pi$ it equals $\frac{\varphi(x)-\varphi(k\pi)}{\sin x}$ (except the point $x = k\pi$). Since the function $\sin x$ has zero of order 1 at the point $x = k\pi$, according to problem 14.23 the function $\frac{\varphi(x)-\varphi(k\pi)}{\sin x}$, which is defined by continuity at the point $x = k\pi$, is infinitely differentiable in the neighborhood of this point. Therefore the function ψ, which is defined by continuity at the points $x = k\pi$, $k \in \mathbb{Z}$, belongs to $C^\infty(\mathbb{R})$. It is clear that ψ has a bounded support, hence $\psi \in \mathcal{D}(\mathbb{R})$.

Let y be a solution of the equation $\sin x \cdot y = 0$. Since

$$\varphi(x) = \psi(x)\sin x + \sum_{k\in\mathbb{Z}} \varphi(k\pi)\xi(x - k\pi), \quad x \in \mathbb{R},$$

then $\langle y(x), \varphi(x)\rangle = \langle y(x), \psi(x)\sin x\rangle + \sum\limits_{k\in\mathbb{Z}} \varphi(k\pi)\langle y(x), \xi(x - k\pi)\rangle$. Denote

$$c_k = \langle y(x), \xi(x - k\pi)\rangle, \quad k \in \mathbb{Z}.$$

Since $\langle y(x), \psi(x)\sin x\rangle = \langle \sin x \cdot y(x), \psi(x)\rangle = 0$, then

$$\langle y(x), \varphi(x)\rangle = \sum_{k\in\mathbb{Z}} c_k\varphi(k\pi) = \left\langle \sum_{k\in\mathbb{Z}} c_k\delta_{k\pi}, \varphi(x)\right\rangle.$$

Hence $y = \sum\limits_{k\in\mathbb{Z}} c_k\delta_{k\pi}$ (the series converges in $\mathcal{D}'(\mathbb{R})$). This function satisfies the equation for any $c_k \in \mathbb{C}$, $k \in \mathbb{Z}$. Indeed, since the operation of multiplication by

$\alpha(x) \in C^\infty(\mathbb{R})$ is continuous on $\mathcal{D}'(\mathbb{R})$, then $\sin x \cdot \sum\limits_{k\in\mathbb{Z}} c_k\delta_{k\pi} = \sum\limits_{k\in\mathbb{Z}} c_k(\sin x\, \delta_{k\pi})$. It is easy to verify that $\sin x\, \delta_{k\pi} = 0$, $k \in \mathbb{Z}$. Therefore the series converges to 0.

(2) $y_{\text{gen}} = \sum\limits_{k\in\mathbb{Z}} c_k\delta_{(k+\frac{1}{2})\pi}$.

14.53 According to problem 14.7 we have $u^{(m)} = \sum\limits_{k=0}^{n-1} c_k\delta_0^{(k)}$, where $c_0, \ldots, c_{n-1}$ are some complex numbers. It remains to find a partial solution of the equation and the general solution of the homogeneous equation $u^{(m)} = 0$.

14.55 By the Fubini's theorem

$$F[f * g](y) = \frac{1}{\sqrt{2\pi}} \int\limits_{\mathbb{R}} e^{iyx} \left(\int\limits_{\mathbb{R}} f(t)g(x-t)dt \right) dx =$$

$$= \frac{1}{\sqrt{2\pi}} \int\limits_{\mathbb{R}} e^{iyt} f(t) \left(\int\limits_{\mathbb{R}} e^{iy(x-t)} g(x-t)dx \right) dt = \sqrt{2\pi}\, F[f](y)F[g](y).$$

14.57° $\mathcal{E}(x) = \theta(x)e^{-ax}$, $u(x) = ce^{-ax} + \int\limits_{-\infty}^{x} e^{-a(x-y)}\varphi(y)dy$.

14.58 (1) $\mathcal{E}(x) = \theta(x)\sinh x$, $u(x) = c_1e^x + c_2e^{-x} + \int\limits_{-\infty}^{x} \sinh(x-y)\varphi(y)dy$; (2) $\mathcal{E}(x) = \theta(x)\frac{x^{n-1}}{(n-1)!}e^{ax}$, $u(x) = \sum\limits_{k=0}^{n-1} c_kx^ke^{ax} + \int\limits_{-\infty}^{x} \frac{(x-y)^{n-1}}{(n-1)!}e^{a(x-y)}\varphi(y)dy$; (3) $\mathcal{E}(x) = \frac{1}{4}\theta(x)(1-e^{-4x})$, $u(x) = c_1 + c_2e^{-4x} + \frac{1}{4}\int\limits_{-\infty}^{x} \left(1-e^{-4(x-y)}\right)\varphi(y)dy$; (4) $\mathcal{E}(x) = xe^x\theta(x)$, $u(x) = c_1e^x + c_2xe^x + \int\limits_{-\infty}^{x} (x-y)e^{x-y}\varphi(y)dy$; (5) $\mathcal{E}(x) = \theta(x)(e^{-x} - e^{-2x})$, $u(x) = c_1e^{-x} + c_2e^{-2x} + \int\limits_{-\infty}^{x} (e^{-(x-y)} - e^{-2(x-y)})\varphi(y)dy$; (6) $\mathcal{E}(x) = \theta(x)e^{2x}\sin x$,

$$u(x) = e^{2x}(c_1\sin x + c_2\cos x) + \int\limits_{-\infty}^{x} e^{2(x-y)} \cdot \sin(x-y)\varphi(y)dy;$$

(7) $\mathcal{E}(x) = \frac{1}{3}\theta(x)\big(e^x - e^{-\frac{x}{2}}\cos\frac{\sqrt{3}x}{2} - \sqrt{3}e^{-\frac{x}{2}}\sin\frac{\sqrt{3}x}{2}\big)$,

$$u(x) = c_1e^x + c_2e^{-\frac{x}{2}}\cos\frac{\sqrt{3}x}{2} + c_3e^{-\frac{x}{2}}\sin\frac{\sqrt{3}x}{2} +$$

$$+\frac{1}{3}\int\limits_{-\infty}^{x}\Big(e^{x-y} - e^{-\frac{x-y}{2}}\cos\frac{\sqrt{3}(x-y)}{2} - \sqrt{3}e^{-\frac{x-y}{2}}\sin\frac{\sqrt{3}(x-y)}{2}\Big)\varphi(y)dy;$$

(8) $\mathcal{E}(x) = \frac{1}{2}\theta(x)(e^x-1)^2$, $u(x) = c_1 + c_2e^x + c_3e^{2x} + \frac{1}{2}\int\limits_{-\infty}^{x}(e^{x-y}-1)^2\varphi(y)dy$;

(9) $\mathcal{E}(x) = \frac{1}{2}\theta(x)(\sinh x - \sin x)$,

$$u(x) = c_1e^x + c_2e^{-x} + c_3\cos x + c_4\sin x +$$

$$+\frac{1}{2}\int\limits_{-\infty}^{x}(\sinh(x-y) - \sin(x-y))\varphi(y)dy;$$

(10) $\mathcal{E}(x) = \frac{1}{2}\theta(x)\cdot(x\cosh x - \sinh x)$,

$$u(x) = c_1e^x + c_2xe^x + c_3e^{-x} + c_4xe^{-x} +$$

$$+\frac{1}{2}\int\limits_{-\infty}^{x}\big((x-y)\cosh(x-y) - \sinh(x-y)\big)\varphi(y)dy.$$

14.60 (1) $\mathcal{E}(x,t) = \frac{\theta(x-t)(1+x^2)}{1+t^2}$. According to problem 14.59 we have $\mathcal{E}(x,t) = \theta(x-t)z(x,t)$, where $z = z(x,t)$ for each fixed $t \in \mathbb{R}$ is the solution of the Cauchy problem $z' - \frac{2x}{1+x^2}z = 0$, $z(t) = 1$. This is a separable differential equation. The solution is $z = z(x,t) = \frac{1+x^2}{1+t^2}$, $x, t \in \mathbb{R}$.

(2) $\mathcal{E}(x,t) = \theta(x-t)e^{x^2-t^2}$; (3) $\mathcal{E}(x,t) = \theta(x-t)e^{t^3-x^3}$; (4) $\mathcal{E}(x,t) = \theta(x-t)e^{\cos t-\cos x}$;

(5) $\mathcal{E}(x,t) = \theta(x-t)e^{\sin t-\sin x}$; (6) $\mathcal{E}(x,t) = \theta(x-t)e^{2\cos x^2-2\cos t^2}$;

(7) $\mathcal{E}(x,t) = \theta(x-t)e^{\frac{\arctan^2 x-\arctan^2 t}{2}}$; (8) $\mathcal{E}(x,t) = \theta(x-t)e^{x^4-x-t^4+t}$;

(9) $\mathcal{E}(x,t) = \theta(x-t)\exp(\frac{2^x-2^t}{\ln 2})$; (10) $\mathcal{E}(x,t) = \theta(x-t)\exp\big(\frac{3^{t^3}-3^{x^3}}{\ln 3}\big)$.

14.61° (1), (3) Yes. (2), (4) No.

14.64° Use item (3) of problem 14.19 and inclusion $\mathcal{D}(\mathbb{R}) \subset \mathcal{S}(\mathbb{R})$.

14.65° (1) Yes. (2), (3) No.

14.71 (2) Verify that $\langle f, \varphi \rangle = -\int\limits_{-\infty}^{+\infty} \sin e^x \varphi'(x)dx,\ \varphi \in \mathcal{S}(\mathbb{R})$.

14.73 (1) $e^{-\frac{y^2}{2}}$; (2) $iye^{-\frac{y^2}{2}}$; (3) $-(y^2-1)e^{-\frac{y^2}{2}}$; (4) $-i(y^3-3y)e^{-\frac{y^2}{2}}$. Apply item (3) of problem 14.72° several times.

14.74 Since $F^4 = I$ according to problem 14.14, then item (3) of problem 11.15* implies that $\sigma(F) \subset \{\lambda \in \mathbb{C} \mid \lambda^4 = 1\} = \{1, i, -1, -i\}$. Use problem 14.73 to verify that $e^{-\frac{x^2}{2}}$, $xe^{-\frac{x^2}{2}}$, $(2x^2-1)e^{-\frac{x^2}{2}}$, and $(2x^3-3x)e^{-\frac{x^2}{2}}$ are eigenfunctions which correspond to the eigenvalues $1, i, -1$, and $-i$, respectively.

14.75 (1) $\frac{1}{\sqrt{2\pi}(a-iy)}$; (2) $\frac{\sqrt{2}a}{\sqrt{\pi}(a^2+y^2)}$; (3) $\frac{2\sqrt{2}iay}{\sqrt{\pi}(a^2+y^2)^2}$; (4) $e^{-a|y|}$; (5) $\frac{e^{iay}-1}{\sqrt{2\pi}iy}$; (6) $\frac{\sqrt{2}\sin ay}{\sqrt{\pi}y}$; (7) $e^{-\frac{y^2+a^2}{2}}\cosh ay$; (8) $ie^{-\frac{y^2+a^2}{2}}\sinh ay$.

14.79 (1) $\frac{i}{\sqrt{2\pi}}\mathcal{P}\frac{1}{y} + \sqrt{\frac{\pi}{2}}\delta_0$. Note that $F[\theta(x)e^{-ax}] \xrightarrow{\mathcal{S}'(\mathbb{R})} F[\theta(x)],\ a \to 0+$, and use item (1) of problem 14.75 and problem 14.32. (2) $-\frac{i}{\sqrt{2\pi}}\mathcal{P}\frac{1}{y} + \sqrt{\frac{\pi}{2}}\delta_0$. (3) $i\sqrt{\frac{2}{\pi}}\mathcal{P}\frac{1}{y}$. (4) $\sqrt{\frac{2}{\pi}}(\mathcal{P}\frac{1}{y})' = -\sqrt{\frac{2}{\pi}}\mathcal{P}\frac{1}{y^2}$. Note that $|x| = x \operatorname{sign} x$ and use item (1) of problem 14.76° and item (2) of problem 14.42. (5) $i\sqrt{\frac{\pi}{2}}\operatorname{sign} y$. (6) $-\sqrt{\frac{\pi}{2}}|y|$. Use items (3), (4) and the formula $F^{-1}[f] = F[f(-x)],\ f \in \mathcal{S}'(\mathbb{R})$.

14.80 (1) For $g \in L_2([0,+\infty))$ put $f_1(x) = g(|x|)$ and $f_2(x) = g(|x|)\operatorname{sign}(x), x \in \mathbb{R}$. Verify that $F[f_1](x) = \sqrt{\frac{2}{\pi}}(Ag)(|x|)$ and $F[f_2](x) = i\sqrt{\frac{2}{\pi}}(Bg)(|x|)\operatorname{sign}(x)$. Hence

$$\frac{4}{\pi}\|Ag\|^2_{L_2([0,+\infty))} = \|F[f_1]\|^2_{L_2(\mathbb{R})} = \|f_1\|^2_{L_2(\mathbb{R})} = 2\|g\|^2_{L_2([0,+\infty))},$$

therefore $\|A\| = \sqrt{\frac{\pi}{2}}$. Similarly $\|B\| = \sqrt{\frac{\pi}{2}}$.

(2) $\sigma(A) = \sigma_p(A) = \sigma(B) = \sigma_p(B) = \left\{\pm\sqrt{\frac{\pi}{2}}\right\}$. For $g \in L_2([0,+\infty))$ in the notations from the solution of item (1) we have $F^2[f_1](x) = \frac{2}{\pi}(A^2g)(|x|)$ and $F^2[f_2](x) = -\frac{2}{\pi}(B^2g)(|x|)\operatorname{sign}(x)$. Since $F^2[f_1] = f_1$ and $F^2[f_2] = -f_2$ due to problem 14.14, it follows that $\frac{2}{\pi}A^2 = \frac{2}{\pi}B^2 = I$.

14.81 (1) For all $h \in L_2(\mathbb{R})$ by the Fubini's theorem and the Cauchy–Schwarz inequality we have

$$\int_{\mathbb{R}} |h(x)| \left(\int_{\mathbb{R}} |f(x-y)g(y)|dy \right) dx = \int_{\mathbb{R}} |h(x)| \left(\int_{\mathbb{R}} |f(t)g(x-t)|dt \right) dx =$$

$$= \int_{\mathbb{R}} |f(t)| \left(\int_{\mathbb{R}} |g(x-y)h(x)|dx \right) dt \leq \|f\|_1 \cdot \|g\|_2 \cdot \|h\|_2.$$

Hence $f * g \in L_2(R)$ and $\|f * g\|_2 \leq \|f\|_1 \cdot \|g\|_2$.

(2) In view of item (1) the operators $A_1 g = F[f * g]$ and $A_2 g = \sqrt{2\pi} F[f] \cdot F[g]$ are continuous in $L_2(\mathbb{R})$. Thus it is enough to verify that $A_1 g = A_2 g$ for g from a dense subset of $L_2(R)$. It remains to note that the set $L_1(R) \cap L_2(R)$ is dense in $L_2(\mathbb{R})$ and $F[f * g] = \sqrt{2\pi} F[f] \cdot F[g]$, $g \in L_1(\mathbb{R})$, according to problem 14.55.

14.82 $\sigma(A) = \{\sqrt{2\pi} F[f](x) \mid x \in \mathbb{R}\}$. Put $Bg = F[g]$ and $\widetilde{A}g = \sqrt{2\pi} F[f] \cdot g$, $g \in L_2(\mathbb{R})$. It follows from problem 14.81 that $A = B^{-1}\widetilde{A}B$. Hence the operators $A, \widetilde{A} \in \mathcal{L}(L_2(\mathbb{R}))$ are similar (see problem 11.35), so $\sigma(A) = \sigma(\widetilde{A})$. Note that $\sqrt{2\pi} F[f] \in C(\mathbb{R})$, because $f \in L_1(\mathbb{R})$.

14.84 (1) $\delta_{(1,1)} - \delta_{(0,1)} - \delta_{(1,0)} + \delta_{(0,0)}$; (2) $-2\delta_{(1,0)} + 2\delta_{(0,1)} - 2\delta_{(-1,0)} + 2\delta_{(0,-1)}$. Let γ be the boundary of the square S, which is counterclockwise oriented. Verify that by Green's theorem for all $\varphi \in \mathcal{D}(\mathbb{R}^2)$ we have

$$\langle f''_{yy} - f''_{xx}, \varphi \rangle = \int_S (\varphi''_{yy} - \varphi''_{xx}) dx dy = \int_\gamma (-\varphi'_y dx - \varphi'_x dy) =$$

$$= -2\varphi(1,0) + 2\varphi(0,1) - 2\varphi(-1,0) + 2\varphi(0,-1).$$

14.85 (1), (4) Yes. (2), (3) No.

14.86 (3) Consider as an example $\mu(A) = \int_A e^{\|x\|} dx$, $A \in \mathcal{B}(\mathbb{R}^m)$.

References

1. Abramovich, Y.A., Aliprantis, C.D.: Problems in Operator Theory, American Mathematical Society (2002)
2. Akhiezer, N.I., Glazman, I.M.: Theory of linear operators in Hilbert space, Dover Publ., New York (1993)
3. Antonevich, A.B., Knyazev, P.N., Radyno, Y.V.: Problems and exercises on functional analysis [in Russian], Vyshcha shkola, Minsk (1978)
4. Antonevich, A.B., Mazel, M.H., Radyno, Y.V.: Functional analysis and integral equations [in Russian], Belarusian State University (2011)
5. Bachman, G., Narici, L.: Functional analysis. Dover Publications, Inc. Mineola, New York (2000)
6. Balakrishnan, A.V.: Applied Functional Analysis: Applications of Mathematics (Vol. 3). Springer Science & Business Media (2012)
7. Banach S. Théorie des opérations linéaires, Monograf. Mat, Warsaw (1932) Extended Ukrainian translation: A course in functional analysis (linear operations) [in Ukrainian], Radyanska shkola, Kyiv (1948)
8. Berezansky, Yu.M., Sheftel, Z.G., Us, G.F.: Functional Analysis, V. 1. Birkhäuser, Basel – Boston – Berlin, (1996)
9. Birman, M.Sh., Solomjak, M.Z.: Spectral theory of selfadjoint operators in Hilbert space, Reidel, Dordrecht (1987)
10. Bogachev, V.I., Smolyanov, O.G.: Real and Functional Analysis, Springer-Verlag, Berlin (2020)
11. Conway, J.B.: A course in functional analysis. 2nd edition, Graduate texts in mathematics, v.96, Springer (2019)
12. Dunford, N., Schwartz, J.T.: Linear operators, Part 1: General Theory, John Wiley & Sons (1988)
13. Gorodetskii, V.V., Nagnibida, N.I., Nastasiev, P.P.: Methods of solving problems in functional analysis [in Ukrainian], Kyiv (1997)
14. Gorodetskii, V.V., Drin, Ya.M., Nagnibida, N.I. Generalized functions. Methods of solving problems [in Ukrainian], Knygy-XXI, Chernivtsi (2011)
15. Halmos, P.: A Hilbert Space Problem Book, Springer-Verlag New York (1982)
16. Hille, E., Phillips, R.S.: Functional analysis and semi-groups (Vol. 31). American Mathematical Soc. (1996)
17. Griffel, D.H.: Applied functional analysis. Dover Publications, Inc. Mineola, New York (2002)
18. Kadets, V.M.: A course in functional analysis and measure theory, Springer, Cham (2018)
19. Kantorovich, L.V., Akilov, G.P.: Functional analysis, 2nd edition, Elsevier (2016)

V. Brayman et al., *Functional Analysis and Operator Theory*, Problem Books in Mathematics, https://doi.org/10.1007/978-3-031-56427-7

20. Kirillov, A.A., Gvishiani, A.D.: Theorems and problems in functional analysis, Springer-Verlag, New York – Berlin (1982)
21. Kolmogorov, A.N., Fomin, S.V.: Introductory real analysis. Dover, New York (1975)
22. Kubrusly, C.S.: Hilbert space operators: a problem solving approach, Birkhäuser (2003)
23. Lusternik, L.A., Sobolev, V.J.: Elements of functional analysis, Gordon and Breach Science Publishers (1968)
24. Mortad, M.H.: An Operator Theory Problem Book, World Scientific Publishing Company (2018)
25. Muscat, J.: Functional Analysis: An Introduction to Metric Spaces, Hilbert Spaces, and Banach Algebras, Springer, Cham (2014)
26. Reed, M., Simon, B.: Methods of Modern Mathematical Physics. I. Functional analysis, 2nd edition, Academic Press, New York (1980)
27. Riesz, F., Sz.-Nagy, B.: Functional analysis. Reprint of the 1955 original, Dover Books on Advanced Mathematics, Dover (1995)
28. Rudin, W.: Functional Analysis, 2nd edition, International Series in Pure and Applied Mathematics, McGraw–Hill, New York (1991)
29. Simon, B.: Real analysis. A comprehensive course in analysis, Part 1. Amer. Math. Soc., Providence, Rhode Island (2015)
30. Simon, B.: Operator theory. A comprehensive course in analysis, Part 4. Amer. Math. Soc., Providence, Rhode Island (2015)
31. Torchinsky, A.: Problems in real and functional analysis, Amer. Math. Soc., Providence, Rhode Island (2015)
32. Trenogin, V.A., Pisarevsky, B.M., Soboleva, T.S.: Problems and exercises in functional analysis [in Russian], Nauka, Moscow (1984)
33. Vladimirov, V.S.: Equations of mathematical physics [in Russian], Mir, Moscow (1984)
34. Vladimirov, V.S. (Ed.): A collection of problems on the equations of mathematical physics, Springer Verlag, Berlin (1986)
35. Willem, M.: Functional analysis: Fundamentals and applications. Springer Nature (2023)
36. Yosida, K.: Functional analysis. Springer Science & Business Media (2012)
37. Zeidler, E.: Applied functional analysis: main principles and their applications (Vol. 109). Springer Science & Business Media (2012)

GPSR Compliance

The European Union's (EU) General Product Safety Regulation (GPSR) is a set of rules that requires consumer products to be safe and our obligations to ensure this.

If you have any concerns about our products, you can contact us on ProductSafety@springernature.com

In case Publisher is established outside the EU, the EU authorized representative is:

Springer Nature Customer Service Center GmbH
Europaplatz 3
69115 Heidelberg, Germany

Batch number: 08237905

Printed by Printforce, the Netherlands